TODAY'S TECHNICIAN ™

Shop Manual for
Automatic Transmissions
and Transaxles

Fifth Edition

TODAY'S TECHNICIAN ™

SHOP MANUAL FOR
AUTOMATIC TRANSMISSIONS
AND TRANSAXLES

FIFTH EDITION

JACK ERJAVEC

DELMAR
CENGAGE Learning™

Australia • Brazil • Japan • Korea• Mexico • Singapore • Spain • United Kingdom • United States

Today's Technician™: Automatic Transmissions and Transaxles, 5th Edition

Jack Erjavec

Vice President, Career and Professional
Editorial: Dave Garza

Director of Learning Solutions: Sandy Clark

Executive Editor: David Boelio

Managing Editor: Larry Main

Senior Product Manager: Matthew Thouin

Editorial Assistant: Jillian Borden

Vice President, Career and Professional
Marketing: Jennifer McAvey

Executive Marketing Manager:
Deborah S. Yarnell

Marketing Manager: Katie Hall

Associate Marketing Manager: Mark Pierro

Production Director: Wendy Troeger

Production Manager: Mark Bernard

Content Project Manager: Cheri Plasse

Art Director: Benj Gleeksman

Library of Congress Control Number: 2010924965

ISBN-13: 978-1-4354-8104-6
ISBN-10: 1-4354-8104-6

Delmar
5 Maxwell Drive
Clifton Park, NY 12065-2919
USA

Cengage Learning is a leading provider of customized learning solutions with office locations around the globe, including Singapore, the United Kingdom, Australia, Mexico, Brazil and Japan. Locate your local office at:
international.cengage.com/region

Cengage Learning products are represented in Canada by Nelson Education, Ltd.

For your lifelong learning solutions, visit **delmar.cengage.com**

Visit our corporate website at **cengage.com.**

Notice to the Reader

Publisher does not warrant or guarantee any of the products described herein or perform any independent analysis in connection with any of the product information contained herein. Publisher does not assume, and expressly disclaims, any obligation to obtain and include information other than that provided to it by the manufacturer. The reader is expressly warned to consider and adopt all safety precautions that might be indicated by the activities described herein and to avoid all potential hazards. By following the instructions contained herein, the reader willingly assumes all risks in connection with such instructions. The publisher makes no representations or warranties of any kind, including but not limited to, the warranties of fitness for particular purpose or merchantability, nor are any such representations implied with respect to the material set forth herein, and the publisher takes no responsibility with respect to such material. The publisher shall not be liable for any special, consequential, or exemplary damages resulting, in whole or part, from the readers' use of, or reliance upon, this material.

Printed in the United States of America
3 4 5 6 XX 16 15 14 13

CONTENTS

Contents

PHOTO SEQUENCES

JOB SHEETS

JOB SHEETS

Thanks to the support the Today's Technician Series has received from those who teach automotive technology, Delmar Cengage Learning, the leader in automotive related textbooks, is able to live up to its promise to provide new editions of the series every few years. We have listened and responded to our critics and our fans and present this new updated and revised fifth edition. By revising this series on a regular basis, we can respond to changes in the industry, changes in technology, changes in the certification process, and to the ever-changing needs of those who teach automotive technology.

We also listened to instructors when they said something was missing or incomplete in the last edition. We responded to those and the results are included in this fifth edition.

The Today's Technician Series, by Delmar Cengage Learning, features textbooks that cover all mechanical and electrical systems of automobiles and light trucks. Principally the individual titles correspond to the certification areas for 2009 areas of ASE (National Institute for Automotive Service Excellence) certification.

Additional titles include remedial skills and theories common to all of the certification areas and advanced or specific subject areas that reflect the latest technological trends.

This new edition, like the last, was designed to give students a chance to develop the same skills and gain the same knowledge that today's successful technician has. This edition also reflects the changes in the guidelines established by the National Automotive Technicians Education Foundation (NATEF) in 2008.

The purpose of NATEF is to evaluate technician training programs against standards developed by the automotive industry and recommend qualifying programs for certification (accreditation) by ASE (National Institute for Automotive Service Excellence). Programs can earn ASE certification upon the recommendation of NATEF. NATEF's national standards reflect the skills that students must master. ASE certification through NATEF evaluation ensures that certified training programs meet or exceed industry-recognized, uniform standards of excellence.

The technician of today and for the future must know the underlying theory of all automotive systems and be able to service and maintain those systems. Dividing the material into two volumes, a Classroom Manual and a Shop Manual, provides the reader with the information needed to begin a successful career as an automotive technician without interrupting the learning process by mixing cognitive and performance learning objectives into one volume.

The design of Delmar's Today's Technician Series was based on features that are known to promote improved student learning. The design was further enhanced by a careful study of survey results, in which the respondents were asked to value particular features. Some of these features can be found in other textbooks, while others are unique to this series.

Each Classroom Manual contains the principles of operation for each system and subsystem. The Classroom Manual has discussions on design variations of key components used by the different vehicle manufacturers. It also looks into emerging technologies that will be standard or optional features in the near future. This volume is organized to build upon basic facts and theories. The primary objective of this volume is to allow the reader to gain an understanding of how each system and subsystem operates. This understanding is necessary to diagnose the complex automobiles of today and tomorrow. Although the basics contained in the Classroom Manual provide the knowledge needed for diagnostics, diagnostic procedures appear only in the Shop Manual. An understanding of the underlying theories is also a requirement for competence in the skill areas covered in the Shop Manual.

A coil-ring–bound Shop Manual covers the "how-to's." This volume includes step-by-step instructions for diagnostic and repair procedures. Photo Sequences are used to illustrate some of the common service procedures. Other common procedures are listed and are accompanied with fine line drawings and photos that allow the reader to visualize and conceptualize the finest details of the procedure. This volume also contains the reasons for performing the procedures, as well as when that particular service is appropriate.

The two volumes are designed to be used together and are arranged in corresponding chapters. Not only are the chapters in the volumes linked together, the contents of the chapters are also linked. This linking of content is evidenced by marginal callouts that refer the reader to the chapter and page that the same topic is addressed in the other volume. This feature is valuable to instructors. Without this feature, users of other two-volume textbooks must search the index or table of contents to locate supporting information in the other volume. This is not only cumbersome, but also creates additional work for an instructor when planning the presentation of material and when making reading assignments. It is also valuable to the students, with the page references they also know exactly where to look for supportive information.

Both volumes contain clear and thoughtfully selected illustrations. Many of which are original drawings or photos specially prepared for inclusion in this series. This means that the art is a vital part of each textbook and not merely inserted to increase the numbers of illustrations.

The page layout, used in the series, is designed to include information that would otherwise break up the flow of information presented to the reader. The main body of the text includes all of the "need-to-know" information and illustrations. In the wide side margins of each page are many of the special features of the series. Items that are truly "nice-to-know" information such as: simple examples of concepts just introduced in the text, explanations or definitions of terms that are not defined in the text, examples of common trade jargon used to describe a part or operation, and exceptions to the norm explained in the text. This type of information is placed in the margin, out of the normal flow of information. Many textbooks attempt to include this type of information and insert it in the main body of text; this tends to interrupt the thought process and cannot be pedagogically justified. By placing this information off to the side of the main text, the reader can select when to refer to it.

Jack Erjavec

HIGHLIGHTS OF THIS EDITION—CLASSROOM MANUAL

The text was updated throughout, to include the latest developments. Some of these new topics include the various transmission designs used (or planned to be used) in hybrid vehicles, the Lepelletier gear setup (which results in five-, six-, seven-, and eight-speed transmissions), power split units, and various constantly variable transmission designs. There is also more information on current electronic control systems, including the various protocols used for multiplexing.

The first chapter introduces the purpose of automatic transmissions and how they link to the rest of the vehicle. The chapter also describes the purpose and location of the subsystems, as well as the major components of the system and subsystems. The goal of this chapter is to establish a basic understanding for students to base their learning on. All systems and subsystems that are discussed in detail later in the text are introduced and their primary purpose described. The second chapter covers the underlying basic theories of operation as the topic of the text. This is valuable to the student and the instructor because it covers the theories that other textbooks assume the reader knows. All related basic physical, chemical, and thermodynamic theories are covered in this chapter.

The third chapter applies those theories to the operation of an automatic transmission. Great emphasis is placed on hydraulics. The fourth chapter goes deeply into the electronics involved in today's transmissions. This is a chapter that was greatly updated, since the manufacturers are constantly adding more sophisticated electronics to transmissions.

The chapters that follow cover the major components of an automatic transmission and transaxle, such as torque converters, pumps, hydraulic circuits, gears and shafts, and reaction and friction units. The last chapter takes a look at the commonly used transmissions and transaxles. This includes their mechanical, hydraulic, and electronic systems.

Current model transmissions are used as examples throughout the text. Many are discussed in detail. This includes five-, six-, seven-, and eight-speed and constantly variable transmissions. This new edition also has more information on nearly all automatic transmission-related topics. Finally, the art has been updated throughout the text to enhance comprehension and improve visual interest.

HIGHLIGHTS OF THIS EDITION—SHOP MANUAL

Along with the Classroom Manual, the Shop Manual was updated to match current trends. Service information related to the new topics covered in the Classroom Manual is included in this manual. In addition, several new photo sequences were added. The purpose of these detailed photos is to show students what to expect when they perform the same procedure. They also can provide a student with familiarity of a system or type of equipment they may not be able to perform at their school. Although the main purpose of the textbook is not to prepare someone to successfully pass an ASE exam, all of the information required to do so is included in the textbook.

Chapters 1 and 2 cover the need-to-know transmission-related information about tools, safety, and typical services procedures. The first chapter covers safety issues. To stress the importance of safe work habits, one full chapter is dedicated to safety. Included in this chapter are common shop hazards, safe shop practices, safety equipment, and the legislation concerning and the safe handling of hazardous materials and wastes. Chapter 2 covers the basics of things a transmission technician does to earn a living, including basic diagnostics. Also included in this chapter are those tools and procedures that are commonly used to diagnose and service automatic transmissions and transaxles.

Chapters 3 and 4 have been heavily revised and updated. This is due to the many new developments that have occurred in transmission controls.

The rest of the chapters have been thoroughly updated. Much of the updating focuses on the diagnosis and service to new systems, as well as those systems instructors have said they need more help in.

New photo sequences on reprogramming a TCM and overhauling a five- and six-speed transmission have been added. Currently accepted service procedures are used as examples throughout the text. These procedures also served as the basis for new job sheets that are included in the text. Finally, the art has been updated throughout the text to enhance comprehension and improve visual interest.

Features of the Classroom Manual include the following:

Chapter 1

SAFETY

UPON COMPLETION AND REVIEW OF THIS CHAPTER, YOU SHOULD ABLE TO:

- Explain how safety practices are part of professional behavior.
- Explain the basic principles of personal safety, including protective eyewear, clothing, gloves, shoes, and hearing protection.
- Inspect equipment and tools for unsafe conditions.
- Properly work around batteries.
- Understand the importance of safety and accident prevention in an automotive shop.
- Explain the procedures and precautions for safely using tools and equipment.

- Explain the precautions that need to be followed to safely raise a vehicle on a lift.
- Properly lift heavy objects.
- Explain the procedures for responding to an accident.
- Recognize fire hazards.
- Extinguish common types of fires.
- Identify substances that could be regarded as hazardous materials.
- Describe the purpose of the laws concerning hazardous wastes and materials, including the right-to-know laws.

INTRODUCTION

Safety is everyone's job. You should work safely to protect yourself and the people around you. Perhaps the best single safety rule is "Think before you act." Too often people working in shops gamble by working in an unsafe way. Often they win and no one gets hurt, but all it takes is one accident and all of their past winnings are lost. By gambling and, perhaps, saving five minutes, you can lose an eye or a hand. By acting first and thinking last, you can ruin your back and lose your career. Accidents in a shop can be prevented by others and by *you*. Safe work habits also prevent damage to the vehicles and equipment in the shop.

Working on automobiles can be dangerous. It can also be fun and very rewarding. To keep the fun and rewards rolling in, you need to try to prevent accidents by working safely. In an automotive repair shop, there is great potential for serious accidents, simply because of the nature of the business and the equipment used. Through carelessness, the automotive repair industry can be one of the most dangerous occupations.

However, the chances of your being injured while working on a car are close to nil if you learn to work safely and use common sense. Shop safety is the responsibility of everyone in the shop: you, your fellow students or employees, and your employer or instructor. Everyone must work together to protect the health and welfare of all who work in the shop.

> It has been said that 50 percent of all shop accidents could have been prevented by a single individual: the technician.

> An **accident** is something that happens unintentionally.

> A vacuum is best defined as any pressure lower than atmospheric pressure.

AUTHOR'S NOTE: Transmission pumps run at very close tolerances. They are very likely to wear if the ATF is dirty or broken down. If the pump wears, fluid movement and pressure will decrease. This could cause poor transmission performance. This is just another reason why an automatic transmission needs to be taken care of by its owner and why you need to work on these things in a clean area.

> **Shop Manual**
> Chapter 6, page 308

> Some gear-type pumps use a wear plate instead of a pump cover to seal the gears in the housing.

> A **crescent** is a half-moon shaped part that isolates the inlet side of the pump from the outlet side of the pump.

Gear-Type Pumps

External tooth gear-type pumps consist of two gears in mesh to cause fluid flow. The gears rotate on their own shafts and in opposite directions. The gears are assembled in a housing that surrounds and totally encloses the gears. The shafts are sealed in the housing by bushings and the housing is normally sealed with a cover. In order for the pump to create low inlet pressures, the pump housing must be sealed.

One pump gear may be driven by the torque converter and drives the other gear. As the gears rotate, the gear teeth move in and out of mesh. As the teeth move out of mesh, inlet oil is trapped between the gear teeth and the walls of the housing. The trapped oil is carried around with the teeth until the gears again mesh. At this time the meshing of the teeth forces the oil out. The continuous release of trapped oil provides for flow as the fluid is pushed out of the pump's outlet port. The meshing of the gears also forms a seal that stops the fluid from moving out of the inlet port. Atmospheric pressure on the fluid ensures that the gap between the gear teeth will be refilled with fluid.

This type of pump is made with close tolerances. Excessive wear or play between the teeth of the gears or between the gears and the housing or pump cover will reduce the output and efficiency of the pump.

Another common type of gear pump is the **crescent** pump, which also uses two gears. One gear of this pump has internal teeth and the other has external teeth. The smaller gear with external teeth is in mesh with one part of the larger gear. In the gap where the teeth are not meshed is a crescent-shaped separator (Figure 6-39). The small gear is driven by the torque converter and it drives the larger gear. The gears' teeth mesh tightly together. As the gears rotate, the teeth mesh and then separate. As they separate, a low pressure is created between the gear teeth. The fluid is pushed by atmospheric pressure into this void until it is full.

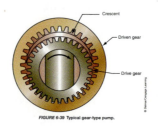

FIGURE 6-39 Typical gear-type pump.

COGNITIVE OBJECTIVES

These objectives define the contents of the chapter and define what the student should have learned upon completion of the chapter.

Each topic is divided into small units to promote easier understanding and learning.

CROSS-REFERENCES TO THE SHOP MANUAL

Reference to the appropriate page in the Shop Manual is given whenever necessary. Although the chapters of the two manuals are synchronized, material covered in other chapters of the Shop Manual may be fundamental to the topic discussed in the Classroom Manual.

TERMS TO KNOW DEFINITIONS

Many of the new terms are pulled out into the margin and defined.

MARGINAL NOTES

These notes add "nice-to-know" information to the discussion. They may include examples or exceptions, or may give the common trade jargon for a component.

AUTHOR'S NOTES

This feature includes simple explanations, stories, or examples of complex topics. These are included to help students understand difficult concepts.

The transmission's input shaft is supported by bushings in the stator support inside the torque converter. There is no mechanical link between the output of the engine and the input of the transmission. The fluid connects the power from the engine to the transmission. The combined weight of the fluid, torque converter, and flexplate serves as the flywheel for the engine.

Torque Converter Construction

A typical torque converter consists of three elements sealed in a single housing: the impeller, the turbine, and the stator. The impeller is the drive member of the unit and its fins are attached directly to the converter cover. Therefore, the impeller is the input device for the converter and always rotates at engine speed.

The turbine is the converter's output member and is coupled to the transmission's input shaft (Figure 6-7). The turbine is driven by the fluid flow from the impeller and always turns at its own speed. The fins of the turbine face toward the fins of the impeller. The impeller and the turbine have internal fins, but the fins point toward each other.

The stator is the reaction member of the converter (Figure 6-8). This assembly is about one-half the diameter of the impeller or turbine and is positioned between the impeller and turbine. The stator is not mechanically connected to either the impeller or turbine; rather, it fits between the turbine outlet and the inlet of the impeller. All of the fluid returning from the turbine to the impeller must pass through the stator. The stator redirects the fluid leaving the turbine back to the impeller (Figure 6-9). By redirecting the fluid so that it is flowing in the same direction as engine rotation, it allows the impeller to rotate more efficiently, creating torque multiplication.

A BIT OF HISTORY

Variations on the basic three-element torque converter have been used. The 1948 Buick Dynaflow had two impellers, two stators, and one turbine. In 1953 the Twin-Turbine Dynaflow was released, with two turbines, one impeller, and one stator. In 1956, Buick introduced a multiple-turbine torque converter that had a variable pitch stator. By the late 1960s, the industry, including Buick, had returned to the basic three-element converter.

A turbine is a finned wheel-like device that receives fluid from the impeller and forces it back to the stator. The turbine transmits engine torque to the

---

SUMMARIES

Each chapter concludes with a summary of key points from the chapter. These are designed to help the reader review the chapter contents.

TERMS TO KNOW LIST

A list of new terms appears next to the Summary.

REVIEW QUESTIONS

Short answer essay, fill-in-the-blank, and multiple-choice questions are found at the end of each chapter. These questions are designed to accurately assess the student's competence in the objectives stated at the beginning of the chapter.

A BIT OF HISTORY

This feature gives the student a sense of the evolution of the automobile. This feature not only contains nice-to-know information, but also should spark some interest in the subject matter.

SUMMARY

- The typical output devices are solenoids and motors, which cause something mechanical or hydraulic to change.
- The decision to shift or not to shift is based on shift schedules and logic programmed into the memory of the computer.
- Adaptive learning allows the computer to compensate for wear and other events that might occur and cause the normal shift programming to be inefficient.
- The computer may receive information from two different sources: directly from a sensor, or through a bus circuit, which connects all of the vehicle computer systems.
- The two most common protocols for serial communication in a multiplex system are the Programmable Controller Interface Data Bus (PCI Bus) and the Controller Area Network (CAN) Bus.
- If the computer loses source voltage, the transmission will enter into default (limp-in) mode. The transmission will also enter into default mode if the computer senses a transmission failure. While in the default mode, the transmission will operate only in PARK, NEUTRAL, REVERSE, and second gears. The transmission will not upshift or downshift. This allows the vehicle to be operated, although its efficiency and performance are hurt.
- Fluid flow to the various apply devices is directly controlled by the solenoids.
- Pressure switches, which give inputs to the transmission computer, are all located within the solenoid assembly.
- Engine speed, throttle position, temperature, engine load, and other typical engine-related inputs are also used by the computer to determine the best shift points. Many of these inputs are available through multiplexing and are input from the common bus.
- Adaptive learning takes place while the computer reads input and output speeds over 140 times per second.
- The CVI is the measurement of the physical amount of fluid and time required to fill the clutch and stroke the piston.
- The TCM relies on information from the engine control system, transmitted through such sensors as the MAP and TP sensor, as well as information from the transmission to determine the optimum shift timing.
- The transmission control system uses many solenoids for control of operation. One solenoid is used for modulated converter clutch control. Another, the EPC solenoid, is used to control hydraulic pressures throughout the transmission. The remaining solenoids are shift solenoids.
- The EPC solenoid replaces the conventional TV cable setup to provide changes in pressure in response to engine load.
- The pulse width modulated (PWM) solenoid is a normally closed valve installed in the valve body. It controls the position of the TCC apply valve.
- On some models, an operational mode selector switch is located on the center console or instrument panel. This switch allows the driver to select different modes to change transmission upshift characteristics.
- Hybrid vehicles are equipped with special automatic transmissions that allow the engine to run at its most efficient speed, which means increased fuel economy and decreased exhaust emissions.
- Honda hybrids use the IMA system, which places an electric motor between the engine and the transmission.
- Some lower voltage hybrid systems use an integrated starter alternator damper (ISAD) between the engine and the transmission to provide for the stop–start feature.
- The two-mode hybrid system fits into a standard automatic transmission housing and is comprised of two planetary gearsets coupled to two electric motor/generators. This arrangement allows for two distinct modes of hybrid drive operation.

TERMS TO KNOW

Adaptive learning
Belt Alternator Starter (BAS)
Controller Area Network (CAN) Bus
Clutch volume indexes (CVIs)
Integrated Motor Assist (IMA)
Integrated starter alternator damper (ISAD)
Input shaft speed (ISS)
Multiplexing
Output shaft speed (OSS) sensor
Power-split device
Programmable Controller Interface Data Bus (PCI Bus)
Shift schedule
Two-mode hybrid system
Variable pulse width modulation (VPWM)
Variable force solenoid (VFS)
Voltage

144

TERMS TO KNOW
(continued)
Slip yoke
Speed reduction
Spur gear
Sun gear
Thrust bearing
Torque
Torque converter
Traction-lock
Transaxle
Transfer case
Underdrive
Universal joint (U-joint)
Vacuum
Valve body
Viscous clutch

- Current automatic transmissions are fitted with one of three types of filter: a screen filter, paper filter, or felt filter.

REVIEW QUESTIONS

Short-Answer Essays

1. What are the primary purposes of a vehicle's drivetrain?
2. Why does torque increase when a smaller gear drives a larger gear?
3. What mechanisms do most CVTs use to vary gear and speed ratios?
4. Why are transmissions equipped with many different forward gear ratios?
5. What is the primary difference between a transaxle and a transmission?
6. Why are U- and CV joints used in the driveline?
7. What does a differential unit do to the torque it receives?
8. What are the purposes of ATF?
9. What kind of gears are commonly used in today's automotive drivelines?
10. When are ball- or roller-type bearings used?

Fill in the Blanks

1. The main components of the drivetrain are the _____ _____, and _____ _____, and _____ _____.

2. The rotating or turning effort of the engine's crankshaft is called _____ _____.
3. Gears are used to apply torque to other rotating parts of the drivetrain and to _____ torque.
4. Torque is calculated by multiplying the applied force by the _____ from the center of the _____ to the point where the force is exerted.
5. Current automatic transmissions are fitted with one of three types of filter: a _____ filter, _____ filter, or _____ filter.
6. Torque is measured in _____-_____ and _____-_____.
7. Gear ratios are determined by dividing the number of teeth on the _____ gear by the number of teeth on the _____ gear.
8. Reverse gear is accomplished by adding a _____ _____ to a two-gear set.
9. The torque converter assembly comprises a _____, an _____, and a _____.
10. In FWD cars, the transmission and drive axle are located in a single assembly called a _____.
11. In RWD cars, the drive axle is connected to the transmission by a _____.

32

To stress the importance of safe work habits, the Shop Manual also dedicates one full chapter to safety. Other important features of this manual include:

PERFORMANCE-BASED OBJECTIVES

These objectives define the contents of the chapter and define what the student should have learned upon completion of the chapter. These objectives also correspond to the list of required tasks for NATEF certification.

Although this textbook is not designed simply to prepare someone for the certification exams, it is organized around the NATEF task list. These tasks are defined generically when the procedure is commonly followed and specifically when the procedure is unique for specific vehicle models. Imported and domestic model automobiles and light trucks are included in the procedures.

SPECIAL TOOLS LISTS

Whenever a special tool is required to complete a task, it is listed in the margin next to the procedure.

CROSS-REFERENCES TO THE CLASSROOM MANUAL

Reference to the appropriate page in the Classroom Manual is given whenever necessary. Although the chapters of the two manuals are synchronized, material covered in other chapters of the Classroom Manual may be fundamental to the topic discussed in the Shop Manual.

BASIC TOOLS LISTS

Each chapter begins with a list of the basic tools needed to perform the tasks included in the chapter.

TERMS TO KNOW DEFINITIONS

Many of the new terms are pulled out into the margin and defined.

MARGINAL NOTES

These notes add "nice-to-know" information to the discussion. They may include examples or exceptions, or may give the common trade jargon for a component.

PHOTO SEQUENCES

Many procedures are illustrated in detailed photo sequences. These detailed photographs show the students what to expect when they perform particular procedures. They also can provide for the student a familiarity with a system or type of equipment that their school may not have.

Cooler line fittings

FIGURE 3-21 Typical location of cooler line connections on a transmission.

Classroom Manual
Chapter 3, page 83

⚠ **WARNING:** Do not remove the radiator cap on a warm engine. Wait until it cools down before proceeding with a coolant check. The hot coolant can burn you.

4. Check the transmission's fluid for signs of engine coolant. Water or coolant will cause the fluid to appear milky with a pink tint. This is also an indication that the transmission cooler leaks and is allowing engine coolant to enter into the transmission fluid.

The cooler can be checked for leaks by disconnecting and plugging the transmission-to-cooler lines, at the radiator. Then remove the radiator cap to relieve any pressure in the system. Tightly plug one of the ATF line fittings at the radiator. Using the shop air supply with a pressure regulator, apply 50 to 70 psi of air pressure into the cooler at the other cooler line

SERVICE TIP: The cooler lines for some transmissions have a seal inside the attaching bore in the transmission housing. A special tool is required to remove the old seal and install a new one (Figure 3-22). If the transmission is so equipped, make sure you replace the seal if the line is leaking or if the line has been disconnected from the housing.

⚠ **CAUTION:** Make sure the air pressure is regulated down to 50 to 70 psi before conducting this test. Most shops have air line pressures of nearly 150 psi. Pressures this high will damage the radiator or cooler.

CAUTIONS AND WARNINGS

Throughout the text, warnings are given to alert the reader to potentially hazardous materials or unsafe conditions. Cautions are given to advise the student of things that can go wrong if instructions are not followed or if a nonacceptable part or tool is used.

Retainer   Main harness

FIGURE 7-17 To remove the transmission wire in an AA80E transmission, unbolt the retainer and pull the harness out of the case.

free, disconnect the shift lever from the manual valve. Then remove the transmission wire from the case. To do this, the retainer's bolt is removed and the harness pulled out (Figure 7-17).

In an U250E transaxle, the wiring harness must be disconnected and removed, along with the oil strainer (filter) must be unbolted and removed with its O-ring. Then the valve body can be removed.

VALVE BODY SERVICE

SERVICE TIP: Another trick is to use the cardboard sheet included in every gasket set. Fold this sheet in one-inch pleats like an accordion and lay it on your bench with the slick side of the cardboard facing up. Then follow these steps:
1. Take the valves and springs out of the valve body and lay them in the different grooves in the sequence they were removed.
2. Clean the valve body castings.
3. Clean the valves and springs. Do not put them back on the cardboard; rather, put them directly into their bores in the valve body.

CUSTOMER CARE: If while you are disassembling and inspecting the valve body you discover signs of poor transmission maintenance or abuse, make note of it. When you next talk to the customer, explain what you found and what the resulting problem was. Make sure you explain this in an understanding and nonoffensive way.

If previous tests suggest a problem with only one or two valves, start your inspection at those valves. Doing this will not only save you time, but will also reduce the chance of something being misplaced or ruined during a total disassembly. If the transmission had heavily contaminated fluid, the entire valve body should be inspected and cleaned, or replaced.

Disassembly

A valve body contains many valves. These valves are typically held in their bores by a plug or cover plate. These must be removed to gain access to the valves. The cover plates are bolted or screwed to the valve body (Figure 7-18). Plugs can be held in place in a number of different ways. All of them can be removed by pressing the plug slightly into the bore.

Begin disassembly by removing the manual shift valve from the valve body (Figure 7-19). Then, remove the pressure regulator retaining screws while keeping one hand around the spring retainer and adjusting screw bracket. Remove the pressure regulator valve. Then remove all of the valves and springs from the valve body. Make sure that you keep all springs and other parts with their associated valves.

It is important to keep track of the placement and position of the springs on each valve. Doing this will make reassembly easy and will ensure you are doing things correctly. One way of doing this is to draw a diagram indicating where the valve was and the position of the spring on the valve. Many technicians use a digital camera or the camera on a cell phone to

342

SERVICE TIPS

Whenever a shortcut or special procedure is appropriate, it is described in the text. These tips are generally those things commonly done by experienced technicians.

CUSTOMER CARE

This feature highlights those little things a technician can do or say to enhance customer relations.

JOB SHEET   16

Name _____   Date _____

TESTING A MAP SENSOR

Upon completion of this job sheet, you should be able to test a manifold absolute pressure sensor in a variety of ways.

ASE Correlation

This job sheet is related to the ASE Automatic Transmission and Transaxle Test's Content Area *General Transmission and Transaxle Diagnosis*.
Task: Diagnose mechanical and vacuum control systems; determine necessary action.

Tools and Materials
Hand-operated vacuum pump   DMM
Lab scope

Describe the vehicle being worked on:
Year _____ Make _____ VIN _____
Model _____

Procedure	Task Completed
1. If the MAP sensor produces an analog voltage signal, follow this procedure.	☐
2. With the ignition switch on, backprobe the 5-volt reference wire.	☐
3. Connect a voltmeter from the reference wire to ground. The reading is _____ volts.	
4. If the reference wire is not supplying the specified voltage, what should be checked next?	
5. With the ignition switch on, connect the voltmeter from the sensor ground wire to the battery ground.	☐
6. What is the measured voltage drop? _____ volts	
7. What does this indicate?	
8. Backprobe the MAP sensor signal wire and connect a voltmeter from this wire to ground with the ignition switch on.	☐
9. What is the measured voltage? _____ volts	
10. What does this indicate?	

225

JOB SHEETS

Located at the end of each chapter, the Job Sheets provide a format for students to perform procedures covered in the chapter. A reference to the ASE Task addressed by the procedure is referenced on the Job Sheet.

CASE STUDIES

Case Studies concentrate on the ability to properly diagnose the systems. Beginning with Chapter 3, each chapter ends with a case study in which a vehicle has a problem, and the logic used by a technician to solve the problem is explained.

CASE STUDY

A customer brought her late-model Chevy Silverado to a shop. Her concern was the truck sometimes stalled when it came to a stop. The technician verified the concern and began diagnosis with a check of the engine. He found nothing unusual. He then checked the TSBs available for the truck and found several. Most were related to a stuck torque converter clutch solenoid; however, there were other possible causes. He began by unplugging the TCC solenoid to see if the problem was still there. If the solenoid is stuck and disconnected, the concern should disappear which is what it did. The technician knew the solenoid could be the cause but the concern could also be caused by dirt in the valve body or a bad signal from the PCM.

He raised the vehicle on a lift with the driving wheels off the ground. He checked the ground and power feed to the solenoid and found it to be within specifications. He then connected a test light across terminals A and D at the transmission. With the engine running while the transmission is Drive, he observed the test light. The vehicle was then slowly accelerated to about 60 mph (97 km/hr). The test light should have illuminated if the PCM was sending a signal to the solenoid. In this case, the test light did light.

While the vehicle was moving at this speed, he applied the brake pedal. This should cause the TCC to disengage, it did not. He therefore moved to check the brake pedal switch and its circuit. During the initial inspection, he found a potential source of the concern. A connector in the wiring harness near the brake switch was damaged. The connection was very loose. To verify that this was the cause of the problem, he taped the connector tightly together. He then repeated his tests and found the TCC disengaging when the brake pedal was pressed down. He then replaced the connector and reconnected everything he had disconnected for his tests. During a road test, he accelerated and stopped several times. Never get the truck stall. His conclusion was the poor connection caused an intermittent problem that would sometimes not turn off the TCC solenoid.

TERMS TO KNOW
Ballooning
Heat exchanger
Mineral spirits

TERMS TO KNOW LIST

A list of new terms appears after the case study.

ASE-STYLE REVIEW QUESTIONS

Each chapter contains ASE-style review questions that reflect the performance-based objectives listed at the beginning of the chapter. These questions can be used to review the chapter as well as to prepare for the ASE certification exam.

ASE-Style Review Questions

1. *Technician A* says a ballooned torque converter can cause damage to the oil pump.
   Technician B says a ballooned torque converter is caused by excessive pressure in the torque converter.
   Who is correct?
   A. A only   C. Both A and B
   B. B only   D. Neither A nor B

2. While checking torque converter endplay:
   Technician A says the torque converter must be installed in the transmission

3. While servicing a variable displacement vane-type pump:
   Technician A says the pump rotor, vanes, and slide have selective sizes and may destroy the pump if the correct ones are not used.
   Technician B says the outer edge of the vanes should be flat.
   Who is correct?
   A. A only   C. Both A and B
   B. B only   D. Neither A nor B

4. *Technician A* says the gears in a gear-type oil pump should be replaced if the outer edges of the teeth are ...mps are

315

ASE Challenge Questions

1. *Technician A* says a lower-than-specified stall speed may be caused by a faulty stator one-way clutch.
   Technician B says a higher-than-specified stall speed may be caused by faulty clutch packs.
   Who is correct?
   A. A only   C. Both A and B
   B. B only   D. Neither A nor B

2. *Technician A* says a shudder after the converter clutch engages could be caused by a damaged or missing clutch check ball.
   Technician B says driveline or converter shudder can be isolated by disconnecting the converter's clutch solenoid.
   Who is correct?
   A. A only   C. Both A and B
   B. B only   D. Neither A nor B

3. The backside of a torque converter is found to be wet.
   Technician A says it may be caused by excessive torque converter hub runout.
   Technician B says it can be caused by insufficient input shaft endplay.
   Who is correct?

4. A customer complains that her vehicle's engine seems to surge when she is driving at about 48 mph.
   Technician A says there may be a problem with the torque converter clutch.
   Technician B says a faulty vehicle speed sensor may cause this problem.
   Who is correct?
   A. A only   C. Both A and B
   B. B only   D. Neither A nor B

5. *Technician A* says a plugged transmission cooler or cooler lines may cause the vehicle to stall when the transmission is shifted into a forward gear.
   Technician B says a plugged transmission cooler or cooler lines may overheat the converter.
   Who is correct?
   A. A only   C. Both A and B
   B. B only   D. Neither A nor B

ASE CHALLENGE QUESTIONS

Each technical chapter ends with five ASE challenge questions. These are not more review questions, rather they test the student's ability to apply general knowledge to the contents of the chapter.

ASE PRACTICE EXAMINATION

A 50 question ASE practice exam, located in the appendix, is included to test students on the contents of the Shop Manual.

APPENDIX A

ASE PRACTICE EXAMINATION

Final Exam Automatic Transmission/Transaxle A2

1. Which of the following is the *least* likely cause for a buzzing noise from a transmission?
   A. Improper fluid level or condition
   B. Defective oil pump
   C. Defective flexplate
   D. Damaged planetary gearset

2. A vehicle experiences engine flare in low gear only.
   Technician A says the torque converter lockup clutch is slipping.
   Technician B says the transmission oil pump is not providing the required pressure.
   Who is correct?
   A. A only   C. Both A and B
   B. B only   D. Neither A nor B

3. The results of a hydraulic pressure test are being discussed:
   Technician A says low idle pressure may be caused by a defective exhaust gas recirculation system.
   Technician B says low neutral and park pressures may indicate a fluid leakage past the clutch and servo seals.
   Who is correct?
   A. A only   C. Both A and B
   B. B only   D. Neither A nor B

4. The vehicle creeps in neutral.
   Technician A says a too high engine idle speed could be the cause.
   Technician B says a too tight clutch pack may be the problem.
   Who is correct?
   A. A only   C. Both A and B
   B. B only   D. Neither A nor B

5. *Technician A* says overtorqued valve body fasteners may cause a lack of engine braking in manual low.
   Technician B says a lack of engine braking in manual third may be caused by a faulty overrunning clutch.
   Who is correct?
   A. A only   C. Both A and B
   B. B only   D. Neither A nor B

6. The transmission's output shaft and its sealing components are being discussed:
   Technician A says the shaft and all of its sealing components must be replaced if nicks and scratches are found in the shaft's sealing area.
   Technician B says all of the shaft's seals and rings must be replaced during a rebuild.
   Who is correct?
   A. A only   C. Both A and B
   B. B only   D. Neither A nor B

7. The vehicle will only upshift to second at full throttle. This could be caused by any of the following EXCEPT:
   A. Clogged oil passages   C. Bad clutch pack
   B. Low fluid level   D. Open upshift switch

8. The vehicle will not move in any gear.
   Technician A says a misadjusted TV cable could be the cause.
   Technician B says leakage at the oil pump and/or valve body could cause this condition.
   Who is correct?
   A. A only   C. Both A and B
   B. B only   D. Neither A nor B

9. Sensors are being discussed:
   Technician A says most speed sensors are AC generators.
   Technician B says most speed sensors use a stationary magnet, rotor, and a voltage sensor.
   Who is correct?
   A. A only   C. Both A and B
   B. B only   D. Neither A nor B

10. *Technician A* says the PCM monitors the amount of voltage generated by the speed sensor to calculate the vehicle's speed.
    Technician B says the output of a speed sensor is pulsed as an on/off voltage signal when displayed on a DSO.
    Who is correct?
    A. A only   C. Both A and B
    B. B only   D. Neither A nor B

539

SUPPLEMENTS

INSTRUCTOR RESOURCES

The Instructor Resources DVD is a robust ancillary that contains all preparation tools to meet any instructor's classroom needs. It includes chapter outlines in PowerPoint with images, video clips and animations that coincide with each chapter's content coverage, chapter tests in ExamView with hundreds of test questions, a searchable Image Library with all photos and illustrations from the text, theory-based Worksheets in Word that provide homework or in-class assignments, the Job Sheets from the Shop Manual in Word, a NATEF correlation chart, and an Instructor's Guide in electronic format.

WEBTUTOR ADVANTAGE

Newly available for this title and to the Today's Technician™ Series is the *WebTutor Advantage* for Blackboard and Angel online course management systems. The *WebTutor for Today's Technician: Automatic Transmissions & Transaxles, 5e* will include presentations in PowerPoint with video clips and animations, end-of-chapter review questions, pre-tests and post-tests, worksheets, discussion springboard topics, job sheets, and more. The *WebTutor* is designed to enhance the classroom and shop experience, engage students, and help them prepare for ASE certification exams.

REVIEWERS

The author and publisher would like to extend a special thanks to the individuals who reviewed this text and offered their invaluable feedback:

Timothy Belt
University of Northwestern Ohio
Lima, OH

Eric Harper
American River College
Sacramento, CA

Thomas L. Corban
University of Northwestern Ohio
Lima, OH

Michael Ronan
Alfred State College
Alfred, NY

Chapter 1

SAFETY

UPON COMPLETION AND REVIEW OF THIS CHAPTER, YOU SHOULD ABLE TO:

- Explain how safety practices are part of professional behavior.

- Explain the basic principles of personal safety, including protective eyewear, clothing, gloves, shoes, and hearing protection.

- Inspect equipment and tools for unsafe conditions.

- Properly work around batteries.

- Understand the importance of safety and accident prevention in an automotive shop.

- Explain the procedures and precautions for safely using tools and equipment.

- Explain the precautions that need to be followed to safely raise a vehicle on a lift.

- Properly lift heavy objects.

- Explain the procedures for responding to an accident.

- Recognize fire hazards.

- Extinguish common types of fires.

- Identify substances that could be regarded as hazardous materials.

- Describe the purpose of the laws concerning hazardous wastes and materials, including the right-to-know laws.

INTRODUCTION

Safety is everyone's job. You should work safely to protect yourself and the people around you. Perhaps the best single safety rule is "Think before you act." Too often people working in shops gamble by working in an unsafe way. Often they win and no one gets hurt, but all it takes is one accident and all of their past winnings are lost. By gambling and, perhaps, saving five minutes, you can lose an eye or a hand. By acting first and thinking last, you can ruin your back and lose your career. Accidents in a shop can be prevented by others and by *you*. Safe work habits also prevent damage to the vehicles and equipment in the shop.

Working on automobiles can be dangerous. It can also be fun and very rewarding. To keep the fun and rewards rolling in, you need to try to prevent accidents by working safely. In an automotive repair shop, there is great potential for serious accidents, simply because of the nature of the business and the equipment used. Through carelessness, the automotive repair industry can be one of the most dangerous occupations.

However, the chances of your being injured while working on a car are close to nil if you learn to work safely and use common sense. Shop safety is the responsibility of everyone in the shop: you, your fellow students or employees, and your employer or instructor. Everyone must work together to protect the health and welfare of all who work in the shop.

> It has been said that 50 percent of all shop accidents could have been prevented by a single individual: the technician.

> An *accident* is something that happens unintentionally.

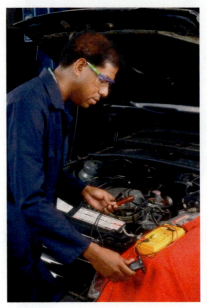

FIGURE 1-1 Proper dress prevents injuries.

PERSONAL SAFETY

Personal safety involves those precautions you take to protect yourself from injury. These include wearing protective gear (Figure 1-1), dressing for safety, working professionally, and correctly handling tools and equipment.

When you have neat work habits, you display a professional attitude. Actually, neat habits are also safe habits. Cleaning up spills and keeping equipment and tools out of the path of others prevent accidents. True professionals take time to clean their tools and work area. With a professional attitude, you do not clown around in the shop, do not throw items in the shop, and do not create an unsafe condition for the sake of saving time. Rather than ignoring basic safety rules to save time, a professional saves time by performing effective diagnostics and repair procedures.

Professionals also take pride in their work, treat customers and their vehicles with respect, and try to stay current with all technical, safety, and environmental concerns.

Eye Protection

Your eyes can become infected or permanently damaged by many things in a shop. Some procedures, such as grinding, result in tiny particles of metal and dust being thrown off at very high speeds. These metal and dirt particles can easily get into your eyes, causing scratches or cuts on your eyeball. Pressurized gases and liquids escaping a hose or hose fitting can spray a great distance. If these chemicals get into your eyes, they can cause blindness. Dirt and sharp bits of corroded metal can easily fall into your eyes while you are working under a vehicle.

Eye protection should be worn whenever you are exposed to these risks. To be safe, you should wear it whenever you are working in the shop. There are many types of eye protection (Figure 1-2). Safety glasses have lenses made of safety glass and some sort of side protection. Regular prescription glasses should not be worn as a substitute for safety glasses. When prescription glasses are worn in a shop, they should be fitted with side shields.

Wearing safety glasses at all times is a good habit to get into. Choose glasses that fit well and feel comfortable. Some shops require that everyone wear eye protection whenever they are in the work area.

Some procedures may require that you wear additional eye protection. For example, when you are cleaning parts with a pressurized spray, you should wear a face shield. The face shield not only gives added protection to your eyes but it also protects the rest of your face.

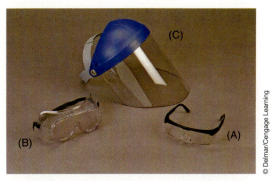

FIGURE 1-2 Different types of eye protection worn by automotive technicians: (A) safety glasses, (B) goggles, and (C) face shield.

FIGURE 1-3 Using an eye wash bottle from an eye station hanging on the wall.

If chemicals such as battery acid, fuel, or solvents get into your eyes, flush them continuously with clean water. Have someone call a doctor and get medical help immediately.

Many shops have eyewash stations (Figure 1-3) or safety showers that should be used whenever you or someone else has been sprayed or splashed with a chemical.

Clothing

Clothing that hangs freely, such as shirttails, can create a safety hazard and cause serious injury. Nothing you wear should be allowed to dangle in the engine compartment or around equipment. Shirts should be tucked in and buttoned and long sleeves buttoned or carefully rolled up. Your clothing should be well fitted and comfortable but made of strong material. Loose, baggy clothing can easily be caught in moving parts and machinery. Neckties should not be worn. Some technicians prefer to wear coveralls or shop coats to protect their personal clothing.

Long hair and loose, hanging jewelry can create the same type of hazard as loose-fitting clothing. They can get caught in moving engine parts and machinery. If you have long hair, tie it back or tuck it under a cap.

Rings, necklaces, bracelets, and watches should not be worn while working. A ring can rip your finger off, a watch or bracelet can cut your wrist, and a necklace can choke you. This is especially true when working with or around electrical wires. The metal used to make jewelry conducts electricity very well and can easily cause a short, through you, if it touches a bare wire.

Keep your clothing clean. If you spill gasoline or oil on yourself, change that item of clothing immediately. Oil that is against your skin for a prolonged period of time can produce rashes or other allergic reactions. Gasoline can irritate cuts and sores.

Foot Protection

You should also protect your feet. Tennis and jogging shoes provide little protection if something falls on your foot. Boots or shoes made of leather or a material that approaches the strength of leather offer much better protection from falling objects. There are many designs of safety shoes and boots that have steel plates built into the toe and **shank** to protect your

An electrical short is basically an alternative path for the flow of electricity.

The **shank** of a shoe is that portion of the shoe that protects the ball of your foot. It is the narrow part of a shoe between the heel and sole.

feet. Many also have soles that are designed to resist slipping on wet surfaces. Foot injuries are not only quite painful but can also put you out of work for some time.

Hand Protection

Good hand protection is often overlooked. A scrape, cut, or burn can seriously impair your ability to work for many days. A well-fitted pair of heavy work gloves should be worn while grinding, welding, or handling chemicals or high-temperature components. Polyurethane or vinyl gloves should be worn when handling strong and dangerous **caustic** chemicals. These chemicals can easily burn your skin.

Many technicians wear thin, surgical-type latex gloves whenever they are working on vehicles (Figure 1-4). These offer little protection against cuts but do offer protection against disease and grease buildup under and around your fingernails. These gloves are comfortable and are quite inexpensive. Latex surgical-type and nitrile gloves are worn as protection against disease and to keep grease from building up under and around your fingernails. Latex gloves are more comfortable to wear but weaken when they are exposed to gas, oil, or solvents. Nitrile gloves are not as comfortable as latex gloves but they are not affected by gas, oil, or solvents. Your choice of hand protection should be based on what you are doing.

Disease Prevention. When you are ill with something that may be contagious, see a doctor and do not go to work or school until the doctor says there is little chance of someone else contracting the illness from you.

You should also be concerned with and protect yourself and others from bloodborne pathogens. **Bloodborne pathogens** are pathogenic microorganisms that are present in human blood and can cause disease. These pathogens include, but are not limited to, hepatitis B virus (HBV) and human immunodeficiency virus (HIV). For everyone's protection, any injury that causes bleeding should be dealt with as a threat to others. You should avoid contact with the blood of another. If you need to administer some form of first aid, make sure you put on hand protection before you do so. You should also wear gloves and other protection

> A **caustic** material has the ability to destroy or eat through something. Caustic materials are considered extremely corrosive.

FIGURE 1-4 Disposable latex gloves.

© Delmar/Cengage Learning

when handling the item that caused the cut. This item should be sterilized immediately. Most importantly, as with all injuries, report the accident to your instructor or supervisor.

You also need to protect your hands and other body parts when working with pressurized fluids. There is a risk of blood poisoning (sepsis) due to the inadvertent introduction of toxins (solvents, diesel fuel, and so on) into the bloodstream from a pressurized source. The best protection for this is working carefully and wearing heavy gloves.

Ear Protection

Exposure to very loud noise levels for extended periods of time can lead to a loss of hearing. Air wrenches, air hammers, engines running under a load, and vehicles running in enclosed areas can all generate annoying and harmful levels of noise. Simple earplugs or earphone-type protectors should be worn in environments that are constantly noisy.

Respiratory Protection

It is not uncommon for a technician to work with chemicals that have toxic fumes. Air or respiratory masks should be worn whenever you will be exposed to toxic fumes. Cleaning parts with solvents and painting are the most common activities for which respiratory masks should be worn. Masks should also be worn when handling hazardous materials.

FIRE HAZARDS AND PREVENTION

The presence of gasoline is so common that its dangers are often forgotten. A slight spark or an increase in heat can cause a fire or explosion. Gasoline fumes are heavier than air. Therefore, when an open container of gasoline is sitting about, the fumes spill out over the sides of the container. These fumes are more flammable than liquid gasoline and can easily explode.

Never smoke around gasoline or in a shop filled with gasoline fumes. If a vehicle has a gasoline leak or you have caused a leak by disconnecting a fuel line, wipe it up immediately and stop the leak. Make sure that any grinding or welding that may be taking place in the area is stopped until the spill is totally cleaned up and the floor has been flushed with water. The rags used to wipe up the gasoline should be taken outside to dry, then stored in an approved dirty rag container. If vapors are present in the shop, have the doors open and turn on the ventilating system. It takes only a small amount of fuel mixed with air to cause combustion.

Gasoline should always be stored in approved containers (Figure 1-5) and never in a glass bottle or jar. If the glass jar were knocked over or dropped, a terrible explosion could take place.

Diesel Fuel. Diesel fuel is not as **volatile** as gasoline, but should be stored and handled in the same way. It is also not as refined as gasoline and tends to be a very dirty fuel. It normally contains many impurities, including active microscopic organisms that can be highly infectious. If diesel fuel happens to get into an open cut or sore, thoroughly wash it immediately.

Solvents. Cleaning solvents are also not as volatile as gasoline, but they are still **flammable**. These should be stored and treated in the same way as gasoline.

Handle all solvents (and any liquids) with care to avoid spillage. Keep all solvent containers closed, except when pouring. Proper ventilation is very important in areas where volatile solvents and chemicals are used. Solvent and other combustible materials must be stored in approved and designated storage cabinets or rooms with adequate ventilation. Never light matches or smoke near flammable solvents and chemicals, including battery acids.

Rags. Oily or greasy rags can also be a source for fires. These rags should be stored in an approved container (Figure 1-6) and never thrown out with normal trash. Like gasoline, oil is a **hydrocarbon** and can ignite with or without a spark or flame. This ignition without an external source of heat or fire is called **spontaneous combustion.**

CAUTION:
Never siphon gasoline or diesel fuel with your mouth. These liquids are poisonous and can make you sick or fatally ill.

The **volatility** of a substance is how easily the substance vaporizes or explodes.

The **flammability** of a substance is how well the substance supports combustion.

A **hydrocarbon** is a substance composed of hydrogen and carbon molecules.

FIGURE 1-5 Flammable liquids should be stored in safety-approved containers.

FIGURE 1-6 Dirty rags, like gasoline, should be put into a safety container.

Fire Extinguishers

In case of a fire you should know the location of the fire extinguishers and fire alarms in the shop and should also know how to use them. You should also be aware of the different types of fires and the fire extinguishers (Figure 1-7) used to put out these types of fires.

Basically, there are four types of fires: **Class A fires** are those in which wood, paper, and other ordinary materials are burning; **Class B fires** are those involving flammable liquids,

FIGURE 1-7 Different types of fire extinguishers.

TABLE 1-1 Guide to fire extinguisher selection.

	Class of Fire	Typical Fuel Involved	Type of Extinguisher
Class A Fires (green)	**For Ordinary Combustibles** Put out a Class A fire by lowering its temperature or by coating the burning combustibles.	Wood Paper Cloth Rubber Plastics Rubbish Upholstery	Water*[1] Foam* Multipurpose dry chemical[4]
Class B Fires (red)	**For Flammable Liquids** Put out a Class B fire by smothering it. Use an extinguisher that gives a blanketing, flame-interrupting effect; cover whole flaming liquid surface.	Gasoline Oil Grease Paint Lighter fluid	Foam* Carbon dioxide[5] Halogenated agent[6] Standard dry chemical[2] Purple K dry chemical[3] Multipurpose dry chemical[4]
Class C Fires (blue)	**For Electrical Equipment** Put out a Class C fire by shutting off power as quickly as possible and by always using a nonconducting extinguishing agent to prevent electric shock.	Motors Appliances Wiring Fuse boxes Switchboards	Carbon dioxide[5] Halogenated agent[6] Standard dry chemical[2] Purple K dry chemical[3] Multipurpose dry chemical[4]
Class D Fires (yellow)	**For Combustible Metals** Put out a Class D fire of metal chips, turnings, or shavings by smothering or coating with a specially designed extinguishing agent.	Aluminum Magnesium Potassium Sodium Titanium Zirconium	Dry powder extinguishers and agents only

Catridge-operated water, foam, and soda-acid types of extinguishers are no longer manufactured. These extinguishers should be removed from service when they become due for their next hydrostatic prerssure test.

Notes:
(1) Freezes in low temperatures unless treated with antifreeze solution, usually weighs over 20 pounds (9 kg), and is heavier than any other extinguisher mentioned.
(2) Also called ordinary or regular dry chemical (sodium bicarbonate).
(3) Has the greatest initial fire-stopping power of the extinguishers mentioned for class B fires. Be sure to clean residue immediately after using the extinguishers so sprayed surfaces will not be damaged (potassium bicarbonate).
(4) The only extinguishers that fight A, B, and C classes of fires. However, they should not be used on fires in liquefied fat or oil of appreciable depth. Be sure to clean residue immediately after using the extinguisher so sprayed surfaces will not be damaged (ammonium phosphates).
(5) Use with caution in unventilated, confined spaces.
(6) May cause injury to the operator if the extinguishing agent (a gas) or the gases produced when the agent is applied to a fire is inhaled.

such as gasoline, diesel fuel, paint, grease, oil, and other similar liquids; and **Class C fires** are electrical fires. **Class D fires** are a unique type of fire in which the burning material is a metal. An example of this is a burning "mag" wheel. The magnesium used in the construction of the wheel is a flammable metal and will burn brightly when subjected to high heat.

Using the wrong type of extinguisher may cause the fire to grow instead of putting it out. All extinguishers are marked with a symbol or letter to signify what class of fire they were intended for (Table 1-1). You should know the location and rating of each fire extinguisher in the shop before you need one.

Using a Fire Extinguisher

Remember, during a fire, never open doors or windows unless it is absolutely necessary; the extra draft will only make the fire worse. Make sure the fire department is contacted before or during your attempt to extinguish a fire. To extinguish a fire, stand 6 to 10 feet from the fire. Each type of extinguisher requires a slightly different use (Figure 1-8). For most, you should follow the guidelines shown in Photo Sequence 1. Before releasing the agent from the extinguisher, hold the extinguisher firmly in an upright position. Aim the nozzle at the base and use a side-to-side motion, sweeping the entire width of the fire. Stay low to avoid inhaling the smoke. If it gets too hot or too smoky, get out. Remember, never go back into a burning building for anything.

To help remember how to use an extinguisher, remember the word PASS:

Pull the pin from the handle of the extinguisher.

Aim the extinguisher's nozzle at the base of the fire.

FOAM
Solution of aluminum sulfate and bicarbonate of soda

Do not spray the stream into the burning liquid. Allow foam to fall lightly on fire.

CARBON DIOXIDE
Carbon dioxide gas under pressure

Direct discharge as close to fire as possible, first at edge of flames and gradually forward and upward.

DRY CHEMICAL

Direct stream at base of flames. Use rapid left-to-right motion toward flames.

SODA-ACID
Bicarbonate of soda solution and sulfuric acid

Direct stream at base of flames.

© Delmar/Cengage Learning

FIGURE 1-8 Each type of fire extinguisher has a slightly different required operation.

USING A DRY CHEMICAL FIRE EXTINGUISHER

All photos in this sequence are © Delmar/Cengage Learning.

P1-1 Multipurpose dry chemical fire extinguisher.

P1-2 Hold the fire extinguisher in an upright position.

P1-3 Pull the safety pin from the handle.

P1-4 Stand eight feet from the fire. Do not go any close to the fire.

P1-5 Free the hose from its retainer and aim it at the base of the fire.

P1-6 Squeeze the lever while sweeping the hose from side to side. Keep the hose aimed at the base of the fire.

Squeeze the handle.

Sweep the entire width of the fire with the contents of the extinguisher.

If there is not a fire extinguisher handy, a blanket or fender cover may be used to smother the flames. Be careful when doing this as the heat of the fire may burn you and the blanket. If the fire is too great to smother, move everyone away from the fire and call the local fire department. A simple under-the-hood fire can cause the total destruction of the car and the building and can take lives. You must be able to respond quickly and precisely to avoid a disaster.

Never put water on a gasoline fire. The water will just spread the fire—the proper fire extinguisher smothers the flames.

ROTATING PULLEYS AND BELTS

Be very careful around belts, pulleys, wheels, chains, or any other rotating mechanism. When working around an engine's drive belts and pulleys, make sure hands, shop towels, or loose clothing do not come in contact with the moving parts. Hands and fingers can be quickly pulled into a revolving belt or pulley even at engine idle speeds.

The thermostatic switch for an electric cooling fan may also be disconnected to prevent the fan from coming on.

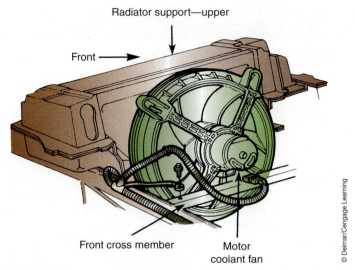

FIGURE 1-9 When possible, disconnect the connector to the cooling fan before reaching into or working near the area around the fan.

 WARNING: Be careful when working around electric engine cooling fans. These fans are controlled by a thermostat and can come on without warning, even when the engine is not running. Whenever you must work around these fans, disconnect the electrical connector to the fan motor (Figure 1-9) before reaching into the area around the fan.

TOOL AND EQUIPMENT SAFETY

When you work with any equipment, make sure you use it properly and that is set up according to the manufacturer's instructions. All equipment should be properly maintained and periodically inspected for unsafe conditions. Frayed electrical cords or loose mountings can cause serious injuries. All electrical outlets should be equipped to allow for the use of three-pronged electrical cords. The third prong allows for a safety ground connection (Figure 1-10). All equipment with rotating parts should be equipped with safety guards that reduce the possibility of the parts coming loose and injuring someone. Do not depend on someone else to inspect and maintain equipment. Check it out before you use it! If you find the equipment unsafe, disconnect it from the power source, put a sign on it to warn others, and notify the person in charge.

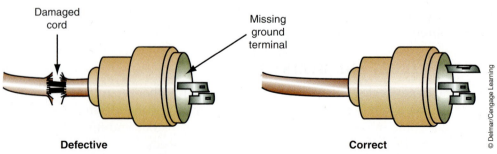

FIGURE 1-10 Make sure all three prongs on an electrical connector are in good condition before plugging in the cord.

Hand Tool Safety

Hand tools should always be kept clean and be used only for the purpose for which they were designed. Oily hand tools can slip out of your hand and cause broken fingers or at least cut or skinned knuckles. Your tools should also be inspected for cracks, broken parts, or other dangerous conditions before you use them.

Knives, chisels, and scrapers must be used in a motion that will keep the point or blade moving away from your body. Always hand a pointed or sharp tool to someone else with the handle toward the person you are handing the tool to.

Power Tool Safety

Power tools are operated by an outside source of power, such as electricity, compressed air, or hydraulic pressure. Safety around power tools is very important. Serious injury can result from carelessness. Always wear safety glasses when using power tools.

If a tool is electrically powered, make sure it is properly grounded. Also, when using electrical power tools, never stand on a wet or damp floor. Disconnect the power source before doing any work on the machine or tool. Before plugging in any electric tool, make sure its switch is in the off position. When you are done using the tool, turn it off and unplug it. Never leave a running power tool unattended.

Before using the tool, check the power cord for bare wires or cracks in the insulation. If the cord is damaged, do not plug it into the wall outlet. Repair the cord before using the tool or use another tool.

When using power equipment on a small part, never hold the part in your hand. Always mount the part in a bench vise or use vise-grip pliers. Never try to use a machine or tool beyond its stated capacity or for operations requiring more than the rated power of the tool.

When working with larger power tools, such as a bench or floor grinding wheel, check the machine and/or the grinding wheels for signs of damage before using them. If a wheel is damaged, it should be replaced and not used. Check the speed rating of the wheel and make sure it matches the speed of the machine. Never spin a grinding wheel at a speed higher than it is rated for.

A safety guard is a protective cover over a moving part. Be sure to place all safety guards in position (Figure 1-11). Although safety guards are designed to prevent injury, you should still wear safety glasses and/or a face shield while using the machine. Make sure there are no people or parts around the machine before starting it. Keep your hands and clothing away from the moving parts. Maintain a balanced stance while using the machine.

Compressed Air Equipment Safety

Compressed air is used to inflate tires, apply paint, and drive tools. Compressed air can be dangerous when it is not used properly. When using compressed air, safety glasses and/or a face shield should be worn. Particles of dirt and pieces of metal, blown by the high-pressure air, can penetrate your skin or get into your eyes.

Before using a compressed air tool, check all hose connections. Pneumatic tools must always be operated at the pressure recommended by the manufacturer.

Always hold an air nozzle or air control device securely when starting or shutting off the compressed air. A loose nozzle can whip suddenly and cause serious injury. Never point an air nozzle at anyone. Never use compressed air to blow dirt from your clothes or hair. Never use compressed air to clean the floor or workbench. Also, never spin bearings with compressed air. If the bearing is damaged, one of the steel balls or rollers might fly out and cause serious injury.

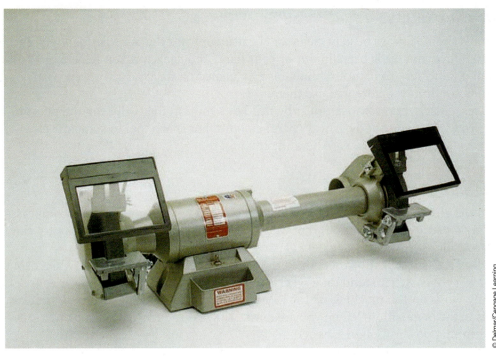

FIGURE 1-11 Before using a bench grinder, make sure all safety guards in place.

© Delmar/Cengage Learning

Lift Safety

Always be careful when raising a vehicle on a lift or a hoist. Adapters and hoist plates must be positioned correctly on twin-post and rail-type lifts to prevent damage to the underbody of the vehicle. There are specific lift points that allow the weight of the vehicle to be evenly supported by the adapters or hoist plates. The correct lift points can be found in the vehicle's service manual. Figure 1-12 shows typical locations for unibody and frame cars. Always

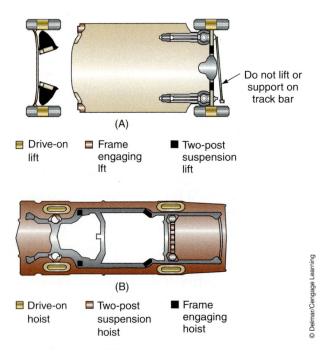

FIGURE 1-12 Typical lift points for (A) unibody and (B) frame/body vehicles.

© Delmar/Cengage Learning

follow the manufacturer's instructions. Before operating any lift or hoist, carefully read the operating manual and follow the operating instructions.

Once the lift supports are properly positioned under the vehicle, raise the lift until the supports contact the vehicle. Then, check the supports to make sure they are in full contact with the vehicle. Shake the vehicle to make sure it is securely balanced, then raise it to the desired working height. Before working under a vehicle, make sure the lift's locking devices are fully engaged.

The Automotive Lift Institute (ALI) is an association concerned with the design, construction, installation, operation, maintenance, and repair of automotive lifts. Their primary concern is safety. Every lift approved by ALI has a safety label. It is a good idea to read through the safety tips included on this label before using a lift.

Jack and Jack Stand Safety

A vehicle can be raised by a hydraulic jack. A handle on the jack is moved up and down to raise part of a vehicle and a valve is turned to release the hydraulic pressure in the jack to lower the part. At the end of the jack is a lifting pad. The pad must be positioned under an area of the vehicle's frame or at one of the manufacturer's recommended lift points. Never place the pad under the floor pan or under steering and suspension components. Always position the jack so the wheels of the vehicle can roll as the vehicle is being raised.

 WARNING: **Never use a lift or jack to move something heavier than it is designed for. Always check the rating before using a lift or jack. If a jack is rated for two tons, do not attempt to use it for a job requiring five tons. It is dangerous for you and the vehicle.**

Safety (jack) stands are placed under a sturdy chassis member, such as the frame or axle housing, to support the vehicle. Once the safety stands (Figure 1-13) are in position, the hydraulic pressure in the jack should be slowly released until the weight of the vehicle is on the stands. Jack stands have a capacity rating. Always use the correct rating of jack stand.

Never move under a vehicle when it is only supported by a jack; rest the vehicle on safety stands before moving under the vehicle. The jack should be removed after the jack stands are set in place. This eliminates potential hazards, such as a jack handle sticking out into a walkway.

© Delmar/Cengage Learning

FIGURE 1-13 Jack or safety stands are placed under the vehicle after it has been raised by a hydraulic jack and before work is done under the vehicle.

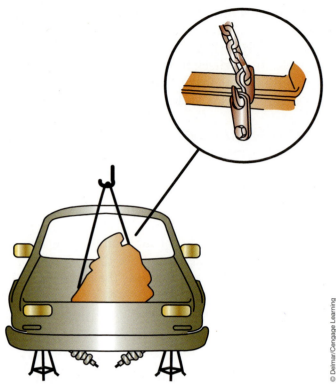

FIGURE 1-14 When using a chain hoist to pull an engine, the attachments to the engine should be secure and strong.

CAUTION:
The active chemical in a battery, the *electrolyte*, is basically sulfuric acid. Sulfuric acid can cause severe skin burns and permanent eye damage, including blindness. If battery acid gets on your skin, wash it off immediately and flush your skin with water for at least five minutes. If the electrolyte gets into your eyes, immediately flush them out with water, then immediately see a doctor. *Never* rub your eyes, just flush them well and go to a doctor. Working with and around batteries is an obvious time to wear safety glasses or goggles.

Chain Hoist and Crane Safety

Heavy parts of the automobile, such as engines, are removed with chain hoists or cranes. (Another term for a chain hoist is *chain fall*.) To prevent serious injury, chain hoists and cranes must be properly attached to the parts being lifted (Figure 1-14). Always use bolts with enough strength to support the object being lifted. Place the chain hoist or crane directly over the assembly. Then, attach the chain or cable to the hoist.

Cleaning Equipment Safety

Cleaning automotive parts can be divided into four basic categories: chemical cleaning, thermal cleaning, abrasive cleaning, and steam cleaning. Regardless of what method you use, you should always follow the precautions given by the manufacturer and you should always wear the recommended protective gear.

BATTERIES

When possible, you should disconnect the vehicle's battery before disconnecting any electrical wire or component. This prevents the possibility of a fire or electrical shock. It also eliminates the possibility of an accidental short, which can ruin the car's electrical system. Disconnect the negative or ground cable first (Figure 1-15), then disconnect the positive cable. Since electrical circuits require a ground to be complete, by removing the ground cable you eliminate the possibility of a circuit accidentally becoming completed. When reconnecting the battery, connect the positive cable first, then the negative.

The hydrogen gases that form in the top of a battery while it is being charged are very explosive. Never smoke or introduce any form of heat around a charging battery. An explosion will not only destroy the battery but may also spray sulfuric acid all over you, the car, and the shop. When connecting a battery charger to a battery, leave the charger off until all of its leads are connected. This will prevent electrical sparks and prevent a possible explosion.

FIGURE 1-15 Always disconnect a battery by removing the negative cable first.

The most dangerous battery is one that has been overcharged. It is hot and has been, or still may be, producing large amounts of hydrogen. Allow the battery to cool before working with or around it. Also, never use or charge a battery that has frozen electrolyte. Extreme amounts of hydrogen may be released from the electrolyte as it thaws.

Working Safely on High-Voltage Systems

Electric drive vehicles (battery operated, hybrid, and fuel cell electric vehicles) have high-voltage electrical systems (from 42 volts to 650 volts). These high voltages can kill you! Most high-voltage circuits are identifiable by size and color. The cables have thicker insulation and are typically colored orange (Figure 1-16). The connectors are also colored orange. On some vehicles, the high-voltage cables are enclosed in an orange shielding or casing. In addition, the high-voltage battery pack and most high-voltage components have "High Voltage" caution labels (Figure 1-17). Be careful not to touch these wires and parts.

Wear insulating gloves, commonly called "lineman's gloves", when working on or around the high-voltage system. These gloves must be class "0" rubber insulating gloves, rated at 1000 volts. Also, to protect the integrity of the insulating gloves, wear leather gloves over the insulating gloves while doing a service.

> ⚠️ **CAUTION:** Always double-check the polarity of the battery charger's connections before turning the charger on. Incorrect polarity can damage the battery or cause it to explode.

FIGURE 1-16 Most high-voltage circuits are identifiable by size and are typically colored orange.

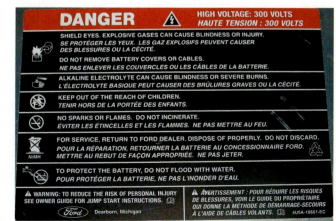

FIGURE 1-17 High-voltage components may have caution notices to alert technicians of the possible dangers.

Make sure they have no tears, holes, or cracks and are dry. Electrons can enter through the smallest of holes in your gloves. The integrity of the gloves should be checked before using them. To check the gloves, blow enough air into each one so they balloon out. Then fold the open end over to seal the air in. Continue to slowly fold that end of the glove toward the fingers. This will compress the air. If the glove continues to balloon as the air is compressed, it has no leaks. If any air leaks out, the glove should be discarded. All gloves, new and old, should be checked before they are used.

Other safety precautions that should always be adhered to when working on an electric drive vehicle are the following:

- Always adhere to the safety guidelines given by the vehicle's manufacturer.
- Obtain the necessary training before working on these vehicles.
- Perform each operation following the procedures defined by the manufacturer.
- Disable or disconnect the high-voltage system before servicing those systems.
- Anytime the engine is running, the generator is producing high voltage and care must be taken to prevent being shocked.
- Systems may have a large capacitor that must be discharged after the high-voltage system has been isolated. Wait for the prescribed amount of time (normally about 10 minutes) before working on or around the high-voltage system.
- After removing a high-voltage cable, cover the terminal with electrical tape.
- Always use insulated tools.
- Alert other technicians that you are working on the high-voltage system with a warning sign such as "high-voltage work: do not touch."
- Always install the correct circuit protection device into a high-voltage circuit.
- Many electric motors have a strong permanent magnet in them; individuals with a pacemaker should not handle these parts.
- When an electric drive vehicle needs to be towed into the shop for repairs, make sure it is not towed on its drive wheels. Doing this will drive the generator(s), which can overcharge the batteries and cause them to explode.
- Always tow these vehicles with the drive wheels off the ground or move them on a flat bed.

If the electric motor is sandwiched between the engine and the transmission, make sure you follow all procedures. The permanent magnet used in the motor is very strong and requires special tools to remove and install it.

VEHICLE OPERATION

When a customer brings a vehicle in for service, certain driving rules should be followed to ensure your safety and the safety of those working around you. For example, before moving a car into the shop, buckle your safety belt. Make sure no one is near, the way is clear, and there are no tools or parts under the car before you start the engine. Check the brakes before putting the vehicle in gear. Then drive slowly and carefully in and around the shop.

When road testing the car, obey all traffic laws. Drive only as far as is necessary to check the automobile and verify the customer's complaint. Never make excessively quick starts, turn corners too quickly, or drive faster than conditions allow.

If the engine must be running while you are working on the car, block the wheels to prevent the car from moving. Place an automatic transmission into park and set the parking (emergency) brake. Never stand directly in front of or behind a running vehicle.

Run the engine only in a well-ventilated area to avoid the danger of poisonous **carbon monoxide (CO)** in the engine exhaust. CO is an odorless but deadly gas. Most shops have an exhaust ventilation system (Figure 1-18); always use it. Connect the hose from the vehicle's tailpipe to the intake for the vent system. Make sure the vent system is turned on before running the engine.

Exhaust contains an odorless, colorless, and deadly gas: **carbon monoxide (CO)**. This poisonous gas gives very little warning to the victim and can kill in just a few minutes.

FIGURE 1-18 When running an engine in a shop, always connect the exhaust to the ventilation system.

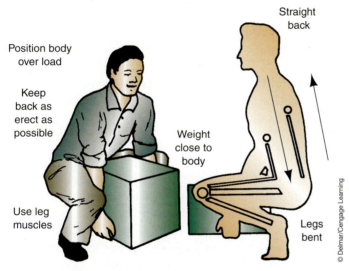

Straight back

Position body over load

Keep back as erect as possible

Weight close to body

Use leg muscles

Legs bent

FIGURE 1-19 When lifting a heavy object, always use your legs and keep the object close to your body.

LIFTING AND CARRYING

When lifting a heavy object like a transmission, use a hoist or have someone else help you. If you must work alone, *always* lift heavy objects with your legs, not your back. Bend down with your legs, not your back, securely hold the object you are lifting, and then stand up, keeping the object close to you (Figure 1-19). Trying to "muscle" something with your arms or back can result in severe damage to your back and may end your career and limit what you do for the rest of your life.

ACCIDENTS

Your entire work area should be kept clean and safe. Any oil, coolant, or grease on the floor can make it slippery. To clean up oil, use commercial oil absorbent. Most shops use absorbent mats in work areas to limit the effects of oil and fluid spills.

Keep all water off the floor. Water is slippery on smooth floors, and electricity flows well through water. Aisles and walkways should be kept clean and wide enough to move through easily. Make sure the work areas around machines are large enough to operate machines safely.

Make sure all drain covers are snugly in place. Open drains or covers that are not flush to the floor can cause toe, ankle, and leg injuries.

Handle all solvents (or any liquids) with care to avoid spillage. Keep all solvent containers closed, except when pouring. Proper ventilation is very important in areas where volatile solvents and chemicals are used. Solvents and other combustible materials must be stored in approved and ventilated storage cabinets or rooms.

Be extra careful when transferring flammable materials from bulk storage. Static electricity can build up enough to create a spark that could cause an explosion. Discard or clean all empty solvent containers. Never light matches or smoke near flammable solvents and chemicals, including battery acids.

Also, accidents can be prevented simply by the way you act. The following are some additional guidelines for working safely in a shop. This list does not include everything you should or shouldn't do; it merely gives some things to think about:

- Never smoke while working on a vehicle or working with any machine in the shop.
- Playing around is not fun when it sends someone to the hospital.
- To prevent serious burns, keep your skin away from hot metal parts such as the radiator, exhaust manifold, tailpipe, catalytic converter, and muffler.
- Always disconnect electric engine cooling fans when working around the radiator. These can turn on without warning and can easily chop off a finger or hand. Make sure you reconnect the fan after you have completed your repairs.
- When working with a hydraulic press, make sure the pressure is applied in a safe manner. It is generally wise to stand to the side when operating the press.
- Properly store all parts and tools by putting them away in a place, where people will not trip over them. This practice not only cuts down on injuries, it also reduces time wasted looking for a misplaced part or tool.

First Aid

If there is an accident, the quicker you respond to it, the less damage there will be. Your supervisor should be immediately informed of all accidents that occur in the shop. The work area should have a first-aid kit (Figure 1-20) for treating minor injuries. Facilities for flushing eyes should also be near or in the shop area. Know where they are. Keep an up-to-date list of emergency telephone numbers clearly posted next to the telephone. These numbers should include a doctor, hospital, and fire and police departments.

It is also a good idea for you to know first aid. The knowledge of how to treat certain injuries can save someone's life. The American Red Cross offers many low-cost but thorough

FIGURE 1-20 A typical-first aid kit and its contents.

© Delmar/Cengage Learning

courses on first aid. You will realize the importance of these classes the first time you have to give first aid to someone or when someone must give it to you.

You should find out if there is a resident nurse in the shop or at the school, and know where the nurse's office is. If there are specific first-aid rules in your school or shop, make sure you are aware of them and follow them.

If someone is overcome by carbon monoxide, get him or her fresh air immediately. Burns should be cooled immediately by rinsing them with water. Whenever there is severe bleeding from a wound, try to stop the bleeding by applying pressure with clean gauze on or around the wound, and get medical help. Never move someone who may have broken bones unless the person's life is otherwise endangered. Moving that person may cause additional injury. Call for medical assistance.

HAZARDOUS MATERIALS

A typical shop contains many potential health hazards for those working in it. These hazards can be classified as:

- Chemical hazards, caused by high concentrations of vapors, gases, or solids in the form of dust.
- Hazardous wastes, which are those substances left behind or result from a process or service.
- Physical hazards, including excessive noise, vibration, pressures, and temperatures.
- Ergonomic hazards, which are conditions that impede normal or proper body position and motion.

There are many federal agencies charged with ensuring safe work environments for all workers. These include the Occupational Safety and Health Administration **(OSHA)**, the Mine Safety and Health Administration (MSHA), and the National Institute for Occupational Safety and Health (NIOSH). These entities, in addition to state and local governments, have instituted regulations that must be understood and followed. Everyone in a shop has the responsibility for adhering to these regulations.

OSHA

OSHA (Occupational Safety and Health Administration) is a branch of the U.S. Department of Labor. This branch was formed to write and enforce operational guidelines for all businesses to ensure that employees work in safe and healthy conditions. The guidelines include standards for cleanliness, air ventilation, fire prevention, emergency measures, equipment condition, and personal protective equipment.

It is the employer's responsibility to provide a place of employment that is free from all recognized safety and health hazards. OSHA regulates all safety and health issues concerning the automotive industry. Businesses may be periodically inspected by OSHA personnel. The owner of the business may be cited for any violations of the standards. The owner will be given a period of time to bring the work area into compliance. If the owner does not correct the situation or has been a repeat violator of the standards, OSHA will fine the owner.

Right-to-Know Laws

Every employee in a shop is protected by "right-to-know" laws. The general intent of these laws is that employers must provide a safe workplace with respect to hazardous materials. All employees must be trained about their rights under the legislation, the nature of the hazardous chemicals in their workplace, the labeling of chemicals, and the information about each chemical listed and described on **Material Safety Data Sheets (MSDSs)**. These sheets (Figure 1-21) are available in writing and/or online from the manufacturers and suppliers of the chemicals. They detail the chemical composition of and precautionary information for all products that can present health or safety hazards. The Canadian equivalents to the MSDS are called Workplace Hazardous Materials Information Systems (WHMIS).

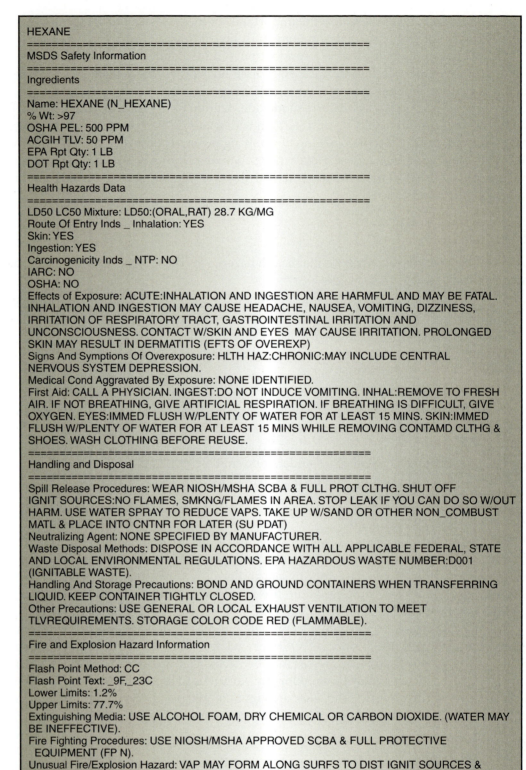

```
HEXANE
============================================================
MSDS Safety Information
============================================================
Ingredients
============================================================
Name: HEXANE (N_HEXANE)
% Wt: >97
OSHA PEL: 500 PPM
ACGIH TLV: 50 PPM
EPA Rpt Qty: 1 LB
DOT Rpt Qty: 1 LB
============================================================
Health Hazards Data
============================================================
LD50 LC50 Mixture: LD50:(ORAL,RAT) 28.7 KG/MG
Route Of Entry Inds _ Inhalation: YES
Skin: YES
Ingestion: YES
Carcinogenicity Inds _ NTP: NO
IARC: NO
OSHA: NO
Effects of Exposure: ACUTE:INHALATION AND INGESTION ARE HARMFUL AND MAY BE FATAL.
INHALATION AND INGESTION MAY CAUSE HEADACHE, NAUSEA, VOMITING, DIZZINESS,
IRRITATION OF RESPIRATORY TRACT, GASTROINTESTINAL IRRITATION AND
UNCONSCIOUSNESS. CONTACT W/SKIN AND EYES  MAY CAUSE IRRITATION. PROLONGED
SKIN MAY RESULT IN DERMATITIS (EFTS OF OVEREXP)
Signs And Symptions Of Overexposure: HLTH HAZ:CHRONIC:MAY INCLUDE CENTRAL
NERVOUS SYSTEM DEPRESSION.
Medical Cond Aggravated By Exposure: NONE IDENTIFIED.
First Aid: CALL A PHYSICIAN. INGEST:DO NOT INDUCE VOMITING. INHAL:REMOVE TO FRESH
AIR. IF NOT BREATHING, GIVE ARTIFICIAL RESPIRATION. IF BREATHING IS DIFFICULT, GIVE
OXYGEN. EYES:IMMED FLUSH W/PLENTY OF WATER FOR AT LEAST 15 MINS. SKIN:IMMED
FLUSH W/PLENTY OF WATER FOR AT LEAST 15 MINS WHILE REMOVING CONTAMD CLTHG &
SHOES. WASH CLOTHING BEFORE REUSE.
============================================================
Handling and Disposal
============================================================
Spill Release Procedures: WEAR NIOSH/MSHA SCBA & FULL PROT CLTHG. SHUT OFF
IGNIT SOURCES:NO FLAMES, SMKNG/FLAMES IN AREA. STOP LEAK IF YOU CAN DO SO W/OUT
HARM. USE WATER SPRAY TO REDUCE VAPS. TAKE UP W/SAND OR OTHER NON_COMBUST
MATL & PLACE INTO CNTNR FOR LATER (SU PDAT)
Neutralizing Agent: NONE SPECIFIED BY MANUFACTURER.
Waste Disposal Methods: DISPOSE IN ACCORDANCE WITH ALL APPLICABLE FEDERAL, STATE
AND LOCAL ENVIRONMENTAL REGULATIONS. EPA HAZARDOUS WASTE NUMBER:D001
(IGNITABLE WASTE).
Handling And Storage Precautions: BOND AND GROUND CONTAINERS WHEN TRANSFERRING
LIQUID. KEEP CONTAINER TIGHTLY CLOSED.
Other Precautions: USE GENERAL OR LOCAL EXHAUST VENTILATION TO MEET
TLVREQUIREMENTS. STORAGE COLOR CODE RED (FLAMMABLE).
============================================================
Fire and Explosion Hazard Information
============================================================
Flash Point Method: CC
Flash Point Text: _9F,_23C
Lower Limits: 1.2%
Upper Limits: 77.7%
Extinguishing Media: USE ALCOHOL FOAM, DRY CHEMICAL OR CARBON DIOXIDE. (WATER MAY
BE INEFFECTIVE).
Fire Fighting Procedures: USE NIOSH/MSHA APPROVED SCBA & FULL PROTECTIVE
  EQUIPMENT (FP N).
Unusual Fire/Explosion Hazard: VAP MAY FORM ALONG SURFS TO DIST IGNIT SOURCES &
FLASH BACK. CONT W/STRONG OXIDIZERS MAY CAUSE FIRE. TOX GASES PRDCED MAY
INCL:CARBON MONOXIDE, CARBON DIOXIDE.
============================================================
```

FIGURE 1-21 A sample of a Material Safety Data Sheet.

CAUTION:

When handling any hazardous material, always wear the appropriate safety protection. Always follow the correct procedures while using the material and be familiar with the information given on the MSDS for that material.

Employees must be familiar with the intended purposes of each substance, the recommended protective equipment, accident and spill procedures, and any other information regarding the safe handling of hazardous materials. This training must be given annually to employees and provided to new employees as part of their job orientation.

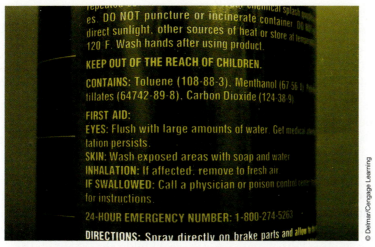

FIGURE 1-22 Carefully read the labels on all chemicals before using them.

An MSDS must include the following information about the product:

- The trade and chemical name of the product.
- The manufacturer of the product.
- All of the ingredients of the product.
- Health hazards such as headaches, skin rashes, nausea, and dizziness.
- The product's physical description; this information may include the product's color, odor, permissible exposure limit (PEL), threshold limit value (TLV), specific gravity, boiling point, freezing point, evaporation data, and volatility rating.
- The product's explosion and fire data, such as its flashpoint.
- The product's reactivity and stability data.
- The product's weight compared to air.
- Protection data, including first aid and proper handling.

Shops must maintain documentation on the hazardous chemicals in the workplace, proof of training programs, records of accidents or spill incidents, satisfaction of employee requests for specific chemical information via the MSDS, and a general right-to-know compliance procedure manual for use within the shop.

All hazardous material must be properly labeled, indicating what health, fire, or reactivity hazard it poses (Figure 1-22) and what protective equipment is necessary when handling each chemical. The manufacturer of the hazardous material must provide all relevant warnings and precautionary information, which must be read and understood by the user before application. Attention to all label precautions is essential for the proper use of the chemical and for prevention of hazardous conditions.

Hazardous Wastes

Many repair and service procedures generate what are known as **hazardous wastes**. Dirty solvents and cleaners are good examples of hazardous wastes. Something is classified as a hazardous waste by the Environmental Protection Agency (EPA) if it is on the EPA list of known harmful materials or has one or more of the following characteristics:

Ignitability. A liquid with a flashpoint below 140°F or a solid that can spontaneously ignite.

Corrosivity. A substance that dissolves metals and other materials or burns the skin.

Reactivity. Any material that reacts violently with water or other materials or releases cyanide gas, hydrogen sulfide gas, or similar gases when exposed to low pH acid solutions. This also includes material that generates toxic mists, fumes, vapors, and flammable gases.

EP (extreme pressure) toxicity. Materials that leach one or more of eight heavy metals in concentrations greater than 100 times primary drinking water standard concentrations.

Reactivity is how easily a substance can cause or be part of a chemical reaction.

A complete EPA list of hazardous wastes can be found in the Code of Federal Regulations. It should be noted that no material is considered hazardous waste until the shop has finished using it and is ready to dispose of it.

In the United States, OSHA regulates the use of many of these materials. The EPA regulates the disposal of hazardous waste. A summary of these regulations follows.

- All businesses that generate hazardous waste must develop a hazardous waste policy.
- Each hazardous waste generator must have an EPA identification number.
- When waste is transported for disposal, a licensed waste hauler must be used to transport and dispose of the waste. A copy of a written manifest (an EPA form) must be kept by the shop.

Regulations on hazardous waste handling and generation have led to the development of equipment that is now commonly found in shops. Examples are thermal cleaning units, closed-loop steam cleaners, waste oil furnaces, oil filter crushers, refrigerant recycling machines, engine coolant recycling machines, and highly absorbent cloths.

 WARNING: **The shop is ultimately responsible for the safe disposal of hazardous wastes, even after the waste leaves the shop. Only licensed waste removal companies should be used to dispose of the waste. Make sure you know what the company is planning to do with the waste. Make sure you have a written contract stating what is supposed to happen with the waste. Leave nothing to chance. In the event of an emergency hazardous waste spill, contact the National Response Center (1-800-424-8802) immediately. Failure to do so can result in a $10,000 fine, a year in jail, or both.**

OSHA and the EPA have other strict rules and regulations that help to promote safety in the auto shop. These are described throughout this text whenever they are applicable.

Maintaining a vehicle involves handling and managing a wide variety of materials and wastes. Some of these wastes can be toxic to fish, wildlife, and humans when improperly managed. No matter the amount of waste produced, it is to the shop's legal and financial advantage to manage waste properly and, even more importantly, to prevent pollution.

Hazardous Waste Disposal

Hazardous wastes must be properly stored and disposed of. It is the shop's responsibility to hire a reputable, financially stable, and state-approved hauler, who will dispose of the shop wastes legally. Select a licensed hazardous waste hauler after seeking recommendations and reviewing the firm's permits and authorizations. If hazardous waste is dumped illegally, your shop may be held responsible.

Always keep hazardous waste separate, properly labeled, and sealed in the recommended containers. The storage area should be covered and may need to be fenced and locked if vandalism could be a problem.

Handling Shop Wastes

The following is a summary of the proper method of preparing and disposing of common hazardous wastes. When handling these classified wastes, make sure to wear the appropriate safety equipment (Figure 1-23). These are general guidelines; always follow the specific state and federal regulations for disposing of these items.

Oil. Recycle oil. Set up equipment, such as a drip table or screen table with a used oil collection bucket, to collect oil dripping off parts. Place drip pans under vehicles that are leaking fluids. Recycle the oil according to local regulations. Do not mix other wastes with used oil, except as allowed by your recycler.

FIGURE 1-23 Wear the appropriate safety equipment when handling hazardous materials and waste.

ATF. Automatic transmission fluid should be collected in a separate container from other lubricants and collected by an approved recycler.

Oil Filters. Drain for at least 24 hours, crush, and recycle used oil filters.

Batteries. Recycle batteries by sending them to a reclaimer or back to the distributor. Store batteries in a water-tight, acid resistant container. Inspect batteries for cracks and leaks when they come in. Treat a dropped battery as if it were cracked. Acid residue is hazardous because it is corrosive and may contain lead and other toxics. Neutralize spilled acid by using baking soda or lime, and dispose of as hazardous material.

Containers. Cap, label, cover, and properly store above ground and outdoors any liquid containers and small tanks within a diked area and on a paved impermeable surface to prevent spills from running into surface or ground water.

Metal Residue from Machining. Collect metal filings when machining metal parts. Keep separate and recycle if possible. Prevent metal filings from falling into a storm sewer drain.

Other Solids. Store materials such as scrap metal, old machine parts, and worn tires under a roof or tarpaulin to protect them from the elements and to prevent the potential to create contaminated runoff. Consider recycling tires by retreading them.

Refrigerants. Recover or recycle refrigerants during the service and disposal of motor vehicle air conditioners and/or refrigeration equipment. You are not allowed to knowingly vent refrigerants into the atmosphere. Recovery or recycling during service must be performed by an EPA-certified technician using certified equipment and following specified procedures.

Solvents. Replace hazardous chemicals with less toxic alternatives that have equal performance. For example, substitute water-based cleaning solvents for petroleum-based solvent degreasers. To reduce the amount of solvent used when cleaning parts, use a two-stage process: dirty solvent followed by fresh solvent. Hire a hazardous-waste management service to clean and recycle solvents. Store solvents in closed containers to prevent evaporation. Evaporation of solvents contributes to ozone depletion and smog formation. Properly label spent solvents and store on drip pans or in diked areas and only with compatible materials.

Liquid Recycling. Collect and recycle coolants from radiators. Store transmission fluids, brake fluids, and solvents containing chlorinated hydrocarbons separately, and recycle or dispose of them properly.

Shop Towels/Rags. Keep waste towels in a closed container marked "contaminated shop towels only." To reduce costs and liabilities associated with disposal of used towels, which can be classified as hazardous waste, investigate using a laundry service that is able to treat the wastewater generated from cleaning the towels.

Waste Storage. Always keep hazardous waste separate, properly labeled, and sealed in the recommended containers. The storage area should be covered and may need to be fenced and locked if vandalism could be a problem. Select a licensed hazardous waste hauler after seeking recommendations and reviewing the firm's permits and authorizations.

SUMMARY

- You have an important role in creating an accident-free work environment. Your appearance and work habits will go a long way toward preventing accidents. Common sense and concern for others will result in fewer accidents.
- It's everyone's job to periodically check for safety hazards in a shop.
- Dressing safely for work is very important. This includes snug-fitting clothing, eye and ear protection, protective gloves, strong shoes, and caps to cover long hair.
- Safety glasses should be worn whenever you are working under a vehicle or when you are using machining equipment, grinding wheels, chemicals, compressed air, or fuels.
- There are two areas of housekeeping you are responsible for: your work area and the rest of the shop.
- Dirty and oily rags should be stored in approved containers.
- All equipment should be inspected for safety hazards before being used.
- Gasoline and diesel fuel are highly flammable and should be kept in approved safety cans.
- It is important to know when to use each of the various types of fire extinguishers. When fighting a fire, aim the nozzle at the base and use a side-to-side sweeping motion.
- Fire, heat, and sparks should be kept away from a battery. These could cause the battery to explode.
- Heavy objects should be lifted with your legs, not your back.
- Special care should be taken whenever using power tools, such as impact wrenches, gear pullers, and jacks.
- Always observe all relevant safety rules when operating a vehicle lift or hoist. Jacks, jack stands, chain hoists, and cranes can also cause injury if not operated safely.
- Use care whenever it is necessary to move a vehicle within the shop. Carelessness and playing around can lead to a damaged vehicle and serious injury.
- Carbon monoxide gas is a poisonous gas present in engine exhaust fumes. It must be properly vented from the shop using tailpipe hoses or other reliable methods.
- Accidents can be prevented by not having anything dangle near rotating equipment and parts.
- Accidents can happen, and when they do, you should respond immediately to prevent further injury or damage.
- All employees have the right to know what hazardous materials they are using to perform their job. Most of the needed information is contained on an MSDS.

TERMS TO KNOW

Bloodborne pathogens

Carbon monoxide (CO)

Caustic

Class A fire

Class B fire

Class C fire

Class D fire

Corrosivity

EP toxicity

Flammability

Hazardous waste

Hydrocarbon

Ignitability

Material Safety Data Sheet (MSDS)

OSHA

Reactivity

Shank

Spontaneous combustion

Volatility

- Always strictly follow the service recommendations and precautions given by the manufacturer when working on a hybrid vehicle.
- Right-to-know laws began in 1983 and are designed to protect employees who must handle hazardous materials and waste on the job.
- Hazardous wastes must be properly disposed of. Typically, shops hire full-service waste management firms to remove and dispose of this waste.

REVIEW QUESTIONS

1. While discussing ways to prevent fires:
   Technician A says all dirty and oily rags should be stored in approved containers.

   Technician B says all dirty and oily rags should be kept in a pile until the end of the day, and then they should be moved to a suitable container.

   Who is correct?
   A. A only
   B. B only
   C. Both A and B
   D. Neither A nor B

2. *Technician A* says accidents can be prevented by not having anything dangle near rotating equipment and parts.

   Technician B says accidents can be prevented by having common sense.

   Who is correct?
   A. A only
   B. B only
   C. Both A and B
   D. Neither A nor B

3. While discussing what to do when an accident does occur:
   Technician A says a technician should immediately determine the cause of the accident.

   Technician B says a technician should respond immediately to prevent further injury or damage.

   Who is correct?
   A. A only
   B. B only
   C. Both A and B
   D. Neither A nor B

4. *Technician A* says gasoline spills should be immediately cleaned up.

   Technician B says gasoline should only be stored in approved containers.

   Who is correct?
   A. A only
   B. B only
   C. Both A and B
   D. Neither A nor B

5. *Technician A* says unsafe equipment should have its power disconnected and be marked with a sign to warn others.

   Technician B says that all equipment should be inspected for safety hazards before being used.

   Who is correct?
   A. A only
   B. B only
   C. Both A and B
   D. Neither A nor B

6. *Technician A* says hazardous wastes must be properly disposed of.

   Technician B says shops hire full-service waste management firms to remove and dispose of hazardous wastes.

   Who is correct?
   A. A only
   B. B only
   C. Both A and B
   D. Neither A nor B

7. While discussing ways to create an accident-free work environment:
   Technician A says everyone in the shop should take full responsibility for ensuring safe work areas.

   Technician B says the appearance and work habits of technicians can help prevent accidents.

   Who is correct?
   A. A only
   B. B only
   C. Both A and B
   D. Neither A nor B

8. While discussing why exhaust fumes should be vented outdoors or drawn into a ventilation/filtration system:
   Technician A says exhaust gases contain amounts of carbon monoxide.

   Technician B says carbon dioxide is an odorless, colorless, and deadly gas.

   Who is correct?
   A. A only
   B. B only
   C. Both A and B
   D. Neither A nor B

9. While discussing simple first-aid procedures:

   Technician A says that if someone is overcome by carbon monoxide, you should get him or her fresh air immediately.

   Technician B says burns should be cooled immediately by rinsing them in water.

   Who is correct?

   A. A only
   B. B only
   C. Both A and B
   D. Neither A nor B

10. While discussing a car's electrical system:

    Technician A says you should always disconnect the negative or ground battery cable first, then disconnect the positive cable.

    Technician B says you should always connect the positive cable first, then the negative.

    Who is correct?

    A. A only
    B. B only
    C. Both A and B
    D. Neither A nor B

Name _____ **Date** _____

SHOP SAFETY SURVEY

As a professional technician, safety should be one of your first concerns. This job sheet will increase your awareness of shop safety rules and equipment. As you survey your shop area and answer the questions, you will learn how to evaluate the safeness of your workplace.

Procedure

Task Completed

Your instructor will review your progress throughout this job sheet and should sign off on the sheet when you complete it.

1. Before you begin to evaluate your work area, evaluate yourself. Are you dressed to work safely? ☐ Yes ☐ No
   If no, what is wrong?

   _____

2. Are your safety glasses OSHA approved? ☐ Yes ☐ No
   Do they have protective shields? ☐ Yes ☐ No

3. Walk around the shop and note any area that poses a potential safety hazard or is an area that you should be aware of.

   _____

   _____

   Any true hazards should be brought to the attention of the instructor immediately.

4. Are there safety areas marked around grinders and other machinery?
   ☐ Yes ☐ No

5. What is the line air pressure in the shop? _____ psi
   What should it be? _____ psi

6. Where are the tools stored in the shop?

   _____

   _____

7. If you could, how would you improve the tool storage area?

   _____

   _____

8. What types of hoists are used in the shop?

   _____

   _____

9. Ask your instructor to demonstrate the proper use of a hoist.

10. Where is the first aid kit(s) kept in the work area?

    _____

    _____

11. What is the shop's procedure for dealing with an accident?

_____

_____

12. Have your instructor supply you with a vehicle's make, model, and year. Using the appropriate service manual, find the location of the correct lift points for that vehicle. On the rear of this sheet, draw a simple figure showing where these lift points are.

Instructor's Response _____

_____

_____

Name _____ **Date** _____

Working Safely around Air Bags

Upon completion of this job sheet, you should be able to work safely around and with air bag systems.

Tools and Materials

A vehicle with air bags
Service manual for this vehicle
Component locator for this vehicle
Safety glasses
A DMM

Describe the vehicle being worked on.

Year _____ Make _____ VIN _____
Model _____

Procedure

Task Completed

1. Locate the information about the air bag system in the service manual. How are the critical parts of the system identified in the vehicle?

   _____
   _____
   _____
   _____

2. List the main components of the air bag system and describe their locations.

   _____
   _____
   _____
   _____
   _____
   _____
   _____
   _____

3. There are some very important guidelines to follow when working with and around air bag systems. These are listed below with some key words left out. Read through these guidelines and fill in the blanks with the correct words.

   a. Wear _____ _____ when servicing an air bag system and handling an air bag module.

   b. Wait at least _____ minutes after disconnecting the battery before beginning any service. The reserve _____ module is capable of storing enough energy to deploy the air bag for up to _____ minutes after battery voltage is lost.

c. Always handle all _____ and other components with extreme care. Never strike or jar a sensor, especially when the battery is connected. This can cause deployment of the air bag.

d. Never carry an air bag by its _____ or _____, and, when carrying it, always face the trim and air bag _____ from your body. When placing a module on a bench, always face the trim and air bag _____.

e. Deployed air bags may have a powdery residue on them. _____ is produced by the deployment reaction and is converted to _____ when it comes in contact with the moisture in the atmosphere. Although it is unlikely that chemicals will still be on the bag, it is wise to wear _____ _____ and _____ when handling a deployed air bag. Immediately wash your hands after handling a deployed air bag.

f. A live air bag must be _____ before it is disposed of. A deployed air bag should be disposed of in a manner consistent with the _____ and manufacturer's procedures.

g. Never use a battery- or AC-powered (alternating-current) _____, _____, or any other type of test equipment on the system unless the manufacturer specifically says to. Never probe with a _____ _____ for voltage.

Instructor's Response _____

_____

_____

Name _____ Date _____

DISCONNECTING THE HIGH-VOLTAGE CIRCUIT ON A HYBRID VEHICLE

Upon completion of this job sheet, you will be able to locate and safely disconnect the high-voltage circuit on a hybrid vehicle.

ASE Correlation

This job sheet is related to the ASE Automatic Transmission and Transaxle Test's Content Area *In-Vehicle Transmission and Transaxle Repair.*

> Identify the location of hybrid vehicle high-voltage circuit disconnect (service plug) and safety procedures.
>
> Identify high-voltage circuits of electric or hybrid electric vehicle and related safety precautions.

Tools and Materials

Service manual

Vinyl tape for insulation

Insulated gloves that are dry and are
   not cracked, ruptured, torn, or
   damaged in any way

Hybrid vehicle

Goggles or safety glasses with side shields

Describe the vehicle being worked on:

Year _____ Make _____ Model _____

VIN _____ Engine type and size _____

Procedure

Task Completed

_____

⚠ **WARNING: Unprotected contact with any electrically charged ("hot" or "live") high-voltage component could cause serious injury or death.**

1. Describe how this vehicle is identified as a hybrid.

   _____

   _____

   _____

2. After referring to the vehicle's owner's manual, briefly describe its operation.

   _____

   _____

   _____

   _____

3. At how many volts are the batteries rated?

   _____

4. Where are the batteries located?

   _____

   _____

5. How are the high-voltage cables labeled and identified?

   _____

   _____

6. What is used to provide short-circuit protection in the high-voltage battery pack?

   _____

   _____

7. What isolates the high-voltage system from the rest of the vehicle when the vehicle is shut off?

   _____

   _____

8. Using the vehicle's service or owner's manual as a guide, list at least five precautions that must be adhered to when servicing this hybrid vehicle.

   _____

   _____

   _____

   _____

   _____

9. If the vehicle has been in an accident and some electrolyte from the batteries has leaked out, how should you clean up the spill?

   _____

   _____

   _____

   _____

10. Look under the hood and describe the components that are visible.

    _____

    _____

    _____

    _____

11. Without touching the high-voltage cables or components, describe the routing of these cables. Include what they appear to be connected from and to.

    _____

    _____

    _____

    _____

WARNING: Never assume that a hybrid vehicle is shut off simply because it is silent. Make sure the ignition key is in your pocket and not in the ignition switch.

12. Describe the procedure for totally isolating the high-voltage system from the vehicle. This often involves the removal of a service plug. Be sure to include the location of this plug in your description of the procedure.

_____

_____

_____

_____

_____

_____

_____

_____

_____

_____

13. Make two signs saying, "Working on High-Voltage Parts. Do not touch!" Attach one to the steering wheel, and set the other one near the parts you are working on. ☐

14. After the service plug has been removed, how long should you wait before working around or on the high-voltage system?

_____

_____

Instructor's Response _____

_____

_____

Chapter 2

SPECIAL TOOLS AND PROCEDURES

UPON COMPLETION AND REVIEW OF THIS CHAPTER, YOU SHOULD BE ABLE TO:

- Describe the use of common diagnostic tools for automatic transmissions and transaxles.

- Describe the use of common pneumatic, electrical, and hydraulic power tools found in an automotive service department.

- Describe some of the special tools used to service automatic transmissions and the driveline.

- List the basic units of measure for length, volume, and mass in the main measuring systems.

- Identify the major measuring instruments and devices used by technicians.

- Explain what common measuring instruments and devices measure and how to use them.

- Describe the proper procedure for measuring with a micrometer.

- Describe the measurements normally taken by a technician while working on a vehicle's drivetrain.

- Describe the different sources for service information that are available to technicians.

- Describe the different types of fasteners used in the automotive industry.

- Describe the requirements for ASE certification as an automotive technician and a master auto technician.

INTRODUCTION

Many different special tools and testing and measuring devices are used to service automatic transmissions. The most commonly used tools are the topic of the following paragraphs. Although you need to be more than familiar with, and will be using, common hand tools, they are not part of this discussion. You should already know what they are and how to use and care for them.

This chapter also discusses some of the everyday service procedures performed by an automatic transmission technician and the basic duties and responsibilities of all automotive technicians.

DIAGNOSTIC TOOLS

Automatic transmission repair requires much more than simply taking a unit apart and putting it back together. Perhaps your most important task when confronted with an automatic transmission problem is to properly identify its cause. Often the problem can be corrected without tearing down the unit. Many tools are available to help with diagnostics. These are discussed here. In addition, basic guidelines for diagnostics are presented.

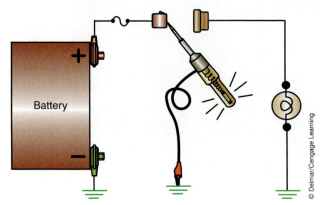

FIGURE 2-1 The bulb of a test light will illuminate if there is power at a connector and the light is properly grounded.

Circuit Tester

Circuit testers are used to identify short and open circuits in any electrical circuit. Low-voltage testers are used to troubleshoot 6- to 12-volt circuits. A **circuit tester**, commonly called a test light, looks like a stubby ice pick. Its handle is transparent and contains a light bulb. A probe extends from one end of the handle and a ground clip and wire from the other end. When the ground clip is attached to a good ground and the probe touched to a live connector, the bulb in the handle will light up (Figure 2-1). If the bulb does not light, voltage is not available at the connector.

 WARNING: Do not use a conventional 12-volt test light to diagnose components and wires in modern electronic systems. The current draw of these test lights may damage computers and system components. High-impedance test lights are available for diagnosing electronic systems.

A self-powered test light is called a continuity tester. It is used on non-powered circuits. It looks like a regular test light, except that it has a small internal battery. When the ground clip is attached to the negative side of a component and the probe touched to the positive side, the lamp will light if there is continuity in the circuit. If an open circuit exists, the lamp will not light. Do not use any type of test light or circuit tester to diagnose automotive air bag systems.

Voltmeter

A voltmeter has two leads: a red positive lead and a black negative lead. The red lead should be connected to the positive side of the circuit or component. The black should be connected to ground or to the negative side of the component. Voltmeters should be connected across the circuit being tested (Figure 2-2).

The voltmeter measures the voltage available at any point in an electrical system. A voltmeter can also be used to test voltage drop across an electrical circuit, component, switch, or connector. A voltmeter can also be used to check for proper circuit grounding.

Ohmmeter

An ohmmeter measures the resistance to current flow in a circuit. In contrast to the voltmeter, which uses the voltage available in the circuit, the ohmmeter is battery powered. The circuit being tested must be open (Figure 2-3). If the power is on in the circuit, the ohmmeter will be damaged.

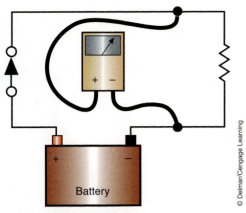

FIGURE 2-2 A voltmeter is used to measure available voltage and voltage drops.

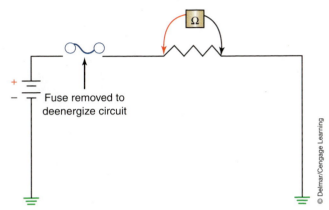

Fuse removed to deenergize circuit

FIGURE 2-3 Always connect an ohmmeter to a component or circuit *after* the power has been disconnected.

The meter sends current through the component and determines the amount of resistance based on the voltage dropped across the load. The scale of an ohmmeter reads from zero to infinity. A zero reading means there is no resistance in the circuit and may indicate a short in a component that should show a specific resistance. An infinite reading indicates a number higher than the meter can measure. This usually is an indication of an open circuit.

Ohmmeters are also used to trace and check wires or cables. Assume that one wire of a four-wire cable is to be found. Connect one probe of the ohmmeter to the known wire at one end of the cable and touch the other probe to each wire at the other end of the cable. Any evidence of resistance, such as meter needle deflection, indicates the correct wire. Using this same method, you can check a suspected defective wire. If resistance is shown on the meter, the wire is sound. If no resistance is measured, the wire is defective (open). If the wire is okay, continue checking by connecting the probe to other leads. Any indication of resistance indicates that the wire is shorted to one of the other wires and that the harness is defective.

Ammeter

An ammeter measures current flow in a circuit. The ammeter must be placed into the circuit or in series with the circuit being tested (Figure 2-4). Normally, this requires disconnecting a wire or connector from a component and connecting the ammeter between the wire or connector and the component. The red lead of the ammeter should always be connected to the side of the connector closest to the positive side of the battery and the black lead should be connected to the other side.

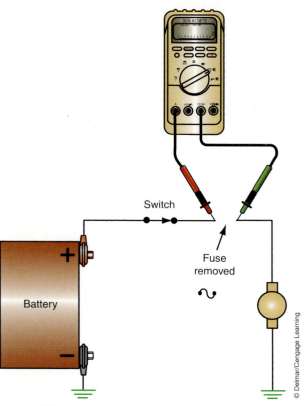

Switch

Battery

Fuse
removed

© Delmar/Cengage Learning

FIGURE 2-4 An ammeter should always be connected in series with the circuit.

© Delmar/Cengage Learning

FIGURE 2-5 An inductive pickup for measuring amperes. This tool is often called a current probe.

It is much easier to test current using an ammeter with an inductive pickup. The pickup clamps around the wire or cable being tested (Figure 2-5). These ammeters measure amperage based on the magnetic field created by the current flowing through the wire. This type of pickup eliminates the need to separate the circuit to insert the meter.

Because ammeters are built with very low internal resistance, connecting them in series does not add any appreciable resistance to the circuit. Therefore, an accurate measurement of the current flow can be taken.

Digital Multimeters

It's not necessary for a technician to own separate meters to measure volts, ohms, and amperes; a multimeter can be used instead. Top-of-the-line multimeters are multifunctional. Most test volts, ohms, and amperes in both DC and AC. Usually, there are several test ranges provided for each of these functions. In addition to these basic electrical tests, multimeters also test engine rpm, duty cycle, pulse width, diode condition, frequency, and even temperature. The technician selects the desired test range by turning a control knob on the front of the meter.

Multimeters are available with either analog or digital displays, but the most commonly used multimeter (Figure 2-6) is the **digital volt/ohmmeter (DVOM)**, which is often referred to as a **digital multimeter (DMM)**. There are several drawbacks to using analog-type meters for testing electronic control systems. Many electronic components require very precise test results. Digital meters can measure volts, ohms, or amperes in tenths and hundredths. Another problem with analog meters is their low internal resistance (input impedance). The low input impedance allows too much current to flow through circuits and should not be used on delicate electronic devices.

Digital meters, on the other hand, have a high input impedance, usually at least 10 megohms (10 million ohms). Metered voltage for resistance tests is well below 5 volts, reducing the risk of damage to sensitive components and delicate computer circuits. A high-impedance digital

© Delmar/Cengage Learning

FIGURE 2-6 A digital multimeter.

Prefix	Symbol	Relation to basic unit
Mega	M	1,000,000
Kilo	k	1000
Milli	m	.001 or $\frac{1}{1000}$
Micro	μ	.000001 or $\frac{1}{1,000,000}$
Nano	n	0.000000001
Pico	p	0.000000000001

© Delmar/Cengage Learning

FIGURE 2-7 The various symbols used to define values shown on a DMM.

multimeter must be used to test the voltage of some components and systems such as an oxygen (O_2) sensor circuit. If a low-impedance analog meter is used in this type of circuit, the current flow through the meter is high enough to damage the sensor.

DMMs have either an "auto range" feature, in which the appropriate scale is automatically selected by the meter, or they must be set to a particular range. In either case, you should be familiar with the ranges and the different settings available on the meter you are using. To designate particular ranges and readings, meters display a prefix before the reading or range. If the meter has a setting for mAmps, that means the readings will be given in milli-amps or 1/1000th of an ampere. Ohmmeter scales are expressed as a multiple of tens or use the prefix K or M. K stands for Kilo or 1000. A reading of 10K ohms equals 10,000 ohms. An M stands for Mega or 1,000,000. A reading of 10M ohms equals 10,000,000 ohms (Figure 2-7). When using a meter with an auto range, make sure you note the range being used by the meter. There is a big difference between 10 ohms and 10,000,000 ohms.

After the test range has been selected, the meter is connected to the circuit in the same way as if it were an individual meter.

When using the ohmmeter function, the DMM will show a zero or close to zero when there is good continuity. If the continuity is very poor, the meter will display an infinite reading. This reading is usually shown as a blinking "1.000," a blinking "1," or an "OL." Before taking any measurement, calibrate the meter. This is done by holding the two leads together and adjusting the meter reading to zero. Not all meters need to be calibrated; some digital meters automatically calibrate when a scale is selected. On meters that require calibration, it is recommended that the meter be zeroed after changing scales.

Multimeters may also have the ability to measure duty cycle, pulse width, and frequency. All of these represent voltage pulses caused by the turning on and off of a circuit or the increase and decrease of voltage in a circuit. **Duty cycle** is a measurement of the amount of time something is on compared to the time of one cycle and is measured in a percentage.

Pulse width is similar to duty cycle except that it is the exact time something is turned on and is measured in milliseconds. When measuring duty cycle, you are looking at the amount of time something is on during one cycle.

The number of cycles that occur in one second is called the **frequency**. The higher the frequency, the more cycles occur in a second. Frequencies are measured in Hertz. One Hertz is equal to one cycle per second.

Some DMMs have a pressure transducer as an attachment. These transducers can be used in place of a pressure gauge. The transducer converts pressure into an analog electrical signal. Pressure transducers are normally available with three types of electrical output; millivolt, amplified voltage, and low milliamps.

Lab Scopes

An **oscilloscope** is a visual voltmeter. An oscilloscope converts electrical signals to a visual image representing voltage changes over a specific period of time. This information is displayed in the form of a continuous voltage line called a **waveform** pattern or trace.

An upward movement of the voltage trace on an oscilloscope screen indicates an increase in voltage, and a downward movement of this trace represents a decrease in voltage. As the voltage trace moves across an oscilloscope screen, it represents a specific length of time (Figure 2-8).

The size and clarity of the displayed waveform is dependent on the voltage scale and the time reference selected. Most scopes are equipped with controls that allow voltage and time interval selection. It is important, when choosing the scales, to remember that a scope displays voltage over time.

Dual-trace oscilloscopes can display two different waveform patterns at the same time. This makes cause and effect analysis easier.

Graphing Multimeter

One of the latest trends in diagnostic tools is a **graphing multimeter (GMM)**. These meters display readings over time, similar to a lab scope. The graph displays the minimum and maximum readings on a graph, as well as displaying the current reading. By observing the graph, a technician can notice any undesirable changes during the transition from a low reading to a high reading, or vice versa. These glitches are some of the more difficult problems to identify without a graphing meter or a lab scope.

Scan Tools

The introduction of computer-controlled systems brought with it the need for tools capable of troubleshooting electronic control systems. There are a variety of scan tools available today that do just that. A **scan tool** is a microprocessor designed to communicate with the vehicle's computer (Figure 2-9). Connected to the computer through diagnostic connectors, a scan

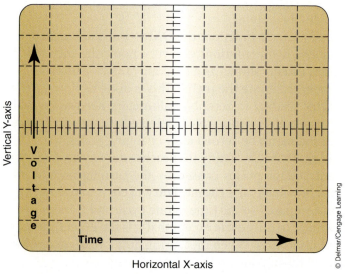

FIGURE 2-8 A scope displays changes in voltage over time.

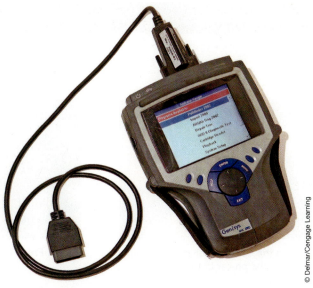

FIGURE 2-9 A scan tool is used to diagnose electronic control systems.

tool can access trouble codes, run tests to check system operations, and monitor the activity of the system (To see how a scan tool is connected to a car, see Photo Sequence 2). Trouble codes and test results are displayed on an LED screen, or printed out on the scanner printer.

Scan tools are capable of testing many onboard computer systems, such as climate controls, transmission controls, engine computers, antilock brake computers, air bag computers, and suspension computers, depending on the year and make of the vehicle and the type of scan tester. In many cases, the technician must select the computer system to be tested with the scanner after it has been connected to the vehicle.

The scan tool is connected to specific diagnostic connectors on the vehicle. Some manufacturers have one diagnostic connector. This connects the data wire from each computer to a specific terminal in this connector. Other manufacturers have several diagnostic connectors on each vehicle, and each of these connectors may be connected to one or more computers. The scan tool must be programmed for the model year, make of vehicle, and type of engine.

With second-generation onboard diagnostic system (OBD-II), the diagnostic connectors are located in the same place on all vehicles. Also, any scan tool designed for OBD-II will work on all OBD-II systems; therefore, the need to have designated scan tools or cartridges is eliminated. Most OBD-II scan tools have the ability to store, or "freeze" data during a road test, and then play back this data when the vehicle is returned to the shop.

There are many different scan tools available. Some are a combination of other diagnostic tools, such as a lab scope and graphing multimeter. These may have the following capabilities:

- Retrieve diagnostic trouble codes (DTCs)
- Monitor system operational data
- Reprogram the vehicle's electronic control modules
- Perform systems diagnostic tests
- Display appropriate service information, including electrical diagrams
- Display Technical Service Bulletins (TSBs)
- Troubleshooting instructions
- Easy tool updating through a personal computer (PC)

The vehicle's computer sets trouble codes when a voltage signal is entirely out of its normal range. The codes help technicians identify the cause of the problem when this is the case. If a signal is within its normal range but is still not correct, the vehicle's computer will not display a trouble code. However, a problem may still exist. Current and historic data is also available on most scan tools (Figure 2-10). This data allows a technician to look at what the system is experiencing when there is or is not a DTC.

CONNECTING A SCAN TOOL

All photos in this sequence are © Delmar/Cengage Learning.

P2-1 Connect the scan tool to the data link connector (DLC) with the ignition off.

P2-2 Turn the scan tool on by depressing the power button.

P2-3 Press ENTER to access the main menu on the tool.

P2-4 Select diagnostics.

P2-5 Select the correct year of the vehicle.

P26 Select the correct model of vehicle.

P2-7 After the vehicle information has been entered, turn on the ignition. You may choose to review data, retrieve DTCs, or activate certain outputs for testing purposes.

P2-8 When you are finished with your diagnostics, clear all stored DTCs and take the vehicle on a road test. Make sure to turn the ignition off whenever you disconnect or connect the scan tool from the vehicle.

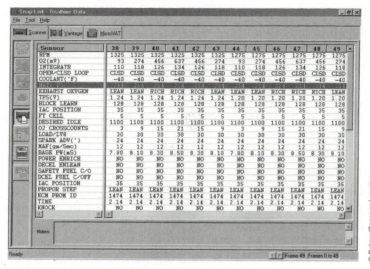

FIGURE 2-10 This is an example of the data that can be viewed with a scan tool. In this example, the information captured by the scan tool was sent to a personal computer.

Some scan tools work directly with a PC through uncabled communication links, such as Bluetooth. Others use a Personal Digital Assistant (PDA). These are small hand-held units that allow you to read diagnostic codes (DTCs) and monitor the activity of sensors. System tests help to determine what service the vehicle requires. Most of these scan tools also have the ability to do the following:

- Perform system and component tests
- Report test results of monitored systems
- Exchange files between a PC and PDA
- View and print files on a PC
- Print DTC/Freeze frame
- Generate emissions reports
- IM/Mode 6 information
- Display related TSBs
- Display full diagnostic code descriptions
- Observe live sensor data
- Update the scan tool as a manufacturer's interfaces change

Computer Memory Saver

Memory savers are an external power source used to maintain the memory circuits in electronic accessories and the engine, transmission, and body computers when the vehicle's battery is disconnected. The saver is plugged into the vehicle's cigar lighter outlet. It can be powered by a 9- or 12-volt battery.

Breakout Box

On some vehicles and with some scan tools, it may be necessary to use a breakout box to monitor serial data. The breakout box is connected in series with Engine control module (ECM)/Powertrain control module (PCM) and the vehicle's wiring harness to the control unit. Often the breakout box comes with many different connectors so it can be installed in a variety of vehicles.

Any wire connected to the ECM/PCM can be monitored through the breakout box. A switch on the box allows a technician to select any pin at the computer's connectors. The scan tool will then display the activity at that pin. Most breakout boxes allow for a display of more than one pin at the same time.

Vacuum Gauge

Measuring intake manifold vacuum is a way to determine the condition of an engine. This is important when diagnosing automatic transmissions. A transmission responds to engine load and engine load affects engine vacuum. As load increases, engine vacuum decreases. When an engine is worn, its ability to create a vacuum decreases. This means a poorly running engine will be seen by the transmission as having a load always on it. This results in poor shift quality and inefficient shift timing.

Manifold vacuum is tested with a vacuum gauge. Vacuum is formed on a piston's intake stroke. As the piston moves down, it lowers the pressure of the air in the cylinder if the cylinder is sealed. This lower cylinder pressure is called *engine vacuum.* Atmospheric pressure will force air into the cylinder when the intake valve of the cylinder with a vacuum opens. This is how the cylinder can be filled with air. The reason atmospheric pressure enters is that whenever there are both low and high pressure, the high pressure will always move toward the low pressure.

Vacuum is measured in inches of mercury (in./Hg) and in kilopascals (kPa) or millimeters of mercury (mm/Hg).

To measure vacuum, a flexible hose on the vacuum gauge is connected to a source of manifold vacuum, either on the manifold or at a point below the throttle plates. Sometimes this requires removing a plug from the manifold and installing a special fitting.

The test is made with the engine cranking (running). A good vacuum reading is typically at least 16 in./Hg. However, a reading of 15 to 20 in./Hg (50 to 65 kPa) is normally acceptable. Since the intake stroke of each cylinder occurs at a different time, the production of vacuum occurs in pulses. If the amount of vacuum produced by each cylinder is the same, the vacuum gauge will show a steady reading. If one or more cylinders are producing different amounts of vacuum, the gauge will show a fluctuating reading.

Hydraulic Pressure Gauge Set

An automatic transmission is a hydraulic machine. The amount of hydraulic pressure and the direction of fluid flow are the keys to proper operation. Therefore, observing the pressure and pressure changes is an excellent way to examine the operation of a transmission.

A common diagnostic tool for automatic transmissions is a hydraulic pressure gauge. A pressure gauge measures pressure in pounds per square inch (psi) or kilopascals (kPa). The gauge is normally part of a kit that contains various fittings and adapters (Figure 2-11). Most transmissions have at least one external fitting that allow for the connection of the pressure gauge to the interior of the transmission. Vehicle manufacturers also publish pressure charts so the results of the test can be compared to specifications.

Logical Diagnostics

The true measure of a good technician is their ability to find and correct the cause of a problem. Service manuals and other information sources will guide you through the diagnosis and repair of a problem. However, those guidelines will not always lead you to the exact cause of the problem. To do this you must use your knowledge and take a logical approach while troubleshooting. **Diagnosis** is not guessing and it is more than following a series of interrelated steps in order to find the solution to a specific problem. Diagnosis is a way of looking at systems that are not functioning the way they should and finding out why. It is knowing how the system should work and deciding if it is working correctly. Through an understanding of the purpose and operation of the system, you can accurately diagnose problems.

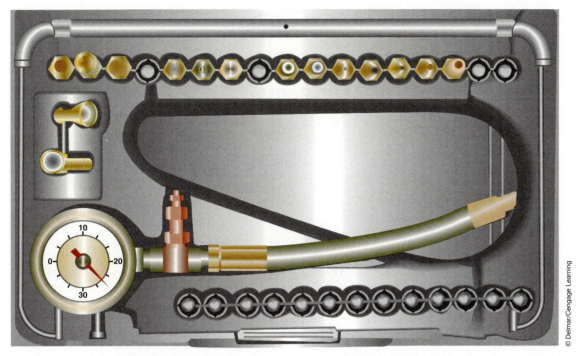

FIGURE 2-11 A hydraulic pressure gauge set with various adapters.

Most good technicians use the same basic diagnostic approach. Because this is such a logical approach, it can quickly lead to the cause of a problem. Logical diagnosis follows these steps:

1. Gather information about the problem.
2. Verify that the problem exists.
3. Thoroughly define what the problem is and when it occurs.
4. Research all available information and knowledge to determine the possible causes of the problem.
5. Isolate the problem by testing.
6. Continue testing to pinpoint the cause of the problem.
7. Locate and repair the problem, then verify the repair.

HYBRID TOOLS

A hybrid vehicle is an automobile and as such is subject to many of the same problems as a conventional vehicle. Most systems in a hybrid vehicle are diagnosed in the same way as well. However, a hybrid vehicle has unique systems that require special procedures and test equipment. It is imperative to have good information before attempting to diagnose these vehicles. Also, make sure you follow all test procedures precisely as they are given.

CAT-III DMM

An important diagnostic tool is a DMM. However, this is not the same DMM used on a conventional vehicle. The meter used on hybrids and other electric-powered vehicles should be classified as a category III (CAT III) meter (Figure 2-12). There are basically four categories for low-voltage electrical meters, each built for specific purposes and to meet certain standards. Low voltage, in this case, means voltages less than 1000 volts. The categories define how safe a meter is when measuring certain circuits. The standards for the various categories are defined by the American National Standards Institute (ANSI), the International Electrotechnical Commission (IEC), and the Canadian Standards Association (CSA). A CAT III meter is

FIGURE 2-12 Only meters with this symbol should be used on high-voltage vehicles.

required for testing hybrid vehicles because of the high voltages, three-phase current, and the potential for high transient voltages. Transient voltages are voltage surges or spikes that occur in AC circuits. To be safe, you should have a CAT III-1000 Volts meter. A meter's voltage rating reflects its ability to withstand transient voltages. Therefore, a CAT III-1000 Volts meter offers much more protection than a CAT III meter rated at 600 volts.

Insulation Resistance Tester

Another important tool is an insulation resistance tester. These can check for voltage leakage through the insulation of the high-voltage cables. Obviously, no leakage is desired and any leakage can cause a safety hazard as well as damage to the vehicle. Minor leakage can also cause hybrid system-related driveability problems. This meter is not one commonly used by automotive technicians, but should be for any one who might service a damaged hybrid vehicle, such as doing body repair. This should also be a CAT III meter and may be capable of checking resistance and voltage of circuits like a DMM.

To measure insulation resistance, system voltage is selected at the meter and the probes placed at their test position. The meter will display the voltage it detects. Normally, resistance readings are taken with the circuit deenergized unless you are checking the effectiveness of the cable or wire insulation. In this case, the meter is measuring the insulation's effectiveness and not its resistance.

 WARNING: The probes for all meters used on high-voltage systems should have safety ridges or finger positioners. These help prevent physical contact between your fingertips and the meter's test leads.

TORQUE WRENCHES

Torque is the twisting force used to turn a fastener against the friction between the threads and between the head of the fastener and the surface of the component. The fact that practically every vehicle and engine manufacturer publishes a list of torque recommendations is ample proof of the importance of using proper amounts of torque when tightening nuts or bolts. The amount of torque applied to a fastener is measured with a torque-indicating or torque wrench.

A **torque wrench** is basically a ratchet or breaker bar with some means of displaying the amount of torque exerted on a bolt when pressure is applied to the handle (Figure 2-13). Torque wrenches are available with the various drive sizes. Sockets are inserted onto the drive and then placed over the bolt. As pressure is exerted on the bolt, the torque wrench indicates the amount of torque.

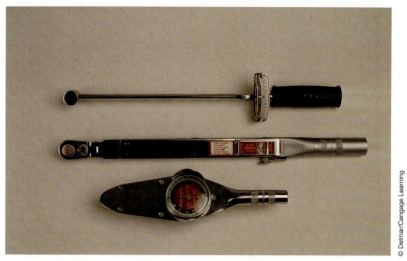

FIGURE 2-13 The basic types of torque-indicating wrenches.

The common types of torque wrenches are available with inch-pound and foot-pound increments:

- A beam torque wrench is not highly accurate. It relies on a beam metal that points to the torque reading.
- A "click"-type torque wrench clicks when the desired torque is reached. The handle is twisted to set the desired torque reading.
- A dial torque wrench has a dial that indicates the torque exerted on the wrench. The wrench may have a light or buzzer that turns on when the desired torque is reached.
- A digital readout type displays the torque and is commonly used to measure turning effort, as well as for tightening bolts. Some designs of this type of torque wrench have a light or buzzer that turns on when the desired torque is reached.

POWER TOOLS

Power tools make a technician's job easier. They operate faster and with more torque than hand tools. However, power tools require greater safety measures. Power tools do not stop unless they are turned off. Power is furnished by air (on pneumatic tools), electricity, or hydraulic fluid.

Pneumatic tools are typically used by technicians because they have more torque, weigh less, and require less maintenance than electric power tools. However, electric power tools tend to cost less than pneumatic ones. Electric power tools can be plugged into most electric wall sockets, but to use a pneumatic tool, you must have an air compressor and an air storage tank. Cordless electrical tools are becoming more popular because they work without a cord or hose.

Air Wrenches

An impact wrench (Figure 2-14) uses compressed air or electricity to hammer or impact a nut or bolt loose or tight. Light-duty impact wrenches are available in three drive sizes, $\frac{1}{4}$, $\frac{3}{8}$, and $\frac{1}{2}$ inch, and two heavy-duty sizes, $\frac{3}{4}$ and 1 inch.

 WARNING: Impact wrenches should not be used to tighten critical parts or parts that may be damaged by the hammering force of the wrench.

CAUTION:
Carelessness or mishandling of power tools can cause serious injury. Make sure you know how to operate a tool before using it.

CAUTION:
The sockets designed for impact wrenches are constructed of thicker but softer steel than other sockets, allowing them to withstand the force of the impact. Ordinary sockets must not be used with impact wrenches. They will crack or shatter because of the force and can cause injury.

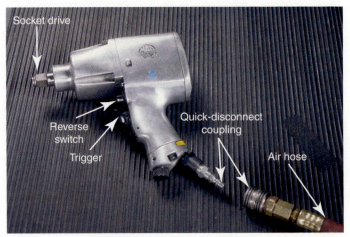

FIGURE 2-14 An air-impact wrench.

Air ratchets are often used during disassembly or reassembly work to save time. Because the ratchet turns the socket without an impact force, these wrenches can be used on most parts and with ordinary sockets. Air ratchets usually have a $\frac{1}{4}$- or $\frac{3}{8}$-inch drive. Air ratchets are not torque sensitive; therefore, a torque wrench should be used on all fasteners after snugging them up with an air ratchet.

Bench Grinder

This electric power tool is generally bolted to a workbench. A bench grinder should have safety shields and guards. Always wear face protection when using a grinder. A bench grinder is classified by wheel size. Six- to ten-inch wheels are the most common in auto repair shops. Three types of wheels are available with this bench tool.

- Grinding wheel. Used for a wide variety of grinding jobs, from sharpening cutting tools to deburring.
- Wire wheel brush. Used for general cleaning and buffing, removing rust and scale, paint removal, deburring, and so forth.
- Buffing wheel. For general purpose buffing, polishing, and light cutting.

Shop Light

Adequate light is necessary when working under and around automobiles. A shop light can be battery powered (like a flashlight) or need to be plugged into a wall socket. Some shops have lights that pull down from a reel suspended from the ceiling. Shop lights use either an incandescent bulb or fluorescent tube. Because incandescent bulbs can pop and burn, it is highly recommended that you only use fluorescent bulbs. Take extra care when using a shop light. Make sure the cord does not get caught in a rotating object. The bulb or tube is surrounded by a cage or enclosed in clear plastic to prevent accidental breaking and burning.

Blowgun

Blowguns are used for air-testing components (Figure 2-15) and blowing off parts during cleaning. Never point a blowgun at yourself or someone else. A blowgun snaps into one end of an air hose and directs airflow when a button is pressed. Always use an OSHA-approved air blowgun. Before using a blowgun, be sure it has not been modified to eliminate air-bleed holes on the side.

FIGURE 2-15 Two examples of OSHA-approved blowguns.

LIFTING TOOLS

Lifting tools are necessary tools for many transmission service procedures. Typically, these tools are provided by the shop and are not the property of a technician. Correct operating and safety procedures should always be followed when using lifting tools.

Jacks

Jacks are used to raise a vehicle off the ground. They are available in two basic designs and in a variety of sizes. The most common jacks are hydraulic floor jacks. They are classified by the weights they can lift: $1\frac{1}{2}$, 2, and $2\frac{1}{2}$ tons, and so on. These jacks are controlled by moving the handle up and down. The other type of portable floor jack uses compressed air. Pneumatic jacks are operated by controlling air pressure at the jack.

Safety Stands

When a vehicle is raised by a jack, it should be supported by **safety stands**. Never work under a car with only a jack supporting it; always use safety stands. Hydraulic seals in the jack can let go and allow the vehicle to drop. Service manuals note the proper locations for jacking and supporting a vehicle while it is raised from the ground. Always follow those guidelines.

Safety stands are commonly called **jack stands**.

Hydraulic Lift

The hydraulic floor lift is the safest lifting tool. It is able to raise the vehicle high enough to allow you to walk and work under it. Various safety features prevent a hydraulic lift from dropping if a seal does leak or if air pressure is lost. Before lifting a vehicle, make sure the lift is correctly positioned.

Some technicians use pneumatic floor and transmission jacks. These jacks typically can raise the vehicle higher than hydraulic units. Before using this type of jack, make sure the unit is sealed and there are no air leaks.

Portable Crane

Often, to remove and install a transmission, the engine will be moved out of the vehicle with the transmission. To remove or install an engine, a portable crane is used. A crane uses hydraulic pressure that is converted to a mechanical advantage and lifts the engine from the

Engine hoists are often referred to as "cherry pickers."

vehicle. To lift an engine, attach a pulling sling or chain to the engine. Some engines have eye plates for use in lifting. If there are none, the sling must be bolted to the engine. The sling-attaching bolts must be large enough to support the engine and must thread into the block a minimum of $1\frac{1}{2}$ times the bolt diameter. Connect the crane to the chain. Raise the engine slightly and make sure the sling attachments are secure. Carefully lift the engine out of its compartment.

Lower the engine close to the floor so the transmission and torque converter can be removed from the engine, if necessary.

Transmission Jacks

Transmission jacks are designed to help you remove a transmission from under the vehicle. The weight of the transmission makes it difficult and unsafe to remove it without much assistance or a transmission jack. These jacks fit under the transmission (Figure 2-16) and are typically equipped with hold-down chains. These chains are used to secure the transmission to the jack. The transmission's weight rests on the jack's saddle.

Transmission jacks are available in two basic styles. One is used when the vehicle is raised by a hydraulic jack and sitting on jack stands. The other style is used when the vehicle is raised on a lift.

Transaxle Removal and Installation Equipment

The removal and replacement (R&R) of transversely mounted engines may require tools not required for removing an RWD engine. The engines of some FWD vehicles are removed by lifting them from the top. Others must be removed from the bottom; this procedure requires

FIGURE 2-16 A typical transmission jack.

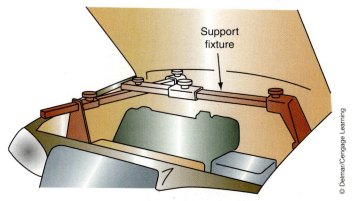

FIGURE 2-17 An engine holding fixture for engine and transaxle removal in a FWD vehicle.

different equipment. Make sure you follow the instructions given by the manufacturer and use the appropriate tools and equipment. The required equipment varies with manufacturer and vehicle model; however, most accomplish the same thing.

To remove the engine and transmission from under the vehicle, the vehicle must be raised. A crane or support fixture is used to hold the engine and transaxle assembly in place while the assembly is being readied for removal. When everything is set for removal of the assembly, the crane is used to lower the assembly onto a cradle. The cradle is similar to a hydraulic floor jack and is used to lower the assembly further so it can be rolled out from under the vehicle. The transaxle can be separated from the engine once it has been removed from the vehicle.

When the transaxle is removed as a single unit, the engine must be supported while it is in the vehicle before, during, and after transaxle removal. Special fixtures (Figure 2-17) mount to the vehicle's upper frame or suspension parts. These supports have a bracket that is attached to the engine. With the bracket in place, the engine's weight is now on the support fixture and the transmission can be removed.

Transmission/Transaxle Holding Fixtures

Special holding fixtures should be used to support the transmission or transaxle after it has been removed from the vehicle (Figure 2-18). These holding fixtures may be standalone units or may be bench mounted, allowing the transmission to be easily repositioned during repair work.

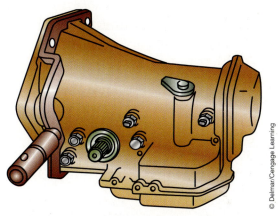

FIGURE 2-18 A typical holding fixture for a transaxle after it has been removed from the vehicle.

A meter is 1/10,000,000 of the distance from the North Pole to the Equator, or 39.37 inches.

MEASURING TOOLS

Many of the procedures discussed in this manual require exact measurements of parts and clearances. Accurate measurements require the use of precision measuring devices that are designed to measure things in very small increments. Measuring tools are delicate instruments and should be handled with great care. Never strike, pry, drop, or force these tools. Also, make sure you clean them before and after every use.

There are many different measuring devices used by automotive technicians. This chapter will only cover those that are commonly used to service transmissions and other driveline components.

Machinist's Rule

The **machinist's rule** looks very much like an ordinary ruler. Each edge of this basic measuring tool is divided into increments based on a different scale. A typical machinist's rule based on the imperial system of measurement may have scales based on $\frac{1}{8}$-, $\frac{1}{16}$-, $\frac{1}{32}$-, and $\frac{1}{64}$-inch intervals. Of course, metric machinist's rules are also available. Metric rules are usually divided into 0.5-mm and 1-mm increments (Figure 2-19).

Some machinist's rules may be based on decimal intervals. These are typically divided into $\frac{1}{10}$-, $\frac{1}{50}$-, and $\frac{1}{1000}$-inch (0.1, 0.03, and 0.01) increments. Decimal machinist's rules are very helpful when measuring dimensions that are specified in decimals; they make such measurements much easier.

Vernier Caliper

A **vernier caliper** can make inside, outside, or depth measurements. It is marked in both British imperial and metric divisions called a vernier scale. A vernier scale consists of a stationary scale and a movable scale, in this case the vernier bar to the vernier plate. The length is read from the vernier scale.

A vernier caliper has a movable scale that is parallel to a fixed scale (Figure 2-20). These precision measuring instruments are capable of measuring outside and inside diameters and most will even measure depth. Vernier calipers are available in both imperial and metric scales. The main scale of the caliper is divided into inches; most measure up to six inches. Each inch is divided into 10 parts, each equal to 0.100 inch. The area between the 0.100 marks is divided into four. Each of these divisions is equal to 0.025 inches (Figure 2-21).

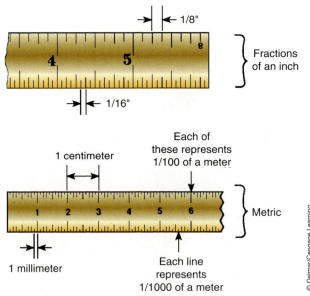

FIGURE 2-19 The graduations on an USCS and metric machinist's rule.

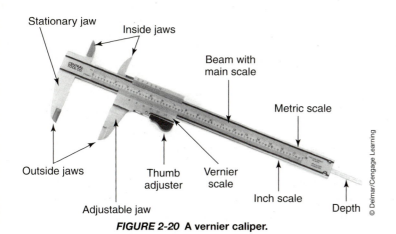

Stationary jaw
Inside jaws
Beam with main scale
Metric scale
Outside jaws
Thumb adjuster
Vernier scale
Adjustable jaw
Inch scale
Depth

FIGURE 2-20 A vernier caliper.

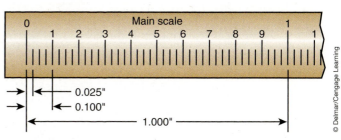

0   Main scale   1   1

0.025"
0.100"
1.000"

FIGURE 2-21 The main scale of a caliper is divided into inches; each inch is divided into 10 parts, each equal to 0.100 inch. The area between the 0.100 marks is divided into four divisions which are equal to 0.025 inches.

The vernier scale has 25 divisions, each one representing 0.001 inch. Measurement readings are taken by combining the main and the vernier scales. At all times, only one division line on the main scale will line up with a line on the vernier scale (Figure 2-21). This is the basis for accurate measurements.

To read the caliper, locate the line on the main scale that lines up with the zero (0) on the vernier scale. If the zero lined up with the 1 on the main scale, the reading would be 0.100 inches. If the zero on the vernier scale does not line up exactly with a line on the main scale, then look for a line on the vernier scale that does line up with a line on the main scale.

Dial Caliper

The **dial caliper** (Figure 2-22) is an easier-to-use version of the vernier caliper. Imperial calipers commonly measure dimensions from 0 to 6 inches. Metric dial calipers typically measure from 0 to 150 mm in increments of 0.02 mm. The dial caliper features a depth scale, bar scale, dial indicator, inside measurement jaws, and outside measurement jaws.

The main scale of an imperial dial caliper is divided into one-tenth (0.1) inch graduations. The dial indicator is divided into one-thousandth (0.001) inch graduations. Therefore, one revolution of the dial indicator needle equals one-tenth inch on the bar scale.

A metric dial caliper is similar in appearance but the bar scale is divided into 2-mm increments. Additionally, on a metric dial caliper, one revolution of the dial indicator needle equals 2 mm.

Both English and metric dial calipers use a thumb-operated roll knob for fine adjustment. When you use a dial caliper, always move the measuring jaws backward and forward to center

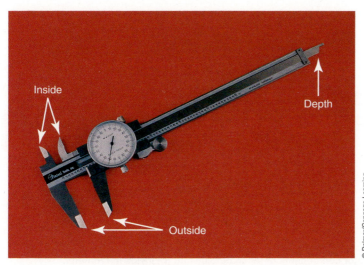

Inside
Depth
Outside

FIGURE 2-22 A dial caliper.

the jaws on the object being measured. Make sure the caliper jaws lay flat on or around the object. If the jaws are tilted in any way, you will not obtain an accurate measurement.

Although dial calipers are precision measuring instruments, they are only accurate to plus or minus two-thousandths (± 0.002) of an inch. Micrometers are preferred when extremely precise measurements are desired.

Micrometers

The **micrometer** is used to measure linear outside and inside dimensions. Both outside and inside micrometers are calibrated and read in the same way. Measurements on both are taken with the measuring points of the tool in contact with the surfaces being measured.

The major components and markings of a micrometer include the frame, anvil, spindle, locknut, sleeve, sleeve numbers, sleeve long line, thimble marks, thimble, and ratchet (Figure 2-23). Micrometers are calibrated in either inch or metric graduations and are available in a range of sizes.

Photo Sequence 3 covers the basics for measuring with and reading a micrometer. Remember, always use the appropriately sized micrometer for the object being measured. Typically, they measure an inch; therefore, the range covered by one micrometer would be 0–1 inch; another would measure 1–2 inches, and so on.

To measure small objects with an outside micrometer, open the jaws of the tool and slip the object between the spindle and the anvil. While holding the object against the anvil, turn the ratchet using your thumb and forefinger until the spindle contacts the object. The ratchet provides only enough pressure on the object to allow it to just fit between the tips of the anvil and spindle. The object should slip through with only a very slight resistance. When a satisfactory feel is reached, lock the micrometer.

Since each graduation on the sleeve represents 0.025 inches, begin reading the measurement by counting the visible lines on the sleeve and multiply that number by 0.025. The graduations on the thimble assembly define the area between the lines on the sleeve; therefore, the number indicated on the thimble should be added to the measurement shown on the sleeve. The sum is the outside diameter of the object.

Micrometers are available to measure in 0.0001 (ten-thousandths) of an inch. Use this type of micrometer if the specifications call for this much accuracy.

To measure larger objects, select a micrometer of the proper size, hold the frame of the micrometer, and slip it over the object. Turn the thimble while continuing to slip the micrometer over the object until you feel a very slight resistance. Rock the micrometer from side to side while doing this to make sure the spindle cannot be closed any farther. Then, lock the micrometer and take a measurement reading.

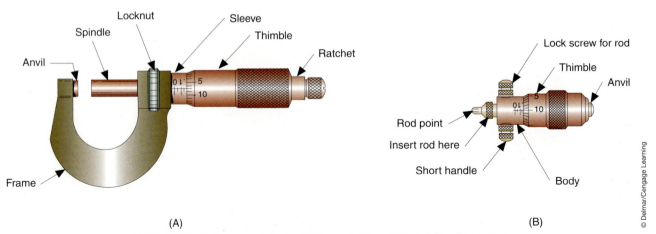

(A)   (B)

FIGURE 2-23 Major components of (A) an outside and (B) inside micrometer.

© Delmar/Cengage Learning

TYPICAL PROCEDURE FOR USING A MICROMETER

All photos in this sequence are © Delmar/Cengage Learning.

P3-1 Micrometers can be used to measure the diameter and thickness of many objects. A common example for a transmission tech is a selective snap ring.

P3-2 Because the thickness of the snap ring is less than one inch, a zero- to one-inch micrometer will be used to measure the snap ring.

P3-3 Each graduation on the sleeve represents 0.025 inch. To read a measurement on a micrometer, begin by counting the visible lines on the sleeve, then multiply 0.025 by that number.

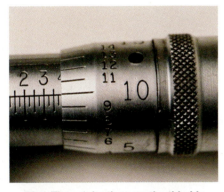

P3-4 The graduations on the thimble assembly define the area between the lines on the sleeve. The number indicated on the thimble is added to the measurement shown on the sleeve.

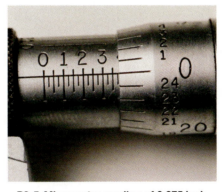

P3-5 Micrometer reading of 0.375 inch.

P3-6 Position the micrometer over the snap ring and slowly close the micrometer around the snap ring until a slight drag is felt while moving the micrometer back and forth over the snap ring.

P3-7 To prevent the reading from changing while you move the micrometer away from the stem, use your thumb to activate the lock lever.

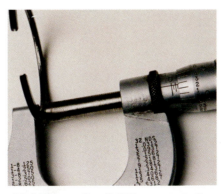

P3-8 This reading (0.078) represents the thickness of the snap ring.

P3-9 Some micrometers are able to measure in 0.0001-inch increments. Use this type of micrometer when the specifications call for this degree of accuracy.

Some technicians use a digital micrometer, which is easier to read. These tools do not have the various scales described here; rather, the measurement is displayed and read directly off the micrometer (Figure 2-24).

A metric micrometer (Figure 2-25) is read in the same way, except the graduations are expressed in the metric system of measurement. Each number on the sleeve represents five millimeters (mm) [0.005 meter (m)]. Each of the 10 equal spaces between each number, which have index lines alternating above and below the horizontal line, represents 0.5 mm (five-tenths of a millimeter). Therefore, one revolution of the thimble changes the reading one space on the sleeve scale (0.5 mm). The beveled edge of the thimble is divided into 50 equal divisions with every fifth line numbered: 0, 5, 10 … 45. Since one complete revolution of the thimble advances the spindle 0.5 mm, each graduation on the thimble is equal to one hundredth of a millimeter. As with the inch-graduated micrometer, the separate readings are added together to obtain the total reading.

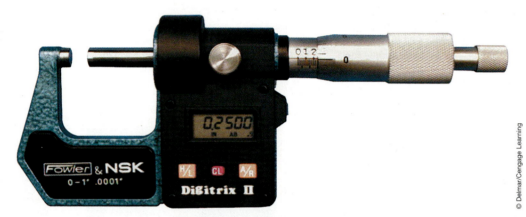

FIGURE 2-24 A digital micrometer.

© Delmar/Cengage Learning

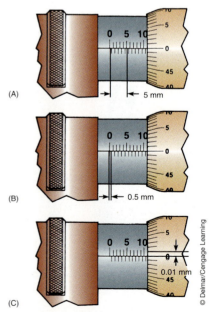

© Delmar/Cengage Learning

FIGURE 2-25 Reading a metric micrometer: (A) 5 mm plus (B) 0.5 mm plus (C) 0.01 mm equals 5.51 mm.

Inside micrometers can be used to measure the inside diameter of a bore. To do this, place the tool inside the bore and extend the measuring surfaces until each end touches the bore's surface. If the bore is large, it might be necessary to use an extension rod to increase the micrometer's range. These extension rods come in various lengths. The inside micrometer is read in the same manner as an outside micrometer.

A depth micrometer is used to measure the distance between two parallel surfaces. The sleeves, thimbles, and ratchet screws operate in the same way as other micrometers. Likewise, depth micrometers are read in the same way as other micrometers.

If a depth micrometer is used with a gauge bar, it is important to keep both the bar and the micrometer from rocking. Any movement of either part will result in an inaccurate measurement.

Telescoping Gauge. Telescoping gauges are used for measuring bore diameters and other clearances (Figure 2-26). They may also be called snap gauges. They are available in sizes ranging from fractions of an inch through six inches. Each gauge consists of two telescoping plungers, a handle, and a lock screw. Snap gauges are normally used with an outside micrometer.

To use the telescoping gauge, insert it into the bore and loosen the lock screw. This will allow the plungers to snap against the bore. Once the plungers have expanded, tighten the lock screw. Then, remove the gauge and measure the expanse with a micrometer.

Small-Hole Gauge. A small-hole or ball gauge works just like a telescoping gauge. However, it is designed for small bores. After it is placed into the bore and expanded, it is removed and measured with a micrometer. Like the telescoping gauge, the small hole gauge consists of a lock, handle, and expanding end. The end expands or retracts by turning the gauge handle.

Feeler Gauge

A feeler gauge is a thin strip of metal or plastic of known and closely controlled thickness. Several of these metal strips are often assembled together as a feeler gauge set that looks like a pocket-knife (Figure 2-27). The desired thickness gauge can be pivoted away from others for convenient use. A steel feeler gauge pack usually contains strips or leaves of 0.002–0.010-inch thickness (in steps of 0.001 inch) and leaves of 0.012–0.024-inch thickness (in steps of 0.002 inch).

A feeler gauge can be used by itself to measure piston clearance, fluid pump clearances, and other distances. It can also be used with a precision straightedge to check the flatness of a sealing surface.

Straightedge. A straightedge is no more than a flat bar machined to be totally flat and straight. Any surface that should be flat can be checked with a straightedge and feeler gauge

© Delmar/Cengage Learning

FIGURE 2-26 A telescoping gauge is used to measure the diameter of a bore.

FIGURE 2-27 A typical feeler gauge pack.

set. The straightedge is placed across and at angles on the surface. At any low points on the surface, a feeler gauge can be placed between the straightedge and the surface. The size of the gauge that fills in the gap determines the amount of warpage or distortion.

Dial Indicator

The **dial indicator** (Figure 2-28) is calibrated in 0.001-inch (one-thousandth–inch) increments. Metric dial indicators are also available. Both types are used to measure movement. Common uses of the dial indicator include measuring endplay (Figure 2-29), clutch pack clearances, and flexplate runout. Dial indicators are available with various face markings and measurement ranges to accommodate many measuring tasks.

A **dial indicator** is a measuring instrument that measures changes in dimension as the indicator is moved across an object or when the object is moved over the stem of the indicator.

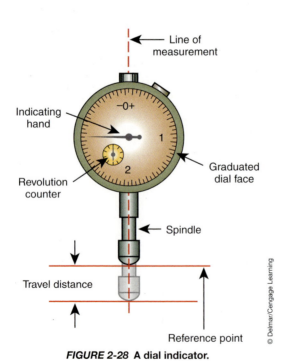

FIGURE 2-28 A dial indicator.

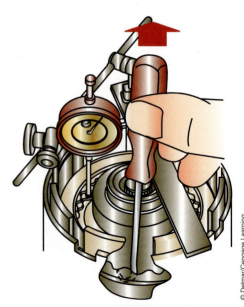

FIGURE 2-29 A dial indicator is often used in automatic transmission service. This setup is measuring the clearance of a multiple-friction disc pack.

To use a dial indicator, position the indicator rod against the object to be measured. Then, push the indicator toward the work until the indicator needle travels far enough around the gauge face to permit movement to be read in either direction. Move the object to one extreme of its travel, then zero the indicator needle on the gauge. Always be sure the range of the dial indicator is sufficient to allow the amount of movement required by the measuring procedure. For example, never attempt to use a one-inch indicator on a component that will move two inches.

TRANSMISSION TOOLS AND EQUIPMENT

Many different tools and equipment are used to service transmissions and drivelines. NATEF has identified many of these and has said an automatic transmission technician must know what they are and how and when to use them. The tools and equipment listed by NATEF and others are covered in the following discussion.

Presses

Many transmission and driveline repairs require the use of a powerful force to assemble or disassemble parts that are **press-fit** together. Removing and installing axle and final drive bearings, universal joint replacement, and transmission assembly work are just a few of the examples. Presses can be hydraulic, electric, air, or hand driven. Capacities range up to 150 tons of pressing force, depending on the size and design of the press. Smaller arbor and C-frame presses can be bench or pedestal mounted, while high-capacity units are freestanding or floor mounted.

Bushing and Seal Pullers and Drivers

Another commonly used group of special tools comprise the various designs of bushing and seal drivers and pullers. Pullers are either a threaded or a slide hammer–type tool (Figure 2-30). Always make sure you use the correct tool for the job; bushings and seals are easily damaged if the wrong tool or procedure is used. Car manufacturers and specialty tool companies work closely together to design and manufacture special tools required to repair cars. Most of these special tools are listed in the appropriate service manuals.

CAUTION:
Always wear safety glasses when using a press.

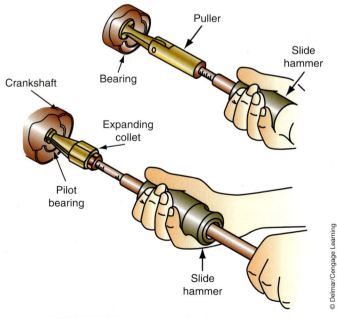

FIGURE 2-30 Examples of slide handle pullers.

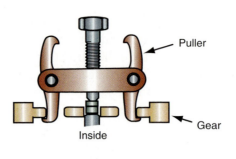

Inside

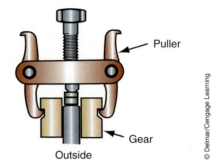

Outside

© Delmar/Cengage Learning

FIGURE 2-31 Examples of the jaws of a gear and bearing puller.

Gear and Bearing Pullers

Many tools are designed for a specific purpose. An example of a special tool is a gear and bearing puller (Figure 2-31). Many gears and bearings have a slight interference fit when they are installed on a shaft or in a housing. Something that has a press-fit has an interference fit. For example, the inside diameter of a bore is 0.001 inch smaller than the outside diameter of a shaft; when the shaft is fitted into the bore it must be pressed in to overcome the 0.001-inch interference fit. This press-fit prevents the parts from moving on each other. The removal of these gears and bearings must be done carefully to prevent damage to the gears, bearings, or shafts. Prying or hammering can break or bind the parts. A puller with the proper jaws and adapters should be used to remove gears and bearings. Using the proper puller, the force required to remove a gear or bearing can be applied with a slight and steady motion.

Retaining (Snap) Ring Pliers

Often, a transmission technician will run into many different styles and sizes of retaining or snap rings that hold subassemblies together or keep them in a fixed location (Figure 2-32). Using the correct tool to remove and install these rings is the only safe way to work with them. All transmission and driveline technicians should have an assortment of retaining ring pliers.

Clutch Tools

Multiple-disc clutch packs contain springs that must be depressed to disassemble and assemble a clutch pack. The spring compressors (Figure 2-33) used to do this must be able to evenly compress the springs while allowing the technician to remove or install a retainer. Often these tools are specifically designed for a particular clutch assembly; however, some universal tools are available.

Special Tool Sets

Vehicle manufacturers and specialty tool companies work closely together to design and manufacture special tools required to repair transmissions. Most of these special tools are listed in the appropriate service manuals and are part of each manufacturer's essential tool kit.

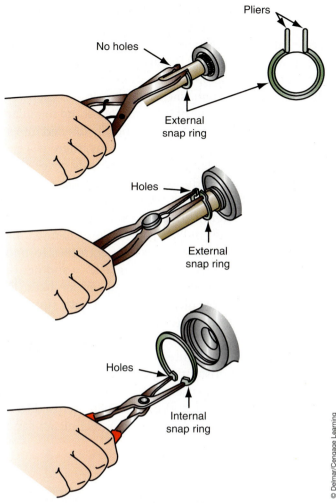

FIGURE 2-32 Internal and external snap ring pliers.

© Delmar/Cengage Learning

FIGURE 2-33 A spring compressor used to remove the snap ring on a multiple-disc clutch pack.

© Delmar/Cengage Learning

SERVICE MANUALS

Perhaps the most important tools you will use are service manuals. Service manuals are a necessary part of transmission and driveline service. There is no way a technician can remember all of the procedures and specifications needed to repair an automobile correctly. Thus, a good technician relies on service manuals and other sources for this information. Good information plus knowledge allows a technician to fix a problem with the least bit of frustration

Service manuals are sometimes simply called *shop manuals.*

and at the lowest expense to the customer. Service manuals also provide drawings and photographs that show where and how to perform certain procedures on the particular vehicle you are working on. Special tools or instruments required are listed and shown. Precautions are also given to prevent injury or damage to parts.

The primary source of repair and specification information for any car, van, or truck is the manufacturer. The manufacturer publishes service manuals each year for every vehicle it builds. Because of the enormous amount of information, some manufacturers publish more than one manual per year per car model. They are typically divided into sections based on the major systems of the vehicle. In the case of transmissions, there is a section for each transmission that may be in the vehicle. Manufacturers' manuals cover all repairs, adjustments, specifications, detailed diagnostic procedures, and special tools required.

The same information that is available in service manuals is now commonly found electronically on compact disks (CD-ROMs), digital video disks (DVDs), and the Internet. A single compact disk can hold a quarter million pages of text, eliminating the need for a huge library to contain all of the printed manuals. Using electronics to find information is also easier and quicker. The disks are normally updated quarterly and not only contain the most recent service bulletins but also engineering and field service fixes. DVDs can hold more information than CDs; therefore, fewer disks are needed with systems that use DVDs. The CDs and DVDs are inserted into a computer (Figure 2-34). All a technician needs to do is enter vehicle information and then move to the appropriate part or system. The appropriate information will then appear on the computer's screen.

Online data can be updated instantly and requires no space for physical storage (Figure 2-35). These systems are easy to use and the information is quickly accessed and displayed. The computer's keyword, mouse, and/or light pen are used to make selections from the screen's menu. Once the information is retrieved, a technician can read it off the screen or print it out and take.

Technical Service Bulletins

Since many technical changes occur on specific vehicles each year, manufacturers' service manuals need to be constantly updated. Updates are published as service bulletins (often referred to as **Technical Service Bulletins** or **TSBs**) that show the changes in specifications

FIGURE 2-34 Service information, today, is widely available on CDs and DVDs.

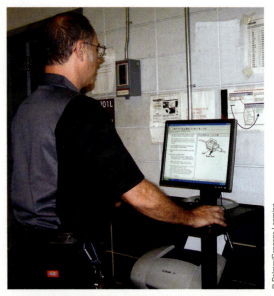

FIGURE 2-35 A technician using internet based service information.

and repair procedures during the model year. These changes do not appear in the service manual until the next year. The car manufacturer provides these bulletins to dealers and repair facilities on a regular basis. The manufacturer also supplies an updated version of the service manual on a CD or DVD.

Automotive manufacturers also publish a series of technician reference books. The publications provide general instructions on how to service and repair their vehicles using their recommended techniques.

General and Specialty Repair Manuals

Some service manuals are published by independent companies rather than manufacturers. However, these publishers pay for and get most of their information from the carmakers. They may present component information, diagnostic steps, repair procedures, and specifications for several car makes in one book. Information is usually condensed and is more general in nature than the manufacturer's manuals. The condensed format allows for more coverage in less space and, therefore, is not always specific. Several model years or car makes may be discussed in one book.

Many of the larger parts manufacturers have excellent guides to the various parts they manufacture or supply. They also provide updated service bulletins on their products. Other sources for up-to-date technical information are trade magazines and trade associations.

Flat-Rate Manuals

Flat-rate manuals contain standards for the length of time a specific repair is supposed to require. Normally, they also contain a parts list with approximate or exact prices of parts. They are excellent for making cost estimates and are published by the manufacturers and independents. Most often, flat-rate manuals are contained in the shop's shop management software or are included in the service information available with the service manuals on CD or DVD.

Using a Service Manual

Although manuals from different publishers vary in presentation and arrangement of topics, all service manuals are easy to use after you become familiar with their organization. Most shop manuals are divided into a number of sections, each covering different aspects of the vehicle. The beginning sections commonly provide vehicle identification and basic maintenance information. The remaining sections deal with each different vehicle system in detail and include diagnostic, service, and overhaul procedures. Each section has an index indicating more specific areas of information.

To use a service manual:

1. Select the appropriate manual for the vehicle being serviced.
2. Use the table of contents to locate the section that applies to the work being done.
3. Use the index at the front of that section to locate the required information.
4. Carefully read the information and study the applicable illustrations and diagrams.
5. Follow all of the required steps and procedures given for that service operation.
6. Adhere to all of the given specifications and perform all measurement and adjustment procedures with accuracy and precision.

Using Electronic Service Data

Today most technicians rely on the Internet to find specific repair information. This is due to the fact that service information can be obtained very quickly and easily. Although much of the information on the web is the same as that found on DVDs, the information on the internet can be easily updated. With DVDs, the provider must send out updated versions to the service facility. This takes time and will only happen periodically, not when an update is necessary. Another advantage of the internet is all of the information is there and there is no need to scan through DVDs to find what you need. A technician simply inputs what he or she needs, and a search engine finds the information. Also, to simplify things, the information is divided by specific systems.

The data available from most electronic service data providers includes the following information for each system of a specific vehicle:

- Adjustments
- Component locators
- Description and operation
- Diagnostic flowcharts
- Diagnostic trouble codes
- Diagrams
- Fluid types and volumes
- Maintenance schedules
- OEM wiring diagrams
- Parts and labor information
- Recall notices
- Required tools and equipment
- Service and repair procedures
- Service precautions
- Short cuts
- Specifications
- Technical Service Bulletins
- Testing and inspection procedures
- User tips

Vehicle Identification

Before performing any service, it is important for you to know exactly what type of vehicle you are working on. The best way to do this is to refer to the **vehicle's identification number** (**VIN**). The VIN (Figure 2-36) is found on a plate behind the lower corner of the driver's side of the windshield, as well as at other locations on the vehicle. The VIN is made up of 17 characters and contains all pertinent information about the vehicle. The use of the 17-character alphanumeric (number and letter) code became mandatory beginning with 1981 vehicles and is used by all domestic and foreign manufacturers of vehicles.

Each character of a VIN has a particular purpose. The first character identifies the country where the vehicle was manufactured, for example:

- 1 or 4—U.S.A.
- 2—Canada
- 3—Mexico
- J—Japan

- K—Korea
- S—England
- W—Germany

The second character identifies the manufacturer, for example:

- A—Audi
- B—BMW
- C—Chrysler
- D—Mercedes Benz
- F—Ford

- G—General Motors
- H—Honda
- N—Nissan
- T—Toyota

The third character identifies the vehicle type or manufacturing division (passenger car, truck, bus, and so on). The fourth through eighth characters identify the features of the vehicle, such as the body style, vehicle model, and engine type.

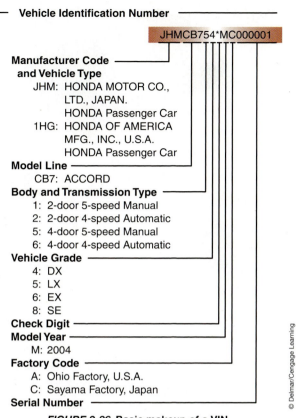

FIGURE 2-36 Basic makeup of a VIN.

The ninth character is used to identify the accuracy of the VIN and is a check digit. The 10th character identifies the model year, for example:

- S—1995
- V—1997
- W—1998
- Y—2000
- 1—2001

- 3—2003
- 5—2005
- 7—2007
- 9—2009

The 11th character identifies the plant where the vehicle was assembled and the 12th to 17th characters identify the production sequence of the vehicle as it rolled off the manufacturer's assembly line.

The specifics needed to further decode the characters of the VIN can be found in a service manual for the vehicle.

Transmission and Driveline Identification. The VIN will give you a general description of the transmission and driveline; however, to obtain the correct transmission specifications and other information, you must first identify the transmission you are working on. The transmission code (Figure 2-37) can be interpreted through information given in the service manual. In addition, the manual may also help you identify the transmission through appearance, casting numbers, or markings on the housing.

Hotline Services

Hotline services provide answers to service concerns by telephone. Manufacturers provide help by telephone for technicians in their dealerships. Subscription hotline services enable independents to get repair information by phone. Some manufacturers also have a phone modem system that can transmit computer information from the car to another location. The vehicle's diagnostic link is connected to the modem. The technician in the service bay runs a

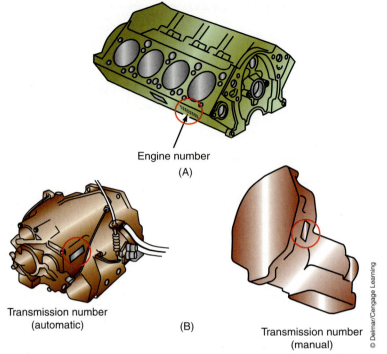

Engine number
(A)

Transmission number
(automatic)

(B)

Transmission number
(manual)

© Delmar/Cengage Learning

FIGURE 2-37 Every manufacturer locates the ID of individual engines and transmission in a specific place, always refer to the service information to determine what transmission you are working with. In this case, the engine serial number (A) is stamped on the cylinder block of the engine, and the transaxle serial number (B) is stamped on the housing.

test sequence on the vehicle. The system downloads the latest updated repair information on that particular model of car. If that does not repair the problem, a technical specialist at the manufacturer's location will review the data and propose a repair.

iATN

The International Automotive Technician's Network (iATN) is comprised of a group of thousands of professional automotive technicians from around the world. The technicians in this group exchange technical knowledge and information with other members. The web address for this group is http://www.iatn.net

FASTENERS

Fasteners are things used to secure or hold parts of something together. Many types and sizes of fasteners are used by the automotive industry. Each fastener is designed for a specific purpose and condition. One of the most commonly used types of fastener is the threaded fastener. Threaded fasteners include bolts, nuts, screws, and similar items that allow a technician to install or remove parts easily.

Threaded fasteners are available in many sizes, designs, and threads. The threads can be either cut or rolled into the fastener. Rolled threads are 30 percent stronger than cut threads. They also offer better fatigue resistance because there are no sharp notches to create stress points. Fasteners are made to imperial or metric measurements. There are four classifications for the threads of imperial fasteners: Unified National Coarse (UNC), Unified National Fine (UNF), Unified National Extrafine (UNEF), and Unified National Pipe Thread (UNPT or NPT). Metric fasteners are also available in fine and coarse threads.

Coarse threads are used for general-purpose work, especially where rapid assembly and disassembly is required. Finely threaded fasteners are used where greater holding force is necessary. They are also used where greater resistance to vibration is desired.

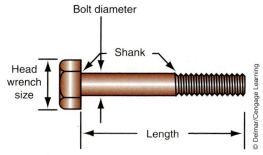

FIGURE 2-38 The basic dimensions of a bolt.

Bolts have a head on one end and threads on the other. Bolts are identified by defining the head size, shank diameter, thread pitch, length (Figure 2-38), and grade.

Studs are rods with threads on both ends. Most often, the threads on one end are coarse while the other end has fine threads. One end of the stud is screwed into a threaded bore. A hole in the part to be secured is fitted over the stud and held in place with a nut screwed over the stud. Studs are used when the clamping pressures of a fine thread are needed and a bolt will not work. If the material the stud is being screwed into is soft (such as aluminum) or granular (such as cast iron), fine threads will not withstand a great amount of pulling force on the stud. Therefore, a coarse thread is used to secure the stud in the work piece and fine threaded nut is used to secure the other part to it. Doing this combines the clamping force of fine threads and the holding power of coarse threads.

Nuts are used with other threaded fasteners when a fastener is not threaded into a piece of work. Many different designs of nuts are found on today's cars. The most common one is the hex nut, which is used with studs and bolts and is tightened with a wrench.

Setscrews are used to prevent rotary motion between two parts, such as a pulley and shaft. Setscrews are either headless, requiring an Allen wrench or screwdriver to loosen and tighten them, or have a square head.

Bolt Identification

The **bolt head** is used to loosen and tighten the bolt; a socket or wrench fits over the head and is used to screw the bolt in or out. The size of the bolt head varies with the diameter of the bolt and is available in imperial and metric wrench sizes. Many confuse the size of the head with the size of the bolt. The size of a bolt is determined by the diameter of its shank. The size of the bolt head determines what size wrench is required.

Bolt diameter is the measurement across the major diameter of the threaded area or across the **bolt shank**. The length of a bolt is measured from the bottom surface of the head to the end of the threads.

The **thread pitch** of a bolt in the imperial system is determined by the number of threads that are in one inch of the threaded bolt length and is expressed in number of threads per inch. A UNF bolt with a $\frac{3}{8}$-inch diameter would be a $\frac{3}{8} \times 24$ bolt. It would have 24 threads per inch. Likewise, a $\frac{3}{8}$-inch UNC bolt would be called a $\frac{3}{8} \times 16$.

The distance, in millimeters, between two adjacent threads determines the thread pitch in the metric system. This distance will vary between 0.8 and 2.0 and depends on the diameter of the bolt. The lower the number, the closer the threads are placed and the finer the threads are.

The bolt's tensile strength, or *grade*, is the amount of stress or stretch it is able to withstand before it breaks. The type of material the bolt is made of and the diameter of the bolt determine its grade. In the imperial system, the tensile strength of a bolt is identified by the number of radial lines (**grade marks**) on the bolt's head. More lines mean higher tensile strength (Table 2-1). Count the number of lines and add two to determine the grade of a bolt.

> The abbreviation *SI* stands for the International System of weights and measures. The SI system is normally called the *metric system*.

TABLE 2-1 STANDARD BOLT STRENGTH MARKINGS

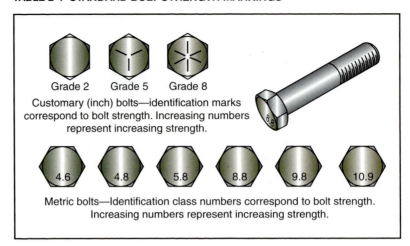

Grade 2   Grade 5   Grade 8

Customary (inch) bolts—identification marks correspond to bolt strength. Increasing numbers represent increasing strength.

4.6   4.8   5.8   8.8   9.8   10.9

Metric bolts—Identification class numbers correspond to bolt strength. Increasing numbers represent increasing strength.

A property class number on the bolt head identifies the grade of a metric bolt. This numerical identification comprises two numbers. The first number represents the tensile strength of the bolt. The higher the number, the greater the tensile strength. The second number represents the yield strength of the bolt. This number represents how much stress the bolt can take before it is not able to return to its original shape without damage. The second number represents a percentage rating. For example, a 10.9 bolt has a tensile strength of 1000 MPa (145,000 psi) and a yield strength of 900 MPa (90 percent of 1000). A 10.9 metric bolt is similar in strength to a SAE grade 8 bolt.

Nuts are graded to match their respective bolts, for example, a grade 8 nut must be used with a grade 8 bolt. If a grade 5 nut were used, a grade 5 connection would result. Grade 8 and critical applications require the use of fully hardened flat washers. These will not dish out when torqued as soft washers will.

Bolt heads can pop off because of **fillet** damage. The fillet is the smooth curve where the shank flows into the bolt head. Scratches in this area introduce stress to the bolt head, causing failure. Removing any burrs around the edges of holes can protect the bolt head. So can placing flat washers with their rounded, punched side against the bolt head and their sharp side against the work surface.

Fatigue breaks are the most common type of bolt failure. A bolt becomes fatigued from working back and forth when it is too loose. Undertightening the bolt causes this problem. Bolts can also be broken or damaged by overtightening, being forced into a nonmatching thread, or bottoming out, which happens when the bolt is too long.

Tightening Bolts

Any fastener is nearly worthless if it is not as tight as it should be. When a bolt is properly tightened, it will be "spring loaded" against the part it is holding. This spring effect is caused by the stretch of the bolt when it is tightened. Normally a properly tightened bolt is stretched to 70 percent of its elastic limit. The elastic limit of a bolt is that point of stretch where the bolt will not return to its original shape when it is loosened. Not only will an overtightened or stretched bolt not have sufficient clamping force, but it will also have distorted threads. The stretched threads will make it more difficult to screw and unscrew the bolt or a nut on the bolt. Always check the service manual to see if there is a torque specification for a bolt before tightening it. If there is, use a torque wrench and tighten the bolt properly.

BASIC GEAR ADJUSTMENTS

While a drivetrain is in operation, the gears, shafts, and bearings are subjected to loads and vibrations. Because of this, the drivetrain must normally be adjusted for the proper fit between parts. These adjustments require the use of precision measuring tools. Three basic adjustments are made when reassembling a unit or when a problem suggests that readjustment is necessary. Adjusting the clearance or play between two gears in mesh is referred to as *adjusting the backlash.* Endplay adjustments limit the amount of end-to-end movement of a gear shaft. Preload is an adjustment made to put a load on an assembly to offset the loads the assembly will face during operation.

Backlash in Gears

Backlash is the clearance between two gears in mesh (Figure 2-39). Excessive backlash can be caused by worn gear teeth, improper meshing of teeth, or bearings that do not support the gears properly. Excessive backlash can result in severe impact on the gear teeth from sudden stops or direction changes of the gears, which can cause broken gear teeth and gears. Insufficient backlash causes excessive overload wear on the gear teeth and can cause premature gear failure.

Backlash is measured with a dial indicator mounted so that its stem is in line with the rotation of the gear and perpendicular to the angle of the teeth. The gear is moved in both directions while the other gear it meshes with is held. The amount of movement on the dial indicator equals the amount of backlash present. The proper placement of shims on a gear shaft is the normal procedure for making backlash adjustments.

Endplay in Gears and Shafts

Endplay refers to the measurable axial (end-to-end) looseness of a bearing. Endplay is always measured in an unloaded condition. To check endplay, a dial indicator is mounted against the side of a gear or the end of a shaft (Figure 2-40). The gear shaft is then pried in both directions and the readings noted. The difference between the two readings is the amount of endplay. Shims or adjusting nuts are widely used to adjust endplay.

End clearance is often referred to as **endplay** and is the measurable axial (end-to-end) looseness of a bearing, assembly, or shaft.

Choosing Selective Washers

A common procedure is the checking and correcting of subassembly end clearances. When these dimensions are correct, the transmission will have good shift characteristics and will wear normally. Look at the forward clutch assembly in Figure 2-41. The clearances in this clutch pack are critical and are controlled by the retaining snap ring. The specifications for

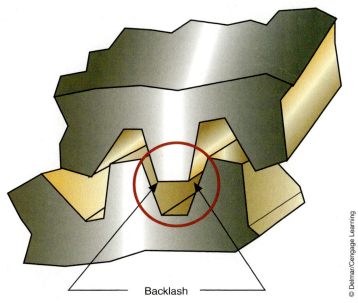

Backlash

© Delmar/Cengage Learning

FIGURE 2-39 Backlash is the clearance between two gears in mesh.

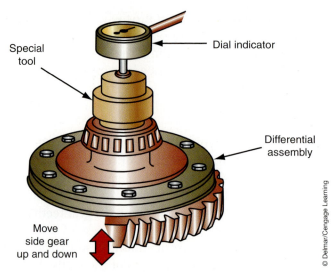

FIGURE 2-40 This setup measures the endplay of a final drive gear assembly.

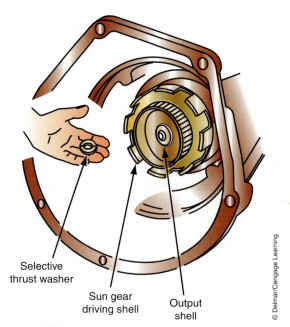

FIGURE 2-41 A selective thrust washer is used between this front and rear multiple-friction disc assembly.

If the part number on a selective snap ring or washer can be read, its thickness can be determined by checking the number against specifications in the service manual.

this clearance are 0.046–0.068 inch (1.17–1.72 mm) and the snap rings are available in four different thicknesses: 0.060–0.064 inch (1.52–1.62 mm), 0.074–0.078 inch (1.87–1.98 mm), 0.086–0.106 inch (2.23–2.69 mm), and 0.102–0.106 inch (2.59–2.69 mm). To determine what size snap ring should be used, the end clearance of the pack needs to be measured.

To do this, assemble the pack with the existing snap ring. Position a dial indicator on the clutch pack's pressure plate. Zero the dial indicator. Now push down on the clutch pack, release the pressure, and zero the indicator again.

Using the required special tool, lift up on the clutch pack until the pressure plate is fully seated against the retaining snap ring. Now read the dial indicator. The reading is the end clearance of the pack with the existing snap ring. If the measurement is not within specifications, a different thickness snap ring must be used. For example, if the end clearance measurement is 0.074 inches, a thicker snap ring must be installed. To determine how thick, remove the existing snap ring and measure its thickness. With a micrometer, the thickness was found to be 0.060 inches. By adding the measured clearance to the snap ring thickness, you know the clearance of the pack. In this case, that clearance is 0.134 inches. To determine what thickness snap ring is best to use, subtract the specification from this total clearance figure. Since the specification is a range, subtract the extremes of the range from the clearance (0.134−0.046 = 0.088 and 0.134−0.068 = 0.066). What you find is that the correct snap ring is also within a range. Since the snap rings are available in four different sizes, the thickness that best meets the need is the one that should be used. In this case, a snap ring with a thickness of 0.074−0.078 inch is the best choice.

The endplay of or clearance between the shafts, clutches, and gears in some transmissions is set by the use of selective thrust washers or shims. The procedure for determining the correct size is the same as followed when selecting a selective snap ring.

Preloading of Geartrains

When normal operating loads are great, geartrains are often preloaded to reduce the deflection of parts. The amount of **preload** is specified in shop manuals and must be correct for the design of the bearings and the strength of the parts. If bearings are excessively preloaded, they will heat up and fail. When bearings are set too loose, the gears or shaft will wear rapidly due to the great amounts of deflection it will experience. Geartrains are preloaded by shims, thrust washers, adjusting nuts, or double-race bearings. Preload adjustments are normally checked by measuring turning effort or torque with a torque wrench (Figure 2-42).

> **Preload** is a fixed amount of pressure constantly applied to a component.

WORKING AS AN AUTO TECH

To be a successful automotive technician you need to have good training and a desire to succeed, and be committed to becoming a good technician and a good employee. A good employee works well with others and strives to make the business successful. The required training is not just in the automotive field. Good technicians need to have good reading, writing, and math skills. These skills will allow you to better understand and use the material found in service manuals and textbooks, and provide you with the basics for good communications with customers and others.

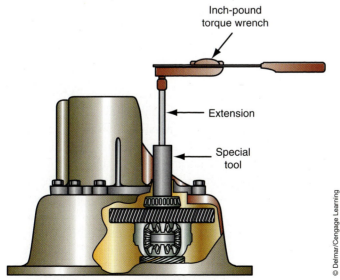

FIGURE 2-42 Turning torque is an indication of bearing preload and is checked with a torque wrench, normally a pound-inch wrench.

Repair Orders

A repair order (RO) is written for every vehicle brought into the shop for service. Repair orders may also be called service orders or work orders. They contain information about the customer, the vehicle, and the customer's concern or request; an estimate of the cost of the services; and when the services should be completed. Repair orders are legal documents that are used for many other purposes, such as payroll and general recordkeeping. Legally, an RO protects the shop and the customer. An RO is signed by the customer, who in doing so authorizes the service and accepts the terms noted on the RO. The customer, however, is protected against being charged more than the estimate given on the RO, unless he or she later authorizes a higher amount. Some states allow shops to be within 10 percent of the estimate, while others hold the shop to the amount that was estimated.

Today, most service facilities have shop management software in their computers. The information needed to complete an RO is input on the computer's keyboard. The software package helps in the estimation of repair costs. The software also takes information from the RO and saves it in various files, each defined by its purpose.

Guidelines for Estimating Repair Costs. For legal reasons and to establish good customer relations, the projected repair costs must be calculated with as much accuracy as possible. To do this, do the following:

1. Make sure you have the correct information about the vehicle.
2. Always use the correct labor and parts guide or database for the specific vehicle.
3. Locate the exact service for that vehicle in the guide or database.
4. Using the guidelines provided in the guide or database, choose the proper time allocation listed for the service.
5. Multiply the allocated time by the shop's hourly flat rate.
6. Using the information given in the guide or database, identify the parts that will be replaced for that service.
7. Locate the cost of the parts in the guide or database or in the catalogs used by the shop.
8. Repeat the process for all other services required or requested by the customer.
9. Multiply the time allocations by the shop's hourly flat rate.
10. Add all of the labor costs together; this sum is the labor estimate for those services.
11. Add the cost of all the parts together; this sum is the estimate for the parts required for the services.
12. Add the total labor and parts costs together. If the shop charges a standard fee for shop supplies, add it to the labor and parts total. This sum is the cost estimate to present to the customer.

Compensation

Technicians are typically paid according to their abilities. Most often, new or apprentice technicians are paid by the hour. While being paid they are learning the trade and the business. Time is usually spent working with a master technician or doing low-skilled jobs. As an apprentice learns more, he or she can earn more and take on more complex jobs. Once technicians have demonstrated a satisfactory level of skills, they can go on flat rate.

Flat rate is a pay system in which a technician is paid for the amount of work they do. Each job has a flat-rate time. Pay is based on that time, regardless of how long it took to complete the job. To explain how this system works, let us look at a technician who is paid $15.00 per flat rate. If a job has a flat-rate time of 3 hours, the technician will be paid $45.00 for the job, regardless of how long it took to complete it. Experienced technicians beat the flat-rate time nearly all of the time. Their weekly pay is based on the time "turned in," not on the time spent. If the technician turns in 60 hours of work in a 40-hour workweek, he or she actually earns $22.50 for each hour worked. However, if he or she turns in only 30 hours in the 40-hour week, the hourly pay is $11.25.

Flat-rate manuals contain standards for the length of time a specific repair is supposed to require.

The flat-rate system favors good technicians who work in a shop that has a large volume of work. The use of flat-rate times allows for more accurate repair estimates to the customers. It also rewards skilled and productive technicians.

Employer–Employee Relationships

Being a good employee requires more than learning the skills for the job. When you begin a job, you enter into a business agreement with your employer. When you become an employee, you sell your time, skills, and efforts. In return, your employer pays you for these resources.

As part of the employment agreement, your employer also has certain responsibilities:

- *Instruction and supervision*—You should be told what is expected of you. A supervisor should observe your work, tell you if it is satisfactory, and offer ways to improve your performance.
- *A clean, safe place to work*—An employer should provide a clean and safe work area as well as a place for personal cleanup.
- *Wages*—You should know how much you are to be paid, what your pay will be based on, and when you will be paid before accepting a job.
- *Fringe benefits*—When you are hired, you should be told what benefits, such as paid vacations and employer contributions to health insurance and retirement plans, you can expect.
- *Opportunity*—This means you should be given a chance to succeed and possibly advance within the company.
- *Fair treatment*—All employees should be treated equally, without prejudice or favoritism.

On the other side of this business transaction, employees have responsibilities to their employers. Your obligations as an employee include the following:

- *Regular attendance*—A good employee is reliable. Businesses cannot operate successfully unless their workers are on the job.
- *Following directions*—As an employee, you are part of a team. Doing things your way may not serve the best interests of the company.
- *Responsibility*—You must be willing to answer for your behavior and work. You need to also realize that you are legally responsible for the work you do.
- *Productivity*—Remember that you are paid for your time as well as your skills and effort.
- *Loyalty*—Loyalty is expected. By being loyal, you will act in the best interests of your employer, both on and off the job.

Another responsibility you have as an employee is good customer relations. Learn to listen and communicate clearly. Be polite and organized, particularly when dealing with customers. Always be as honest as you possibly can.

Look and present yourself as a professional, which is what automotive technicians are. Professionals are proud of what they do and they show it. Always dress and act appropriately and watch your language, even when you think no one is near.

Respect the vehicles on which you work. They are important to the lives of your customers. Always return a vehicle to the owner in a clean, undamaged condition. Remember, a car is the second largest expense a customer has. Treat it that way. It does not matter if you like the car. It belongs to the customer; treat it respectfully.

Explain the repair process to the customer in understandable terms. Whenever you are explaining something to a customer, make sure you do this in a simple way but without making the customer feel stupid. Always show the customers respect and be courteous to them. Not only is this the right thing to do, but it also leads to loyal customers.

ASE CERTIFICATION

The National Institute for Automotive Service Excellence (**ASE**) has established a voluntary certification program for automotive, heavy-duty truck, auto body repair, and engine machine shop technicians. In addition to these programs, ASE also offers individual testing in the areas of parts, alternate fuels, and advanced engine performance. This certification system combines voluntary testing with on-the-job experience to confirm that technicians have the skills needed to work on today's more complex vehicles. ASE recognizes two distinct levels of service capability—the automotive technician and the master automotive technician. The master automotive technician is certified by ASE in all eight major automotive systems. An automotive technician may have certification in only certain areas.

To become ASE certified, a technician must pass one or more tests that stress diagnostic and repair problems. One of these areas is Automatic Transmissions and Transaxles. This test covers the diagnosis and repair of transmissions, transaxles, torque converters, and electronic control systems. You will find content on all of these topics in this manual, as well as the accompanying Classroom Manual.

After passing at least one exam and providing proof of two years of hands-on work experience, the technician becomes ASE certified. Retesting is necessary every five years to remain certified. A technician who passes one examination receives an automotive technician shoulder patch. The master automotive technician patch is awarded to technicians who pass all eight of the basic automotive certification exams (Figure 2-43).

You may receive credit for one of the two years by substituting relevant formal training in one, or a combination, of the following:

- High school training. Three years of training may be substituted for one year of experience.
- Post–high school training. Two years of post–high school training in a public or private trade school, technical institute, community or four-year college, or apprenticeship program may be counted as one year of work experience.
- Short courses. For shorter periods of post–high school training, you may substitute two months of training for one month of work experience.
- You may receive full credit for the experience requirement by satisfactorily completing a three- or four-year apprenticeship program.

Each certification test consists of 40 to 80 multiple-choice questions. The questions are written by a panel of technical service experts, including domestic and foreign vehicle manufacturers, repair and test equipment and parts manufacturers, working automotive technicians, and automotive instructors. All questions are pretested and quality-checked on a national sample of technicians before they are included in the actual test. Many test questions

FIGURE 2-43 My badge that says I am a certified Master Auto Technician.

force the student to choose between two distinct repair or diagnostic methods. The knowledge and skills needed to pass the tests includes:

- Basic technical knowledge—What is it? How does it work? This requires knowing what is in a system and how the system works. It also calls for knowing the procedures and precautions to be followed in making repairs and adjustments.
- Repair knowledge and skill—What is a likely source of a problem? How do you fix it? This requires you to understand and to apply generally accepted procedures and precautions for inspecting, disassembling, rebuilding, replacing, or adjusting components within a particular system.
- Testing and diagnostic knowledge and skill—How do you find what is wrong? How do you know you have corrected a problem? This requires that you be able to recognize that a problem does exist and to know what steps should be taken to identify the cause of the problem.

For further information on the ASE certification program, write to National Institute for Auto Service Excellence (ASE), 101 Blue Seal Drive, S.E., Suite 101, Leesburg, VA 20175 or go to http://www.asecert.org

SUMMARY

- *Diagnosis* means finding the cause or causes of a problem. It requires a thorough understanding of the purpose and operation of the various automotive systems. Diagnostic charts found in service manuals can aid in diagnostics.
- Scan tools and vacuum and pressure gauges are common diagnostic tools for automatic transmission technicians.
- Torque wrenches are used to tighten fasteners to a specified torque.
- Always use the correct tool, in the correct way, for the job.
- Many special tools are available for special purposes. These tools are not normally part of a basic tool set but are purchased on an as-needed basis.
- Dial indicators are used to measure movement and are commonly used to measure the backlash and endplay of a set of gears.
- A micrometer can be used to measure the outside diameter of shafts and the inside diameter of holes. It may be calibrated in either imperial or metric graduations.
- The primary source of repair and specification information for any vehicle is the manufacturer's service manual. Updates are published as service bulletins. These include changes made during the model year, which will not appear in the manual until the following year.
- Flat-rate manuals and software are ideal for making cost estimates. Published by manufacturers and independent companies, they contain figures showing how long specific repairs should take to complete, as well as a list of the necessary parts and their prices.
- *Gear backlash* reflects how tightly the teeth of two gears mesh.
- Whenever bolts are replaced, they should be replaced with exactly the same size, grade, and type as was installed by the manufacturer.
- Besides learning technical and mechanical skills, service technicians must learn to work as part of a team. As an employee, you will have certain responsibilities to your employer and your customers.
- Customer relations are an extremely important part of doing business. Professional, courteous treatment of customers and their vehicles is a must.
- The National Institute for Automotive Service Excellence (ASE) actively promotes professionalism within the industry. Its voluntary certification program for automotive technicians and master auto technicians helps guarantee a high level of quality service.

TERMS TO KNOW

ASE
Backlash
Blowgun
Bolt head
Bolt shank
Circuit tester
Diagnosis
Diagnostic trouble codes (DTCs)
Dial caliper
Dial indicator
Digital multimeter (DMM)
Digital volt/ohmmeter (DVOM)
Duty cycle
Endplay
Feeler gauge
Fillet
Flat rate
Frequency
Grade marks
Graphing multimeter (GMM)
Jack (Safety) stands
Machinist's rule
Micrometer
Oscilloscope
Preload
Press-fit
Pulse width
Repair order
Scan tool
Technical Service Bulletin (TSB)
Thread pitch
Torque wrench
Vehicle identification number (VIN)
Vernier caliper
Waveform

1. While discussing automotive fasteners:

   Technician A says bolt sizes are listed by their appropriate wrench size.

   Technician B says whenever bolts are replaced, they should be replaced with exactly the same size and type that was installed by the manufacturer.

   Who is correct?

   A. A only   C. Both A and B

   B. B only   D. Neither A nor B

2. While adjusting the end clearance of a clutch with a specified clearance of 0.044–0.051 inches and a measured clearance of 0.075 inches with the existing thrust washer thickness of 0.055 inches:

   Technician A uses a washer with a thickness between 0.079–0.086 inches to correct the end clearance.

   Technician B says the proper thickness for the washer is determined by adding the measured clearance to the thickness of the existing washer, then subtracting the specifications from that total.

   Who is correct?

   A. A only   C. Both A and B

   B. B only   D. Neither A nor B

3. *Technician A* says dial indicators are commonly used to measure the backlash of a set of gears.

   Technician B says dial indicators are commonly used to measure the endplay of a set of gears.

   Who is correct?

   A. A only   C. Both A and B

   B. B only   D. Neither A nor B

4. *Technician A* says one of the most important tools for a technician is a shop or service manual.

   Technician B says technicians must constantly update their tools in order to work on newer vehicles.

   Who is correct?

   A. A only   C. Both A and B

   B. B only   D. Neither A nor B

5. *Technician A* says that after an individual passes a particular ASE certification exam, he or she is certified in that test area.

   Technician B says all of the questions on an ASE certification exam are written as " Technician A and Technician B" questions.

   Who is correct?

   A. A only   C. Both A and B

   B. B only   D. Neither A nor B

6. While discussing the purpose of a torque wrench:

   Technician A says they are used to tighten fasteners to a specified torque.

   Technician B says they can be used to provide added leverage while loosening or tightening a bolt.

   Who is correct?

   A. A only   C. Both A and B

   B. B only   D. Neither A nor B

7. While discussing the purpose of micrometers:

   Technician A says micrometers are used to measure the diameter of an object.

   Technician B says outside micrometers are used to measure the outside diameter of an object, while inside micrometers are used to measure the inside diameter.

   Who is correct?

   A. A only   C. Both A and B

   B. B only   D. Neither A nor B

8. *Technician A* says that gear backlash is a statement of how much movement a gear shaft has.

   Technician B says gear backlash is a statement of how tightly the teeth of two gears mesh.

   Who is correct?

   A. A only   C. Both A and B

   B. B only   D. Neither A nor B

9. *Technician A* looks at the ninth digit of the VIN to determine the exact year of the vehicle.

   Technician B says the information about the vehicle's engine can be found in the fourth through seventh characters of the VIN.

   Who is correct?

   A. A only   C. Both A and B

   B. B only   D. Neither A nor B

10. *Technician A* says finely threaded fasteners are used where greater resistance to vibration is desired.

    Technician B says coarse threads are used for general-purpose work, especially where rapid assembly and disassembly are required.

    Who is correct?

    A. A only   C. Both A and B

    B. B only   D. Neither A nor B

Name _____ Date _____

IDENTIFICATION OF SPECIAL TOOLS

Upon completion of this job sheet, you should be able to use the service manual to determine what special tools are required to properly service a particular transmission.

Tools and Materials

Appropriate service manual

Procedure

_____

Task Completed

Your instructor will assign you a transmission type. You will use the service manual to identify and describe the special tools recommended for servicing this transmission. You will also identify any tools your shop has that can be used in place of the factory-specified tools.

1. The transmission assigned to you is:

   Transmission Model _____

   from: year _____ make _____ model _____

   The vehicle's VIN is: _____

2. The manual you will use to find the information:

   _____

3. List the special tools referenced in the overhaul section of the service manual for this transmission (include in your list what part or service each of the tools is used for):

   _____
   _____
   _____
   _____
   _____
   _____
   _____

4. List the special tools that are available for your use in the shop:

   _____
   _____
   _____
   _____
   _____
   _____
   _____

5. List the available tools that can be used in place of the specified tools:

_____

_____

_____

_____

_____

_____

Instructor's Response _____

_____

_____

Name _____ **Date** _____

FILLING OUT A WORK ORDER

Upon completion of this job sheet, you will be able to prepare a service work order based on customer input, vehicle information, and service history.

ASE Correlation

This job sheet is related to the ASE Automatic Transmission and Transaxle Test's Content Area *In-Vehicle Transmission and Transaxle Repair.*
Task: Complete work order to include customer information, vehicle identifying information, customer concern, related service history, cause, and correction.

Tools and Materials

An assigned vehicle or the vehicle of your choice
Service work order or computer-based shop management package
Parts and labor guide

Work Order Source:

Describe the system used to complete the work order. If a paper repair order is being used, describe the source.

_____

_____

Procedure

Task Completed

1. Prepare the shop management software for entering a new work order or obtain a blank paper work order. ☐

2. Enter customer information, including name, address, and phone numbers onto the work order. ☐

3. Locate and record the vehicle's VIN. ☐

4. Enter the necessary vehicle information, including year, make, model, engine type and size, transmission type, license number, and odometer reading. ☐

5. Does the VIN verify that the information about the vehicle is correct? _____

6. Normally, you would interview the customer to identify his or her concerns. However to complete this job sheet, assume the customer desires to have the fluid and filter changed in the transmission. ☐

7. The history of service to the vehicle can often help diagnose problems as well as indicate possible premature part failure. Gathering this information from the customer can provide some of this information. For this job sheet assume the vehicle has not had a transmission problem and was not recently involved in a collision. Service history is further obtained by searching files based on customer name, VIN, and license number. Check the files for any related service work. ☐

8. Search for Technical Service Bulletins on this vehicle that may relate to the customer's concern.

9. Based on the customer's concern, service history, TSBs, and your knowledge, what is the likely cause of this concern?

_____

_____

☐

10. Enter this information onto the work order.

11. Prepare to make a repair cost estimate for the customer. Identify all parts that may need to be replaced to correct the concern. List these here.

_____

_____

12. Describe the task(s) that will be necessary to replace the part.

_____

_____

☐

13. Using the parts and labor guide, locate the cost of the parts that will be replaced and enter the cost of each item onto the work order at the appropriate place for creating an estimate.

14. Now, locate the flat-rate time for work required to correct the concern. List each task with its flat-rate time.

_____

_____

☐

15. Multiply the time for each task by the shop's hourly rate and enter the cost of each item onto the work order at the appropriate place for creating an estimate.

☐

16. Many shops have a standard amount they charge each customer for shop supplies and waste disposal. For this job sheet, use an amount of $10 for shop supplies.

☐

17. Add the total costs and insert the sum as the subtotal of the estimate.

18. Taxes must be included in the estimate. What is the sales tax rate and does it apply to both parts and labor, or just one of these?

_____

_____

☐

19. Enter the appropriate amount of taxes to the estimate, then add this to the subtotal. The end result is the estimate to give the customer.

20. By law, how accurate must your estimate be?

_____

_____

21. Generally speaking, the work order is complete and is ready for the customer's signature. However, some businesses require additional information; make sure you enter that information on the work order. On the work order, there is a legal statement that defines what the customer is agreeing to. Briefly describe the contents of that statement.

_____

_____

Instructor's Response _____

_____

_____

Name _____ **Date** _____

GATHERING VEHICLE INFORMATION

Upon completion of this job sheet, you will be able to gather service information about a vehicle and its automatic transmission or transaxle.

ASE Correlation

This job sheet is related to the ASE Automatic Transmission and Transaxle Test's Content Area *In-Vehicle Transmission and Transaxle Repair.*

Tasks: Research applicable vehicle and service information, such as transmission/transaxle operation, vehicle service history, service precautions, and Technical Service Bulletins. Locate and interpret vehicle and major component identification numbers (VIN, vehicle certification labels, calibration labels).

Tools and Materials

Appropriate service manuals
Computer

Describe the vehicle being worked on:

Year _____ Make _____ Model _____
VIN _____

Procedure

1. Using the service manual or other information source, describe what each letter and number in the VIN for this vehicle represents.

    _____

    _____

2. Locate the Vehicle Emissions Control Information (VECI) label and describe where you found it.

    _____

    _____

3. Summarize the information you found on the VECI label.

    _____

    _____

4. While looking in the engine compartment or under the vehicle, locate the identification tag on the transmission or transaxle. Describe where you found it.

    _____

    _____

5. Summarize the information contained on this label.

_____

_____

_____

6. Using a service manual or electronic database, locate the information about the vehicle's automatic transmission. List the major components of the control system and describe the basic operation parameters.

_____

_____

_____

_____

_____

7. Using a service manual or electronic database, locate and record all service precautions regarding the automatic transmission noted by the manufacturer.

_____

_____

_____

_____

8. Using the information that is available, locate and record the vehicle's service history.

_____

_____

_____

_____

9. Using the information sources that are available, summarize all Technical Service Bulletins for this vehicle that relate to the automatic transmission and its control system.

_____

_____

_____

_____

Instructor's Response _____

_____

_____

Chapter 3

DIAGNOSIS, MAINTENANCE, AND BASIC ADJUSTMENTS

BASIC TOOLS
Technician's basic tool set
Appropriate service manual

UPON COMPLETION AND REVIEW OF THIS CHAPTER, YOU SHOULD BE ABLE TO:

- Listen to the driver's complaint, road test the vehicle, then determine the needed repairs.
- Diagnose unusual fluid usage, level, and condition problems.
- Replace automatic transmission fluid and filters.
- Diagnose noise and vibration problems.
- Diagnose electronic, mechanical, and vacuum control systems.
- Inspect, replace, and align powertrain mounts.
- Inspect, adjust, and replace vacuum modulator, lines, and hoses.
- Diagnose mechanical and vacuum control systems; determine needed repairs.

- Perform oil pressure tests; determine needed repairs.
- Inspect, adjust, and replace manual valve shift linkage.
- Inspect, adjust, and replace cables or linkages for throttle valve (TV) kickdown and accelerator pedal.
- Inspect and replace external seals and gaskets while the transmission is in the vehicle.
- Inspect, repair, and replace extension housing.
- Inspect and replace speedometer drive gear, driven gear, and retainers while the transmission is in the vehicle.
- Inspect and replace parking pawl, shaft, spring, and retainer while the transmission is in the vehicle.
- Adjust bands.

Automatic transmissions are used in many rear-wheel-drive and four-wheel-drive vehicles. Automatic transaxles are most used on front-wheel-drive vehicles. The major components of a transaxle are the same as those in a transmission, except the transaxle assembly includes the final drive and differential gears. An automatic transmission/transaxle shifts automatically without the driver moving a gearshift or pressing a clutch pedal.

Because of the many similarities between a transmission and a transaxle, most of the diagnostic and service procedures are similar. Therefore, all references to a transmission apply equally to a transaxle unless otherwise noted. This rule also holds true for the questions on the certification test for automatic transmissions and transaxles.

Transmissions are strong and typically trouble-free units that require little maintenance. Maintenance normally includes fluid checks and scheduled linkage adjustments and oil changes.

Nearly all current automatic transmissions have electronic controls. These controls work with the hydraulic and mechanical systems of the transmission to provide reliable and efficient operation. Because of the variety of different electronic systems used by the manufacturers and the complexity of some of these systems, electrical and electronic transmission systems are covered in their own chapter (Chapter 4). The content of this chapter applies to all transmissions, whether they have electronic controls or not.

Some of the procedures noted in this chapter are for both in-vehicle and out-of-vehicle service.

SERVICE TIP:
Whenever you are diagnosing or repairing a transaxle or transmission, make sure you refer to the appropriate service manual before you begin.

IDENTIFICATION

Prior to beginning diagnostics or any service on a transmission, its type must be properly identified. Proper identification ensures that the correct procedures will be followed, the correct replacement parts will be installed, and the correct information from Technical Service Bulletins can be retrieved. Correct identification of a transmission may not be possible just by looking at it or assuming that a certain transmission is always used in a particular model vehicle. Some transmission technicians rely on the shape of the oil pan to identify the transmission. This often is a worthwhile technique, but the shape may only indicate the basic model and not the exact design. Some transmissions have more than one pan; this feature can also be used to help identify the transmission. For example, a Ford AX4N transaxle has a bottom pan and a side cover pan. This feature is not a perfect way to identify this transaxle, however, because the AXOD-E looks just like it but is a very different transmission. Sometimes the location or number of electrical connectors can be used. The 45RFE transmission can be identified by a 23-way electrical connector on the driver's side of the transmission; other Chrysler transmissions do not have this connector.

The only positive way to identify the exact design and model of transmission is by locating its identification numbers. The transmission identification number (TIN) is typically listed on a sticker affixed to a late-model transmission (Figure 3-1). These stickers include a bar code and give information about the manufacturer of the transmission, the actual build date, the transmission part number, and the exact model of transmission.

Some examples of the location of and the information contained on the identification tags of common transmissions follow.

On Chrysler FWD vehicles, the transaxle can be identified by a transaxle identification number (TIN) stamped on a boss, located on the transaxle housing just above the oil pan flange. On RWD vehicles, the TIN is stamped on a pad on the left side of the transmission case's oil pan flange. In addition to the TIN, each transmission carries an assembly part number that must be referenced when ordering transaxle replacement parts. Transmission operation requirements are different for each vehicle and engine combination. Some internal parts will differ between models. Always refer to the seven-digit part number for positive transmission identification when replacing parts.

Late-model Ford RWD vehicles can be identified by an identification code letter found on the lower line of the vehicle certificate label under "TR." This label is attached to the left (driver's) side door lock post. A RWD transmission can be identified by a metal tag attached to the transmission by the lower extension housing retaining bolt. The tag on a transaxle is attached to the valve body cover and gives the transmission model number, line shift code, build date code, and assembly and serial numbers (Figure 3-2).

FIGURE 3-1 An example of a transmission identification sticker. These are typically in an easily viewed spot.

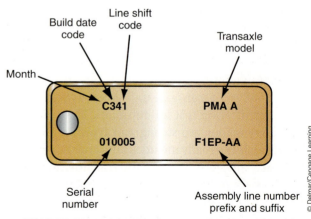

FIGURE 3-2 A typical identification tag for a Ford Motor Company automatic transmission.

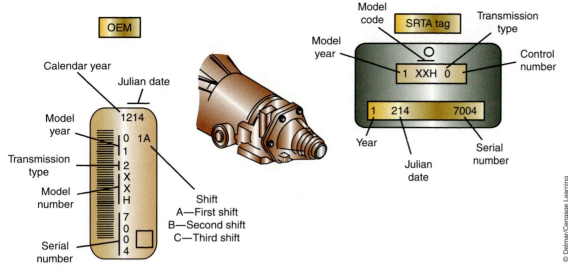

FIGURE 3-3 Location of and information contained on a General Motors' transaxle identification plate.

Early Ford RWD models can be identified by an identification tag located under the lower intermediate servo cover bolt or attached to the lower extension housing retaining bolt. A number appearing after the suffix indicates internal parts in the transmission have been changed after initial production startup. The top line of the tag shows the transmission model number and build date code.

The transaxle in General Motors FWD vehicles can be identified by an identification number stamped on a plate attached to the rear face of the transaxle (Figure 3-3). The transmission/transaxle model is printed on the Service Parts Identification label located inside the vehicle.

The transmission in early RWD General Motors products can be identified by the production number, which is located on the ID plate attached to the right side of the case near the modulator. The production number consists of a year code, a two-character model code, and a build-date code.

The transmission in late-model RWD GM products can be identified by a letter code contained in the identification number. The ID number is stamped on the transmission case above the oil pan rail on the right rear side. The identification number contains information that must be used when ordering replacement parts.

The transaxle in many Hondas and Acuras can be identified by an identification number stamped on a metal pad on top of the transmission. The first two characters indicate the transmission model.

Aisin Warner transmissions, which are used by Jeep, Isuzu, and Volvo, can be identified by a plate attached to a side of the transmission case. The plate shows the transmission model number and serial number.

The transmission used in some models from Audi, Porsche, and Volkswagen can be identified by numbers cast into the top rear of the case. The transaxle model code is identified by figures stamped into the torque converter housing. These figures consist of the model code and build-date code. Some models have code letters and the date of manufacture stamped into the machined flat on the bellhousing rim. The valve body may also be stamped on its machined boss. The valve body identification tag is secured with valve body mounting screws. A torque converter code letter is stamped on the side of the attaching lug. Some of these model vehicles use a Mercedes transmission. These units have their model stamped on the right pan rail of the case.

Models that use Borg Warner transmissions have an identification number stamped on a plate attached to the torque converter housing near the throttle cable and distributor.

Jatco transmissions, which are manufactured by the Japan Automatic Transmission Company, are used by Chrysler, Mitsubishi, Mazda, and Nissan. The model may be identified by a stamped metal plate attached to the right side of the transmission case. The model code appears on the second line and the serial number is on the bottom line.

Some Mazda, Mercedes-Benz, Toyota, and Subaru (Gunma) transmissions are best identified by the 11th character in the VIN, which is located at the top left of the instrument panel and on the transaxle flange on the exhaust side of the engine or the driver's door post. Most Mercedes-Benz transmissions have the identification number stamped into the pan rail on the right side of the case.

BMW, Peugeot, and some models of Volvo use ZF transmissions, which have an identification plate fixed to the left side of the transmission case. The lower left series of numbers on the plate indicate the number of gears, type of controls, type of gears, and torque capacity.

DIAGNOSTICS

The complexities involved in the operation of an automatic transmission (Figure 3-4) can make diagnosis quite complicated. This is especially true if a technician does not have a thorough understanding of the operation of a normally working transmission.

Automatic transmission problems are usually caused by one or more of the following conditions:

- Poor engine performance
- Poor maintenance
- Problems in the hydraulic system
- Abuse resulting in overheating
- Mechanical malfunctions

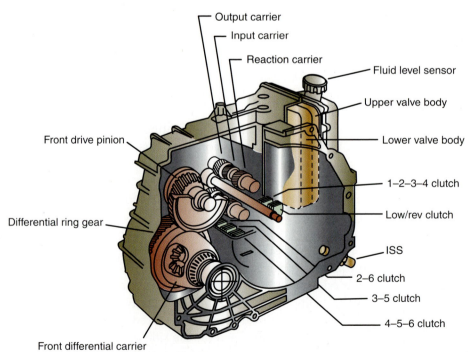

FIGURE 3-4 Basic components of a 6T70 transaxle.

© Delmar/Cengage Learning

- Electronic failures
- Improper adjustments

Diagnosis should begin by checking the condition of the fluid and its level, conducting a thorough visual inspection, and by checking the various control systems. On late-model vehicles, the initial inspection should include a check of the electronic control system. These control systems vary with the type of transmission and the manufacturer and require specific diagnostic and service procedures. Therefore, a separate chapter for these procedures is included in this book.

Poor engine performance can have a drastic affect on transmission operation. Low engine vacuum will cause a vacuum modulator to sense a load condition when it actually is not present. This will cause delayed and harsh shifts. Delayed shifts can also result from the action of the TV assembly, if the engine runs so badly that the throttle pedal must be frequently pushed to the floor to keep it running or to get it going.

Engine performance can also affect torque converter clutch operation. If the engine is running too poorly to maintain a constant speed, the converter clutch will engage and disengage at higher speeds. The customer complaint may be that the converter chatters, however the problem may be the result of engine misses.

If the vehicle has an engine performance problem, the cause should be found and corrected before any conclusions on the transmission are made. A quick way to identify if the engine is causing shifting problems is to connect a vacuum gauge to the engine and take a reading while the engine is running. The gauge should be connected to intake manifold vacuum, anywhere below the throttle plates. A normal vacuum gauge reading is steady and at 17 in.Hg. The rougher the engine runs, the more the gauge readings will fluctuate. The lower the vacuum readings, the more severe the problem.

Diagnosing a problem should follow a systematic procedure to eliminate every possible cause that can be corrected without removing the transmission. In order to properly diagnose a problem, you must totally understand the customer's concern or complaint. Make sure you know all you can about the conditions that exist when the problem occurs, such as the following:

- Cold or hot vehicle operating temperatures
- Cold or hot outside temperatures
- Loaded or unloaded vehicle
- City or highway driving
- Type of terrain
- Upshifting
- Downshifting
- Particular gear ranges
- Coasting
- Braking

CUSTOMER INTERVIEW

In order to properly diagnose a problem, you must totally understand the customer's concern or complaint. It is essential that you gather as much information from the customer as possible. The more details you have about the problem, the more accurate your diagnosis will be. Keep in mind that the problem may not be a transmission problem; it may be an engine or drivetrain problem. Gathering as much information as you can to describe the problem will help you decide if the problem is a transmission problem.

Get a good description of the problem from the customer. Find out when the problem was first noticed and if the problem is evident now.

INSPECTION

Conducting a thorough inspection of the transmission and the rest of the vehicle is extremely important. Inspect the battery and all transmission-related wiring and connectors. Check the wheels and tires of the vehicle. Make sure they are the correct size for the vehicle; a change in tire diameter can affect transmission operation. Check the overall condition of the tires because excessive wear or damage can cause symptoms that the customer will blame on the transmission. Also, identify any and all noninstalled electrical accessories. Make sure these have been properly installed. If these are incorrectly connected to the system or connected into the control module's harness, the system will not operate properly.

Your inspection should also include a careful look at the transmission's fluid, the physical appearance of the transmission, the transmission's cooling system, and the engine and transmission mounts. All warning lamps in the instrument panel should be checked. If any of these remain on after the engine is started, the reason should be identified and corrected before you continue with your diagnosis. Lastly, use a scan tool to retrieve any fault codes that are held in the computer's memory.

Fluid Check

Your diagnosis should continue with a fluid check. The transmission and torque converter cannot work properly if the fluid level is not correct. To check the ATF level, make sure the vehicle is on a level surface. Before removing the dipstick, wipe all dirt off the filler or dipstick tube and the dipstick handle.

On most automobiles, the ATF level can be checked accurately only when the transmission is at operating temperature. The temperature of the transmission or transaxle is a vital factor when checking ATF levels. Fluid-level checking can be misleading because some transmissions have bimetallic elements that block fluid flow to the transmission cooler until a predetermined temperature is reached. The bimetallic element keeps the fluid inside the transmission hot. Also, if the vehicle has just been driven for a long trip or has been pulling a trailer, the fluid needs to cool before you can get an accurate reading.

Most manufacturers recommend that the engine should be running while you check the fluid level. Others recommend that the engine should not be running. Always refer to the vehicle's service manual to identify the correct procedure. Also, make sure that the parking brake is engaged and take all necessary safety precautions while working under the hood.

When checking the fluid level, make sure the vehicle is on a level surface. If the transmission has a dipstick, wipe all dirt off the protective disc and the dipstick handle. Remove the dipstick and wipe it clean with a lint-free white cloth or paper towel. Reinsert the dipstick, remove it again, and note the reading. Markings on a dipstick indicate ADD levels, and on some models, FULL levels for cool, warm, or hot fluid (Figures 3-5 and 3-6). Always check the fluid level at least twice. If there are inconsistent readings, inspect the transmission's vent assembly to make sure it is clean and not clogged.

If the fluid level is low or off the crosshatch section of the dipstick, the problem could be external fluid leaks. Check the transmission case, oil pan, and cooler lines for evidence of leaks.

Low fluid levels may also be caused by poor maintenance. The level needs to be checked on a regular basis. If the transmission is equipped with a vacuum modulator, low fluid level may result from a bad diaphragm in the modulator. Engine vacuum will draw the fluid into the cylinders, where the engine will burn it. This problem may be found quickly by looking for fluid in the vacuum hose at the modulator. There should be none.

Classroom Manual

Chapter 3, page 82

If the dipstick has a "hot" level marked on it, this mark should be used only when the dipstick is hot to the touch.

SERVICE TIP: Some manufacturers recommend that a cold check of the fluid be taken to make sure the transmission has enough fluid to operate safely until a hot check can be made. The cold check should be followed by a hot check as soon as possible. Use the cold check to check fluid level when the fluid temperature is between 80°F and 90°F (27°C and 32°C). The hot check is the most accurate way to check the fluid level.

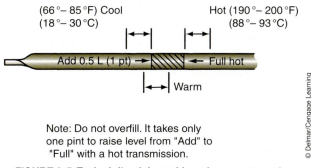

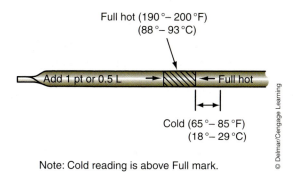

FIGURE 3-5 Typical dipstick markings for an automatic transmission.

Note: Do not overfill. It takes only one pint to raise level from "Add" to "Full" with a hot transmission.

Note: Cold reading is above Full mark.

FIGURE 3-6 Typical dipstick markings for an automatic transaxle.

Low fluid levels can cause a variety of problems. Air can be drawn into the oil pump's inlet circuit and mixed with the fluid. This will result in aerated fluid, which causes slow pressure buildup, and low pressures, which causes slippage between shifts. Air in the pressure-regulator valve will cause a buzzing noise when the valve tries to regulate pump pressure. A low fluid level can also cause delayed shifting and slipping, which leads to overheating and accelerated clutch and band wear.

Excessively high fluid levels can also cause **aeration**. As the planetary gears rotate in high fluid levels, air can be forced into the fluid. Aerated fluid can foam, overheat, and oxidize. All of these problems can interfere with normal valve, clutch, and servo operation. Foaming may be evident by fluid leakage from the transmission's vent.

> **CUSTOMER CARE: Customers should be made aware that fluid level and condition should be checked at least every six months. Temperature fluctuations from summer to winter can cause a thermal breakdown of ATF. Even high-quality fluids can experience breakdown as a result of these frequent and extreme temperature changes. It is said that nearly 90 percent of all transmission failures are caused by fluid breakdown or oxidation.**

The condition of the fluid should be checked while checking the fluid level. Examine the fluid carefully. The normal color of ATF is pink or red. If the fluid has a dark brownish or blackish color or a burned odor, the fluid has been overheated. A milky color indicates that water has mixed with the fluid, possibly from engine coolant leaking into the transmission's cooler in the radiator. If there is any question about the condition of the fluid, drain out a sample for closer inspection.

If there is evidence that there is water or moisture in the fluid, the transmission must be completely disassembled and the following parts should be cleaned or replaced:

- The torque converter
- All internal and external seals
- All transmission fluid filters
- All clutches and bands that have friction material
- All solenoids

After the transmission is reassembled, the transmission's fluid cooler(s) and its hoses and tubes should be flushed and cleaned.

After checking the ATF level and color, wipe the dipstick on absorbent white paper and look at the stain left by the fluid. Dark particles are normally band or clutch material, while

CAUTION:
If the fluid level is so low that it doesn't appear on the dipstick or is at the very bottom of the dipstick, the vehicle should not be driven until the level is brought up to normal. Driving with a very low level can cause internal transmission damage.

Aeration is the process of mixing air with a liquid.

Oxidation occurs when something is mixed with oxygen to produce an oxygen-containing compound. Oxidation is also the term given to the chemical breakdown of a substance or compound caused by its combination with oxygen.

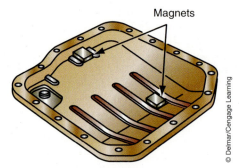

Magnets

FIGURE 3-7 **This oil pan has two magnets to collect metal shavings that may be suspended in the fluid.**

© Delmar/Cengage Learning

silvery metal particles are normally caused by the wearing of the transmission's metal parts. If the dipstick cannot be wiped clean, it is probably covered with varnish, which results from fluid oxidation. Varnish will cause the spool valves to stick, leading to improper shifting speeds. Varnish or other heavy deposits indicate the need to change the transmission's fluid and filter.

Sometimes it is difficult to accurately define the condition of the fluid. At these times, it is wise to examine the fluid and the residue in the oil pan. In most cases, this should be done after a road test. Typically, the oil pan will be the collecting point for anything suspended in the fluid. It is quite common to find some residue in the pan, even with a fresh transmission. However, an examination of the residue will help decide whether the transmission has an internal problem or not. Also, many transmission pans have magnets (Figure 3-7) in them to collect metal shavings that may be floating in the fluid. Check the magnets because excessive metal particles indicate internal damage. The permanent magnet-type input and output speed sensors can also be removed and inspected. If metal particles have gathered on the sensors, an internal problem may be indicated. If the fluid is heavily contaminated, the transmission must be disassembled and all parts, including the torque converter and cooling system, must be thoroughly cleaned. If the fluid is clean and there is only a small amount of residue in the fluid, the transmission problem is most likely caused by something external to the transmission.

If the fluid level is low and shows no signs of contamination, the transmission fluid level can be brought up to the desired level. However, if the transmission has a problem that has yet to be diagnosed, it is best to top off the fluid after road testing the vehicle. In general, it is best not to make any repairs or adjustments before the road test. Doing so may alter the symptoms reported by the customer.

If fluid needs to be added, add it in ½ pint (0.25L) increments through the transmission fluid filler tube. Do not overfill the transmission fluid. Pour the fluid in slowly and carefully to avoid spilling it on the transmission and/or engine. Clean up any spills immediately; ATF can damage components in the engine compartment.

Final Drives. Some transaxles have a separate oil reservoir for the final drive assemblies. The fluid level in these units is done with a separate check. Normally, the fluid level is checked by removing the filler plug and observing the level of the fluid. The fluid should be level with the hole for the filler plug. Often, there is an additional plug at the bottom of the unit to allow for draining of the fluid.

Transmissions without a Dipstick. Many late-model transmissions do not have a dipstick, and the fluid level is checked in the same way as a manual transmission. The dipstick and filler tube were removed from these transmissions to prevent overfilling. Research has found that many transmission failures were caused by overfilling and/

or using the wrong fluid. Without a dipstick, it is difficult to check the fluid level and condition. These transmissions have a vent/fill cap typically located on the side of the transmission. Some also have a drain plug in the bottom of the pan. In addition, these transmissions are fitted with a fluid level sensor that will inform the driver when the fluid level is dangerously low.

To check the fluid level, the transmission must be warm and the vehicle must be on a level surface or raised on a lift. It is important that all four wheels are raised, because just lifting the front will put the vehicle on a slant and you will be unable to get an accurate reading. Find the fill plug and remove it. A small amount of fluid should leak out of the vent/fill opening if the fluid is at the correct level. If fluid does not come out, add fluid until it does.

Some vehicles and the types of transmissions without a dipstick include the following:

- 2004–05 Cadillac Catera w/5L40/5L50E
- 2004–05 Chevrolet Aveo w/Aisin 81-40LE
- 1997 Chevrolet Cavalier, Cobalt w/4T40/45E
- 2005 Chevrolet Equinox w/AF33
- 2005 Chrysler 300 3.5L w/42RLE or NAG-1
- 2005 Ford cars and light trucks w/5R55N/S/W
- 2005 Lincoln Navigator 5.4L w/ZF-6SHP-26
- 2004 Mazda MPV w/5F31J
- 2004 Mazda Miata w/N4AEL
- 2004–05 Saturn Ion w/AF23
- 2004–05 Saturn Vue w/4 or 5-speed

Transmission Fluids

ATF is dyed red so that it is not easily confused with engine oil. The exact fluid that should be used in an automatic transmission depends on the transmission design and the year the transmission was built. It is very important that the correct type of ATF be used. Always refer to the service or owner's manual for the correct type of fluid to use. Some transmission dipsticks are also marked with the type of ATF required.

Although there are many types of ATF available, the following are the most common:

- **ATF+3 (Chrysler Specification MS-7176E)**—This is a fluid formulated for Chrysler automatic transmissions where ATF+, ATF+2, ATF+3 is recommended.
- **TYPE F**—This fluid is typically recommended for Ford and some imported vehicle automatic transmissions built prior to the 1977 model year as well as some 1977 through 1982 models. (Don't assume that all Ford vehicles use type F, they don't and it has been a long time since they did!)
- **Dexron® VI/Mercon®**—This fluid is sometimes referred to as multi-purpose ATF because it is recommended for all GM and Ford automatic transmissions (since 1983) requiring Dexron or Mercon transmission fluids. It also is suitable for most Mercedes-Benz passenger car automatic transmissions.
- **Multi-Vehicle ATF**—This ATF is specially formulated to meet the requirements of a broad range of automatic transmission specifications. It can be safely used in most U.S. vehicles but should not be used in a few pre-1986 vehicles where Type F fluids were specified, vehicles requiring Dexron VI, or some recent vehicles equipped with continuous variable transmissions (CVTs).

Some transmissions require the use of fluids not mentioned here. CVTs require a fluid that is much different than that used in automatic transmissions. Always use the fluid recommended by the manufacturer. The use of the wrong fluid may cause the transmission to operate improperly and/or damage the transmission.

CAUTION:
Make sure you always install the correct fluid in a transmission. The use of the wrong fluid can cause poor shifting or torque converter shudder and may cause damage to the transmission. It is important to note that often manufacturers recommend a special fluid for their transmissions. This is especially true of CVTs.

SPECIAL TOOLS

Large drain pan
Clean white rags
Inch-pound
torque wrench

Fluid Changes

> **CUSTOMER CARE:** Abusive driving can overheat a transmission and cause fluid oxidation and breakdown. Inform your customers that they should always stay within the recommended towing load for their vehicle. Also tell them to avoid excessive rocking back and forth when they are stuck in snow or mud.

The transmission's fluid and filter should be changed whenever there is an indication of oxidation or contamination. Photo Sequence 4 shows the correct procedure for performing an automatic transmission fluid and filter change. Periodic fluid and filter changes are also part of the preventive program for most vehicles. The frequency of this service depends on the conditions that the transmission normally operates under. Severe usage requires that the fluid and filter be changed every 15,000 miles. Severe usage is defined as:

1. More than 50 percent operation in heavy city traffic during hot weather (above 90°F).
2. Police, taxi, or commercial-type operation or trailer towing.

The mileage interval that a manufacturer will recommend for a fluid and filter change will also depend on the type of transmission. For example, some General Motors transmissions use aluminum valves in their valve body. Since aluminum is softer than steel, aluminum valves are less tolerant of abrasives and dirt in the fluid. Therefore, to prolong the life of the valves, more frequent fluid changes are recommended.

> **CUSTOMER CARE:** Older transmissions did not require as frequent fluid and filter changes as do the later-model vehicles. Customers should be made aware of the recommended frequency and the reason for this change. In older vehicles, both the engine and transmission ran cooler, which extended the life of transmission fluid. These transmissions operated at 175°F and the transmission fluid lasted about 100,000 miles before oxidizing. However, fluid life is halved with every 20° its temperature rises above 175°F. For example, fluid life expectancy drops to 50,000 miles at 195°F and just 25,000 miles at 215°F, which is the temperature at which most newer transmissions run.

Classroom Manual

Chapter 3, page 82

 WARNING: Be careful when draining the transmission fluid. It can be very hot and will tend to adhere to your skin, causing severe burns.

Change the fluid only when the engine and transmission are at normal operating temperatures. On most transmissions, you must remove the oil pan to drain the fluid. Some transmission pans on recent vehicles include a drain plug. A filter or screen is normally attached to the bottom of the valve body. Filters are made of paper or fabric and are held in place by screws, clips, or bolts (Figure 3-8). Filters should be replaced, not cleaned.

To drain and refill a typical transmission, the vehicle must be raised on a hoist. After the vehicle is safely in position, place a drain pan with a large opening under the transmission's oil pan. Then loosen the pan bolts and tap the pan at one corner to break it loose. Fluid will begin to drain from around the pan. When all fluid is drained, remove the oil pan.

TYPICAL PROCEDURE FOR PERFORMING AN AUTOMATIC TRANSMISSION FLUID AND FILTER CHANGE

All photos in this sequence are © Delmar/Cengage Learning.

P4-1 Raise the vehicle to a good working height and safely positioned on the life.

P4-2 Place the large-diameter oil drain pan under the transmission pan.

P4-3 Loosen all of the pan bolts except three at one end. This will allow some fluid to drain out.

P4-4 Support the pan with one hand and remove remaining bolts to remove the pan. Pour the fluid in the pan into the drain pan.

P4-5 Inspect the residue in the pan for indications of transmission problems. Then remove the old pan gasket and wipe the pan clean with a lint-free rag.

P4-6 Unbolt the filter from the transmission.

P4-7 Compare the replacement gasket and filter with the old ones to make sure the replacements are the right ones for this application.

P4-8 Install the new filter and tighten the attaching bolts to specifications. Then lay the new gasket over the sealing surface of the pan.

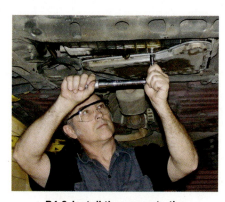

P4-9 Install the pan onto the transmission. Install and tighten the bolts to specifications. Then lower the vehicle and pour new fluid into the transmission. Run the engine to circulate the new fluid, then turn it off and raise the vehicle. Check for fluid leaks.

FIGURE 3-8 Transmission fluid filters are attached to the transmission case by screws, bolts, retaining clips, and/or by the pickup tube.

© Delmar/Cengage Learning

After draining the fluid, carefully remove the pan. There will be some fluid left in the pan. Be prepared to dump it into the drain pan. Check the bottom of the pan for deposits. A small amount of blackish deposits from clutches and bands is normal. However, an excess amount along with a burnt smell indicates burned frictional linings. If the pan has goldish brown coating, the old fluid is oxidized. If there are many metal particles in the fluid or coating the pan, this is evidence of worn gears, washers, or bushings. Normally, the metal is gold looking, it is brass or bronze. These particles come from worn bushings or thrust washers. If the particles are steel or iron, they are from the gearsets, springs, or needle bearings. Plastic debris can be from a spring retainer or thrust washer.

Any abnormal contamination or debris means the transmission is worn and should be overhauled. If the residue appears normal, clean the oil pan and its magnet.

Filters. Normally, the filter should be changed every 60,000–100,000 miles. The filters are coarse, meaning they can pass small debris through them. These particles can damage the transmission. It is best to replace the filter every time the fluid is changed or after the system has been flushed. If the old filter is kept in place during flushing, it may be able to trap some of the contaminants that were loosened during flushing. It is important to note that some transmissions do not have a filter, rather a screen is used to stop debris from circulating through the transmission.

Remove the filter (Figure 3-8), cut it open, and inspect it. Use a magnet to determine if metal particles are steel or aluminum. Steel particles indicate severe internal transmission wear or damage. If the metal particles are aluminum, they may be part of the torque-converter stator. Some torque converters use phenolic plastic stators, therefore metal particles found in these transmissions must be from the transmission itself. Filters are always replaced, whereas screens are cleaned. Screens are removed in the same way as filters. Clean the screen with fresh solvent and a stiff brush.

Some transmissions have a screw-on filter (Figure 3-9). These are similar to a conventional engine oil filter. Make sure the replacement filter is tightened properly and that the sealing gasket from the old filter is not stuck to the transmission case. Also apply some clean ATF to the seal before tightening the filter.

Other transmissions have a filter plate installed above the internal control module/solenoid assembly (Figure 3-10). Some transmissions also have an inline filter connected in the fluid line going to the fluid cooler (Figure 3-11).

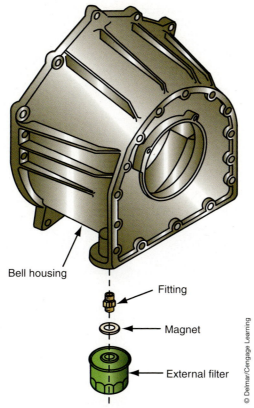

Bell housing

Fitting

Magnet

External filter

© Delmar/Cengage Learning

FIGURE 3-9 An external oil filter mounted to the bellhousing.

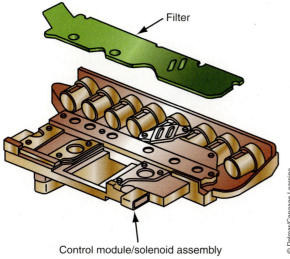

Filter

Control module/solenoid assembly

© Delmar/Cengage Learning

FIGURE 3-10 This control module/solenoid assembly is mounted to the side of the transmission and has a separate filter.

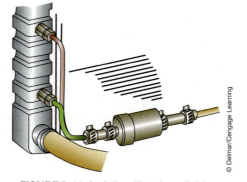

© Delmar/Cengage Learning

FIGURE 3-11 An inline filter for a fluid cooler.

Refilling the Assembly. Remove any traces of the old pan gasket on the case housing. Then, install a new filter and gasket on the bottom of the valve body and tighten the retaining bolts to the specified torque. If the filter is sealed with an O-ring, make sure it is properly installed.

Remove any traces of the old pan gasket from the oil pan and transmission housing. Make sure the mounting flange of the oil pan is not distorted and bent. Oil pans are typically made of stamped steel. The thin steel tends to become distorted around the attaching boltholes.

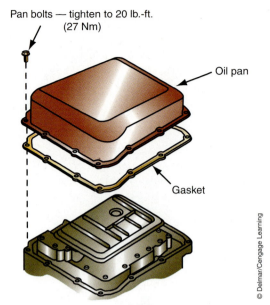

Pan bolts — tighten to 20 lb.-ft. (27 Nm)

Oil pan

Gasket

© Delmar/Cengage Learning

FIGURE 3-12 Always tighten the pan bolts to the correct torque specification.

CAUTION:

Make sure you carefully check the specifications. The required torque is often given in pounds-inch. You can very easily break the bolts or damage the transmission case if you tighten the bolts to pounds-feet.

Some transaxles vent through the dipstick handle or tube.

These distortions can prevent the pan from fitting tightly against the transmission case. To flatten the pan, place the mounting flange of the pan on a block of wood and flatten one area at a time with a ball-peen hammer.

Verify that the new gasket is correct by comparing the old with the new. Then, re-install the pan using the gasket or sealant recommended by the manufacturer. Tighten the pan retaining bolts to the specified torque (Figure 3-12); often this specification is given in inch-pounds rather than foot-pounds.

Transmission gaskets should not be installed with any type of liquid adhesive or sealant, unless specifically noted by the manufacturer. If any sealer gets into the valve body, severe damage can result. Also, sealant can clog the oil filter. If a gasket is difficult to install, a thin coating of transmission assembly lube or petroleum jelly can be used to hold the gasket in place while assembling the parts.

One type of gasket that presents unique installation problems is the cork-type gasket. These gaskets tend to change shape and size with changes in humidity. If a cork gasket is slightly larger than it should be, soak it in water and lay it flat on a warm surface. Allow it to dry before installing it. If the gasket is slightly smaller than required, soak it in warm water prior to installation. Another way to make a cork gasket grow is to lay in on a flat clean and hard surface, then, with a hammer, strike it all the way around until it is the correct size.

Whenever you are using a cork gasket to create a seal between two parts, make sure you properly tighten the two surfaces together. Tighten the attaching bolts or nuts, in a staggered pattern, to the specified torque so that the gasket material is evenly squeezed between the two surfaces. If too much torque is applied, the gasket may split.

With the pan securely in place, lower the vehicle. Now, pour a little less than the required amount of fluid into the transmission through the dipstick tube. Always use the recommended type of ATF. The wrong fluid will alter the shifting characteristics of the transmission. For example, if type F fluid is used in a transmission designed for Dexron type fluid, the shifting will be harsher.

Start the engine and allow it to idle for at least one minute. Then, with the parking and service brakes applied, move the gear selector lever momentarily to each position, ending in park. Recheck the fluid level and add a sufficient amount of fluid to bring the level to about of an inch below the "add" mark.

Run the engine until it reaches normal operating temperature. Then recheck the fluid level. It should be in the "hot" region on the dipstick. Make sure the dipstick is fully seated into the dipstick tube opening. This will prevent dirt from entering the transmission.

Using a Fluid Evacuator. The use of a fluid evacuator (Figure 3-13) makes it unnecessary to loosen and remove the oil pan. This makes the job simpler and certainly less messy. The evacuator draws the old fluid out of the transmission through the filler tube. Fluid evacuators create a vacuum in their reservoir. This low pressure is either created with an air compressor or with a hand pump. The fluid in the transmission is at atmospheric pressure, therefore when it is exposed to the low pressure, it will move toward it. To use a typical evacuator tool, perform the following steps:

1. Bring the ATF's temperature up to its normal operating temperature.
2. Turn on the evacuator to create a vacuum in its reservoir.
3. After a suitable vacuum is present, insert the tool's tube into the filler tube as far as it can go.
4. Open the valve on the evacuator and collect the fluid in a container.
5. Properly dispose of the old fluid.
6. Remove the tool's tube from the filler tube.
7. Refill the transmission with new ATF.

Flushing. Draining the transmission does not remove all of the old fluid for the unit. Much fluid remains inside the torque converter, the transmission cooler, and the cooler lines. It is said that a drain and refill service replaces only 40 percent of the old contaminated ATF. In

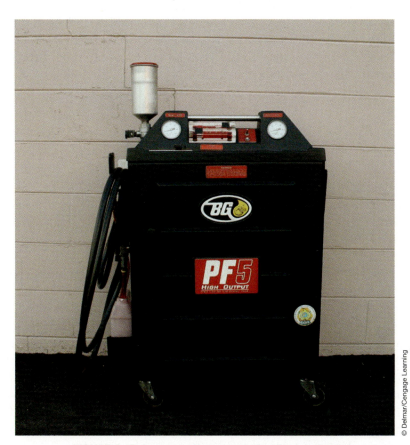

FIGURE 3-13 An automatic transmission fluid exchanger

© Delmar/Cengage Learning

Case porosity is caused by tiny holes that are formed by trapped air bubbles during the casting process.

Some transmissions do not use a gasket; instead, the pan is sealed with RTV.

SERVICE TIP:
On transmissions with a governor, after starting the engine, quickly move the gear selector to reverse. Many transmissions do not feed fluid to the governor when they are in reverse. By quickly placing the transmission in reverse, the chance of dirt entering the governor is greatly decreased.

SERVICE TIP:
Many technicians recommend that the transmission not be flushed if the old fluid smells or looks burnt or if it has flakes in it. These indicate a need to overhaul the transmission. Flushing will make the transmission worse than it was.

fact, when you drain the fluid through a drain plug, you end up draining only 20 percent of the fluid. In both cases, the new fluid mixes with the old and reduces the advantages of having new fluid. Also, the fresh ATF can often loosen sludge and varnish deposits leading to shift problems and even transmission failure.

To remove most of the old fluid and fill the system with fresh lubricant, the transmission should be flushed or the fluid totally exchanged. Normally, flushing also involves the use of a chemical that will dissolve varnish and other contaminants. Some manufacturers also recommend the use of a conditioner added to the new fluid after flushing. The conditioners may extend the service life of the ATF life, revitalize seals and 0-rings, and maintain smooth shifting.

Flushing is done with a flushing tool. Some flushing tools have a pressure gauge that can be used to detect a restriction in the hydraulic circuits, filter, cooler, or lines and hoses. If the pressure reading is less than 8 psi, there is a restriction in the fluid path.

There are two basic types of flushing tools: a cooler line flush machine and a pump inlet flush machine. The cooler line flush machine connects into the line from the transmission to the transmission cooler. This type of machine may also be called a fluid exchange unit. The outlet from the machine is connected to the cooler. This means the fluid moves from the transmission into the machine and then to the cooler. The machine has a chamber separated by a diaphragm. The area above the diaphragm is filled with fresh fluid. With the engine running, fluid flows from the transmission into the lower part of the chamber. As the old fluid fills its portion within the chamber, it pushes up on the diaphragm and forces new fluid into the cooler. This continues until most of the old fluid is replaced by new fluid.

The pump inlet flush machine connects to the intake of the oil pump. The oil pan and filter are removed and a drain pan is set under the transmission. The machine provides fresh fluid to the pump inlet and as the fluid passes through the transmission it leaks from the bottom of the transmission into the drain pan. The new fluid has little chance of mixing with the old and the system is filled with fresh oil. This machine wastes a bit of fluid, however. It will cycle through about 20 quarts of fluid to flush out 15 quarts of old fluid. After the system is flushed, the filter and pan are reinstalled and the unit is filled with fluid.

FLUID LEAKS

Continue your diagnostics by conducting a quick and careful visual inspection. Check all drivetrain parts for looseness and leaks. If the transmission fluid was low or if there was no fluid, raise the vehicle and carefully inspect the transmission for signs of leakage. Leaks are often caused by defective gaskets or seals. Common sources of leaks are the oil pan seal, rear cover, final drive cover (on transaxles), extension housing, speedometer drive gear assembly, and electrical switches mounted into the housing (Figure 3-14). The housing itself may have a porosity problem, allowing fluid to seep through the metal. Case porosity may be repaired using an epoxy-type sealer.

During your inspection, check the area around the filler tube for signs of leakage. This type of leak is often corrected by replacing the seal or grommet between the tube and the housing. Also, check the fluid lines and hoses at the transmission and the oil cooler. At times, these leaks can be corrected by tightening the fittings. However, do not overtighten them; this can cause damage to the fittings or housing. If the fittings are not loose, the problem is usually corrected by installing new seals at the connections.

Sometimes the source of the leak is difficult to find. A careful cleaning of the transmission and the area surrounding it may help. After the transmission is cleaned, run the engine and with the brake pedal depressed. Move the gear selector through its gear ranges; this circulates the fluid throughout the transmission. Turn off the engine and inspect the transmission. The source of the leak may be now apparent.

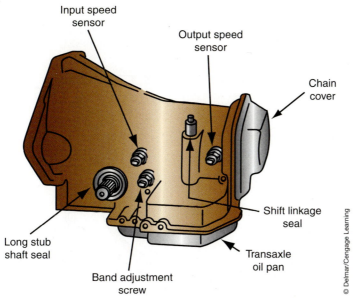

Input speed sensor

Output speed sensor

Chain cover

Shift linkage seal

Long stub shaft seal

Band adjustment screw

Transaxle oil pan

© Delmar/Cengage Learning

FIGURE 3-14 **Possible sources of fluid leaks on this transaxle.**

If the leak is still not found, a fluorescent dye can be added to the fluid. With these products, the leak will be evident when an ultraviolet light passes by the source of the leak. The dye will also help to determine if ATF is leaking into the engine's cooling system. The following procedure is the typical way to conduct this type of leak check. Always follow the instructions of the manufacturer of the dye.

1. Add the recommended amount of dye to the transmission fluid.
2. Raise the vehicle and clean off all traces of oil from the top and bottom of the torque converter housing, the transmission case, the rear of the engine, and the engine's oil pan. Use compressed air to dry all the surfaces after they have been cleaned.
3. Lower the vehicle. Then start the engine and allow it to run until it reaches normal operating temperature.
4. Raise the vehicle with the engine still running and carefully use the ultraviolet light to find evidence of a leak.
5. Continue to run the engine until the probable source of leakage can be determined. Sometimes this process is sped up by running the engine at fast idle and occasionally shifting between drive and reverse.
6. When the leak is discovered, lower the vehicle and turn off the engine.

Oil Pan

A common cause of fluid leakage is the seal of the oil pan to the transmission housing. If there are signs of leakage around the rim of the pan, retorquing the pan bolts may correct the problem. If tightening the pan does not correct the problem, the pan must be removed and a new gasket installed (Figure 3-12). Make sure the sealing surface of the pan's rim is flat and capable of providing a seal before reinstalling it.

Torque Converter

Torque converter problems can be caused by a leaking converter (Figure 3-15). This type of problem may be the cause of customer complaints of slippage and a lack of power. To check the converter for leaks, remove the converter access cover and examine the area around the torque converter shell. An engine oil leak may be falsely diagnosed as a converter leak. The color of engine oil is different from transmission fluid and may help identify the true source

Classroom Manual

Chapter 3, page 83

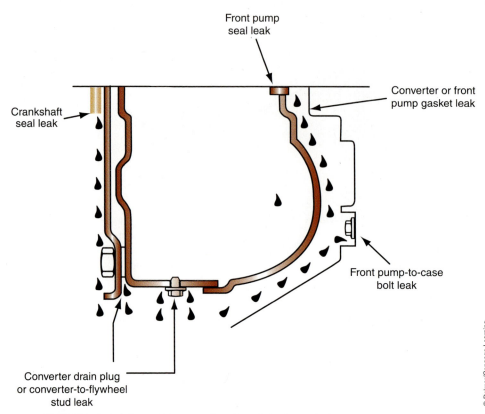

Front pump
seal leak

Converter or front
pump gasket leak

Crankshaft
seal leak

Front pump-to-case
bolt leak

Converter drain plug
or converter-to-flywheel
stud leak

© Delmar/Cengage Learning

FIGURE 3-15 By determining the direction of fluid travel, the cause of a fluid leak around the torque converter can be identified.

of the leak. However, if the oil or fluid has absorbed much dirt, both will look the same. An engine leak typically leaves an oil film on the front of the converter shell, whereas a converter leak will cause the entire shell to be wet. If the transmission's oil pump seal is leaking, only the back side of the shell will be wet. If the converter is leaking or damaged, it should be replaced.

If the torque converter hub seal is leaking, it should be replaced. To do this, the transmission must be removed. Normally, a slide hammer and a seal removal and installation tool designed for the transmission are required to replace the seal.

Once the transmission has been removed, the torque converter is pulled out of the unit. Then the seal is pulled from the oil pump. Be careful not to damage the area where the seal is seated.

Position the new hub seal and drive it into position with the seal installation tool. Then install the torque converter and the transmission.

Pump Seal and Gasket. If the fluid leak appears to be at the center of the torque converter, the oil pump seal or gasket may be leaking. If this is the case, the transmission must be removed to install new seals and gaskets.

Differential/Axle Seals. If oil is leaking from around where the axle shaft enters a transaxle, the seal is probably worn or damaged. Before proceeding to replace the seal, check the service information to see if the final drive unit has a separate oil reservoir from the transaxle. If it does, identify the type of fluid that must be used in the final drive.

To replace the seal, remove the axle shaft according to the procedure outlined in the service information. Once the shaft has been removed, use the proper tool to remove the seal, often a slide hammer-type puller is used (Figure 3-16).

CAUTION:
Be careful not to damage the housing while removing the seal.

FIGURE 3-16 Using a slide hammer puller to remove an axle seal from the transaxle case.

© Delmar/Cengage Learning

To install a new seal, position the new seal on the installation tool designed for the transaxle. Then install the seal. Make sure it is squarely and securely positioned in its bore. Apply a light coat of lubricant on the end of the axle shaft and insert it into the transaxle housing.

Then, check the fluid level and add fluid as necessary. Never overfill the unit as that can cause damage to the transaxle.

Extension Housing

An oil leak stemming from the mating surfaces of the extension housing and the transmission case may be caused by loose bolts. To correct this problem, tighten the bolts to the specified torque. Also check for signs of leakage at the rear of the extension housing. Fluid leaks from the seal of the extension housing can be corrected with the transmission in the car. Often, the cause for the leakage is a worn extension housing bushing, which supports the sliding yoke of the driveshaft. Normally, bushing wear or damage is caused by faulty universal joints. When the driveshaft is installed, the clearance between the sliding yoke and the bushing should be minimal. If the clearance is satisfactory, a new oil seal will correct the leak. If the clearance is excessive, the repair requires that a new seal and a new bushing be installed. If the seal is faulty, the transmission vent should be checked for blockage.

Classroom Manual

Chapter 3, page 93

Rear Oil Seal and Bushing Replacement

Procedures for the replacement of the rear oil seal and bushing vary little with each car model. General procedures for the replacement of the oil seals and bushings follow.

To replace the rear seal:

1. Remove the driveshaft.
2. Remove the old seal from the extension housing (Figure 3-17).
3. Lubricate the lip of the seal, then install the new seal in the extension housing.
4. Install the driveshaft.

To replace the rear bushing and seal:

1. Remove the driveshaft from the car. Inspect the yoke for wear and replace it if necessary.
2. Insert the appropriate puller tool into the extension housing until it grips the front side of the bushing.
3. Pull the seal and bushing from the housing.
4. Using the correct tools, drive a new bushing into the extension housing.
5. Install a new seal in the housing (Figure 3-18).
6. Install the driveshaft.

SPECIAL TOOLS

Slide hammer
Seal remover tool
Bushing removing tool
Seal driver
Bushing driver

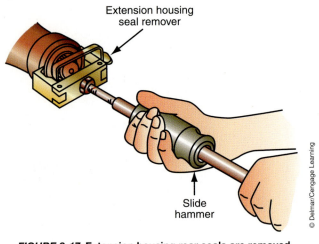

Extension housing
seal remover

Slide
hammer

© Delmar/Cengage Learning

FIGURE 3-17 Extension housing rear seals are removed with a slide hammer and a special removal tool.

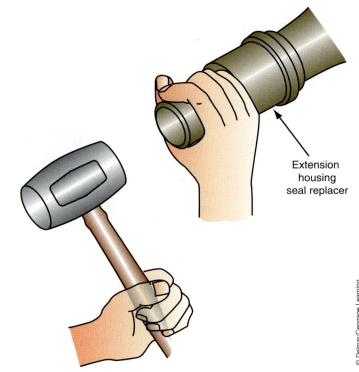

Extension
housing
seal replacer

© Delmar/Cengage Learning

FIGURE 3-18 Rear seals are installed into the extension housing with the proper driver.

Speed Sensors

Classroom Manual

Chapter 3, page 95

Speed sensors can be the source of an oil leak. Typically, a transmission has two speed sensors fitted into the housing (Figure 3-19): the input shaft speed sensor (ISS; sometimes referred to as the TSS for turbine speed sensor) and the output shaft speed sensor (OSS).

Both of these sensors can be serviced with the transmission in the vehicle. Most are retained by a bolt that must be torqued to specifications. To remove them, put the gear selector in neutral and raise the vehicle. Then disconnect the connector to the sensor and remove the retaining bolt. The sensor can then be pulled from the housing. To install the sensor, apply a light coat of clean ATF on the seal and push it into place, then install and tighten the retaining bolt.

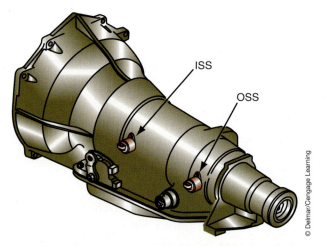

ISS

OSS

© Delmar/Cengage Learning

FIGURE 3-19 The location of the ISS and OSS in a typical transmission.

Electrical Connections

A visual inspection of the transmission should include a careful check of all electrical wires, connectors, and components. This inspection is especially important for electronically controlled transmissions and for transmissions that have a torque converter clutch. All faulty connectors, wires, and components should be repaired or replaced before continuing your diagnosis of the transmission.

Check all electrical connections to the transmission. Faulty connectors or wires can cause harsh or delayed and missed shifts. On transaxles, the connectors can normally be inspected through the engine compartment, whereas they can only be seen from under the vehicle on longitudinally mounted transmissions. To check the connectors, release the locking tabs and disconnect them, one at a time, from the transmission. Carefully examine them for signs of corrosion, distortion, moisture, and transmission fluid. A connector or wiring harness may deteriorate if ATF reaches it. Also check the connector at the transmission. Using a small mirror and flashlight may help you get a good look at the inside of the connectors. Inspect the entire transmission wiring harness for tears and other damage. Road debris can damage the wiring and connectors mounted underneath the vehicle.

Because the operation of the engine and transmission is integrated through the control computer, a faulty engine sensor or connector may affect the operation of both the engine and the transmission. The various sensors and their locations can be identified by referring to the appropriate service information. The engine control sensors that are the most likely to cause shifting problems are the throttle position sensor, MAP sensor, and vehicle speed sensor.

Checking the Transaxle Mounts

The engine and transmission mounts on FWD cars are (Figure 3-20) important to the operation of the transaxle. Any engine movement may change the effective length of the shift and throttle cables and therefore may affect the engagement of the gears. Delayed or missed shifts may result from linkage changes as the engine pivots on its mounts. Problems with transmission mounts may also affect the operation of a transmission on a RWD vehicle, but this type of problem will be less detrimental than the same type of problem on a FWD vehicle. Many shifting and vibration problems can be caused by worn, loose, or broken engine and transmission mounts. Visually inspect the mounts for looseness and cracks. To get a better look at the condition of the mounts, pull up and push down on the transaxle case while watching the mount. If the mount's rubber separates from the metal plate or if the case moves up but not down, replace the mount. If there is movement between the metal plate and its attaching point on the frame, tighten the attaching bolts to an appropriate torque.

Then, from the driver's seat, apply the foot brake, set the parking brake, and start the engine. Put the transmission into a gear and gradually increase the engine speed to about 1500–2000 rpm. Watch the torque reaction of the engine on its mounts. If the engine's reaction to the torque appears to be excessive, broken or worn drivetrain mounts may be the cause.

If it is necessary to replace the transaxle mount, make sure you follow the manufacturer's recommendations for maintaining the alignment of the driveline. Failure to do this may result in poor gear shifting, vibrations, or broken cables. Some manufacturers recommend that a holding fixture or special bolt be used to keep the unit in its proper location.

When removing the transaxle mount, begin by disconnecting the battery's negative cable. Disconnect any electrical connectors that may be located around the mount. It may be necessary to move some accessories, such as the horn, in order to service the mount without damaging some other assembly. Be sure to label any wires you remove to facilitate reassembly.

Install the engine support fixture and attach it to an engine hoist. Lift the engine just enough to take the pressure off of the mounts. Remove the bolts attaching the transaxle mount to the frame and the mounting bracket, then remove the mount.

Classroom Manual
Chapter 3, page 95

Even the slightest amount of corrosion can affect the output of a sensor. Increased resistance will always change the voltage signal of a circuit.

A **throttle position sensor** is most often referred to as a TP sensor.

A **manifold absolute pressure (MAP) sensor**, senses engine vacuum.

Some manufacturers require that the mount bolts be discarded and new ones installed during reassembly.

Classroom Manual
Chapter 3, page 96

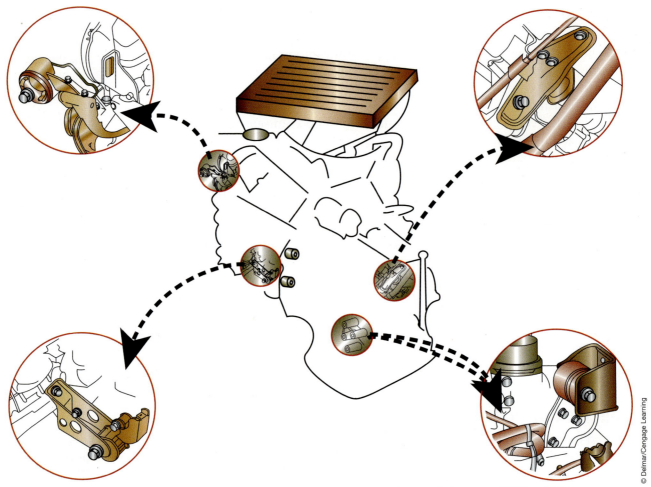

FIGURE 3-20 The various mounts used to position the engine and transaxle in a typical FWD vehicle.

SERVICE TIP:
In the procedure for removing the mount, some manufacturers recommend the use of a special alignment bolt, which is installed in an engine mount. This bolt serves as an indicator of powertrain alignment. If excessive effort is required to remove the alignment bolt, the powertrain must be shifted to allow for proper alignment.

To install the new mount, position the transaxle mount in its correct location on the frame and tighten its attaching bolts to the proper torque. Install the bolts that attach the mount to the transaxle bracket. Prior to tightening these bolts, check the alignment of the mount. Once you have confirmed that the alignment is correct, tighten all loosened bolts to their specified torque. Remove the engine hoist fixture from the engine and reinstall all accessories and wires that may have been removed earlier.

Transmission Cooler and Line Inspection

Transmission coolers are a possible source of fluid leaks. The efficiency of the coolers is also critical to the operation and longevity of the transmission. The vehicle may be equipped with the standard-type cooler in the radiator or may also have an auxiliary cooler. Both should be visually inspected and tested in the same way.

Follow these steps when inspecting the transmission cooler and associated lines and fittings:

1. Check the engine cooling system. The transmission cooler cannot be efficient if the engine's cooling system is defective. Repair all engine cooling system problems before continuing to check the transmission cooler.
2. Inspect the fluid lines and fitting between the cooler and the transmission (Figure 3-21). Check these for looseness, damage, signs of leakage, and wear. Replace any damaged lines. Replace or tighten any leaking fittings.
3. Inspect the engine's coolant for traces of transmission fluid. If ATF is present in the coolant, the transmission cooler leaks.

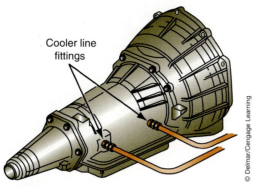

Cooler line
fittings

FIGURE 3-21 Typical location of cooler line connections on a transmission.

Classroom Manual
Chapter 3, page 83

 WARNING: Do not remove the radiator cap on a warm engine. Wait until it cools down before proceeding with a coolant check. The hot coolant can burn you.

4. Check the transmission's fluid for signs of engine coolant. Water or coolant will cause the fluid to appear milky with a pink tint. This is also an indication that the transmission cooler leaks and is allowing engine coolant to enter into the transmission fluid.

The cooler can be checked for leaks by disconnecting and plugging the transmission-to-cooler lines, at the radiator. Then remove the radiator cap to relieve any pressure in the system. Tightly plug one of the ATF line fittings at the radiator. Using the shop air supply with a pressure regulator, apply 50 to 70 psi of air pressure into the cooler at the other cooler line fitting. Look into the radiator. If you see bubbles, the cooler leaks.

Checking Flow. The cooler system should also be checked for proper flow. To do this, make sure there are no fluid leaks. Start the engine and allow the transmission to reach normal operating temperature. Then turn the engine off and check the fluid level. Remove the transmission dipstick and install a funnel into the dipstick tube. Raise the vehicle up on a hoist. Then disconnect the cooler return line at the transmission. Securely fasten a rubber hose to the end of the cooler line and place the other end of the hose into the funnel. Now lower the vehicle and start the engine. Allow the engine to run at a fast idle and observe the flow of fluid into the funnel.

SERVICE TIP:
The cooler lines for some transmissions have a seal inside the attaching bore in the transmission housing. A special tool is required to remove the old seal and install a new one (Figure 3-22). If the transmission is so equipped, make sure you replace the seal if the line is leaking or if the line has been disconnected from the housing.

CAUTION:
Make sure the air pressure is regulated down to 50 to 70 psi before conducting this test. Most shops have air line pressures of nearly 150 psi. Pressures this high will damage the radiator or cooler.

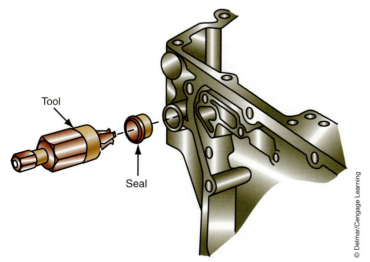

Tool

Seal

FIGURE 3-22 The special tool required to remove a cooler line seal on some transmissions.

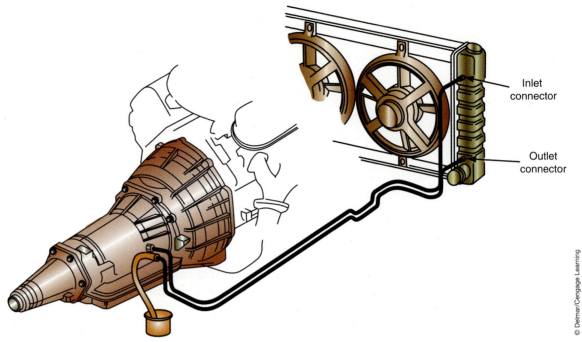

FIGURE 3-23 Checking the flow of fluid through the cooler circuit.

Another way to do this test is with the help of an assistant. Have the assistant pour the same amount of fluid into the filler tube as you are getting to flow into a bucket (Figure 3-23). If you get one pint, have the assistant pour in one pint. By replenishing the fluid, your assistant is ensuring that you will not run out of fluid. Make sure the assistant keeps track of what is being poured in. Continue the test for 30 seconds.

The preferred way to check flow is with a flow meter. A flow meter is preferred because it has restrictions on the return side, which is closer to reality than an open hose. To use a flow meter, connect it into the cooler lines, start the engine, and observe.

Most manufacturers list specifications for transmission cooler flow rates. This is normally expressed in pints per second (such as one pint in 15–20 seconds). You should observe a steady flow of fluid. If the flow is erratic or weak, the cooler may be restricted, the transmission's pump may be weak, or there is an internal leak in the transmission.

CHECKING FOR DTCs

Prior to taking the vehicle on a road test, you should retrieve DTCs. At this stage of the diagnostic procedure, it is important to recognize what the control system is seeing. Although a detailed look at electronic control systems is given in the next chapter (Chapter 4), basic retrieval and interpretation information is necessary now. Connect the scan tool to the appropriate data line connector. Calibrate the tool to the vehicle being tested. Then set it to retrieve the DTCs. Record all recovered DTCs. Clear the codes prior to your test drive. Use the scan tool after the test drive to see if the codes were reset during the test.

It is important that you do not correct the problems indicated by the DTCs before you take the vehicle on a road test. Correcting them may change your results and leave you with altered road test results. However, if the road test verifies the customer's initial complaint and anything else you found wrong, correct these problems before moving on with other repairs. All engine-related and body computer codes should be dealt with before you address the transmission codes. If the system displays only "pass" codes, continue your diagnosis according to the symptoms, using all available information to do this. Remember,

Classroom Manual

Chapter 3, page 79

the control system will only set a DTC when it senses that a sensor or output is not functioning the way it thinks it should. There are many things that can influence the thinking of the computer.

ROAD TESTING THE VEHICLE

All transmission complaints should be verified by road testing the vehicle and attempting to duplicate the customer's complaint. A knowledge of the exact conditions that cause the symptom and a thorough understanding of transmissions will allow you to accurately diagnose problems.

Many problems that appear to be transmission problems may in fact be problems with the engine, the driveshaft, the U-joint or CV joint, the wheel bearings, wheel/tire imbalance, or other conditions. Make sure these are not the cause of the problem before you begin to diagnose and repair a transmission. Diagnosis becomes easy if you think about what is happening in the transmission when the problem occurs. If there is a shifting problem, think about the parts that are being engaged and what these parts are attempting to do.

Test Preparation

Road testing a vehicle to exactly duplicate the symptoms is essential to proper diagnosis. During the road test, the transmission should be operated in all possible modes and its operation noted. The customer's complaint should be verified and any other transmission malfunctions should be noted. Your observations during the road test, your understanding of the operation of a transmission, and the information given in service manuals, will help you identify the cause of any transmission problem. If conducted properly, road testing is diagnosis through a process of elimination.

Before beginning your road test, find and duplicate, from a service manual, the chart (Figure 3-24) which shows the band and clutch application for different gear selector positions. Using these charts will greatly simplify your diagnosis of automatic transmission problems. It is also wise to have a notebook or piece of paper to jot down notes about the operation of the transmission. If the transmission is electronically controlled, take a scan tool on the road test with you. It will be an added convenience if the scan tool has memory or a printer.

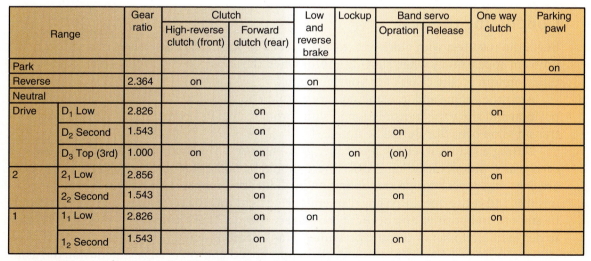

Range		Gear ratio	Clutch		Low and reverse brake	Lockup	Band servo		One way clutch	Parking pawl
			High-reverse clutch (front)	Forward clutch (rear)			Opration	Release		
Park										on
Reverse		2.364	on		on					
Neutral										
Drive	D₁ Low	2.826		on					on	
	D₂ Second	1.543		on			on			
	D₃ Top (3rd)	1.000	on	on		on	(on)	on		
2	2₁ Low	2.856		on					on	
	2₂ Second	1.543		on			on			
1	1₁ Low	2.826		on	on				on	
	1₂ Second	1.543		on			on			

© Delmar/Cengage Learning

FIGURE 3-24 A typical band and clutch application chart. This chart should be referred to during a road test and when determining the cause of any shifting problems.

Symptom - Engagement Concerns	Electrical	Hydraulic/ Mechanical
No Forward	No electrical concerns	**Fluid** • Incorrect level • Condition Shift **Linkage** • Damaged or incorrectly adjusted **Incorrect Pressures** • Low forward clutch pressure, low line pressure **Fluid Filter and Seal Assembly** • Plugged, damaged • Filter seal damaged **Main Controls** • 3-4 shift valve, main regulator valve, manual valve — stuck, damaged • Bolts not tightened to specifications • Gaskets damaged· 2-3 accumulator and seals damaged • Pressure regulator valve **Pump Assembly** • Bolts not tightened to specifications • Porosity/cross leaks/ball missing or leaking, plugged hole • No. 3 and No. 4 seal rings damaged • Gaskets damaged **Forward Clutch Assembly** • Seals, piston damaged • Check balls damaged, missing, mislocated, not seating correctly • Friction elements damaged or worn **One-Way Clutch Assembly (Planetary)** • Worn, damaged or assembled incorrectly **Output Shaft** • Damaged

© Delmar/Cengage Learning

FIGURE 3-25 A portion of a typical manufacturer's diagnosis by symptom chart.

Some manufacturers recommend the use of a symptoms chart or checklist that is provided in their service manuals (Figure 3-25).

Prior to road testing a vehicle, always check the transmission fluid level and condition of oil and correct any engine performance problems. Inspect the transmission for signs of fluid leakage. If leakage is evident, wipe off the leaking oil and dust. Then, note the location of the leaks in your notebook or on the symptoms chart.

Conducting the Test

Begin the road test with a drive at normal speeds to warm the engine and transmission. If a problem appears only when starting and/or when the engine and transmission is cold, record this symptom on the chart or in your notebook.

After the engine and transmission are warmed up, place the shift selector into the DRIVE position and allow the transmission to shift through all of its normal shifts. Any questionable transmission behavior should be noted.

During a road test, check for proper gear engagement as you move the selector lever to each gear position, including PARK. There should be no hesitation or roughness as the gears are engaging. Check for proper operation in all forward ranges, especially the 1–2, 2–3, 3–4 upshifts and converter clutch engagement during light throttle operation. These shifts should be smooth and positive and occur at the correct speeds. These same shifts should feel firmer under medium to heavy throttle pressures. Transmissions equipped with a torque converter clutch should be brought to the specified apply speed and their engagement noted. Again, record the operation of the transmission in these different modes in your notebook or on the diagnostic chart.

Force the transmission to kickdown and record the quality of this shift and the speed at which it downshifts. These will be later compared to the specifications. Manual downshifts should also be made at a variety of speeds. The reaction of the transmission should be noted, as should all abnormal noises, and the gears and speeds at which they occur.

After the road test, check the transmission for signs of leakage. Any new leaks and their probable cause should be noted. Then compare your written notes from the road test to the information given in the service manual to identify the cause of the malfunction. The service manual usually has a diagnostic chart to aid you in this process.

The following problems and their causes are given as examples. The actual causes of these types of problems will vary with the different models of transmissions. Always refer to the appropriate band and clutch application chart while diagnosing shifting problems. Using these will allow you to identify the cause of the shifting problems through the process of elimination.

These are typical causes of common problems:

- If the shift for all forward gears is delayed, a slipping front or forward clutch is indicated.
- If there is a delay or slip when the transmission shifts into any two forward gears, a slipping rear clutch is indicated.
- If there is a shift delay while the transmission was in DRIVE and it shifted into third gear, either the front or rear clutch may be slipping. To determine which clutch is defective, look at the behavior of the transmission when one of the two clutches are not applied. For example, if the rear clutch is not engaged while the transmission is in REVERSE and there was no slippage in REVERSE, the problem of the slippage in third gear is caused by the rear clutch. If there is a delay only during shifts into REVERSE and from second to third gear, the Reverse/High clutch is slipping.
- If there is delayed shifting from first to second gear, there are intermediate clutch or band problems.
- If there is slippage in first gear when the gear selector is in the DRIVE position, but not when first gear is manually selected, the one-way or overrunning clutch may be the cause.

It is important to remember that delayed shifts or slippage may also be caused by leaking hydraulic circuits or sticking spool valves in the valve body. Since the application of bands and clutches are controlled by the hydraulic system, improper pressures will cause shifting problems. Other components of the transmission can also contribute to shifting problems. For example, on transmissions equipped with a vacuum modulator, if upshifts do not occur at the specified speeds or do not occur at all, either the modulator is faulty or the vacuum supply line is leaking.

Abnormal transmission noises and vibrations can be caused by faulty bearings, damaged gears, worn or damaged clutches and bands, or a bad oil pump as well as by contaminated fluid or improper fluid levels. Vibrations can also be caused by torque converter problems. All noises and vibrations that occur during the road test should be noted, as well as when they occur.

Transaxles have unique noises associated with them because of their construction. These noises can result from problems in the differential or drive axles. Use the following guide to determine if a noise is caused by the transaxle or by other parts in the drivetrain.

A knock at low speeds may indicate:

1. Worn drive axle CV joints or U-joints
2. Worn side gear hub counterbore

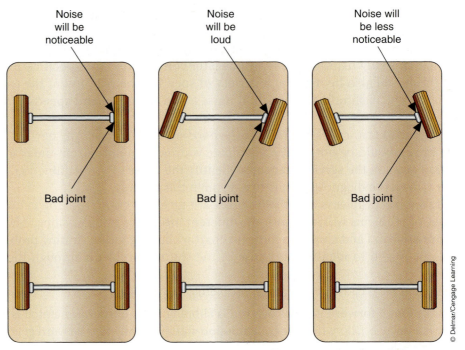

Noise will be noticeable

Noise will be loud

Noise will be less noticeable

Bad joint

Bad joint

Bad joint

FIGURE 3-26 A bad outside CV joint will have a different sound as the vehicle changes direction.

Noise that is most pronounced on turns may indicate:

1. Differential gear noise
2. Dry, worn CV joints

Clunk on acceleration or deceleration may indicate:

1. Loose engine mounts
2. Worn differential pinion shaft or side gear hub counterbore in the case
3. Worn or damaged drive axle inboard CV joints or U-joints

Clicking noise in turns may indicate:

1. Worn or damaged outboard CV joint (Figure 3-26)

Noises and Vibrations

Classroom Manual

Chapter 3, page 66

Most vibration problems are caused by an unbalanced torque converter assembly, a poorly mounted torque converter, or a faulty output shaft. The key to determining the cause of the vibration is to pay particular attention to the vibration in relationship to engine and vehicle speed. If the vibration changes with a change in engine speed, the cause of the problem is most probably the torque converter. If the vibration changes with vehicle speed, the cause is probably the output shaft or the driveline connected to it. The later type of problem can be a bad extension housing bushing or universal joint, which would become worse at higher speeds.

Begin your diagnosis by determining if the cause of the problem is the drive line or the transmission. To do this, put the transmission in gear and apply the foot brakes. If the noise is no longer evident, the problem must be in the drive line or the output of the transmission. A common source of noise is worn wheel bearings. If the noise is still present, the problem must be in the transmission or torque converter.

Noise problems are also best diagnosed by paying a great deal of attention to the speed and the conditions at which the noise occurs. The conditions to pay most attention to are the operating gear and the load on the drive line. If the noise is engine speed related and is present in all gears, including PARK and NEUTRAL, the most probable source of the noise is the oil pump because it rotates whenever the engine is running. However, if the noise is engine related and is present in all gears except PARK and NEUTRAL, the most probable sources of the noise are those parts that rotate in all gears, such as the drive chain, the input shaft, and torque converter.

Noises that only occur when a particular gear is operating must be related to those components responsible for providing that gear, such as a band or clutch. On some RWD vehicles, noise only in first gear can be caused by a defective rear transmission mount. If the noise is vehicle related, the most probable causes are the output shaft and final drive assembly. Often, the exact cause of noise and vibration problems can only be identified through a careful inspection of a disassembled transmission.

Whatever the exact fault in the transmission is the cause of a vibration or noise, the transmission will need to come out and the entire unit checked and carefully inspected. Proper diagnosis prior to disassembling the transmission will identify the specific areas that should be carefully looked at and can prevent unnecessary transmission removal and teardown.

The reverse/high clutch is sometimes called the direct clutch.

Vacuum Modulator

Diagnosing a vacuum modulator begins with checking the vacuum at the line or hose to the modulator (Figure 3-27). The modulator should be receiving engine manifold vacuum. If it does, there are no vacuum leaks in the line to the modulator. Check the modulator itself for leaks with a hand-held vacuum pump (Figure 3-28). The modulator should be able to hold approximately 18 in./Hg. If transmission fluid is found when you disconnect the line at the modulator, the vacuum diaphragm in the modulator is leaking and the modulator should be replaced. If the vacuum source, vacuum lines, and vacuum modulator are in good condition but shift characteristics indicate a vacuum modulator problem, the modulator may need adjustment.

Classroom Manual

Chapter 3, page 88

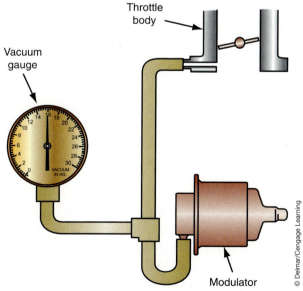

FIGURE 3-27 The correct hookup for connecting a vacuum gauge to vacuum modulators.

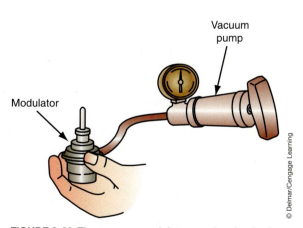

FIGURE 3-28 The vacuum modulator can be checked for leakage and action by activating it with a handheld vacuum pump and observing the vacuum gauge and the action of the modulator.

If no engine vacuum was found at the modulator, check the vacuum line from the engine to the modulator. If the engine is running it is very unlikely that it would have zero vacuum, so some vacuum should be present unless the line or fittings leak.

If the vacuum to modulator was low, check engine vacuum. Run the normal engine vacuum tests to determine where the problem is. A common problem for low vacuum is a restricted exhaust.

A vacuum drop test is a very important test. It checks the soundness of the vacuum sources and connecting lines and hoses. With the engine running, foot on the brake pedal, and the transmission in gear, quickly press the gas pedal to the floor then release it. The reading on the vacuum gauge should fall to zero and return to about 17 in./Hg. immediately. If it falls slowly or goes down only to about 5 in. there is a restriction. A restriction is also noted by slow return to 17 in.

Most modulators must be removed to be adjusted, if it is adjustable. However, there are some that have an external adjustment. This adjustment allows for fine-tuning of modulator action. To remove a vacuum modulator from the transmission, loosen the retaining clamp and bolt. Some units are screwed into the transmission case. While pulling the modulator out of the housing, be careful not to lose the modulator actuating pin, which may fall out as the modulator is removed. Use a hand held vacuum pump with a vacuum gauge and the recommended gauge pins to adjust the modulator according to specifications.

DIAGNOSE HYDRAULIC AND VACUUM CONTROL SYSTEMS

The best way to identify the exact cause of the problem is to use the results of the road test, logic, and oil circuit charts for the transmission being worked on. Before doing this, however, you should always check all sources for information about the symptom. Also, make sure you check the basics: trouble codes in the computer, fluid level and condition, leaks, and mechanical and electrical connections. Using the oil circuits, you can trace problems to specific valves, servos, clutches, and bands.

The basic oil flow is the same for all transmissions. The oil pump supplies the fluid flow that is used throughout the transmission. Fluid from the pump always goes to the pressure-regulating valve. From there, the fluid is directed to the manual shift valve. When the gear selector is moved, the manual valve directs the fluid to other valves and to the apply devices. By following the flow of the fluid on the oil circuit chart, you can identify which valves and apply devices should be operating in each particular gear selector position. Through a process of elimination, you can identify the most probable cause of the problem.

Most often, the valve body can be removed for service without removing the entire transmission. However, on some models, the transmission must be removed.

Mechanical and vacuum controls can also contribute to shifting problems. The condition and adjustment of the various linkages and cables should be checked whenever there is a shifting problem. If all checks indicate that the problem is either an apply device or in the valving, an air pressure test can help identify the exact problem. Air pressure tests are also performed during disassembly to locate leaking seals and during reassembly to check the operation of the clutches and servos.

An air pressure test is conducted by applying clean, moisture-free air, at approximately 40 psi, through a rubber-tipped air nozzle. With the valve body removed, direct air pressure to the case holes that lead to the servo, accumulator, and clutch apply passages if possible (Figure 3-29). Cover the vent hole of the circuit being tested with a clean, lint-free shop towel; this will catch any fluid that may spray out. You should clearly hear or feel a dull thud that indicates the action of a holding device. If a hissing noise is heard, a seal is probably leaking in that circuit (Figure 3-30). If you cannot hear the action of the servo or clutch, apply air

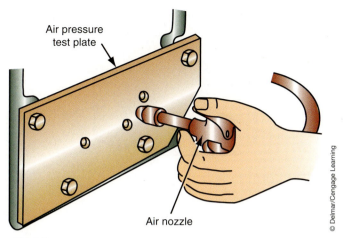

Air pressure test plate

Air nozzle

FIGURE 3-29 Using air pressure through a test plate to test hydraulic circuits.

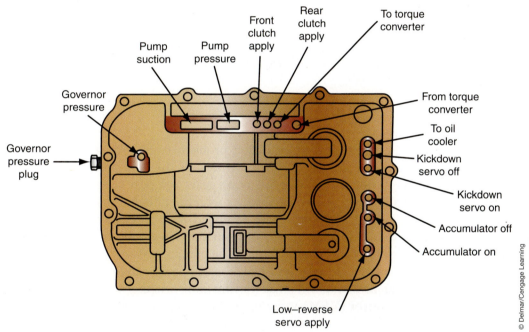

Pump suction

Pump pressure

Front clutch apply

Rear clutch apply

To torque converter

Governor pressure

From torque converter

To oil cooler

Kickdown servo off

Kickdown servo on

Accumulator off

Accumulator on

Governor pressure plug

Low–reverse servo apply

FIGURE 3-30 Air testing points on a typical transmission.

pressure again and watch the assembly to see if it is reacting to the air. A servo or accumulator should react immediately to the release of the apply air. If it does not, something is making it stick. Repair or replace the apply devices if they do not operate normally.

Air pressure can normally be directed to the following circuits through the appropriate hole for each: front clutch, rear clutch, kickdown servo, low servo, and reverse servo. Some manufacturers recommend the use of a specially drilled plate and gasket, which is bolted to the transmission case (Figure 3-31). This plate not only clearly identifies which passages to test but also seals off the other passages. Air is applied directly through the holes in the plate.

SERVICE TIP:
Whenever there is a shifting problem that is not readily diagnosed, check the service manual and Technical Service Bulletins before making assumptions about the problem or getting frustrated. Sometimes the cause of a problem is something that you would least expect. For example, a single, and hard-to-diagnose, problem in a 4L30 transmission can cause it to have no movement, forward movement in neutral, no forward movement, or no reverse gear. All of these complaints may be caused by a defective center support. The center support is made of two pieces, the aluminum housing and the steel sleeve, which is pressed into the center of the support. Grooves machined into the sleeve provide channels for several oil paths through the support. If the sleeve rotates, the potential for several problems exists. The most common complaint is a bind-up in reverse. It is not recommended that the sleeve be removed and a new one installed; rather, the entire center support assembly should be replaced.

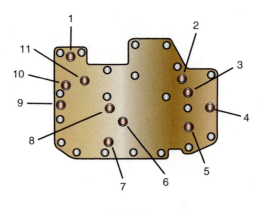

LEGEND

1. Converter bypass
2. Direct clutch
3. Forward clutch
4. 2D3 accumulator top
5. 2D3 accumulator bottom
6. Reverse servo
7. Overdrive servo apply
8. Overdrive servo release
9. Intermediate clutch
10. Reverse clutch
11. 1D2 accumulator apply

© Delmar/Cengage Learning

FIGURE 3-31 An example of a transmission air test plate.

Classroom Manual

Chapter 3, page 92

PRESSURE TESTS

If you cannot identify the cause of a transmission problem from your inspection or road test, a pressure test should be conducted. This test measures the fluid pressure of the different transmission circuits during the various operating gears and gear selector positions. The number of hydraulic circuits that can be tested varies with the different makes and models of transmissions. However, most transmissions are equipped with pressure taps that allow the pressure test equipment to be connected to the transmission's hydraulic circuits (Figures 3-32 and 3-33). Pressure testing checks the operation of the oil pump, pressure regulator valve, throttle valve, and vacuum modulator system (if the vehicle is equipped with one), plus the governor assembly. Pressure tests can be conducted on all transmissions whether their pressures are regulated by a **variable force motor (VFM)**, vacuum modulator, or through conventional valving.

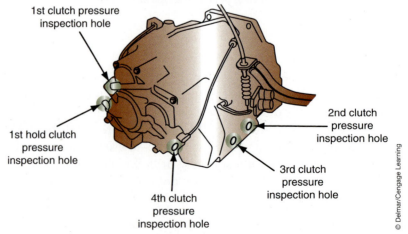

FIGURE 3-32 Pressure taps on the outside of a typical Honda transaxle case.

© Delmar/Cengage Learning

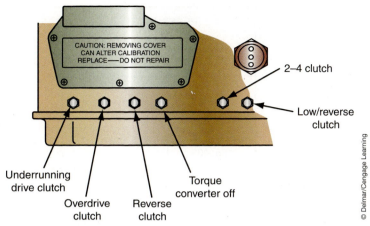

FIGURE 3-33 Pressure taps on the outside of a typical Chrysler transaxle.

A pressure test has its greatest value when the transmission does not shift smoothly or when the shift timing is wrong. Both of these problems may be caused by excessive line pressure, which can be verified by a pressure test.

Conducting a Pressure Test

Before conducting a pressure test on an electronic automatic transmission, correct and clear all trouble codes retrieved from the system. Also make sure the transmission fluid level and condition is okay and that the shift linkage is in good order and properly adjusted. To conduct a pressure test, use a tachometer, two pressure gauges (electronic gauges if available), the correct manufacturer specifications (Figure 3-34), and (for vehicles equipped with a vacuum modulator) a vacuum gauge and a hand held vacuum pump.

The pressure test is best conducted with three pressure gauges, but two will work. Two of the gauges should read up to 400 psi and the other to 100 psi. The two 400-psi gauges are usually used to check mainline and an individual circuit, such as mainline and direct or forward circuits. If a circuit is 10 psi lower than mainline pressure when they are both tested at exactly the same time, a leak is indicated. This is why it is best to use two 400-psi gauges. A 100-psi gauge may be used on TV and governor circuits. This ensures an observation of the critical pressures from these two circuits. Next to the scan tool, a pressure gauge is the most valuable tool for automatic transmission diagnostics.

The pressure gauges are connected to the pressure taps in the transmission housing and routed so that the driver can see the gauges. The vehicle is then road tested and the gauge readings observed during the following operational modes: slow idle, fast idle, and wide-open throttle (WOT).

 WARNING: **Always stop the engine when connecting and disconnecting the pressure gauges at the transmission. The fluid is under pressure and can easily spray at you and get in your eyes if a hose, fitting, or gauge leaks, so always wear safety goggles when using a pressure gauge.**

During the road test, observe the starting pressures and the steadiness of the increases that should occur with slight increases in load. The amount the pressure drops as the transmission shifts from one gear to another should also be noted. The pressure should not drop more than 15 psi between shifts.

SERVICE TIP: When using two or more pressure gauges, be certain to put them on a "Tee" fitting at the same time to calibrate them. The gauges should be calibrated to read the same at 50, 75, 100, 125, and 150 psi.

WOT is a common acronym for wide-open throttle.

Gear Selector Position	Actual Gear	PRESSURE TAPS					
		Underdrive Clutch	Overdrive Clutch	Reverse Clutch	Torque Converter Clutch Off	2/4 Clutch	Low/Reverse Clutch
PARK* 0 mph	PARK	0–2	0–5	0–2	60–110	0–2	115–145
REVERSE* 0 mph	REVERSE	0–2	0–7	165–235	50–100	0–2	165–235
NEUTRAL* 0 mph	NEUTRAL	0–2	0–5	0–2	60–110	0–2	115–145
L# 20 mph	FIRST	110–145	0–5	0–2	60–110	0–2	115–145
3# 30 mph	SECOND	110–145	0–5	0–2	60–110	115–145	0–2
3# 45 mph	DIRECT	75–95	75–95	0–2	60–90	0–2	0–2
OD# 30 mph	OVERDRIVE	0–2	75–95	0–2	60–90	75–95	0–2
OD# 50 mph	OVERDRIVE WITH TCC	0–2	75–95	0–2	0–5	75–95	0–2

* Engine speed at 1500 rpm.
\# CAUTION: Both front wheels must be turning at the same speed.

FIGURE 3-34 An oil pressure chart. These pressures are for an engine speed of 1500 rpm.

Any pressure reading not within the specifications indicates a problem (Figure 3-35). Typically, when the fluid pressures are low, there is an internal leak, clogged filter, low oil pump output, or faulty pressure regulator valve. If the fluid pressure increased at the wrong time or the pressure was not high enough, sticking valves or leaking seals are indicated. If the pressure drop between shifts is greater than approximately 15 psi, an internal leak at a servo or clutch seal is indicated. Always check the manufacturer's specifications for maximum dropoff before jumping to any conclusions.

To maximize the usefulness of a pressure test and to be better able to identify specific problems, begin the test by measuring line pressure. Mainline pressure should be checked in all gear ranges and at the three basic engine speeds. It is very helpful for you to make a quick chart such as the one shown in Figure 3-36 to record the pressures during the test.

If the pressure in all operating gears is within specifications at slow idle, the pump and pressure regulator are working fine. If all pressures are low at slow idle, it is likely that there is a problem in the pump, a stuck-open pressure regulator, clogged filter, improper fluid level, or there is an internal pressure leak. To further identify the cause of the problem, check the pressure in the various gears while the engine is at a fast idle.

If the pressures at fast idle are within specifications, the cause of the problem is normally a worn oil pump; however, the problem may be an internal leak or a pressure regulator valve

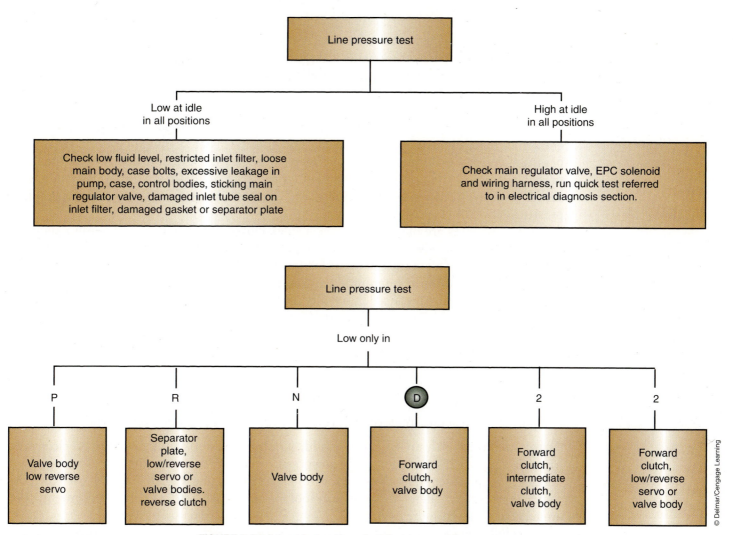

FIGURE 3-35 A troubleshooting chart for abnormal line pressure.

	Slow idle	Fast idle	WOT
P			/////
R			
N			/////
D			
3			
2			
1			

FIGURE 3-36 Pressure chart to aid in diagnostics.

© Delmar/Cengage Learning

that is stuck open. Internal leaks typically are more evident in a particular gear range because that is when ATF is being sent to a particular device through a particular set of valves and passages. If any of these components leaks, the pressure will drop when that gear is selected or when the transmission is operating in that gear.

Further diagnostics can be made by observing the pressure change when the engine is operated at WOT in each gear range. A clogged oil filter will normally cause a gradual pressure drop at higher engine speeds because the fluid cannot pass through the filter fast enough to meet the needs of the transmission and the faster-turning oil pump. If the fluid pressure changed with the increase in engine speed, a stuck pressure regulator is the most probable cause of the problem. A stuck-open pressure regulator may still allow the pressure to build with increases in engine speed, but it will not provide the necessary boost pressures.

If the pressures are high at slow idle, a stuck-closed pressure regulator, a misassembled or stuck open boost valve, a poorly adjusted or damaged shift cable, or a stuck-open throttle valve is indicated. If all of the pressures are low at WOT, pull on the TV cable or disconnect the vacuum hose to the vacuum modulator. If this causes the pressures to be in the normal range, the low pressure is caused by a faulty cable or there is a problem in the vacuum modulator or vacuum lines. If the pressures stay below specifications, the most likely causes of the problem are the pump or the control system.

If all pressures are high at WOT, compare the readings to those taken at slow idle. If they were high at slow idle and WOT, a faulty pressure regulator is indicated. If the pressures were normal at slow idle and high at WOT, the throttle system is faulty. To verify that the low pressures are caused by a weak or worn oil pump, conduct a *reverse stall test*. If the pressures are low during this test but are normal during all other tests, a weak oil pump is indicated.

Transmissions with an EPC Solenoid. On transmissions equipped with an electronic pressure control (EPC) solenoid, only one pressure gauge is needed along with a scan tool. To check the pressure, connect the scan tool and start the engine. Check the fluid level and correct it as necessary. Then check the scan tool for current and stored DTCs. Record any that are present. Then turn the engine off.

Carefully remove the plug for the line pressure test port. Then install the pressure gauge into that bore (Figure 3-37). Access the scan tool transmission output controls for the line PC

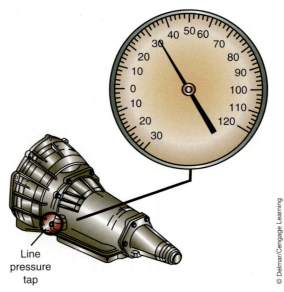

Line pressure tap

FIGURE 3-37 The pressure gauge is installed in the housing at the EPC pressure tap.

© Delmar/Cengage Learning

	Transmission pressure with TP at 1.5 volts and vehicle speed above 8 km/h (5 mph)				
Gear	EPC Tap	Line Pressure Tap	Forward Clutch Tap	Intermediate Clutch Tap	Direct Clutch Tap
1	276–345 kPa (40–50 psi)	689–814 kPa (100–118 psi)	620–745 kPa (90–108 psi)	641–779 kPa (93–113 psi)	0–34 kPa (0–5 psi)
2	310–345 kPa (45–50 psi)	731–869 kPa (106–126 psi)	662–800 kPa (96–116 psi)	689–827 kPa (100–120 psi)	655–800 kPa (95–116 psi)
3	341–310 kPa (35–45 psi)	620–758 kPa (90–110 psi)	0–34 kPa (0–5 psi)	586–724 kPa (85–105 psi)	551–689 kPa (80–100 psi)

© Delmar/Cengage Learning

FIGURE 3-38 A pressure chart for a transmission equipped with an EPC.

solenoid. Start the engine. To have accurate line pressure readings, this procedure must be performed at least three times. Also, keep in mind that normally a scan tool is only able to control the line PC solenoid in PARK and NEUTRAL with the engine running below 1500 rpm.

With the scan tool, increase and decrease the line PC solenoid in increments of approximately 15 psi (100 kPa). Between each of the increments allow the pressure to stabilize. Note the pressure readings on the scan tool and compare them to the readings on the pressure gauge. If the readings vary greatly, diagnose the cause.

Once the readings are noted, turn the engine off and remove the pressure gauge. Make sure to tighten the test plug properly.

EPC readings can also be taken during the road test. Make sure you meet the conditions recommended by the manufacturer before coming to any conclusion about the condition of the solenoid. If the EPC pressure is not within specifications (Figure 3-38), follow the instructions for checking the solenoid.

Governors

If the pressure tests suggest that there is a governor problem, the governor should be removed, disassembled, cleaned, and inspected. Some governors are mounted internally and the transmission must be removed to service the governor, whereas others can be serviced by removing the extension housing or oil pan, or by detaching an external retaining clamp and then removing the unit.

A faulty governor or governor drive gear system typically causes improper shift points. However, some electronically controlled transmissions do not rely on the hydraulic signals from a governor; rather, they rely on the electrical signals from these sensors. Sensors, such as speed and load sensors, signal to the transmission's computer when gears should be shifted. Faulty electrical components or loose connections can also cause improper shift points.

LINKAGES

Many transmission problems are caused by improper adjustment of the linkages. All transmissions have either a cable or a rod-type gear selector linkage. Some transmissions also have a throttle valve linkage, while others use an electric switch connected to the throttle to control forced downshifts.

Normal operation of a *neutral safety switch* provides a quick check for the adjustment of the gear selector linkage. To do this, move the selector lever slowly until it clicks into the park

The gear selector linkage is often referred to as the manual linkage because it controls the manual shift valve.

position. Turn the ignition key to the start position. If the starter operates, the park position is correct. After checking the park position, move the lever slowly toward the neutral position until the lever drops at the end of the N stop in the selector gate. If the starter also operates at this point, the gearshift linkage is properly adjusted. This quick test also tests the adjustment of the neutral safety switch. If the engine does not start in either or both of these positions, the neutral safety switch or the gear selector linkage needs adjustment or repair. Also, the starter should not operate in any other gear selector position. If it does, adjust the linkage.

 WARNING: Since you must work under the vehicle to adjust most shift linkages, make sure you properly raise and support the vehicle before working under it. Also, wear safety glasses or goggles while working under the vehicle.

Neutral Safety Switch

Most neutral safety switches are combinations of a neutral safety switch and a backup lamp switch. Others have separate units for these two distinct functions. When the reverse light switch is not part of the neutral safety switch, it seldom needs adjustment, as it is directly activated by the transmission (Figure 3-39) or the gear selector linkage. A neutral safety switch allows current to pass from the starter switch to the starter when the lever is placed in the P or N position. Some neutral safety switches are nonadjustable. If these prevent starting in P or N and the gear selector linkage is correctly adjusted, they should be replaced.

A voltmeter can be used to check the switch for voltage when the ignition key is turned to the start position with the shift lever in P or N. If there is no voltage, the switch should be adjusted or replaced.

To adjust a typical neutral safety switch:

1. Place the shift lever in neutral.
2. Loosen the attaching bolts for the switch.
3. Using an aligning pin, move the switch until the pin falls into the hole in its rotor.
4. Tighten the attaching bolts.
5. Recheck the switch for continuity. If no voltage is present, replace the switch.

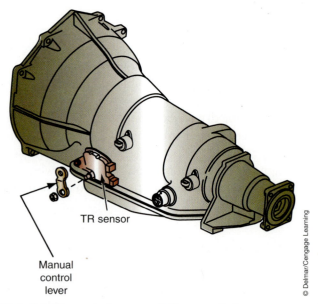

TR sensor

Manual control lever

© Delmar/Cengage Learning

FIGURE 3-39 A digital transmission range (TR) sensor that serves as a neutral start switch, back-up lamp switch, and as an input to the computer. It is located on the side of the transmission and is moved by the shift linkage.

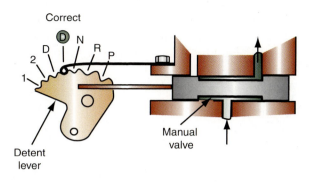

Correct

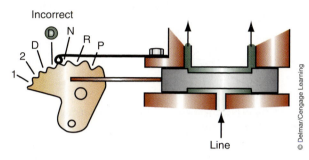

Incorrect

Manual valve

Detent lever

Line

© Delmar/Cengage Learning

FIGURE 3-40 Incorrect linkage adjustments may cause the manual shift valve to be positioned improperly in its bore and cause slipping during gear changes.

Gear Selector Linkage

A worn or misadjusted gear selector linkage will affect transmission operation. The transmission's manual shift valve must completely engage the selected gear (Figure 3-40). Partial manual shift valve engagement will not allow the proper amount of fluid pressure to reach the rest of the valve body. If the linkage is misadjusted, poor gear engagement, slipping, and excessive wear can result. If the linkage is out of adjustment, the manual shift valve will not be in its proper position and can cause internal fluid leaks. This can result in delayed shifting or slippage. The gear selector linkage should be adjusted so the manual shift valve detent position in the transmission matches the selector level detent and position indicator.

To check the adjustment of the linkage, move the shift lever from the park position to the lowest drive gear. Detents should be felt at each of these positions. If the detent cannot be felt in either of these positions, the linkage needs to be adjusted. While moving the shift lever, pay attention to the gear position indicator. Although the indicator will move with an adjustment of the linkage, the pointer may need to be adjusted so that it shows the exact gear after the linkage has been adjusted.

Linkage Adjustment

 WARNING: Always set the parking brake and block the wheels before moving the gear selector through its positions. This will prevent the vehicle from rolling over you or someone else.

To adjust a typical floor-mounted gear selector linkage:

1. Place the shift lever into the drive position.
2. Loosen the locknuts and move the shift lever until drive is properly aligned and the vehicle is in the D range.
3. Tighten the locknut (Figure 3-41).

Classroom Manual
Chapter 3, page 95

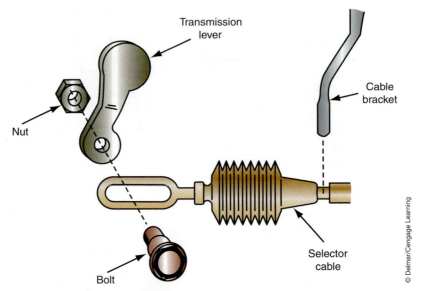

FIGURE 3-41 Gearshift linkages normally attach directly to the shift lever at the transmission and are retained by a locknut.

To adjust a typical cable-type linkage:

1. Place the shift lever into the park position.
2. Loosen the clamp bolt on the shift cable bracket.
3. Make sure the preload adjustment spring engages the fork on the transaxle bracket.
4. By hand, pull the shift lever to the front detent position (park), then tighten the clamp bolt. The shift linkage should now be properly adjusted.
5. If the cable assembly is equipped with an adjuster (Figure 3-42), move the shift lever into the park position and adjust the cable by rotating the adjuster into the lock position. Typically the adjuster will click when the lock is fully adjusted.

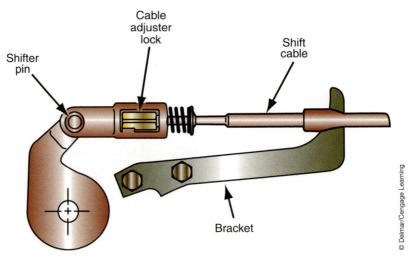

FIGURE 3-42 A shift cable adjuster lock assembly.

To adjust a typical rod-type linkage:

1. Loosen or disconnect the shift rod at the shift lever bracket.
2. Place the gear selector into park and the manual shift valve lever into the park detent position.
3. With both levers in position, tighten the clamp on the sliding adjustment to maintain their relationship. On the threaded type of linkage adjustment, lengthen or shorten the connection as needed. On some vehicles, you may need to adjust the neutral safety switch after resetting the linkage.

After adjusting any type of shift linkage, recheck it for detents throughout its range. Make sure a positive detent is felt when the shift lever is placed into the park position, as a safety measure. If you are unable to make an adjustment, the levers' grommets may be badly worn or damaged and should be replaced (Figure 3-43). When it is necessary to disassemble the linkage from the levers, the plastic grommets used to retain the cable or rod should be replaced. Use a prying tool to force the cable or rod from the grommet, then cut out the old grommet. Pliers can be used to snap the new grommets into the levers and the cable or rod into the levers.

To check the adjustment of the linkage, make sure the transmission is in PARK when the selector in placed in that position. Also make sure the engine starts in PARK and NEUTRAL,

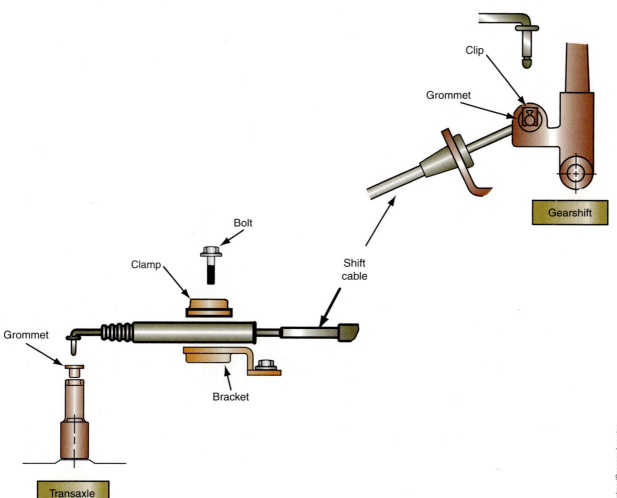

© Delmar/Cengage Learning

FIGURE 3-43 If the shift cable cannot be properly adjusted, the cable is distorted or some of the various grommets or brackets are worn and should be replaced.

The throttle valve linkage is called the TV linkage.

Classroom Manual

Chapter 3, page 90

CAUTION:

It is important that you check the service manual before making adjustments to the throttle or downshift linkages. Some of these linkage systems are not adjustable and if you loosen them in an attempt to adjust them, you may break them.

and only in those positions. Check that the reverse lamps come on when the selector is placed in REVERSE. Check the gear selector indicator to make sure it displays the proper range.

Throttle Valve Linkages

The throttle valve cable connects the throttle pedal movement to the throttle valve in the transmission's valve body. On some transmissions, the *throttle valve (TV) linkage* may control both the downshift valve and the throttle valve. Others use a vacuum modulator to control the throttle valve and a throttle linkage to control the downshift valve. Late-model transmissions may not have a throttle cable. Instead, they rely on electronic sensors and switches to monitor engine load and throttle plate opening. The action of the throttle valve produces throttle pressure (Figure 3-44). Throttle pressure is used as an indication of engine load and influences the speed at which automatic shifts will take place.

A misadjusted TV linkage may result in a throttle pressure that is too low in relation to the amount the throttle plates are open, causing early upshifts or slipping. Throttle pressure that is too high can cause harsh and delayed upshifts, and part-and wide-open throttle downshifts will occur earlier than normal. When adjusting the TV and downshift linkages, always follow the manufacturer's recommended procedures. An adjustment as small as a half-turn can make a big difference in shift timing and feel.

To adjust a typical throttle cable:

1. Run the engine until it has reached normal operating temperature.
2. Loosen the cable mounting bracket or swivel lock screw.
3. Position the bracket so that both bracket alignment tabs are touching the transaxle, then tighten the lock screw to the recommended torque.
4. Release the readjust tab on the cable assembly.
5. Make sure the cable is free to slide all the way toward the engine against its stop, after the readjust or locking tab is released.
6. Move the throttle control lever clockwise against its internal stop, then press the readjust tab downward into its locked position.

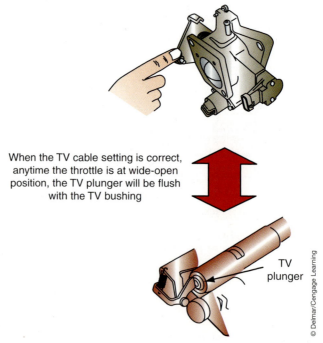

When the TV cable setting is correct, anytime the throttle is at wide-open position, the TV plunger will be flush with the TV bushing

TV plunger

© Delmar/Cengage Learning

FIGURE 3-44 The movement of the throttle plate causes the TV valve to move.

7. At this point, the cable is adjusted and the backlash in the cable is removed.
8. Check the cable for free movement by moving the throttle lever counterclockwise and slowly releasing it to confirm it will return fully clockwise.
9. No lubrication is required for any component of the throttle cable system.
10. Check the adjustment of the cable by conducting a system pressure test.

To adjust a typical throttle lever rod:

1. Run the engine until it has reached normal operating temperature.
2. Loosen the linkage's swivel lock screw.
3. Make sure the swivel is free to slide along the flat end of the throttle rod. Disassemble and clean or repair parts to assure free movement.
4. Hold the transaxle's throttle lever firmly toward the engine and against its internal stop, then tighten the swivel lock screw.
5. The rod is now adjusted and any backlash in the linkage should be taken up by the pre-load spring.
6. Check the adjustment of the cable by conducting a system pressure test.

To adjust a typical downshift linkage:

1. Run the engine until it has reached normal operating temperature.
2. Put the transmission in neutral with the parking brake set and allow the engine to run at its normal idle speed.
3. Using the specified amount of pressure, press down on the downshift rod.
4. Rotate the adjustment screw to obtain the specified clearance between the screw and the throttle arm.

OTHER ADJUSTMENTS

Band Adjustment

If a transmission problem still exists after the shift linkage and throttle pressure cable and rod have been adjusted and all electrical switches and sensors checked, the bands of the transmission may need adjustment. Photo Sequence 5 shows a typical procedure for adjusting a transmission band.

On some transmissions, slippage during shifting can be corrected by adjusting the holding bands. To help identify if a band adjustment will correct the problem, refer to the written results of your road test. Compare your results with the clutch and band application chart in the service manual. If slippage occurs when there is a gear change that requires holding by a band, the problem may be corrected by tightening the band.

On some vehicles, the bands can be adjusted externally with a torque wrench. On others, the transmission fluid must be drained and the oil pan removed. Locate the band-adjusting nut, and then clean off all dirt on and around the nut. Now, loosen the band-adjusting bolt locknut and back it off approximately five turns (Figure 3-45). Use a calibrated inch-pound torque wrench to tighten the adjusting bolt to the specified torque (Figure 3-46). Then, back off the adjusting screw the specified number of turns, and tighten the adjusting bolt locknut while holding the adjusting stem stationary. Reinstall the oil pan with a new gasket and refill the transmission with fluid. If the transmission problem still exists, an oil pressure test or transmission teardown must be done.

Parking Pawl

Any time you have the oil pan off, you should inspect the transmission parts that are exposed. This is especially true of the parking pawl assembly (Figure 3-47). This component is typically not hydraulically activated; rather, the gearshift linkage moves the pawl into position to lock

An automatic transmission tune-up usually includes fluid and filter change, linkage checks, and adjustment of the bands.

Classroom Manual
Chapter 3, page 91

CAUTION:
Do not excessively back off the adjusting stem because the anchor block may fall out of place, making it necessary to remove and disassemble the transmission to fit it back in place.

SPECIAL TOOLS
Large drain pan
Pound-inch torque wrench

TYPICAL PROCEDURE FOR ADJUSTING TRANSMISSION BANDS

All photos in this sequence are © Delmar/Cengage Learning.

P5-1 Make sure you properly identify the transmission before making band adjustments.

P5-2 Locate band-adjusting screw.

P5-3 Loosen adjusting screw locknut without allowing the screw to turn. If the locknut has a fluid seal, do not reuse the locknut. Install a new nut.

P5-4 Loosen the adjusting screw so that the band can relax around the drum and all tension is off of the adjusting screw.

P5-5 Tighten the adjusting screw to the specified torque.

P5-6 Back off the adjusting screw the exact number of turns that is specified in the service manual.

P5-7 Position a wrench on the adjusting screw and over the locknut so that you can tighten the locknut without moving the adjusting screw.

P5-8 Hold the adjusting screw in position and tighten the locknut to the specified torque.

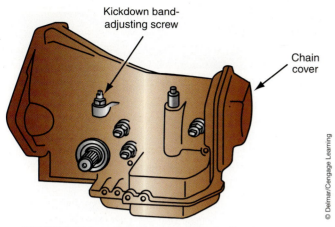

FIGURE 3-45 Location of external band-adjusting screw.

Kickdown band-adjusting screw

Chain cover

© Delmar/Cengage Learning

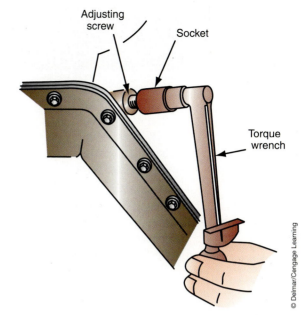

Adjusting screw

Socket

Torque wrench

FIGURE 3-46 Bands are typically adjusted to a specific inch-pound torque setting.

© Delmar/Cengage Learning

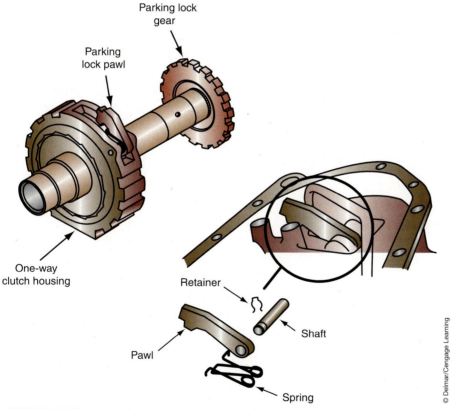

Parking lock gear

Parking lock pawl

One-way clutch housing

Retainer

Shaft

Pawl

Spring

FIGURE 3-47 The entire parking pawl and gear assembly should be carefully inspected for wear or defects.

© Delmar/Cengage Learning

the output shaft of the transmission. Unless the customer's complaint indicates a problem with the parking mechanism, no test will detect a problem here.

Check the pawl assembly for excessive wear and other damage. Also check to see how firmly the pawl is in place when the gear selector is shifted into the park mode. If the pawl can be easily moved out, it should be repaired or replaced.

CASE STUDY

A customer with a late-model Ford pickup equipped with an E4OD transmission complained that every time he hit a bump, the transmission would downshift. Then, when driven on smooth surfaces, the transmission would begin to cycle on its own between third and fourth gears.

The technician originally thought the problem was caused by a bad fourth gear clutch that couldn't hold the planetary gearset in overdrive or by a problem in the lockup torque converter circuit. Beginning the diagnosis with a visual inspection of the transmission and the many sensors that provide information to the E4OD's control module, the technician found faulty signals from the manual lever position (MLP) sensor. These signals explained the erratic shifting, because the computer uses the sensor to determine what gear the transmission should be in and how much modified line pressure to supply.

The technician used the required special tool to realign the sensor with the gear selector, but still found that the resistance readings were out of specifications. The sensor was replaced and the problem of erratic shifting was solved.

ASE-STYLE REVIEW QUESTIONS

1. While diagnosing noises apparently coming from a transaxle assembly:

   Technician A says a knocking sound at low speeds is probably caused by worn CV joints.

   Technician B says a clicking noise heard when the vehicle is turning is probably caused by a worn or damaged outboard CV joint.

   Who is correct?

   A. A only   C. Both A and B

   B. B only   D. Neither A nor B

2. *Technician A* says if the shift for all forward gears is delayed, a slipping front or forward clutch is normally indicated.

   Technician B says a slipping rear clutch is indicated when there is a delay or slip when the transmission shifts into any forward gear.

   Who is correct?

   A. A only   C. Both A and B

   B. B only   D. Neither A nor B

3. While discussing the results of an oil pressure test:

   Technician A says when the fluid pressures are high, internal leaks, a clogged filter, low oil pump output, or a faulty pressure regulator valve are indicated.

   Technician B says if the fluid pressure increased at the wrong time, an internal leak at the servo or clutch seal is indicated.

   Who is correct?

   A. A only   C. Both A and B

   B. B only   D. Neither A nor B

4. *Technician A* says low engine vacuum will cause a vacuum modulator to sense a load condition when it actually is not present, causing delayed and harsh shifts.

   Technician B says poor engine performance can cause delayed shifts through the action of the TV assembly.

   Who is correct?

   A. A only   C. Both A and B

   B. B only   D. Neither A nor B

5. *Technician A* says delayed shifting can be caused by worn planetary gearset members.

   Technician B says delayed shifts or slippage may be caused by leaking hydraulic circuits or sticking spool valves in the valve body.

   Who is correct?

   A. A only   C. Both A and B

   B. B only   D. Neither A nor B

6. While discussing proper band adjustment procedures:

   Technician A says that on some vehicles the bands can be adjusted externally with a torque wrench.

   Technician B says a calibrated pound-inch torque wrench is normally used to tighten the band-adjusting bolt to a specified torque.

   Who is correct?

   A. A only   C. Both A and B

   B. B only   D. Neither A nor B

7. While checking the condition of a car's ATF:

   Technician A says if the fluid has a dark brownish or blackish color or a burned odor, the fluid has been overheated.

   Technician B says if the fluid has a milky color, engine coolant has been leaking into the transmission's cooler.

   Who is correct?

   A. A only   C. Both A and B

   B. B only   D. Neither A nor B

8. While discussing the proper way to diagnose a kickdown switch:

   Technician A says that when the throttle pedal is fully depressed, a click should be heard just before the pedal reaches its travel stop. If the click is not heard, the switch should be replaced.

   Technician B says if the transmission cannot be forced to automatically downshift, the kickdown switch is open and should be replaced.

   Who is correct?

   A. A only   C. Both A and B

   B. B only   D. Neither A nor B

9. While discussing a pressure test:

   Technician A says this test is the most valuable diagnostic check for slippage in one gear.

   Technician B says the test can identify the cause of late or harsh shifting.

   Who is correct?

   A. A only   C. Both A and B

   B. B only   D. Neither A nor B

10. While checking the engine and transmission mounts on a FWD car:

   Technician A says any engine movement may change the effective length of the shift and throttle cables and therefore may affect the engagement of the gears.

   Technician B says delayed or missed shifts are caused by hydraulic problems, not linkage problems.

   Who is correct?

   A. A only   C. Both A and B

   B. B only   D. Neither A nor B

ASE CHALLENGE QUESTIONS

1. *Technician A* says engine lugging during downshifts may be caused by a faulty lockup torque converter.

   Technician B says poor engine performance may result in delay shifts on a TV cable controlled transmission or transaxle.

   Who is correct?

   A. A only   C. Both A and B

   B. B only   D. Neither A nor B

2. *Technician A* says a leaking torque converter may be indicated by the presence of a red or reddish-orange fluid in the bell housing.

   Technician B says transmission oil on the converter's rear shell indicates a leaking torque converter.

   Who is correct?

   A. A only   C. Both A and B

   B. B only   D. Neither A nor B

3. Worn or broken engine/transaxle mounts may cause all of the following EXCEPT:

   A. Delay shifts

   B. Change on the linkage effective length

   C. No PARK gear

   D. Early or late kickdown shifts

4. *Technician A* says a ruptured vacuum modulator may be indicated by the color of the engine's exhaust.

   Technician B says a stuck modulator actuating pin may be indicated by a no-reverse engagement condition.

   Who is correct?

   A. A only   C. Both A and B

   B. B only   D. Neither A nor B

5. Pressure testing reveals low pressure in all gears.

   Technician A says using a reverse stall test could isolate the faulty component.

   Technician B says a reverse stall test will determine whether the low-reverse servo is the malfunctioning component.

   Who is correct?

   A. A only   C. Both A and B

   B. B only   D. Neither A nor B

Name _____ **Date** _____

VISUAL INSPECTION OF AN AUTOMATIC TRANSMISSION

Upon completion of this job sheet, you should be able to conduct a preliminary inspection of the transmission.

Tools and Materials

A vehicle   Safety glasses

A torque wrench   Service manual

Describe the vehicle being worked on:

Year _____ Make _____ VIN _____

Model _____

Model and type of transmission _____

Procedure

_____

Task Completed

1. Service manual referred to:

   _____

2. Check the transmission housing for damage, cracks, and signs of leaks. Comments:

   _____

   _____

   _____

3. Check the slip joint area in the transmission extension housing for leaks. Comments:

   _____

   _____

   _____

4. Check for leaks at everything attached to the transmission. Comments:

   _____

   _____

   _____

5. Check the transmission's linkages for looseness, wear, and damage. Comments:

   _____

   _____

   _____

6. Check any transmission cables for binding, wear, and damage. Comments:

   _____

   _____

7. Check the fluid condition and level. Comments:

_____

_____

8. What are your conclusions from these checks?

_____

_____

Instructor's Response _____

_____

_____

Name _____ Date _____

ROAD TEST A VEHICLE TO CHECK THE OPERATION OF THE AUTOMATIC TRANSMISSION

Upon completion of this job sheet, you should be able to road test a vehicle with an automatic transmission to verify a customer's complaint and begin the diagnostic procedure.

Tools and Materials

A vehicle with an automatic transmission
Service manual
Clean shop rag
Pad of paper and pencil

Describe the vehicle being worked on:

Year _____ Make _____ VIN _____

Model _____

Model and type of transmission _____

Procedure

Task Completed

1. Park the vehicle on a level surface. ☐

2. Wipe all dirt off of the protective disc and the dipstick handle. ☐

3. Start the engine and allow it to reach operating temperature. ☐

4. Remove the dipstick and wipe it clean with a lint-free cloth or paper towel. ☐

5. Reinsert the dipstick, remove it again, and record the reading.

_____

6. Describe the condition of the fluid (color, condition, and smell):

_____

_____

7. What is indicated by the fluid's condition?

_____

_____

8. Find and duplicate the chart from a service manual that shows the band and clutch application for different gear selector positions. Using these charts will greatly simplify your diagnosis of automatic transmission problems. It is also wise to have a notebook or piece of paper to jot down notes about the operation of the transmission.

9. Inspect the transmission for signs of fluid leakage. Comments:

_____

_____

10. Drive the vehicle at normal speeds to warm up the engine and transmission. Describe the behavior of the transmission and torque converter.

_____

_____

11. Place the shift selector into the DRIVE position and allow the transmission to go through all of its normal shifts. Describe the operation of the transmission and torque converter.

_____

_____

12. Check for proper operation in all forward ranges, especially the 1–2, 2–3, and 3–4 upshifts and converter lockup during light throttle operation. Describe the operation.

_____

_____

13. Force the transmission to kickdown and record the quality of this shift and the speed at which it downshifts. Comments:

_____

_____

14. Manually cause the transmission to downshift. How did it react?

_____

_____

15. Record any vibrations or noises that occur during the test drive.

_____

_____

16. Park the vehicle and move the shifter into each gear range. Pay attention to shift quality as each range (including park) is selected. Record the results:

_____

_____

17. Compare your notes with the specifications and shifting chart. What are your conclusions about the transmission and torque converter? What transmission parts could be causing the problem? Use the clutch/band application chart to answer this.

_____

_____

_____

_____

Instructor's Response _____

_____

_____

Name _____ Date _____

SERVICING LINKAGES

Upon completion of this job sheet, you will be able to inspect, adjust, or replace throttle valve and gear selector linkages and cables.

ASE Correlation

This job sheet is related to the ASE Automatic Transmission and Transaxle Test's Content Area *In-Vehicle Transmission/Transaxle Maintenance and Repair.*

Task: Inspect, adjust, and replace manual valve shift linkage, transmission range sensor/ switch, and park/neutral position switch.

Tools and Materials

Basic hand tools
Service information

Describe the vehicle that was assigned to you.

Year _____ Make _____ VIN _____

Model _____

Model and type of transmission _____

Procedure

1. If the gear selector linkage is misadjusted, poor gear engagement, slipping, and excessive wear can result. The gear selector linkage should be adjusted so that the manual shift valve detent position in the transmission matches the selector level detent and position indicator. To check the adjustment of the linkage, move the shift lever from the PARK position to the lowest DRIVE gear. Detents should be felt at each of these positions. If the detent cannot be felt in either of these positions, the linkage needs to be adjusted. Describe your findings.

   _____

   _____

   _____

2. While moving the shift lever, pay attention to the gear position indicator. Although the indicator will move with an adjustment of the linkage, the pointer may need to be adjusted so that it shows the exact gear after the linkage has been adjusted. Describe your findings.

   _____

   _____

   _____

   ⚠️ **CAUTION:** Always set the parking brake before moving the gear selector through its positions.

3. To adjust a typical floor-mounted gear selector linkage,
   a. Place the shift lever into the DRIVE position.
   b. Loosen the locknuts and move the shift lever until DRIVE is properly aligned and the vehicle is in the "D" range.
   c. Tighten the locknut.

Did you have any problems?

_____

_____

_____

4. To adjust a typical cable-type linkage,
   a. Place the shift lever into the PARK position.
   b. Loosen the clamp bolt on the shift cable bracket.
   c. Make sure the preload adjustment spring engages the fork on the bracket. By hand, pull the shift lever to the front detent position (PARK), then tighten the clamp bolt. The shift linkage should now be properly adjusted.

   Did you have any problems?

_____

_____

_____

5. To adjust a typical rod-type linkage,
   a. Loosen or disconnect the shift rod at the shift lever bracket.
   b. Place the gear selector into PARK and the manual shift valve lever into the PARK detent position.
   c. With both levers in position, tighten the clamp on the sliding adjustment to maintain their relationship. On the threaded type of linkage adjustment, lengthen or shorten the connection as needed.

   Did you have any problems?

_____

_____

_____

6. On some vehicles, you may need to adjust the neutral safety switch after resetting the linkage. Did you need to do this? Why?

_____

_____

_____

7. After adjusting any type of shift linkage, recheck it for detents throughout its range. Make sure a positive detent is felt when the shift lever is placed into the PARK position, as a safety measure. Describe your findings.

_____

_____

_____

8. If you are unable to make an adjustment, the levers' grommets may be badly worn or damaged and should be replaced. When it is necessary to disassemble the linkage from the levers, the plastic grommets used to retain the cable or rod should be replaced. Use a prying tool to force the cable or rod from the grommet, and then cut out the old

grommet. Pliers can be used to snap the new grommets into the levers and the cable or rod into the levers. What did you find?

_____

_____

_____

9. The throttle valve cable connects the movement of the throttle pedal movement to the throttle valve in the transmission's valve body. On some transmissions, the throttle linkage may control both the downshift valve and the throttle valve. Others use a vacuum modulator to control the throttle valve and a throttle linkage to control the downshift valve. Late-model transmissions may not have a throttle cable; instead, they rely on electronic sensors and switches to monitor engine load and throttle plate opening. The action of the throttle valve produces throttle pressure. What does this transmission use to relay throttle pressure to the valve body?

_____

_____

_____

Instructor's Response _____

_____

_____

Name _____ Date _____

PRESSURE TESTING A TRANSMISSION

Upon completion of this job sheet, you should be able to conduct a pressure test on a transmission.

ASE Correlation

This job sheet is related to the ASE Automatic Transmission and Transaxle Test's Content Area *General Transmission and Transaxle Diagnosis.*

Task: Perform pressure tests; determine necessary action.

Tools and Materials

A vehicle with an automatic transmission   Vacuum gauge

Hoist   T-fitting and vacuum hose

Tachometer   Lint-free shop towels

Pressure gauges   Service manual

Describe the vehicle that was assigned to you:

Year _____ Make _____ VIN _____

Model _____

Model and type of transmission _____

Procedure

Task Completed

1. Describe the pressure specifications and conditions that are listed in the service manual.

   _____

   _____

   _____

   _____

2. Start the engine and allow the engine and transmission to reach normal operating temperature. Then turn off the engine.   ☐

3. Connect the tachometer to the engine if the vehicle is not equipped with one.   ☐

4. Have a fellow student sit in the vehicle to operate the throttle, brakes, and transmission during this test.   ☐

5. Raise the vehicle on the hoist to a comfortable working height.   ☐

6. Use the service manual to locate the pressure test ports. Describe their locations.

   _____

   _____

   _____

7. Connect the pressure gauges to the appropriate service ports.

☐ 8. If the transmission has a vacuum modulator, use the T-fitting and vacuum hose to connect the vacuum gauge into the modulator circuit.

☐ 9. Start the engine. Have your helper move the gearshift selector into the first test position.

☐ 10. Run the engine at the specified speed. Move the gear selector as required by the specifications.

☐ 11. Observe and note the pressure and vacuum readings in the various gear ranges.

_____

_____

12. Move the pressure gauges to the appropriate test ports for the next range to be tested. Describe this location.

_____

_____

13. Restart the engine and have your helper move the gear selector into the range to be tested and increase the engine's speed to the test point.

☐ 14. Observe and note the pressure and vacuum readings in the various gear ranges.

_____

_____

15. Have your helper slowly apply the vehicle's brakes. Once the engine has returned to idle, turn it off.

16. Repeat this sequence until the recommended test sequence has been completed.

☐ 17. Summarize the results of these tests.

_____

_____

_____

_____

_____

_____

Instructor's Response _____

_____

_____

Name _____   **Date** _____

AIR PRESSURE TESTING

Upon completion of this job sheet, you should be able to conduct an air pressure test on a transmission.

ASE Correlation

This job sheet is related to the ASE Automatic Transmission and Transaxle Test's Content Area *Off-Vehicle Transmission/Transaxle Repair Friction and Reaction Units.*

Tasks: Air test the operation of clutch and servo assemblies.

Tools and Materials

An automatic transmission or transaxle on a bench, or
  a vehicle with an automatic transmission
Compressed air and rubber tipped air nozzle
Pound-inch torque wrench

Eye protection
Air pressure testing plate
Service manual
Lint-free shop towels

Describe the transmission that was assigned to you:

Year _____ Make _____ VIN _____

Model _____

Model and type of transmission _____

Procedure

Task Completed

1. If this job sheet will be completed on a transmission installed in a vehicle, raise the vehicle on a hoist. Then drain the transmission fluid and remove the oil pan, oil filter, and valve body.

   ☐

2. If the job sheet will be completed with the transmission on a bench, remove the valve body.

   ☐

3. Refer to the service manual and identify the fluid passages required for testing.

   ☐

4. If you have an air pressure testing plate, install it with the appropriate gaskets and tighten the plates to the manufacturer's specifications. The specified torque is
   _____ .

5. Apply air pressure to the servo and clutch passages and listen for 5–10 seconds. Describe the sounds you hear and record what they indicate.

   _____

   _____

   _____

6. Release the air pressure and reapply pressure. Listen and record the results.

   _____

   _____

Task Completed

☐

7. Remove the testing plate.

8. Summarize the results of this test.

_____

_____

_____

_____

Instructor's Response _____

_____

_____

Name _____ Date _____

TRANSMISSION COOLER INSPECTION AND FLUSHING

Upon completion of this job sheet, you should be able to inspect, test, and flush the cooler system for an automatic transmission.

ASE Correlation

This job sheet is related to the ASE Automatic Transmission and Transaxle Test's Content Area *In-Vehicle Transmission and Transaxle Repair.*

Tasks: Check condition of engine cooling system: inspect, test, flush, and replace transmission cooler, lines, and fittings.

Tools and Materials

A vehicle with an automatic transmission or transaxle	Various lengths of rubber hose
Line wrenches	Drain pan
Tubing cutter	Compressed air and air nozzle
Tubing flaring kit	Supply of clean solvent or mineral spirits
ATF pressure tester	Lint-free shop towels
	Service manual

Describe the vehicle that was assigned to you:

Year _____ Make _____ VIN _____

Model _____

Model and type of transmission _____

Procedure

Task Completed

1. Remove the transmission dipstick and describe the condition, smell, and color of the fluid.

   _____

   _____

2. Open the radiator cap (after the engine is cooled down) and check the coolant for traces of ATF. Also check the cap and gasket for signs of ATF. Record your findings.

   _____

   _____

3. Replace the radiator cap. ☐

4. Inspect the metal lines and fittings to and from the transmission cooler. Look for damage and signs of leakage. Describe your findings.

   _____

   _____

5. Summarize the results of your inspection.

_____

_____

☐   6. Place a drain pan under the fittings that connect the cooler lines to the radiator.

☐   7. Disconnect both cooler lines from the radiator. Plug or cap the lines from the transmission.

☐   8. Plug or cap one fitting at the radiator.

☐   9. Remove the radiator cap.

☐   10. Hold the compressed air nozzle tightly against the open fitting.

☐   11. Apply air pressure through the fitting (no more than 75 psi).

12. Check the coolant in the radiator for signs of air bubbles or air movement. Describe your findings.

_____

_____

☐   13. Unplug the other fitting and cooler lines.

☐   14. Reconnect the cooler lines to the radiator.

15. Summarize the results of the leak test.

_____

_____

_____

_____

☐   16. Set the parking brake and start the engine. Allow the engine and transmission to reach normal operating temperature before proceeding. Turn off the engine.

☐   17. Remove the transmission dipstick. Place a funnel in the dipstick tube.

☐   18. Raise the vehicle on a hoist.

☐   19. Disconnect the cooler return line at the point where it enters the transmission case.

☐   20. Attach a rubber hose to the disconnected cooler return line.

☐   21. Lower the vehicle and place the end of the hose into the funnel.

☐   22. Start the engine and run it at about 1000 rpm.

23. Observe the flow of fluid into the funnel. Describe the flow and summarize what this indicates.

_____

_____

_____

_____

☐   24. Turn off the engine.

☐   25. Raise the vehicle and remove the rubber hose from the return line.

26. Reconnect the return line. ☐

27. Lower the vehicle and reinstall the dipstick. ☐

The remainder of this job sheet covers a procedure for flushing the transmission cooler. Check with your instructor before proceeding.

1. Raise the vehicle on a hoist. ☐

2. Place a drain pan under the engine's radiator (or place it under the external transmission cooler if the vehicle is so equipped). ☐

3. Disconnect both cooler lines at the ends of the lines and at the radiator or cooler. ☐

4. Clean the fittings at the ends of the lines and at the radiator or cooler. ☐

5. Connect a rubber hose to the inlet fitting at the radiator. ☐

6. Place the other end of the hose into the drain pan. ☐

7. Attach a hose funnel to the cooler return fitting. Install a funnel in the other end of this hose. ☐

8. Pour small amounts of mineral spirits into the funnel. Observe the flow and the condition of the fluid moving into the drain pan. What does it look like?

_____

_____

9. If there is little flow, apply some air pressure to the hose connected to the return line. ☐

10. Continue pouring mineral spirits into the funnel until the fluid entering the drain pan is clear. ☐

11. Disconnect the cooler lines at the transmission and place the drain pan at one end of the lines. ☐

12. Install the rubber hose and funnel to the other end of the lines. ☐

13. Pour mineral spirits through the lines until the flow is clear. Describe your findings.

_____

_____

14. Pour ATF through the cooler lines to remove the mineral spirits. ☐

15. Remove all hoses and reconnect the cooler lines to the transmission and radiator. ☐

16. Check the transmission's fluid level and correct it if necessary. ☐

Instructor's Response _____

_____

_____

Chapter 4

ELECTRICAL AND ELECTRONIC SYSTEM DIAGNOSIS AND SERVICE

UPON COMPLETION AND REVIEW OF THIS CHAPTER, YOU SHOULD BE ABLE TO:

- Diagnose electronic control systems and determine needed repairs.

- Conduct a road test to determine if a fault is electrical or hydraulic.

- Use common electrical test instruments to locate problems.

- Conduct preliminary checks on EAT systems and determine needed repairs or service.

- Retrieve trouble codes from common electronically controlled automatic transmissions and determine needed repairs or service.

- Follow the prescribed diagnostic procedures according to DTC or symptom and determine needed repairs or service.

- Perform converter clutch system tests and determine needed repairs or service.

- Inspect, test, and replace electrical/electronic switches.

- Inspect, test, and replace electrical/electronic sensors.

- Inspect, test, bypass, and replace electrical/electronic solenoids.

- Inspect, test, adjust, or replace transmission-related electrical/electronic components.

One of the first tasks during diagnosis of an electronically controlled transmission is to determine if the problem is caused by the transmission or by electronics. To determine this, the transmission must be observed to see if it responds to commands given by the computer. Identifying whether the problem is related to the transmission or is electrical will determine what steps need to be followed to diagnose the cause of the problem.

An electronically controlled transmission (Figure 4-1) will operate efficiently, only if the commands it receives from the computer are valid. This holds true even if the hydraulic and mechanical parts of the transmission are normal. All diagnostics should begin with a scan tool to check for trouble codes in the system's computer. After the received codes are addressed, you can begin a more detailed diagnosis of the system and transmission. Your next step may be to manually activate the shift solenoids by connecting a jumper wire to them or by using a transmission tester that allows you to manually activate the solenoids. Prior to doing this, the wiring to the solenoids should be studied to determine if the computer activates them by supplying voltage to them or by completing the ground circuit. Also you need to know what gear certain solenoids are activated in. This information can be found in the service manual.

BASIC TOOLS
Basic mechanic's tool set
Scan tool
DMM
Jumper wires
Appropriate service manual

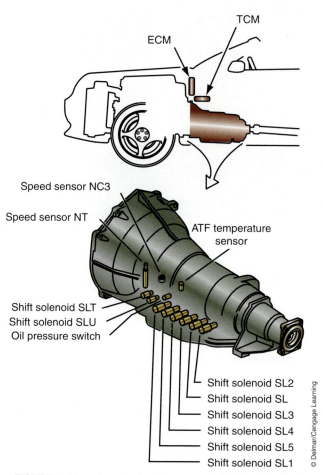

FIGURE 4-1 The key electronic components of an AA80E transmission.

Speed sensor NC3
Speed sensor NT
ATF temperature sensor
Shift solenoid SLT
Shift solenoid SLU
Oil pressure switch
Shift solenoid SL2
Shift solenoid SL
Shift solenoid SL3
Shift solenoid SL4
Shift solenoid SL5
Shift solenoid SL1

ECM
TCM

© Delmar/Cengage Learning

Classroom Manual

Chapter 4, page 117

EAT is a commonly used acronym for electronic automatic transmission.

A **TSB** is a Technical Service Bulletin in which the latest fixes and updates for a component or system are found.

BASIC EAT TESTING

The best way to diagnose an electronically controlled transmission is to approach solving the problem in a logical way.

It is important that you totally understand what the complaint or problem is before you venture in and try to find the cause. This may include an interview or road test with the customer to thoroughly define the complaint and to identify when and where the problem occurs.

Since many EAT problems are caused by the basics, it is wise to conduct all of the preliminary checks required for a nonelectronically controlled transmission. In addition, you should conduct a thorough inspection of the electronic system. Whenever diagnosing a transmission, remember that an engine problem can and will cause the transmission to act abnormally.

Often, accurately defining the problem and locating related information in TSBs and other materials can indicate the cause of the problem. No manufacturer makes a perfect vehicle or transmission, so when the manufacturer recognizes common occurrences of a problem, it will issue a statement regarding the way to fix the problem. Also, for many DTCs and symptoms the service information will give a simple diagnostic chart or path for identifying the cause of the problem. These are designed to be followed step by step and will lead to a conclusion if you follow the path that corresponds exactly to the symptom. Check all available information before moving on in your diagnostics.

Sometimes the symptom will not match any of those described in the service information. This doesn't mean it is time to guess. It means it is time to clearly identify what is working correctly. By eliminating those circuits and components that are working correctly from the list of possible causes of the problem, you can identify what may be causing the problem and what should be tested further.

Although the first steps in troubleshooting include retrieving diagnostic codes, there are problems that will not be made evident by a code. These problems may be solved with the diagnostic charts or pure logic. However, this logic must be based on a thorough understanding of the transmission and its controls.

GUIDELINES FOR DIAGNOSING EATs

Classroom Manual
Chapter 4, page 117

1. Make sure the battery has at least 12.6 volts before troubleshooting the transmission.
2. Check all fuses and identify the cause of any blown fuses.
3. Compare the wiring to all suspected components against the colors given in a service information.
4. When testing electronic circuits, always use a high-impedance test light or DMM (digital multimeter).
5. If an output device is not working properly, check the power circuit to it.
6. If an input device is not sending the correct signal back to the computer, check the reference voltage it is receiving and the voltage it is sending back to the computer.
7. Compare the voltages in and out of a sensor with the voltages the computer is sending out and receiving.
8. Before replacing a computer, check the solenoid isolation diodes according to the procedures outlined in the service manual.
9. Make sure computer wiring harnesses do not run parallel with any high-current wires or harnesses. The magnetic field created by the high current may induce a voltage in the computer harness. You should also be aware that antenna cables and CB radios can cause interference.
10. Take necessary precautions to prevent the possibility of static discharge while working with electronic systems.
11. While checking individual components, always check the voltage drop of the ground circuits. This is becoming more and more important because newer cars contain less material that conducts electricity well.
12. Make sure the ignition is off whenever you disconnect or connect an electronic device in a circuit.
13. All sensors should be checked in cold and hot conditions.
14. All wire terminals and connections should be checked for tightness and cleanliness.
15. Use electrical cleaning spray to clean all connectors and terminals.
16. Use dielectric grease at all connections to prevent future corrosion.
17. If you must break through the insulation of a wire to take an electrical measurement, make sure you tightly tape over the area after you are finished testing.

Electrostatic Discharge

Some manufacturers mark certain components and circuits with a code or symbol to warn technicians that they are sensitive to electrostatic discharge (Figure 4-2). Static electricity can destroy or render a component useless.

When handling any electronic part, especially those that are static sensitive, follow the guidelines below to reduce the possibility of electrostatic build up on your body and the inadvertent discharge to the electronic part. If you are not sure whether a part is sensitive to static, treat it as if it were.

1. Always touch a known good ground before handling the part. This should be repeated while handling the part and again after sliding across a seat, sitting down from a standing position, or walking a distance.
2. Avoid touching the electrical terminals of the part unless you are instructed to do so in the written service procedures. It is good practice to keep your fingers off all electrical terminals as the oil from your skin can cause corrosion. Use an anti-static strap, if one is available.

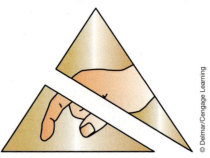

FIGURE 4-2 A manufacturer's warning symbol to show a circuit or component is sensitive to electrostatic discharge (ESD).

3. When you are using a voltmeter, always connect the negative meter lead first.
4. Do not remove a part from its protective package until it is time to install the part.
5. Before removing the part from its package, ground yourself and the package to a known good ground on the vehicle.
6. When replacing a programmable read-only memory (PROM) unit, ground your body by putting a metal wire around your wrist and connect the wire to a known good ground.

Although diagnostic trouble codes (DTCs) are very helpful during diagnosis, they also can become a stumbling block. This is especially true if more than one DTC is present. If two or more codes are present, you should look at the relationship of the codes to identify if they could have a common cause. Reacting to the individual codes may move your troubleshooting efforts beyond the common cause.

It is also important to remember that codes can be set by out-of-range signals. This doesn't mean the sensor or sensor circuit is bad. It could mean the sensor is working properly but there is a mechanical or hydraulic problem causing the abnormal signals. Not only can internal transmission problems cause codes to be set; so can basic electrical problems. Problems such as loose connections, broken wires, corrosion, and poor grounds will affect the signals in that circuit.

BASIC ELECTRICAL TESTING

Although the emphasis of this chapter is electronically controlled transmissions, basic electrical checks must be covered first. Many of the preliminary checks involve the use of common electrical test equipment. The proper use of test equipment is also necessary to test and identify the exact cause of problems after diagnostic codes are retrieved.

Ammeters

Measuring current flow can provide a summary of the activity in a circuit. If the current flow is lower than expected, there is some extra resistance in the circuit. This resistance can be the result of corrosion, frayed wires, or a loose connection. When current is higher than expected, a short or low resistance is indicated. A normally operating circuit will have the expected amount of current flow.

There are two main types of ammeters and each requires a different hook-up into the circuit. Prior to using an ammeter in a circuit, make sure the meter is capable of handling the current in the circuit. Circuits like the starting motor circuit have very high amounts of current and only specially designed meters are able to withstand the current flow of these circuits. Often ammeters have ranges of measurement. Always select the proper range prior to connecting the meter.

One type of ammeter has two leads, a positive lead and a negative lead. In order for this type of ammeter to accurately read current flow, the meter must be connected in series to the circuit (Figure 4-3). The circuit should be disconnected from its power source, then opened by disconnecting a connector in the circuit and inserting the red lead in the connector half

The static electricity that can be generated by a vinyl seat is 20,000 volts. It only takes 35 volts to destroy any electronic automotive device.

Classroom Manual
Chapter 4, page 103

Normally the positive lead is red and the negative lead is black.

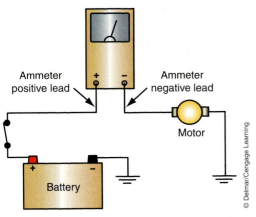

FIGURE 4-3 An ammeter is always connected in series with the circuit. This allows circuit current to flow through the meter.

FIGURE 4-4 An inductive pickup for an ammeter.

coming from the positive side of the battery and inserting the black lead into the other half. The circuit is now completed through the meter and, when activated, current will flow through the circuit and the meter, which will indicate the amount.

Inductive ammeters are not connected into series with the circuit. This type of meter utilizes an inductive pick-up, which monitors the magnetic field formed by current flowing through a conductor (Figure 4-4). These meters normally have three leads: positive, negative, and inductive. The positive and negative leads are connected to their appropriate posts of the battery. The inductive lead is clamped around a conductor in the circuit. When the circuit is activated, current flow can be read on the meter. Meters used to measure current flow in the starting and charging systems normally are the inductive type.

Voltmeters

Electrical circuits and components cannot operate properly if the proper amount of voltage is not available to them. Not only must the battery be able to deliver the proper amount of voltage, but the proper amount of voltage must be available to operate the intended component. As current flows through a resistance, voltage is lost or dropped. In circuits where undesired resistance is present, the voltage available for the circuit is decreased. Because of these conditions, voltage is measured at the source, at the electrical loads, and across the loads and circuits.

Connecting the positive lead to any point within a circuit and connecting the negative lead to a ground will measure the voltage at the point where the positive lead was connected. If battery voltage is being measured, the positive lead is connected to the positive terminal of the battery and the negative lead to the negative post or to a clean spot on the common ground. The voltmeter will read the potential difference between the two points and display that difference on its scale. In the case of a 12-volt battery, the potential at the positive post is 12 volts and the potential at the negative is zero. Therefore the meter will read 12 volts.

The positive lead can be connected anywhere within the circuit and it will measure the voltage present at that point. To measure the amount of available voltage, the negative lead can be connected to any point in the common ground circuit. To measure the amount of voltage drop across any wire or component of the circuit, the positive lead is connected to the battery side of the component and the negative lead is placed directly to the other side of the component. For example, to measure the voltage drop of a light bulb in a simple circuit, connect the positive lead to the battery or power lead at the bulb and the other lead to the ground side of the bulb (Figure 4-5). The meter will read the amount of voltage drop across the bulb when the circuit is activated. If no other resistances are present in the circuit, the amount of voltage drop will equal the amount of source voltage. However, if there is a resistance present in the ground connection, the bulb will have a voltage drop that is less than battery voltage.

SERVICE TIP:
Always wipe the meter leads off with a damp cloth after checking a battery. This will prevent the battery acid from damaging the ends of the leads.

Classroom Manual
Chapter 4, page 103

Battery voltage is typically called *source voltage*.

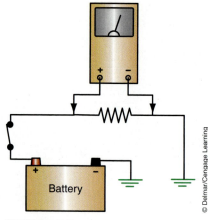

FIGURE 4-5 To measure voltage drop, the voltmeter is connected in parallel or across the component being tested.

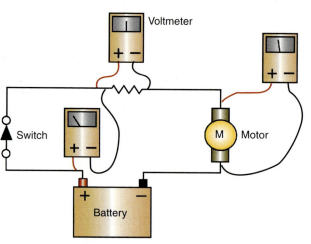

FIGURE 4-6 Voltmeter connections for measuring voltage drop across various parts of a circuit.

If the ground is bad, the voltage drop across the ground will equal the source voltage minus the amount of voltage dropped by the light. In order to measure voltage drop and available voltage at various points within a circuit, the circuit must be activated (Figure 4-6).

Ohmmeters

An ohmmeter is frequently used to test for circuit completeness or continuity. If a circuit is complete, the meter will display a low resistance, whereas if the circuit is open, a large amount of resistance and an infinite reading will be displayed.

Most ohmmeters have a variety of scales to measure within. A technician chooses between scales of 1, 100, or 10,000 ohms. The anticipated resistance determines the scale to be selected. After the scale has been selected, the meter must be zeroed in that scale. To do this, the leads of the meter are held together and the needle or number display is adjusted until it reads zero. Typically, an ohmmeter is equipped with an adjustment control to zero the meter.

To measure the resistance of a circuit or component, disconnect the power from the circuit or remove the component from the circuit (Figure 4-7). With the meter zeroed in the appropriate scale, connect one of the meter's leads to the power side of the component and the other lead to the ground side. Make sure you are not holding the metal leads while taking measurements. If you do this, you will become a parallel circuit with the component

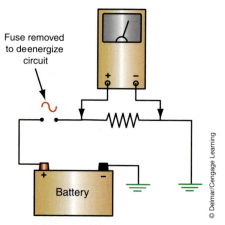

Fuse removed to deenergize circuit

FIGURE 4-7 Using an ohmmeter to measure resistance. Note that the circuit's fuse has been removed.

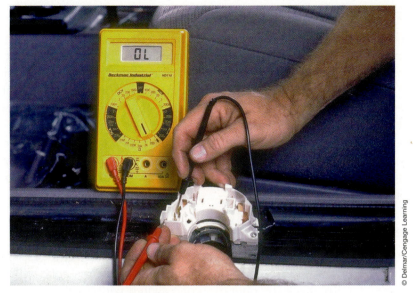

FIGURE 4-8 A digital ohmmeter showing an infinite reading.

or circuit you are checking and you will get inaccurate results. Always hold the leads with the plastic handle. The meter will display the amount of resistance present between the two points. If the resistance is greater than the selected scale, the meter will show a reading of infinity (Figure 4-8). When this reading results, the technician should move up a scale on the meter. Whenever the scales have been changed, the meter must be zeroed for that range prior to measuring the resistance. If subsequent scale changes result in continued readings of infinity, it can be assumed that there is no continuity between the two measured points.

Ohmmeters can also be used to compare the resistance of a component to the value it should have. Many electrical components have a specified resistance value, which is listed in the service manuals. This resistance value is important, as it controls the amount of voltage dropped and the amount of current that will flow in the circuit. If a component does not have the proper amount of resistance, the circuit will not operate properly.

Prior to testing a component or circuit with an ohmmeter, the service manual should be checked for precautions regarding the impedance of the meter. Ohmmeters supply their own power to measure resistance and often electronic components can be damaged by causing excessive current to flow in them. Using an ohmmeter with the proper impedance will avoid damage to the components. If an impedance specification is not given in the manual, it can be safely assumed that any ohmmeter is appropriate for that circuit.

> Digital multimeters may display "O.L" when there is an infinite measurement.

Test Lights

Often a test light is used to verify the presence of voltage at a particular point. Although the test light cannot give a reading of the amount of voltage present, it can clearly display the presence of voltage. Normally a test light has one wire lead. This is connected to ground. The positive lead is actually a piercing probe that is slipped into the circuit at the desired point to verify voltage. If voltage is present, the bulb in the test light will shine.

A self-powered test light is often used instead of an ohmmeter to test for continuity. A self-powered test light does not rely on the power of the circuit to light its bulb. Rather, it contains a battery for a power supply; when connected across a completed circuit, the bulb will light. This type of test light is connected to the circuit in the same manner as an ohmmeter.

Neither type of test light should be used to test electronic circuits or components. The use of test lights should be limited to checking for battery voltage at various points within an electrical circuit.

A **multimeter** is a meter that combines the functions of a voltmeter, ammeter, and ohmmeter.

A multimeter is often referred to as a VOM (volt–ohm meter).

Impedance is best defined as the operating resistance of an electrical device.

Classroom Manual

Chapter 4, page 103

Multimeters

Ohmmeters, ammeters, and voltmeters can be separate meters or can be combined in one meter called a **multimeter**. Most multimeters can measure DC volts, DC current (amps), AC volts, and ohms. Multimeters have either a needle-type analog display, or a numeric digital display. To use this type of meter, the desired function and the scale are selected prior to connecting the meter. Once the meter is set in a function, the **digital multimeter (DMM)** should be used as an individual function type meter.

A DMM with high **impedance** is normally required for testing electronic components and circuits. The high impedance prevents a surge of high current through the sensitive semiconductors when the tester is connected into a circuit. High current can destroy electronic circuits and components.

Using Meters

Meters can be used to test electrical components, such as switches, loads, and semiconductors. Switches can be tested for operation and for excessive resistance with a voltmeter, test light, or ohmmeter. To check the operation of a switch with a voltmeter or a test light, connect the meter's positive lead to the battery side of the switch. With the negative lead attached to a good ground, voltage should be measured at this point. Without closing the switch, move the positive lead to the other side of the switch. If the switch is open, no voltage will be present at that point. The amount of voltage present at this side of the switch should equal the amount on the other side when the switch is closed. If the voltage decreases, the switch is causing a voltage drop due to excessive resistance. If no voltage is present on the ground side of the switch with it closed, the switch is not functioning properly and should be replaced.

If a switch has been removed from the circuit, it can be tested with an ohmmeter or a self-powered test light. By connecting the leads across the switch connections, the action of the switch should open and close the circuit.

Ohmmeters are used to test semiconductors. Because semiconductors allow current flow only in certain directions, they can be tested by connecting an ohmmeter to their connections. Some digital meters have a feature that allows for accurate testing of diodes. This feature should be used. Analog meters are the preferred instrument for checking diodes because the results are easily observed. By placing the leads of the meter on the semiconductor connections, continuity should be observed. Reversing the meter leads to the same connections should result in different readings of continuity. For example, a diode should show good continuity (low resistance) when the leads are connected to it and poor continuity (high resistance) when the leads are reversed.

In addition to meters and test lights, troubleshooting electrical problems may involve the use of jumper wires, circuit breakers, or short detection devices. A jumper wire is simply a length of wire, normally equipped with alligator clips on either end. These clips allow the wire to be easily and safely connected to various points within a circuit. Jumper wires are primarily used to bypass a particular point within a circuit and should contain an in-line fuse or circuit breaker to prevent damage to the circuit being tested and the wire itself.

ELECTRICAL PROBLEMS

Classroom Manual

Chapter 4, page 106

Normally, the circuits in an automobile have a circuit protection device placed in the common path from the battery. This protection device is usually a fuse or a circuit breaker. They are designed to protect the wires and components from excessive current. When a great amount of current flows through the fuse or breaker, an element will burn out, causing the circuit to be opened and stopping current flow. This action prevents the high current from burning up the wires or components in the circuit.

High current is caused by low resistance. A decrease in the amount of resistance is typically the result of a **short**. A short is best defined as an additional and unwanted path to ground. Most shorts, such as what occurs when a bare wire contacts the frame of the car, create an extremely low resistance parallel branch. A slow-turning motor can also cause low resistance and high current.

A short to ground can be present before the load in the circuit or internally within the load or component. A short can also connect two or more circuits together, causing additional parallel legs and uncontrolled operation of components. An example of a possible result from a wire-to-wire short would be the horn blowing each time the brake pedal is depressed. This could be caused by a wire-to-wire short between the horn and brake light circuits. Shorts are one of the three common types of electrical problems.

A circuit breaker is a protection device that resets itself after it has been tripped by high current. Circuit breakers open due to the heat of high current. While open, the breaker cools and closes again to complete the circuit. When a fuse is blown or burned out by high current, it must be replaced to reactivate the circuit. When diagnosing a circuit with a short, a completed circuit makes locating the fault easier. By inserting a circuit breaker in place of the fuse, the circuit can be activated and still protected.

A number of techniques and test instruments can be used to locate a short. The most common of these is called a *short detector*. This device actually includes a circuit breaker and a magnetic sensing device (Figure 4-9). The latter may be called a Gauss gauge or a compass. The gauge relies on the magnetic field formed by current flow to indicate the location of a short. The cycling of the circuit breaker is indicated by the sweeps of the detector's needle as it is moved along the length of the circuit. When the needle no longer sweeps in response to the breaker, the location of the short can be assumed to be before that point in the circuit. Carefully moving the gauge back through the circuit should locate the exact place of the short.

Another common electrical fault is the **open**. An open causes an incomplete circuit and can result from a broken or burned wire, a loose connection, or a faulty component. If a circuit is open, there is no current flow and the component will not operate. If there is an open in one leg of a parallel circuit, the remaining part of that parallel circuit will operate normally.

To determine the location of an open, a test light or voltmeter is used. By probing along the circuit, the completed path can be traced. The point at which voltage is no longer present is the point at which the open is. If the circuit is open, there will be no current flow. This means that if the open is after the load, source voltage will be measured after the load because there is no voltage drop in the circuit. Often a jumper wire is used to verify or isolate the problem area.

Excessive resistance at a connector (Figure 4-10), internally in a component, or within a wire is also a common electrical problem. High resistance will cause low current flow and the component will not be able to operate normally, if at all.

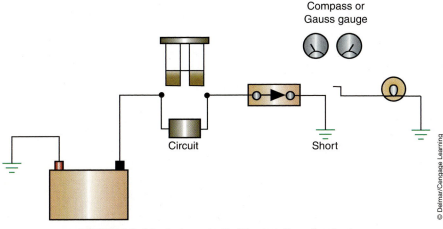

FIGURE 4-9 A typical way to find the location of a short.

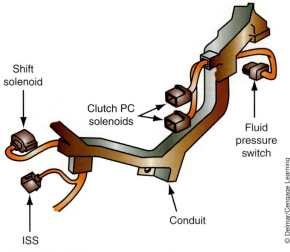

FIGURE 4-10 The various transmission connectors at the transmission should be inspected and checked for high-resistance problems.

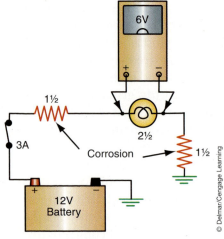

FIGURE 4-11 A simple light circuit with unwanted resistance. Notice the reduced voltage drop across the lamp and the reduced circuit current.

The source of high resistance is easily found by measuring the voltage drop across the circuit. It is important to remember that excessive resistance can occur anywhere in the circuit. The best way to identify the source of high resistance is to divide the affected circuit into three parts: the component, the ground circuit, and the power circuit. If the voltage drop across the component is equal to source voltage (Figure 4-11), there are no other resistances in the circuit. If the voltage drop is less than source voltage, check the voltage drop across the ground circuit. If the voltage drop is more than 0.1 volts, there is high resistance in that part of the circuit. The source can be located by separating that circuit into small measurable parts and then measuring each part. The part that has the excessive voltage drop across it is the source for the high-resistance problem. The power side of the circuit is checked in the same way.

LAB SCOPES

Lab scope is the name given to a small (sometimes handheld) oscilloscope. An oscilloscope is actually a visual voltmeter. Lab scopes have become the diagnostic tool of choice for many good technicians. A scope allows you to see voltage changes over time (Figure 4-12). Voltage is displayed across the screen of the scope as a waveform. By displaying the waveform, the scope shows the slightest changes in voltage. This is a valuable feature for a diagnostic tool.

SPECIAL TOOLS

Lab scope
Assortment of jumper wires

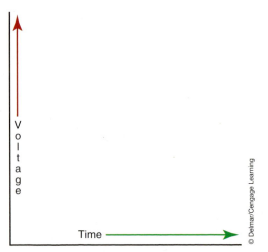

FIGURE 4-12 Voltage and time axes on a scope's screen.

With a scope, precise measurement of voltage is possible. When measuring voltage with an analog voltmeter, the meter only displays the average values at the point being probed. Digital voltmeters simply sample the voltage several times each second and update the meter's reading at a particular rate. If the voltage is constant, good measurements can be made with both types of voltmeters. A scope will display any change in voltage as it occurs. This is especially important for diagnosing intermittent problems.

The screen of a lab scope is divided into small divisions of time and voltage (Figure 4-13). These divisions set up a grid pattern on the screen. The horizontal movement of the wave-form represents time. Voltage is measured with the vertical position of the waveform. Since the scope displays voltage over time, the waveform moves from the left (the beginning of measured time) to the right (the end of measured time). The value of the divisions can be adjusted to improve the view of the voltage waveform. For example, the vertical scale can be adjusted so that each division represents 0.5 volts and the horizontal scale can be adjusted so that each division equals 0.005 seconds (5 milliseconds). This allows you to view small changes in voltage that occur in a very short period of time. The grid serves as a reference for measurements.

Since a scope displays actual voltage, it will display all electrical noises or disturbances that accompany the voltage signal (Figure 4-14). Noise is primarily caused by **radio frequency interference (RFI)**, which may come from the ignition system. RFI is an unwanted voltage signal that rides on a signal. This noise can cause intermittent problems with unpredictable results. The noise causes slight increases and decreases in the voltage. When a computer receives a voltage signal with noise, it will try to react to the minute changes. As a result, the computer responds to the noise rather than the voltage signal.

Electrical disturbances or **glitches** are momentary changes in the signal. These can be caused by intermittent shorts to ground, shorts to power, or opens in the circuit. These problems can occur for only a moment or may last for some time. A lab scope is handy for finding these and other causes of intermittent problems. By observing the voltage signal and wiggling or pulling a wiring harness, any looseness can be detected by a change in the voltage signal.

A **glitch** is best defined as abnormal movement of the waveform.

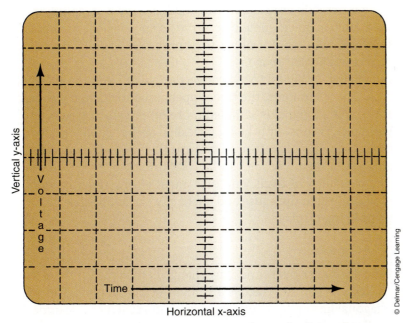

FIGURE 4-13 Grids on a scope screen serve as time and voltage references.

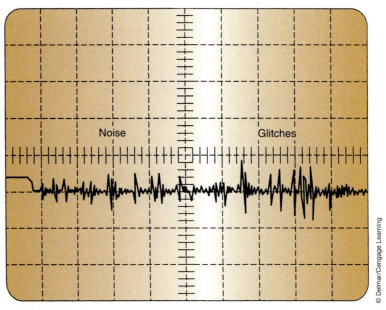

Noise

Glitches

FIGURE 4-14 RFI noise and glitches may appear on the voltage signal.

The update rate is the time it takes the trace from the end of the screen on the right to move back to the left of the screen.

A captured signal is a signal stored in a DSO's memory.

The sweep rate of a scope is simply how fast the electron beam moves across the screen.

Binary numbers are strings of zeros and ones, or ons and offs, which represent a numeric value.

The height of a waveform is called its **amplitude**.

Analog vs. Digital Scopes

Analog scopes show the actual activity of a circuit and are called *real-time scopes*. This simply means that what is taking place at the point being measured or probed is what you see on the screen. Analog scopes have a fast update rate, which allows for the display of activity without delay.

A digital scope, commonly called a **digital storage oscilloscope (DSO)**, converts the voltage signal into digital information and stores it in its memory. Some DSOs send the signal directly to a computer or a printer, or save it to a disk. To help in diagnostics, a technician can "freeze" the captured signal for close analysis. DSOs also have the ability to capture low-frequency signals. Low-frequency signals tend to flicker when displayed on an analog screen. To have a clean waveform on an analog scope, the signal must be repetitive and occur in real time. The signal on a DSO is not quite real time; rather, it displays the signal as it occurred a short time before.

This slight delay is actually very slight. Most DSOs have a sampling rate of one million samples per second. This is quick enough to serve as an excellent diagnostic tool. This fast sampling rate allows slight changes in voltage to be observed. Slight and quick voltage changes cannot be observed on an analog scope.

A DSO uses an analog-to-digital (A/D) converter to digitize the input signal. Since digital signals are based on binary numbers, the trace appears slightly choppy when compared to an analog trace. But the voltage signal is sampled more often, which results in a more accurate waveform. The waveform is constantly being refreshed as the signal is pulled from the scope's memory. Remember, the sampling rate of a DSO can be as high as a million times per second.

Both an analog and a digital scope can be dual trace scopes (Figure 4-15). This means they both have the capability of displaying two traces at one time. By watching two traces simultaneously, you can watch the cause and effect of a sensor, or compare a good or normal waveform to the one being displayed.

Waveforms

A waveform represents voltage over time. Any change in the **amplitude** of the trace indicates a change in the voltage. When the trace is a straight horizontal line, the voltage is constant (Figure 4-16). A diagonal line up or down represents a gradual increase or decrease in voltage. A sudden rise or fall in the trace indicates a sudden change in voltage.

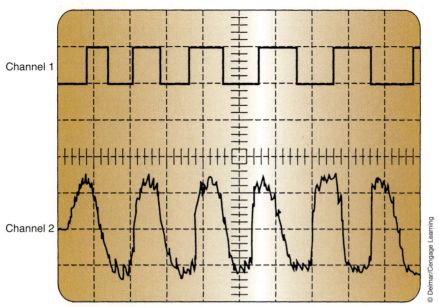

FIGURE 4-15 Some scopes can display two traces at one time; these are called dual-trace scopes.

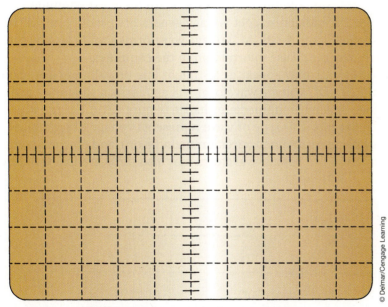

FIGURE 4-16 A constant voltage waveform.

Scopes can display AC and DC voltage either one at a time or simultaneously, as in the case of noise caused by RFI. Noise results from AC voltage riding on a DC voltage signal. The consistent change of polarity and amplitude of the AC signal causes slight changes in the DC voltage signal. A normal AC signal changes its polarity and amplitude over a period of time. The waveform created by AC voltage is typically called a **sine wave** (Figure 4-17). One complete sine wave shows the voltage moving from zero to its positive peak, then moving down through zero to its negative peak and returning to zero. If the rise and fall from positive and negative are the same, the wave is said to be **sinusoidal**. If the rise and fall are not the same, the wave is nonsinusoidal. Therefore, it can be said that all AC voltage waveforms are not sine waves.

A **sine wave** is a waveform of a single frequency alternating current.

The term **sinusoidal** means the wave is shaped like a sine wave.

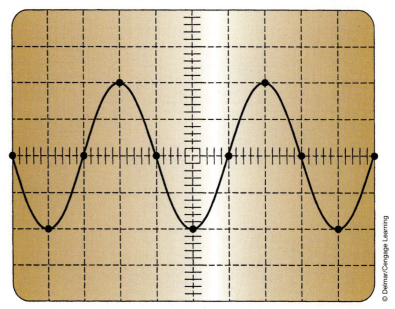

FIGURE 4-17 An AC voltage sine wave.

A **cycle** is one set of changes in a signal that repeats itself several times.

One complete sine wave is a **cycle**. The number of cycles that occur per second is the frequency of the signal. Checking frequency or cycle time is one way of checking the operation of some electrical components. Input sensors are the most common components that produce AC voltage. Permanent magnet voltage generators produce an AC voltage that can be checked on a scope (Figure 4-18). AC voltage waveforms should also be checked for noise and glitches, which may send false information to the computer.

DC voltage waveforms may appear as a straight line or a sloping line showing a change in voltage. Sometimes a DC voltage waveform will appear as a square wave or digital signal, which shows voltage making an immediate change (Figure 4-19). Square waves have straight vertical sides and a flat top. This type of wave represents voltage being applied (circuit being turned on), voltage being maintained (circuit remaining on), and no voltage applied (circuit is turned off). Of course, a DC voltage waveform may also show gradual voltage changes.

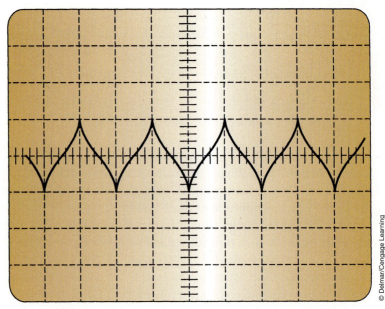

FIGURE 4-18 An AC voltage trace from a typical permanent magnet generator-type pickup or sensor.

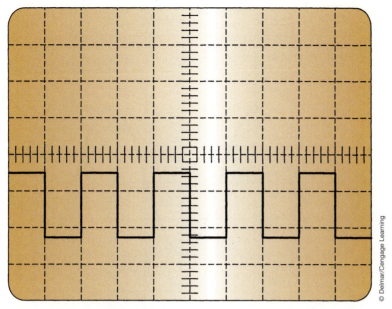

FIGURE 4-19 A typical square wave.

Scope Controls

Depending on the manufacturer and model of the scope, the type and number of its controls will vary. However, nearly all scopes have these: intensity, vertical (y-axis) adjustments, horizontal (x-axis) adjustments, and trigger adjustments. The intensity control is used to adjust the brightness of the trace. This allows for clear viewing regardless of the light around the scope screen.

The vertical adjustment actually controls the voltage displayed. The voltage setting of the scope is the voltage that will be shown per division (Figure 4-20). If the scope is set at 0.5 (500 millivolts), this means a 5-volt signal will need 10 divisions. Likewise, if the scope is set to 1 volt, 5 volts will need only 5 divisions. While using a scope, it is important to set the vertical so that voltage can be accurately read. Setting the voltage too low may cause the waveform to move off the screen, while setting it too high may cause the trace to be flat and unreadable. The vertical position control allows the vertical position of the trace to be moved anywhere on the screen.

Examples of permanent magnet generators are some crankshaft and camshaft sensors, ignition pickup units, vehicle speed sensors, and wheel speed sensors.

Some of the commonly found components powered by digital signals are fuel injectors, mixture-control solenoids, and digital EGR valves.

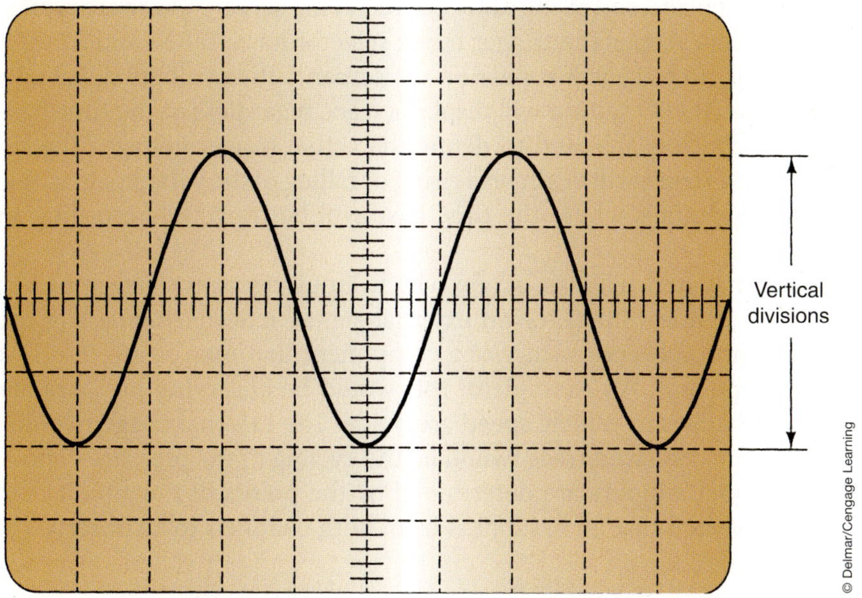

Vertical divisions

FIGURE 4-20 Vertical divisions represent voltage.

163

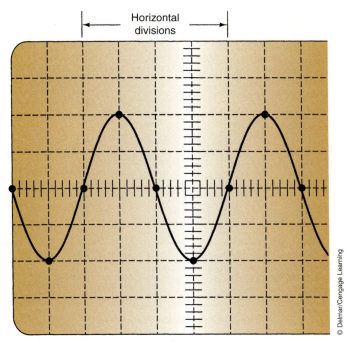

FIGURE 4-21 Horizontal divisions represent time.

The horizontal position control allows the horizontal position of the trace to be set on the screen. The horizontal control is actually the time control of the trace (Figure 4-21). Setting the horizontal control is setting the time base of the scope's sweep rate. If the time per division is set too low, the complete trace may not show across the screen. Also, if the time per division is set too high, the trace may be too crowded for detailed observation. The time per division (TIME/DIV) can be set from very short periods of time (millionths of a second) to full seconds.

The rate at which the trace moves across the display is called the **sweep rate** and is determined by the trigger. The trigger is usually the input signal and allows the trace to be synchronized with that signal.

Trigger controls tell the scope when to begin a trace across the screen. Setting the trigger is important when trying to observe the timing of something. Proper triggering will allow the trace to repeatedly begin and end at the same points on the screen. There are typically numerous trigger controls on a scope. The trigger mode selector has NORM and AUTO positions. In the NORM setting, no trace will appear on the screen until a voltage signal occurs within the set time base. The AUTO setting will display a trace regardless of the time base.

Slope and level controls are used to define the actual trigger voltage. The slope switch determines whether the trace will begin on a rising or falling of the voltage signal (Figure 4-22). The level control determines where the time base will be triggered according to a certain point on the slope.

A trigger source switch tells the scope which input signal to trigger on. This can be Channel 1, Channel 2, line voltage, or an external signal. External signal triggering is very useful when you want to observe a trace of a component that may be affected by the operation of another component. An example of this would be observing pressure control solenoid activity when changes in vehicle speed are made. The external trigger would be voltage change at the VSS. The displayed trace would be the cycling of the pressure control solenoid. Channel 1 and Channel 2 inputs are determined by the points of the circuit being probed. Some scopes have a switch that allows inputs from both channels to be observed at the same time or alternately.

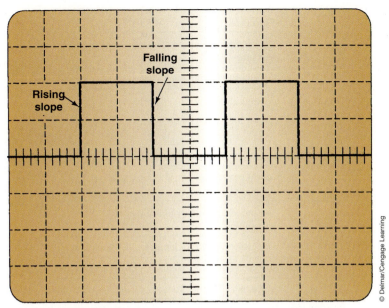

FIGURE 4-22 The trigger can be set to start the trace with a rise or fall of the voltage.

SCAN TOOLS

Scan tools will retrieve fault codes from a computer's memory and digitally display these codes on the tool. A scan tool may also perform many other diagnostic functions depending on the year and make of the vehicle. Some aftermarket scan tools have removable modules that are updated each year. These modules are designed to test the computer systems on various makes of vehicles.

The scan tool must be programmed for the model year, make of vehicle, and type of engine. With some scan tools, this selection is made by pressing the appropriate buttons on the tester, as directed by the digital tester display. On other scan testers, the appropriate memory card must be installed in the tester for the vehicle being tested. Some scan testers have a built-in printer to print test results, while other scan testers may be connected to an external printer.

As automotive computer systems become more complex, the diagnostic capabilities of scan testers continue to expand. Many scan testers now have the capability to store, or "freeze," data into the tester during a road test, and then play back this data when the vehicle is returned to the shop.

Some scan testers now display diagnostic information based on the fault code in the computer memory. Service bulletins published by the manufacturer of the scan tester may be indexed in the tester after the vehicle information is entered in the tester. Other scan testers will display sensor specifications for the vehicle being tested.

A computer will set a DTC when a voltage signal is entirely out of its normal range. If a signal is within its normal range but still not correct for the conditions, the computer will not set a DTC. However, there is still a problem that can cause a customer concern. To help identify the cause of the concern, the signals to and from the computer should be carefully observed. This is done with a scan tool. The tool is set to read **serial data**, which is often displayed in PID values (Figure 4-23).

Although OBD II simplified things, the DTCs retrieved do not reflect the same problems across the different transmission models produced by the different manufacturers.

A **scan tool** is actually a computer designed to communicate with the vehicle's computer.

The communications to and from the computer are commonly referred to as the system's **serial data**.

PID Name	Description of PID
EPC	Commanded Electronic Pressure Control Pressure—in psi
GEAR	Commanded Gear—not actual
LINEDSD	Commanded Line Pressure—in psi
OSS	Input from Output Shaft Speed Sensor—in rpm
RPM	Input from Engine Speed Sensor—in rpm
SSA	Commanded State of Shift Solenoid No. 1—ON or OFF
SSB	Commanded State of Shift Solenoid No. 2—ON or OFF
SS1F	Shift Solenoid No. 1 Circuit Fault—YES or NO
SS2F	Shift Solenoid No. 2 Circuit Fault—YES or NO
TCCACT	Slippage of Torque Converter Clutch—in rpm
TCCCMD	Commanded State of Torque Converter Clutch Solenoid—in %
TCCF	Torque Converter Clutch Solenoid Curcuit Fault—YES or NO
TFT	Transmission Fluid Temperature—in voltage or degrees
TP	Throttle Position—in voltage
TR	Transmission Range Sensor—by position
TRANRAT	Actual Transmission Gear Ratio—by position
TSS	Input from Turbine Shaft Speed Sensor—in rpm

© Delmar/Cengage Learning

FIGURE 4-23 Common transmission PIDs for Ford products.

There are some codes that reflect the same basic problem. For example, a P0705 code relates to a TR range sensor circuit problem. However, each manufacturer may have several other DTCs that describe particular failures, such as Ford's P0708 code, which says a particular circuit for the TR sensor is open. Always consult the service manual when deciphering a DTC.

GENERAL EAT DIAGNOSIS

Some early electronic transmissions are only partially controlled; that is, only the engagement of the converter clutch and third-to-fourth shifting are electronically controlled. Other models feature electronic shifting into all gears, plus electronic control of the TCC (Figure 4-24).

The controls of an EAT direct hydraulic flow by using solenoid-controlled valves. When used to control TCC operation, the solenoid opens a hydraulic circuit to the TCC spool valve, causing the spool valve to move and direct mainline pressure to apply the clutch. Electronically controlled shifting is accomplished in much the same way. Shifting occurs when a solenoid is either turned on or turned off (Figure 4-25). At least two shift solenoids are incorporated into the system and shifting takes place by controlling the solenoids. The desired gear is put into operation through a combination of on and off solenoids.

Classroom Manual

Chapter 4, page 117

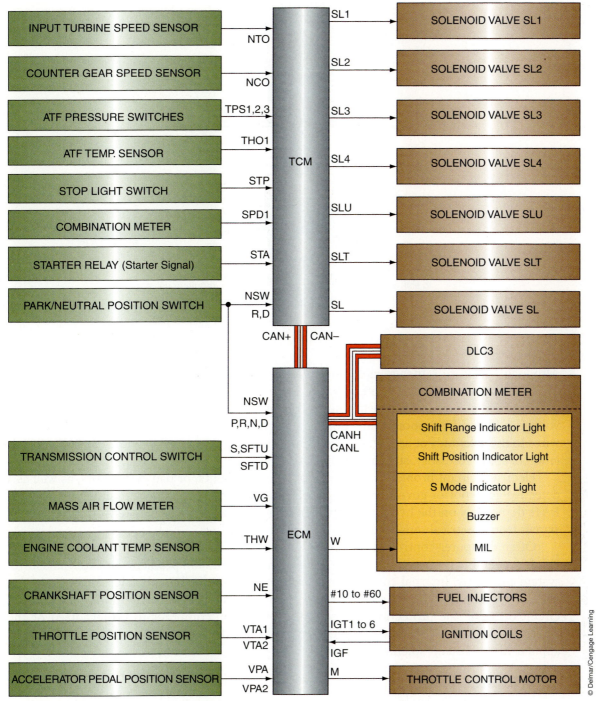

FIGURE 4-24 The components involved in the operation of a typical late-model EAT system.

Several sensors and switches are used to inform the control computer of the current operating conditions. Most of these sensors are also used to calibrate engine performance. The computer then determines the appropriate shift time for maximum efficiency and best feel. The shift solenoids are controlled by the computer, which either supplies power to the solenoids or supplies a ground circuit. The techniques for diagnosing electronic transmissions are basically the same techniques used to diagnose TCC systems.

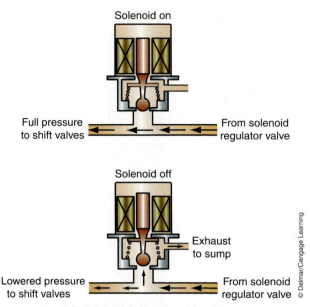

FIGURE 4-25 Shift solenoid action.

PRELIMINARY CHECKS

Although EATs are rather reliable, they have introduced new problems for the automatic transmission technician to contend with. Some of the common problems that affect shift timing and quality, as well as the timing and quality of TCC engagement, are incorrect battery voltage, a blown fuse, poor connections, a defective TP sensor or VSS, defective solenoids, wires to the solenoids or sensors crossed, corrosion at an electrical terminal, or faulty installation of some accessory, such as a cellular telephone.

Electrical circuit problems, faulty electrical components, or bad connectors, as well as a defective governor or governor drive gear assembly, can cause improper shift points. Most EATs do not rely on the hydraulic signals from a governor; rather, they rely on the electrical signals from electrical sensors to determine shift timing.

Often computer-controlled transmissions will start off in the wrong gear. This can happen for several reasons, either because of internal transmission problems or external control system problems. Internal transmission problems can be faulty solenoids or stuck valves. External problems can be the result of a complete loss of power or ground to the control circuit or a fail-safe protection strategy initiated by the computer to protect itself or the transmission from an observed problem. Typically the default gear is simply the gear that is applied when the shift solenoids are off.

Basic System Checks

Troubleshooting a transmission's electronic control system is like troubleshooting any other electronic system; you need to make preliminary checks of the system before moving into specific checks. The first step in a preliminary check is verifying system voltage. With a DMM, check the open-circuit voltage of the battery. Minimum voltage at the battery should be 12.6 volts. If the battery is below this, recharge the battery. If the voltage is still low after recharging, replace it.

Continue by checking the condition of the battery cables. Conduct a voltage drop test across each cable. Make sure the system is activated when doing a voltage drop test. There should be no more than 0.1 volt dropped across the positive or the negative cable. If the voltage drop exceeds that amount, clean or replace the cables.

A visual inspection of the circuit should follow. Carefully look at the entire system and check for damaged or corroded wires, loose connections, and damaged connectors (Figure 4-26). Pay particular attention to the connectors. Make sure there are no bent or broken terminals. If any are present, replace the connector. Also look inside each connector for signs of corrosion.

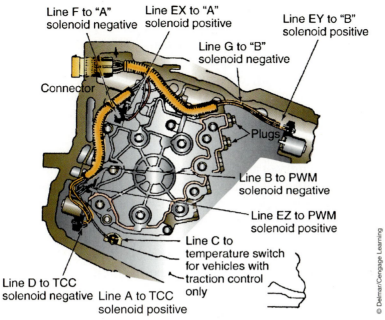

Line F to "A" solenoid negative
Line EX to "A" solenoid positive
Line EY to "B" solenoid positive
Line G to "B" solenoid negative
Connector
Plugs
Line B to PWM solenoid negative
Line EZ to PWM solenoid positive
Line C to temperature switch for vehicles with traction control only
Line D to TCC solenoid negative
Line A to TCC solenoid positive

© Delmar/Cengage Learning

FIGURE 4-26 The position and condition of the transmission wiring harness and its connectors should be carefully checked.

If any is found, clean the wires and terminals. If the corrosion cannot be cleaned, replace the wires or the connector.

Continue your basic inspection by checking the fuse or fuses to the control module. To accurately check a fuse, either test it for continuity with an ohmmeter or check each side of the fuse for power when the circuit is activated.

> **CUSTOMER CARE:** Vehicles equipped with control computers may require a relearn procedure after the battery is disconnected. Many computers memorize and store vehicle operation patterns for optimum driveability and performance. When the vehicle's battery is disconnected, this memory is lost. The computer will use default data until new data from each key start is stored. Customers often complain of driveability problems during the relearn stage because the vehicle acts differently than it did before it was serviced. To reduce the possibility of complaints, the vehicle should be road tested and the correct relearn procedure followed.

If the system has no apparent problems, continue testing the system. The diagnostic procedure for most EAT systems would now include checking the system with a scan tool. The scan tool will display any trouble codes in the system. More importantly to the technician, most scan tools will display serial data. The serial data stream allows you to monitor system sensor and actuator activity during operation. Comparing the test values to the manufacturer's specifications will greatly help in diagnostics.

It is possible that the data displayed by a scan tool is not the actual value. Most computer systems will disregard an input that is well out of range and rely on a default value held in its memory. These default values are hard to recognize and do little for diagnostics; this is why detailed testing with a DMM or lab scope is preferred by many technicians. These test instruments are also used to further test the system after a scan tool has identified a problem.

Road Test

Critical to proper diagnosis of EAT and TCC control systems is a road test. The road test should be conducted in the same way as one for a nonelectronic transmission, except that a scanner tool should also be connected to the circuit to monitor engine and transmission operation.

CAUTION: Remember to disconnect power to the component before checking it with an ohmmeter. Failure to do so can result in the meter being destroyed.

During the road test, the vehicle should be driven in the normal manner. All pressure and gear changes should be noted. Also, the various computer inputs should be monitored and the readings recorded for future reference. Some scan tools have the capability of printing out a report of the test drive. Critical information from the inputs includes engine speed, vehicle speed, manifold vacuum, operating gear, and the time it took to shift gears. If the scanner does not have the ability to give a summary of the road test, you should record this same information after each gear or operating condition change.

Basic Transmission Circuits

To summarize how the circuitry of an electronically controlled transmission interrelates with a computerized engine control system, a summary of the transmission controls for a Ford E4OD follows. This system is similar to the other systems used by Ford, as well as by other manufacturers. The transmission is controlled by the PCM of the engine control system. The computer receives inputs from both the engine and the transmission. Based on this information, the PCM can send signals to operate components in both the engine and transmission. There are many inputs to the PCM from the transmission. The transmission range (TR) sensor informs the PCM what gear has been selected. The vehicle speed sensor, located in the same place as the governor would be, inputs the mph at which the vehicle is traveling. The turbine speed sensor (not found on all models) sends signals to inform the PCM of the speed of the input to the transaxle. The transmission oil temperature sensor monitors the temperature of the ATF. This signal determines whether the PCM should go to a cold-start shift cycle or shift according to its schedule.

There are four nontransmission inputs: the TP sensor, MAF, PIP, and brake on switch. The system uses five solenoids, which are located in the transmission: the electronic pressure control solenoid that regulates transmission operating pressure, a modulated solenoid that controls the converter clutch, and three shift solenoids that control fluid flow to the various holding and clutching devices (Figure 4-27).

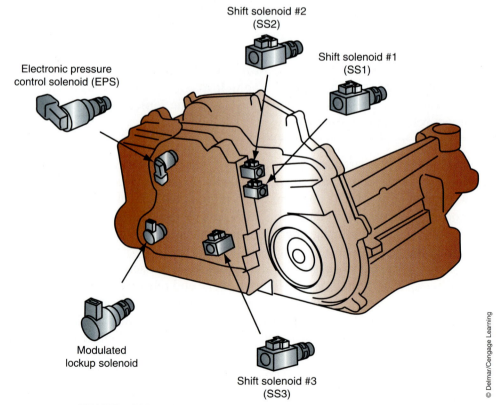

FIGURE 4-27 Location of output solenoids on an AXOD-E transaxle.

© Delmar/Cengage Learning

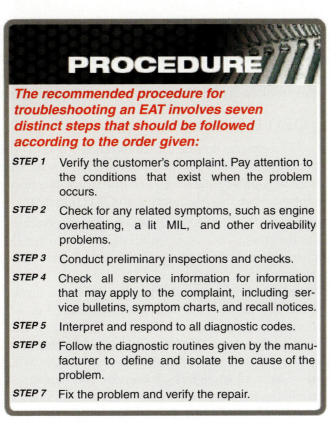

PROCEDURE

The recommended procedure for troubleshooting an EAT involves seven distinct steps that should be followed according to the order given:

STEP 1 Verify the customer's complaint. Pay attention to the conditions that exist when the problem occurs.

STEP 2 Check for any related symptoms, such as engine overheating, a lit MIL, and other driveability problems.

STEP 3 Conduct preliminary inspections and checks.

STEP 4 Check all service information for information that may apply to the complaint, including service bulletins, symptom charts, and recall notices.

STEP 5 Interpret and respond to all diagnostic codes.

STEP 6 Follow the diagnostic routines given by the manufacturer to define and isolate the cause of the problem.

STEP 7 Fix the problem and verify the repair.

© Delmar/Cengage Learning

FIGURE 4-28 A flow chart showing the major steps that should be followed while diagnosing an EAT.

Because the engine and transmission control systems share information, a common component may affect both systems. Nearly all computer control systems have some form of self-diagnostics. During this mode of operation, the computer will display a trouble code that represents a circuit or component it determines is faulty. Part of the basic procedure for diagnosing an EAT is to retrieve the trouble codes from the computer. This information will determine what step you should take next (Figure 4-28).

As you can readily tell, in this system, as well as the others, engine performance problems can cause the transmission to perform unsatisfactorily. Since the computer controls the outputs to the engine and the transmission based on input sensors common to both, it is very important that you perform basic engine checks before spending time diagnosing the transmission. This was important when diagnosing nonelectronically controlled transmissions and is even more important when diagnosing an EAT.

On-Board Diagnostics

The use of **On-Board Diagnostics II (OBD II)** systems has brought computerization to most systems on a vehicle. On-board diagnostic capabilities are incorporated into a vehicle's computer to monitor virtually every component that can affect emission performance. Each component is checked by a diagnostic routine to verify that it is functioning properly. In addition to self-diagnostics, OBD provide a means to check the operation of the control systems.

The system continuously monitors operating conditions for possible system malfunctions. It compares system conditions against programmed parameters. If the conditions fall outside of these limits, it detects a malfunction and sets a trouble code that indicates the portion of the system that has the fault. The setting of some of these codes will also cause the malfunction indicator lamp (MIL) to illuminate. This lamp informs the driver that a problem is present. It is important to remember that not all codes will cause the MIL to illuminate; therefore, an unlit MIL does not mean there are no DTCs in the computer's memory.

Classroom Manual
Chapter 4, page 116

171

A scan tool will display any trouble codes in the system. DTCs are designed to help technicians identify and locate problems in the transmission's system. It is very important that you refer to the service information when interpreting DTCs. When diagnosing an EAT, make sure there are no engine-related codes that could affect the operation of the transmission. If there is an engine problem, fix it before continuing with your diagnosis.

Troubleshooting OBD II Systems

The following steps provide a general outline for troubleshooting OBD II systems. There are slight variations in different years and with different models. Always refer to the manufacturer's information before beginning your diagnosis. The below is a brief explanation of the steps that should be followed when troubleshooting an OBD II system:

1. **Interview the Customer**—Gather as much information as possible from the customer. Ask the customer to describe the driving conditions present when the problem appears. This should include weather, traffic, and speed. The initial steps during diagnosis should verify the customer's concern.

2. **Check the MIL**—The MIL should turn on when the ignition is turned on and the engine is not running. When the engine is started, the MIL should go off. If either of these does not occur, troubleshoot the lamp system before continuing. The MIL is basically an engine malfunction light, but if the TCM detects a problem that may affect emissions, it will send a request over the data bus to the PCM to turn on the MIL lamp.

3. **Visual Inspection**—Check all wires frays, looseness, and damage. Also, check the wiring for burned or chafed spots, pinched wires, or contact with sharp edges or hot exhaust parts. Try not to wiggle the wires while doing this, a wiggle may correct an intermittent problem that may be hard to find later. Check the condition of the battery and all sensors and actuators for physical damage. Check all connections to sensors, actuators, control modules, and ground points. Also check all vacuum hoses for pinches, cuts, or disconnects. Correct any problems.

4. **Connect the Scan Tool**—Make sure the tool is OBD II compliant. The scan tool must have the appropriate connector to fit the DLC on an OBD II system and the proper software for the vehicle being tested. Cable adapters are used to connect a standard scan tool to an OBD II system. The software inserts make the scan tool compatible with the PCM in the vehicle. Make sure the recommended insert for that make and model of vehicle is used in the scan tool. Always follow the instructions given in the scan tool's manual when testing the system.

5. **Check DTC(s) and Freeze Frame Data**—When using a scan tool, an * (asterisk) next to the DTC often indicates there is stored freeze frame data associated with that DTC. During diagnostics, it is helpful to know the conditions present when the DTC was set, such as was the vehicle running or stopped, what was the engine's temperature, and was the air/fuel ratio rich or lean. Print or record all DTC and related information. If there was a no or poor communication DTC, solve that problem before continuing.

6. **Check Service History and Service Publications**—There may be a TSB or other service alert that may have the necessary repair information. Check these and follow the procedures outlined in them before continuing. Service history can give clues about cause of the problem, as the problem may be related to a recent repair.

7. **Clear DTC and Freeze Frame Data**—Using the scan tool, clear all DTCs and freeze frame data (Figure 4-29). Make sure you do this only after you have recorded all retrieved data.

8. **Check DTCs**—Repeat step 4. If there are no DTCs, check the status of the monitors' readiness and the pending codes on the scan tool. Do what is necessary to complete the necessary drive cycles before continuing. For many DTCs, the PCM will enter into the fail-safe mode. This means the PCM has substituted a value to allow the engine to run. Refer to the service information to determine if any DTCs indicate the fail-safe mode,

The Malfunction indicator lamp (MIL) is a warning lamp in a vehicle's instrument panel that lets the driver know when the vehicle's electronic control units detected a problem.

Diagnostic trouble codes (DTCs) are alphanumerical codes generated by the electronic control system to indicate a problem in a circuit or subsystem.

CAUTION:
Do not clear DTCs unless directed by a diagnostic procedure. Clearing DTCs will also clear valuable freeze frame and failure records data.

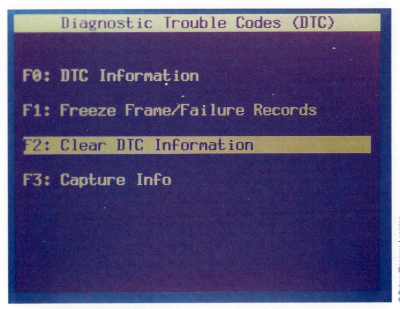

FIGURE 4-29 After all retrieved data are recorded, clear all DTCs and freeze frame data.

if so, follow the appropriate diagnostic steps. Current DTCs indicate a problem that is present. Use the DTC chart to determine what was detected, the probable problem areas, and how to diagnose that DTC.

9. **Basic Inspection**—When the DTC is not confirmed in the previous check, use the problems symptoms chart in the service information to diagnose a problem when no code is displayed but the problem is still occurring.

10. **DTC Chart**—Use the DTC chart to determine what was detected, the probable problem areas, and how to diagnose that DTC (Figure 4-30).

11. **Check for Intermittent Problems**—If the cause of the problem has not yet been determined, proceed to check for an intermittent problem.

12. **Perform Repairs**—Once the cause of the concern has been identified, perform all required services.

13. **Repair Verification Test**—After repairs have been made, check your work by clearing all codes, checking the MIL, and rechecking for codes.

Malfunction Indicator Lamp Operation

The MIL is only used to inform the driver that the vehicle should be serviced soon. It also informs a technician that the computer has set a trouble code. OBD II systems continuously monitor the entire system, switches on a MIL if something goes wrong, and stores a fault code in the PCM when it detects a problem.

When the ignition is initially turned on, the MIL will momentarily flash ON then OFF and remain ON until the engine is running or the computer determines there are no DTCs stored in memory. The TCM monitors all transmission related electrical components and typically sets a DTC if it detects a problem. However, not all set codes illuminate the MIL. The MIL (Figure 4-31) is used to inform the driver of a serious problem. Therefore, DTCs can be set even though the MIL is not illuminated. Many of the diagnostic steps required under OBD II must be performed under specific operating conditions. The software in the TCM determines which tests will be conducted and when they will be run. Based on these tests, or monitors, the TCM will store different types of DTCs, as well as their history. The DTCs remain in memory until a certain number of trips or start cycles have completed and the problem has not re-occurred. The DTCs will also be removed when the battery is disconnected or cleared with the scan tool.

DTC	COMPONENT	DESCRIPTION	CONDITION	SYMPTOM
P0705	Digital TR	Digital TR circuit failure.	This indicates an invalid pattern in TR-D. It is caused by a short to ground or an open in TR4, TR3A, TR2 and/or TR1 circuits. This DTC will not be set by an incorrectly adjusted TR sensor.	An increase in EPC pressure (harsh shifts). Defaults to (D) or D for all gear positions.
P0712	TFT, wiring, PCM	315 °F (157 °C) indicates the TFT circuit is grounded.	The voltage drop across the sensor exceeds the scale set for that temperature.	Firm shift feel.
P0713	TFT, wiring, PCM	−40 °F (−40 °C) indicates the TFT circuit is open.	The voltage drop across the sensor exceeds the scale set for that temperature.	Firm shift feel.
P0721	OSS	The OSS sensor signal is noisy.	The PCM has detected an erratic OSS signal.	Harsh shifts, abnormal shift schedule, no TCC engagement.
P0731	Shift Solenoid A (SSA), Shift Solenoid B (SSB) or internal parts	1st gear error.	No 1st gear.	Incorrect gear selection, depending on failure or mode and manual lever position. Shift errors may also be due to other internal transmission concerns (stuck valves, damaged friction material). Engine rpm could be higher or lower than expected.
P0732	SSA, SSB or internal parts	2nd gear error.	No 2nd gear.	Incorrect gear selection, depending on failure or mode and manual lever position. Shift errors may also be due to other internal transmission concerns (stuck valves, damaged friction material). Engine rpm could be higher or lower than expected.
P0733	SSA, SSB or internal parts	3rd gear error.	No 3rd gear.	Incorrect gear selection, depending on failure or mode and manual lever position. Shift errors may also be due to other internal transmission concerns (stuck valves, damaged friction material). Engine rpm could be higher or lower than expected.
P0734	SSA, SSB or internal parts	4th gear error.	No 4th gear.	Incorrect gear selection, depending on failure or mode and manual lever position. Shift errors may also be due to other internal transmission concerns (stuck valves, damaged friction material). Engine rpm could be higher or lower than expected.
P0963	EPC solenoid	The EPC solenoid is shorted to the power circuit.	The voltage through the EPC solenoid is checked. An error will be noted if tolerance is exceeded.	Maximum EPC pressure, harsh engagements and shifts.
P1760	EPC, wiring, PCM	The EPC solenoid circuit is shorted or the output driver is faulty.	The PCM detected a loss of EPC during operation.	Unexpected reduction in engine torque.

© Delmar/Cengage Learning

FIGURE 4-30 A sample of transmission related DTCs.

© Delmar/Cengage Learning

FIGURE 4-31 An example of a MIL that illuminates to warn the driver that transmission control has been switched to the default mode.

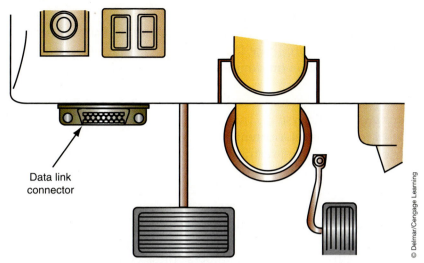

FIGURE 4-32 Standard location for the DLC in an OBD II system.

Data Link Connector

Standards require the **Data Link Connector (DLC)** for OBD II systems to be mounted in the passenger compartment out of sight of vehicle passengers. The standard DLC is a 16-pin connector. The same seven pins are used for the same information, regardless of the vehicle's make, model, and year. The connector is "D"-shaped and has guide keys that allow the scan tool to be only installed one way. Using a standard connector design and by designating the pins, it allows data retrieval with any scan tool designed for OBD II. Some European vehicles meet OBD II standards by providing the designated DLC along with their own connector for their own scan tool.

According to the OBD standards, the DLC must be easily accessible from the driver's seat (Figure 4-32). Therefore, they are located somewhere between the left end of the instrument panel and to the right of the center of the panel. The DLC cannot be hidden behind panels.

The DLC is designed only for scan tool use. You cannot jump across any of the terminals to display codes on an instrument panel or other indicator lamp. Any generic scan tool can be connected to the DLC and can access the diagnostic data stream. The connector pins are arranged in two rows and are numbered consecutively. Seven of the 16 pins have been assigned by the OBD II standard. The remaining nine pins can be used by the individual manufacturers to meet their needs and desires.

Diagnostic Trouble Codes

Diagnostic trouble codes (DTCs) are designed to help technicians identify and locate problems in the transmission's control system. DTCs from pre-OBD II systems can indicate a variety of things. Each manufacturer (and sometimes each vehicle model) used different codes to identify detected problems. It is very important that you refer to the service manual when interpreting DTCs.

Some OBD II codes are universal in that they are the same for all manufacturers and vehicle models. Other codes are manufacturer specific and must be interpreted through reference to the appropriate service information.

DTCs are set by the computer when a signal from a component or circuit is not within the normal operating range for the operating conditions. A code will also be set when there is no signal from an input circuit or to an output device. EATs also have the ability to conduct a self-test. During this test, the sensor circuits are checked and if a problem is found, a DTC is set.

SERVICE TIP:
When a vehicle has a 16-pin DLC, it does not always mean the vehicle is equipped with OBD II.

There are basically two types of DTCs. A hard code is a DTC that represents a problem that is present at the time of retrieval. These are the codes that should be responded first, during diagnostics. Soft codes are those DTCs that are not currently present. These codes can be retrieved and represent an intermittent problem or a problem that existed but is no longer present.

An OBD II DTC is a five-character code with both letters and numbers (Figure 4-33). This is called the alphanumeric system. The first character of the code is a letter. This defines the system where the code was set. Currently there are four possible first character codes: "B" for body, "C" for chassis, "P" for powertrain, and "U" for undefined. The U-codes are designated for future use.

The second character is a number. This defines the code as being a mandated code or a special manufacturer code. A "0" code means that the fault is defined or mandated by OBD II. A "1" code means the code is manufacturer specific. Codes of "2" or "3" are designated for future use.

The third through fifth characters are numbers. These describe the fault. The third character tells you where the fault occurred. The remaining two characters describe the exact condition that set the code. The numbers are organized so that the various codes related to a particular sensor or system is grouped together.

DTCs that relate to transmission faults can be caused by engine or transmission input and/or output devices. These codes may seem to indicate a problem with an input or output circuit but may actually be caused by an internal transmission problem. Remember, codes are set by out-of-range values. Therefore, when the TCM is receiving a too low or high input signal, the cause is not necessarily a bad sensor. The sensor can be fine and a mechanical or

The SAE J2012 standards specify that all DTCs will have a five-digit alphanumeric numbering and lettering system. The following prefixes indicate the general area to which the DTC belongs:

1. P — powertrain
2. B — body
3. C — chassis

The first number in the DTC indicates who is responsible for the DTC definition.

1. 0 — SAE
2. 1 — manufacturer

The third digit in the DTC indicates the subgroup to which the DTC belongs. The possible subgroups are:

0 — Total system
1 — Fuel-air control
2 — Fuel-air control
3 — Ignition system misfire
4 — Auxiliary emission controls
5 — Idle speed control
6 — PCM and I/O
7 — Transmission
8 — Non-EEC power train

The fourth and fifth digits indicate the specific area where the trouble exists. Code P1711 has this interpretation:

P — Powertrain DTC
1 — Manufacturer-defined code
7 — Transmission subgroup
11 — Transmission oil temperature (TOT) sensor and related circuit

© Delmar/Cengage Learning

FIGURE 4-33 An explanation of the five digits in an OBD II trouble code.

hydraulic transmission problem is causing the abnormal signals. Not only can internal transmission problems cause codes to be set, but so can basic electrical problems. Problems such as loose connections, broken wires, corrosion, and poor grounds will affect the signals.

OBD II codes are quite definitive and can lead you to the cause of the problem. For example, there are five codes that relate to the transmission fluid temperature circuit. Code P0710 indicates there is a fault in the sensor's circuit, P0711 indicates the circuit is operating out of the acceptable range, P0712 indicates the circuit lower than normal input signals, P0713 indicates the input signals are higher than normal, and P0714 indicates the circuit appears to have an intermittent problem.

If the TCM is unable to communicate with the PCM, there is a data bus problem. These problems will normally result in poor operation as well as the inability to retrieve DTCs from the TCM. The PCM constantly monitors the data bus and if it is unable to establish communication, it will order a data bus DTC.

Freeze Frame Data

One of the mandated capabilities of OBD II is the "**freeze frame**" or snapshot feature. Although the regulations mandate just emission-related DTCs, manufacturers can choose to include this feature for other systems. With this feature, the PCM takes a snapshot of the activity of the various inputs and outputs at the time the PCM illuminated the MIL or set a DTC. The PCM uses this data for identification and comparison of similar operating conditions if the same problem occurs in the future. This feature is also valuable to technicians, especially when trying to identify the cause of an intermittent problem. The action of sensors and actuators when the code was set can be reviewed. This can be a great help in identifying the cause of a problem. The informations held in freeze frame are actual values; they have not been altered by the adaptive strategy of the PCM.

Once a DTC and the related freeze frame data are stored in memory, they will stay there even if other emission-related DTCs are set. This data can only be removed with a scan tool. When a scan tool is used to erase a DTC, it automatically erases all associated freeze frame data. The data are stored by priority, information related to misfire and fuel control have priority over other DTCs. Fuel system misfires will overwrite any other type of data except for other fuel system misfire data.

Test Modes

All OBD II systems have the same basic test modes and all are accessible with an OBD II scan tool. Always refer to the manufacturer's information when using these test modes for diagnosis.

Mode 1 is the **Parameter Identification (PID)** mode. This mode allows access to current emission related data values of inputs and outputs, calculated values, and system status information. Some PID values are manufacturer specific; others are common to all vehicles. This information is referred to as serial data.

Mode 2 is the freeze frame data access mode. This mode permits access to emission-related data values from specific generic PIDs. The number of freeze frames that can be stored is limited. The Freeze Frame information updates if the condition recurs.

Mode 3 permits scan tools to obtain stored DTCs. The information is transmitted from the PCM to the scan tool following a Mode 3 request. The DTC, its descriptive text, or both can be displayed on the scan tool.

Mode 4 is the PCM reset mode. It allows the scan tool to clear all emission-related diagnostic information from its memory. When this mode is activated, all DTCs, freeze frame data, DTC history, monitoring test results, status of monitoring test results, and on-board test results are cleared and reset.

Mode 5 is the oxygen sensor monitoring test. This mode gives the actual oxygen sensor outputs during the test cycle. These are stored values, not current values that are retrieved in Mode 1. This information is used to determine the effectiveness of the catalytic converter.

Mode 6 is the output state mode (OTM) and can be used to identify potential problems in the noncontinuous monitored systems.

Mode 7 reports the test results for the continuous monitoring systems.

Mode 8 is the request for control of an on-board system test or component mode. It allows the technician to control the PCM, through the scan tool, to test a system. In some cases, the scan tool will only set the conditions for conducting a test and not actually conduct the test. An example of this is the EVAP leak test, which is done with other test equipment.

Mode 9 is the request for vehicle information mode. This mode reports the vehicle's identification number, calibration identification, and calibration verification. This information can be used to see if the most recent calibrations have been programmed into the PCM.

Serial Data

Although the first steps in diagnosis include retrieving DTCs, there are problems that will not be evident by a code. These problems are solved with further testing, symptom charts, or pure logic. This logic must be based on an understanding of the transmission and its controls. It is possible to pinpoint the exact cause of a transmission problem: Monitor the serial data with a scan tool. The serial data stream allows you to monitor system activity during operation. Comparing the observed values to the manufacturer's specifications will greatly help in diagnostics.

To observe the serial data, enter the desired PID into the scan tool. Many PIDs are standard for all OBD II systems, however not all vehicles will support all PIDs and there are manufacturer-defined PIDs that are not part of the OBD-II standard. Manufacturers list and define the various applicable PIDs in their service information. This information should be compared to the observed data. Always refer to the service information and the scan tool's operating instructions for proper identification of normal or expected data. If an item is not within the normal values, record the difference and diagnose that particular item.

It is possible that the data displayed by a scan tool are not the actual value. Most computer systems will disregard inputs that are well out of range and rely on a default value held in its memory. These default values are hard to recognize and do little for diagnostics; this is why detailed testing with a DMM or lab scope is preferred by many technicians. These test instruments are also used to further test the system after a scan tool has identified a problem.

CVI. Chrysler's use of CVI for adjusting line pressure to compensate for wear provides data concerning the transmission's wear. Chrysler was the first manufacturer to offer this type of information. The 41TE's TCM monitored the oil volume necessary to apply a clutch. Based on this, the TCM could adjust the flow rate of the oil by adjusting the pulse width to the shift solenoids. The TCM also monitored the input and output speed sensors to calculate the change rate of a shift and the gear ratio present after the shift. These two items make up the CVI value that can be displayed on a scan tool. The displayed values relate to both volume and time.

Watching the CVI (Figure 4-34) can help diagnose hard shifting or slippage during shifting problems. As a clutch begins to wear or an internal leaks starts, the CVI will rise which

Clutch	When updated	Oil temp.	Proper CVI (64 CVI = 1 cubic inch)
L/R	2–1 or 3–1 manual downshift	>110 °F (43 °C)	82 to 134
2C	3–2 kickdown	>110 °F (43 °C)	25 to 64
OD	2–3 upshift	>110 °F (43 °C)	30 to 64
4C	3–4 upshift	>110 °F (43 °C)	30 to 64
UD	4–3 part throttle kickdown	>110 °F (43 °C)	25 to 92
2–3 OD	After first 2 to 3 shift	>65 °F (18 °C)	Higher than OD

© Delmar/Cengage Learning

FIGURE 4-34 The standard CVI indexes for a Chrysler RFE transmission.

indicates a slightly longer shift transition. A low CVI number indicates a short shift, while a higher number indicates a longer shift. Abnormal CVI reading can be caused by other things besides clutch wear, such as a weak solenoid, faulty pressure regulator, bad TCM, or an internal oil leak. Other Chrysler transmissions have a similar system in their late-model units. These systems are called the Shift Time Adaptation system and the readout on the scan tool is given in torque or strength of the shift.

GM also has a similar system. The scan tool can display shift time and shift error data for every shift. Shift time is the actual time it took to shift; while shift error is the difference between the desired and actual shift times. When a positive value is displayed, the actual shift time took longer than the desired shift time. These values are displayed as TAP (Transmission Adaptive Pressure) cells. Along with these values is a code that signifies the throttle position when the less than ideal shifts occurred. These codes range from 4 to 16; 4 means light throttle, 8 means medium throttle, and 12 and higher results from a wide throttle opening. These values are displayed for each shift. When the TAP cells are within specifications and are very similar, the transmission is operating normally. If one or more gears have a higher value, the clutches or solenoids for those gears may be bad or worn.

Intermittent Faults

An intermittent fault is a fault that is not always present. It may not activate the MIL or cause a DTC to be set. By studying the system and the relationship of each component to another, you should be able to create a list of possible causes for the intermittent problem.

Most intermittent problems are caused by faulty connections or wiring. Refer to a wiring diagram for each of the suspected circuits or components to identify the connections and components in them. That entire circuit should be carefully inspected. Check for burnt or damaged wire insulation, damaged terminals at the connectors, corrosion at the connectors, loose connectors, loose wire terminals, and loose ground wire or straps.

The vehicle can also be taken for a test drive with the scan tool connected. The scan tool will monitor the activity of a circuit while the vehicle is being driven. This allows a look at a circuit's response to changing conditions. The snapshot or freeze frame feature stores the conditions and operating parameters at command or when the PCM sets a DTC. If the snapshot is taken when the intermittent problem is occurring, the problem will be easier to diagnose.

With a scan tool, actuators can be activated and their functionality tested. The results of their operation can be monitored. Also the outputs can be monitored as they respond to changes from the inputs. When an actuator is activated, watch the response on the scan tool. Also listen for the clicking of the relay that controls that output. If no clicking is heard, measure the voltage at the relay's control circuit, there should be a change of more than 4 volts when it is activated.

To locate the source of the problem, a voltmeter can be connected to the suspected circuit and the wiring harness wiggled (Figure 4-35). As a guideline for what voltage should be expected in the circuit, refer to the reference value table in the service manual. If the voltage reading changes with the wiggles, the problem is in that circuit. The vehicle can also be taken for a test drive with the voltmeter connected. If the voltmeter readings become abnormal with changing operating conditions, the circuit being observed is probably has the problem.

Service Information

After retrieving the DTCs, find the description of the DTC in the service information. There is typically more than one possible cause of the problem. The description of the DTC also leads to pinpoint tests. These are designed to guide you through a step-by-step procedure. To be effective, each step should be performed, in the order given, until the problem has been identified.

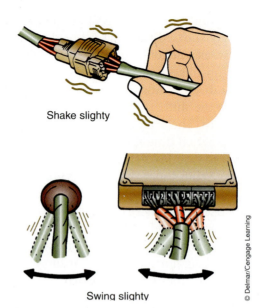

Shake slighty

Swing slighty

© Delmar/Cengage Learning

FIGURE 4-35 By slightly wiggling and moving the wires at a connector, the cause of an intermittent problem may be found.

Make sure to check all available service information related to the DTCs. There may be a TSB related to the code. This may be a recommendation to reprogram the computer's software.

Often, accurately defining the problem and locating related information in TSBs and other materials can identify the cause of the problem. No manufacturer makes a perfect vehicle or transmission and as the manufacturer recognizes common occurrences of a problem, they will issue a statement regarding the fix of the problem. Check all available information before moving on in your diagnostics.

Correct diagnosis depends on correctly interpreting all data collected and performing all subsequent tests properly. The following information will help:

- Instructions for retrieving DTCs and obtaining freeze frames
- Instructions for diagnosing a no communication problem between the system and the scan tool
- Current technical bulletins for the vehicle
- A fail-safe chart that shows what strategy is taken when certain DTCs are set.
- The operating manual for the scan tool
- A DTC chart with the codes and possible problem areas
- A parts locator for the system
- An electrical wiring diagram for the system
- Identification of the various terminals of the PCM
- DTC troubleshooting guide
- Component testing sequences

Reprogramming Control Modules. Often, there will be a TSB that recommends reprogramming of the computer. This is typically called "**flashing**" the computer. When a computer is flashed, the old program is erased and a new one written in. Reprogramming is often necessary when the manufacturer discovers a common concern that can be solved through changing the system's software. New programs are downloaded into the scan tool and then downloaded into the computer through a dedicated circuit.

Each type scan tool has a different procedure for flashing. Always follow the manufacturer's instructions. Some scan tools are connected to a PC and the software is transferred from a CD or a website. Photo Sequence 6 shows a typical procedure for flashing a BCM.

FLASHING A BCM

All photos in this sequence are © Delmar/Cengage Learning.

P6-1 Use the scan tool to retreive the BCM's part number or the vehicle's VIN and record them.

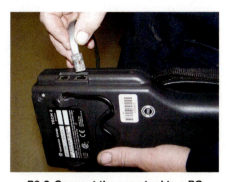

P6-2 Connect the scan tool to a PC that can link the scan tool to the flash software. Some scan tools will connect directly to an internet site.

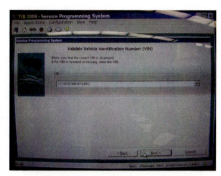

P6-3 Enter the BCM part number in the appropriate field and select "Show Updates" on the menu.

P6-4 Select the desired flash line.

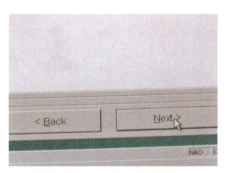

P6-5 Hit the "next" button to begin downloading of the software.

P6-6 Monitor the progress of the downloading of the scan tool.

P6-7 Once downloading is complete, connect a battery charger to the vehicle's battery. Turn it on and maintain about 13.5 volts at the battery.

P6-8 Connect the scan tool to the DLC and turn the scan tool on.

P6-9 Move through the menus on the scan tool until the desired flash screen is shown. Then follow all instructions given on the tool.

Electronic Defaults

When the TCM detects a serious transmission problem or a problem in its circuit (or in some cases an engine control or data bus problem), it may switch to a default, limp-in, or fail-safe mode. Limp-in may also be initiated if the TCM loses its battery power feed. This **default mode** allows for limited driving capabilities and is designed to prevent further transmission damage while allowing the driver to drive with decreased power and efficiency to a service facility for repairs.

While diagnosing a problem in an EAT, always refer to the service information to identify the normal "default" operation of the transmission. You could spend time tracing the wrong problem by not recognizing that the transmission is operating in default.

The capabilities of a transmission while it is in limp-in mode depend on the extent of the fault, the manufacturer, and the model of transmission. When the TCM moves into the limp-in mode, a DTC is set and the transmission will only operate in this mode until the problem is corrected. Examples of operating characteristics during limp-in include the following:

- The transmission locks in third gear when the gear selector is in the Drive position or second gear when it is in a lower position.
- The transmission will remain in whatever gear it was in but will shift into third or second gear and stay there as soon as the vehicle slows down.
- The transmission will only use first and third gears while in the Drive position.
- The transmission will operate only in Park, Neutral, Reverse, and second gears and will not upshift or downshift.

If the malfunction causes improper system operation, the computer may minimize the effects of the malfunction by using fail-safe action. During this, the computer will control a system based on programmed values instead of the input signals. This allows the system to operate on a limited basis instead of shutting down completely.

Symptom-Based Diagnosis

At times, no DTCs are set but a problem exists. To discover the cause of these concerns, the description of the problem or its symptoms should be used to determine what systems or components should be checked. Before diagnosing a problem based on its symptoms, make sure the following:

- The PCM and MIL are operating correctly.
- There are no stored DTCs.
- All data observed on the scan tool are within normal ranges.
- There is communication between the scan tool and the control system.
- There are no TSBs available for the current symptom.
- All of the grounds for the BCM/PCM are sound.
- All vehicle modifications are identified.
- The vehicle's tires are properly inflated and are the correct size.

There is typically a section in the service information dedicated to symptom-based diagnosis. Although a customer may describe a problem in nontechnical terms, you should summarize the concern to match one or more of the various symptoms listed by the manufacturer. Each potential problem area should be checked. It is important to realize that some problems may cause more than one symptom.

Sometimes the symptom will not match any of those described in the service information. This does not mean it is time to guess. It means it is time to clearly identify what is working right. By eliminating those circuits and components that are working correctly from the list of possible causes of the problem, you can identify what may be causing the problem and what should be tested further.

Communication Checks

Performing diagnostic checks on late-model vehicles should include a communications check. If the different control modules are not communicating with each other, there is no way to properly diagnose the systems. When a scan tool is installed, it will try to communicate with every module that could be in the vehicle. If an option is not there, the scan tool will display "No Comm" for that control module. That same message will appear if the module is present but not communicating. Therefore, always refer to the service information to identify what modules should be present before coming to any conclusions.

The system periodically checks itself for communication errors. The different buses send messages to each other immediately after it sends a message. The message between messages checks the integrity of the communication network. All of the modules in the network also receive a message within a specific time. If the message is not received, the control module will set a DTC stating it did not receive the message.

There are three types of DTCs used by CAN buses:

- **Loss of communication**—Loss of communication (and Bus-off) DTCs are set when there is a problem with the communication between modules (Figure 4-36). This could be caused by bad connections, wiring, or the module. Note: In most cases, a lost communication DTC is set in modules other than the module with the communication problem.

DTC	COMPONENT	DESCRIPTION	CONDITION	SYMPTOM
P1657	TCM	TCM communication link error.	Controller area network (CAN) link error detected by PCM.	No engagements, no adaptive or self learning strategy. Limited fuel and spark.
U0073	TCM CAN	TCM communication link error.	CAN link error detected by PCM.	No engagements, no adaptive or self learning strategy.
U0100	TCM CAN	TCM communication link error.	CAN link error detected by PCM.	No engagements, no adaptive or self learning strategy.
U0101	TCM/PCM	TCM communication error.	PCM lost communication with the TCM.	• Engine driveability concerns. • Will turn on MIL and driver warning lamp.
U0121	ABS	TCM communication link error.	PCM/TCM have detected an error in the CAN wheel rpm information from the ABS system.	No engagements, no adaptive or self learning strategy.
U0401	PCM/TCM	Invalid data received from the PCM or engine components.	Data received from the PCM or engine components is not correct for the vehicle operating conditions.	Transmission may enable limp-home strategies or increase pressures. Engine components and PCM may or may not set additional DTCs. P2544, fuel monitor error, ECT sensor failure, and MAF sensor failures may be present. MIL may illuminate.
U0415	Wheel speed sensor	TCM communication link error.	PCM/TCM have detected an error in the CAN wheel rpm information from the ABS system.	No adaptive or self learning strategy.

© Delmar/Cengage Learning

FIGURE 4-36 Examples of some of the DTCs that reflect a problem with network communications.

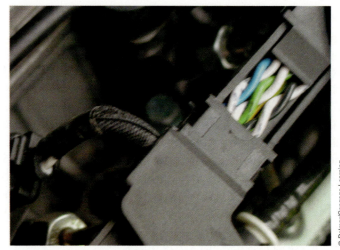

FIGURE 4-37 CAN bus wires are twisted pairs. When making wire repairs, it is important to keep these wires twisted.

- **Signal error**—The control modules can run diagnostics on some input circuits to determine if they are operating normally. If a circuit fails the test, a DTC will set.
- **Internal error**—The modules also run internal checks. If there is a problem, it will set an internal error DTC.

Bus Wire Service. If the bus wire needs repair, due to an open, short, or high resistance, it must not be relocated or untwisted. The twisting serves an extremely important purpose (Figure 4-37). After a bus wire has been repaired by soldering, wrap that part of the wire with vinyl tape. Never run the repair wire in such a way as it bypasses the twisted sections. CAN bus wires are likely to be influenced by noise if you bypass the twisted wires.

TESTING ELECTRONIC CIRCUITS AND COMPONENTS

Diagnosing a computer-controlled system is much more than accessing DTCs. You need to know what to test, when to test it, and how to test it. Most electronic circuits can be checked in the same way as other electrical circuits. However, only high-impedance meters should be used. Further testing may be required to define a particular DTC or when the problem does not set a DTC and when verifying that a component or circuit is faulty.

To pinpoint the exact cause of a transmission problem, you will need to use logic and basic electrical troubleshooting equipment, such as wiring diagrams, diagnostic charts, special transmission testers, and basic electrical testing tools. The operation of some components can be monitored with the scan tool; however, additional tests are normally necessary. These tests include the following:

- **Ohmmeter Checks**—Most sensors and output devices can be checked with an ohmmeter.
- **Voltmeter Checks**—Many sensors, output devices, and their wiring can be diagnosed by checking the voltage to them, and in some cases, from them. Plus voltage drop tests are extremely valuable.
- **Lab Scope Checks**—The activity of sensors and actuators can be monitored with a lab scope or a graphing multimeter. By watching their activity, abnormal behavior can be observed.

 WARNING: Before disconnecting any electronic component, be sure the ignition is turned off. Disconnecting components may cause high induced voltages and computer damage.

Testing Sensors

To monitor conditions, the computer uses a variety of sensors. All sensors perform the same basic function. They detect a mechanical condition (movement or position), chemical state, or temperature and change them into electrical signals that can be used by the PCM.

If a DTC directs you to a faulty sensor or sensor circuit, or if you suspect a faulty sensor, it should be tested. Always follow the correct procedures when testing sensors and other electronic components. Also, make sure you have the correct specifications for each part. Sensors are tested with a DMM, scan tool, and/or lab scope or GMM.

If the preliminary tests pointed to a possible problem in an input circuit, the circuit should be tested. Make sure to check all suspect circuits for resistance problems; conduct voltage drop tests on those circuits. Often, the manufacturers will give specific testing procedures for specific sensors; always follow them.

Some sensors are simple on-off switches. Others are variable resistors that change resistance according to temperature changes. Some are voltage or frequency generators, while others send varying signals according to the rotational speed of a device. Knowing what they are measuring and how they respond to changes are the keys to accurately testing a sensor.

A sensor is any device that sends an input signal to a computer.

Testing Switches

Many different switches are used as inputs or control devices for EATs. Each of the switches serves a different purpose. However, all of these switches either complete or open an electrical circuit. By completing a circuit, these switches either provide a ground for the circuit or connect two wires together. To open the circuit, switches either disconnect the ground from the circuit or disconnect two wires.

Most of the switches are either mechanically or hydraulically controlled. The operation of these switches can be easily checked with an ohmmeter (Figure 4-38). With the meter connected across the switch's leads, there should be continuity or low resistance when the switch is closed and there should be infinite resistance across the switch when it is open. A test light can also be used. When the switch is closed, power should be present at both sides of the switch. When the switch is open, power should be present at only one side.

Most grounding switches react to some mechanical action to open or close. There are some, however, that responds to changes in pressure or temperature. An example of this type of switch is the power steering pressure switch. This switch informs the PCM when power steering pressures reach a particular point. When the power steering pressure exceeds that point, the PCM knows there is an additional load on the engine and will increase idle speed.

A switch can be easily tested with an ohmmeter. Disconnect the connector at the switch. Refer to the wiring diagram to identify the terminals at the switch if there are more than two.

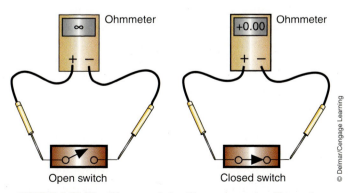

FIGURE 4-38 Checking a switch with an ohmmeter. The meter is placed in parallel with the component after the power is removed from the circuit.

Connect the ohmmeter across the switch's terminals. Perform whatever action is necessary to open and close the switch.

Switches can also be checked with a voltmeter. The signal to the PCM from supply side switches should be 0 volts with the switch open and supply voltage when the switch is close. Using a voltmeter is preferred because it tests the circuit as well as the switch. If less than supply voltage is present with the switch is closed, there is unwanted resistance in the circuit. Expect the opposite readings on a ground side switch.

TR Sensor. The Manual Lever Position or Transmission Range (TR) sensor is a switch that provides information to the computer as to what operating range has been selected by the driver. Based on that information, the computer determines the proper shift strategy for the transmission. The following is a list of problems that may result from a faulty TR Sensor:

- No upshifting
- Slipping out of gear
- High line pressure in transmissions equipped with a pressure control solenoid
- Delayed gear engagements
- Engine starts in other lever positions besides Park and Neutral

Because this switch is open or closed, depending on position, it can be checked with an ohmmeter. By referring to a wiring diagram, you should be able to determine when the switch should be open. Doing this for some switch designs may be a little difficult as the switch assembly may be made up of more than one switch. Then connect the meter across the input and out of the switch. Move the lever into the desired position and measure the resistance (Figure 4-39). An infinite reading is expected when the switch is open. If there is any resistance, the switch should be replaced. The switch should also be replaced if there is some resistance when the switch is closed.

Pressure Switch. The pressure switches (Figure 4-40) used in today's transmissions either complete or open an electrical circuit. Normally, open switches will have no continuity across the terminals until oil pressure is applied to it. Normally, closed switches will have continuity across the terminals until oil pressure is applied to it. Refer to the wiring diagram to determine the type of switch and test the switch with an ohmmeter. There should be continuity when the switch is closed and no continuity when the switch is open. Base your expected results on the type of switch you are testing.

Pressure switches can be checked by applying air pressure to the part that would normally be exposed to oil pressure. When applying air pressure to these switches, check them for leaks. Although a malfunctioning electrical switch will probably not cause a shifting problem, it will, if it leaks. If the switch leaks off the applied pressure in a hydraulic circuit to a holding device, the holding member may not be able to function properly.

When possible, you should check pressure switches when they are installed and controlled by the vehicle. By watching an ohmmeter connected to a governor switch, you can check its spring rating. Connect the ohmmeter to the "A" terminal and ground. Bring the engine's speed up. When the ohmmeter reads about 25 ohms, the governor switch is closed. The speed at which this occurs is the speed required to close the grounding switch. Other grounding type switches can be checked in the same way. However, it is important that you identify if it is a normally open or closed switch before testing.

Temperature Switch. Temperature responding switches operate in the same way. When a particular temperature is reached, the switch opens. This type of switch is best measured by removing it, connecting it to an ohmmeter, and submerging it in heated water. A good temperature responding switch will open (have an infinite reading) when the water temperature reaches the specified amount. If the switch fails this test, it should be replaced.

When possible, you should check pressure switches when they are installed and controlled by the vehicle.

CAUTION:
Always disconnect an electrical switch or component from the circuit or from a power source before testing it with an ohmmeter. Failure to do so will damage the meter.

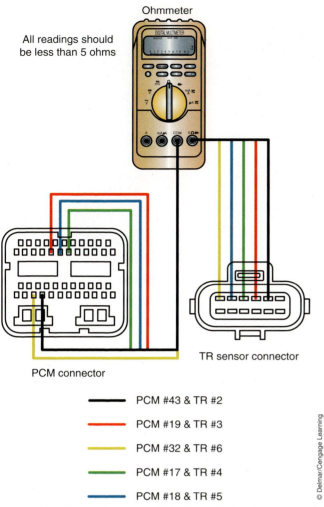

All readings should be less than 5 ohms

Ohmmeter

PCM connector

TR sensor connector

	PCM #43 & TR #2
	PCM #19 & TR #3
	PCM #32 & TR #6
	PCM #17 & TR #4
	PCM #18 & TR #5

FIGURE 4-39 Checking a TR switch at the PCM and TR sensor connectors. The colors mark the different locations for placing the meter's leads to test the switch.

© Delmar/Cengage Learning

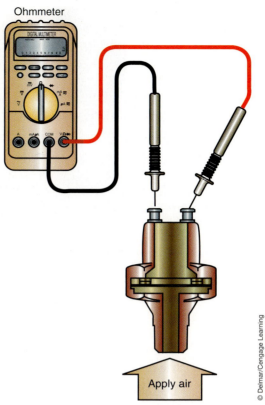

Ohmmeter

Apply air

© Delmar/Cengage Learning

FIGURE 4-40 To check a normally open pressure switch, apply air to the bottom oil passage and check for continuity across the terminals.

Testing Temperature Sensors

Temperature sensors are designed to change resistance with changes in temperature. A temperature sensor is based on a thermistor. Some thermistors increase resistance with an increase of temperature. Others decrease the resistance as temperature increases. The PCM changes the operation of many systems based on temperature. Temperature sensor problems are often caused by wiring faults or loose or corroded connections. Nearly all temperature sensors are NTC thermistors. Typically, the PCM supplies a reference voltage of 5 volts to the sensor. That voltage is changed by the change of the sensor's resistance and is fed back to the PCM. Based on the return voltage, the PCM calculates the exact temperature. When the sensor is cold, its resistance is high, and the return voltage signal is also high. As the sensor warms up, its resistance drops and so does the voltage signal. Many testers are able to show where to place the probes of the tester to check sensors, such as an ECT (Figure 4-41).

Temperature sensor circuits should be tested for opens, shorts, and high resistance. Often, a DTC will be set. Scan tool data should also be looked at. If the observed temperature is the coldest possible, the circuit is open. If the temperature is the highest possible, the circuit has a short. Also, if the connector to the sensor is disconnected, the readings should drop to cold. High resistance problems will cause the PCM to respond to a lower temperature than the actual temperature. This can be verified by using a good thermometer (infra-red is best)

An NTC thermistor is a Negative Temperature Coefficient thermistor and a PTC is a Positive Temperature Coefficient thermistor that reacts in the opposite way as an NTC.

FIGURE 4-41 Many scan tools can show where to place the meter's leads when testing an input or output.

to measure the temperature and compare it to the scan tool readings. There will be a slight difference between the two if the sensor circuit is working properly. Unwanted resistance in the circuit can cause delayed or poor shifting.

DMM Checks. Obviously, these sensors can be checked with an ohmmeter. To do so, remove the sensor. Use an infrared temperature probe to measure its temperature and measure the resistance across it. Compare your reading to the chart of normal resistances given in the service information (Figure 4-42). With the sensor installed, its terminals may be back-probed to connect a voltmeter to the sensor terminals. The sensor should provide the specified voltage drop at any temperature.

The sensors can also be tested by removing them and placing them into a container of water with an ohmmeter connected across the sensor terminals. A thermometer is also placed in the water. When the water is heated, the sensor should have the specified resistance at any temperature. If the sensor does not have the specified resistance, replace the sensor.

Then, disconnect the wiring from the sensor to the computer, connect an ohmmeter from each sensor terminal to the appropriate computer terminal. Both sensor wires should indicate less resistance than specified by the manufacturer. If the wires have high resistance, the wires or wiring connectors must be repaired.

°F	°C	Resistance in ohms
−40 to −4	−40 to −20	967k to 284k
−3 to 31	−19 to −1	284k to 100k
32 to 68	0 to 20	100k to 37k
69 to 104	21 to 40	37k to 16k
105 to 158	41 to 70	16k to 5k
159 to 194	71 to 90	5k to 2.7k
195 to 230	91 to 110	2.7k to 1.5k
231 to 266	111 to 130	1.5k to 0.8k
267 to 302	131 to 150	0.8k to 0.54k

FIGURE 4-42 An example of acceptable resistance readings for a fluid temperature sensor when it is at various temperatures.

WARNING: Never apply an open flame to a temperature sensor for test purposes. This action will damage the sensor.

Lab Scope Checks. Thermistor activity can be monitored with a lab scope. Connect the scope across to the output of the thermistor or temperature sensor. Run the engine and watch the waveform. As the temperature increases, there should be a smooth increase or decrease in voltage. Look for glitches in the signal. These can be caused by changes in resistance or an intermittent open.

Testing Pressure Sensors

Most pressure sensors are piezometric sensors (Figure 4-43). A silicon chip in the sensor flexes with changes in pressures. One side of the chip is exposed to a reference pressure, which is either a perfect vacuum or a calibrated pressure. The other side is the pressure that will be measured. As the chip flexes in response to pressure, its resistance changes. This changes the voltage signal sent to the PCM. The PCM looks at the change and calculates the pressure change.

DMM Checks. Normally, the PCM sends a voltage reference signal to a pressure sensor. With the ignition switch on, backprobe the reference wire and measure the voltage. If the reference wire does not have the specified voltage, check the reference voltage at the PCM. If the voltage is within specifications at the PCM, but low at the sensor, repair the wire. When this voltage is low at the PCM, check the voltage supply wires and ground wires for the PCM. If the wires are good, replace the computer.

Then connect the voltmeter from the sensor's ground wire to the battery ground. If the voltage drop across this circuit exceeds specifications, repair the ground wire from the sensor to the computer. To check the voltage signal of a pressure sensor, turn the ignition switch on and connect a voltmeter to the sensor signal wire. Operate the transmission in a way that should cause the monitored pressure to change. If there was no change in the voltage signal or if it was not what was expected, replace the sensor.

Some pressure sensors use a diaphragm to sense pressure changes. As the diaphragm moves, the resistance of the sensor changes. To test these, remove the sensor and apply varying amounts of air pressure to the port below the diaphragm. Observe the resistance changes with an ohmmeter. Compare your findings to the specifications, if they are available.

Lab Scope Checks. To check a pressure sensor with a lab scope, connect the scope to the sensor's output and a good ground. Operate the transmission in a way that should cause the

> The term *Piezoresistive* represents the characteristic of something that changes resistances in relationship to changes in pressure.

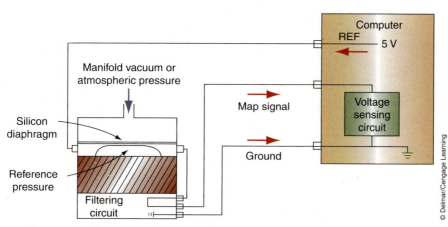

FIGURE 4-43 A piezoresistive sensor is made up of a silicon diaphragm sealed in a quartz plate.

monitored pressure to change. If there was no change in the voltage signal or if it was not what was expected, replace the sensor. If the signal is erratic, the sensor or sensor wires are defective.

Testing Speed Sensors

EATs rely on electrical signals from a speed sensor to control shift timing (Figure 4-44). The use of this type of sensor negates the need for hydraulic signals from a governor. When this sensor fails or sends faulty readings, it can cause complaints that include no overdrive, no converter-clutch engagement, and no upshifts.

There are basically two types of speed sensors, permanent magnetic (PM) generator sensors and Hall-effect switches. Identifying the type of sensor used in a particular application dictates how the sensor should be tested.

If the PCM does not receive a speed signal it will set a DTC, it may also set a code if the signal does not correlate with other inputs. Check the wiring and connectors at the sensor and the control modules. Make sure the connections are tight and not damaged.

DMM Checks. Speed sensors can be checked with an ohmmeter. Most manufacturers list a resistance specification. The resistance of the sensor is measured across the sensor's terminals (Figure 4-45). The typical range for a good sensor is 800–1400 ohms of resistance. If there is no continuity, the sensor is open and should be replaced. Reposition the leads of the meter so that one lead is on the sensor's case and other to a terminal. There should be no continuity in this position. If there is any measurable amount of resistance, the sensor is shorted.

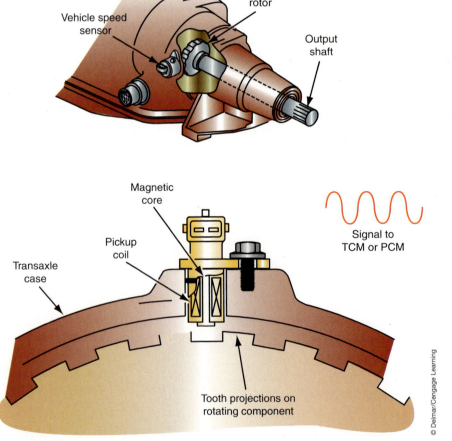

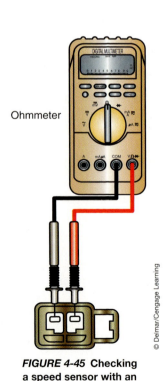

FIGURE 4-44 The vehicle speed sensor will display a pulse each time a tooth of the rotor passes by the sensor.

FIGURE 4-45 Checking a speed sensor with an ohmmeter.

A speed sensor can also be checked with the vehicle on a hoist. The vehicle should be positioned so the drive wheels are free to rotate. Backprobe the sensor's output wire and connect the voltmeter leads from this wire to ground. Select the 20 V AC scale on the meter. Then start the engine. Put the transmission in a forward gear and observe the meter. If the sensor's voltage is not 0.5 volts or more, replace the sensor. If the signal is correct, backprobe the sensor's terminal at the PCM and measure the voltage with the wheels rotating. If 0.5 volts is at this terminal, the trouble may be in the PCM.

 WARNING: **While conducting this procedure, it is possible to damage the CV joints if the drive wheels are allowed to dangle in the air. Place safety stands under the front suspension arms to maintain proper drive axle operating angles. Also, when running the vehicle on a lift make sure you stay clear of the wheels and other rotating parts. If you are caught by something powered by the engine, serious injury can result.**

When 0.5 volts is not available, turn the ignition off and disconnect the wire from the sensor to the PCM. Connect an ohmmeter across the wire. The meter should read 0 ohms. Repeat the test with the leads connected to the sensor's ground and the PCM ground terminal. This wire should also have 0 ohms. If the resistance of these wires is more than specified, repair the wires.

Lab Scope Checks. Magnetic pulse generators can be tested with a lab scope. Connect the scope's leads across the sensor's terminals. The expected pattern is an AC signal, which should be a perfect sine wave when the speed is constant (Figure 4-46). When the speed is changing, the AC signal should change in amplitude and frequency. If the readings are not steady and do not smoothly change with a change in speed, suspect a faulty connector, wiring harness, or sensor.

Hall-Effect Sensors. To test a Hall-effect switch, disconnect its wiring harness. Connect a low-voltage source across the positive and negative terminals of the Hall layer. Then connect a voltmeter across the negative and signal voltage terminals. Insert a metal feeler gauge

A Hall-effect switch produces a voltage pulse when there is the presence of a magnetic field. The Hall-effect is the consequence of moving current through a thin conducting that is exposed to a magnetic field; as a result of this, voltage is produced.

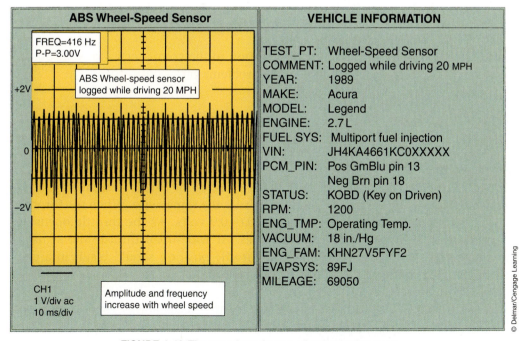

FIGURE 4-46 The waveform from a wheel speed sensor.

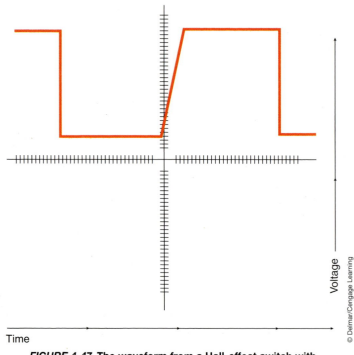

FIGURE 4-47 The waveform from a Hall-effect switch with a bad transistor.

between the Hall layer and the magnet. Make sure the feeler gauge is touching the Hall element. If the sensor is operating properly, the meter will read close to battery voltage. When the feeler gauge blade is removed, the voltage should decrease. On some units, the voltage will drop to near zero. Check the service information to see what voltage you should observe when inserting and removing the feeler gauge.

When observing a Hall-effect sensor on a lab scope, pay attention to the downward and upward pulses. These should be straight. If they appear at an angle (Figure 4-47), this indicates the transistor is faulty causing the voltage to rise slowly. The waveform should be a clean and flat square wave. Any change from a normal trace means the sensor should be replaced.

Testing Throttle Position Sensors

A throttle position (TP) sensor sends information to the computer as to what position the throttle is in. If this sensor fails, the following problems can result:

- No upshifts
- Quick upshifts
- Delayed shifts
- Line-pressure problems with transmissions that have a line pressure control solenoid
- Erratic converter clutch engagement

A basic TP sensor has three wires. One wire carries the 5-volt reference signal, another serves as the ground for the resistor, and the third is the signal wire. When the throttle plates are closed, the signal voltage will be around 0.6–0.9 volts. As the throttle opens, there is less resistance and the voltage signal increases. At wide-open throttle the signal will be approximately 3.5–4.7 volts. All changes in throttle position should result in a change in voltage. Often, the connector terminals for the sensor are gold plated. The plating makes the connector more durable and corrosion resistant.

Be careful, some TP sensors have four wires. The additional wire is connected to an idle switch. Normally, when the switch is closed there will be 0 volts and battery voltage when the switch is open. Check the wiring diagram before deciding if the switch and circuit is good.

Scan Tool Checks. The action of a TP sensor can also be monitored on a scan tool. Compare the position, expressed in a percentage, to the voltage specifications for that throttle position.

A faulty TP sensor may not cause a DTC to be set. The diagnostic capabilities of the PCM must be able to determine if the sensor is working correctly or not. If it does not have this capability, it may not set a code. The PCM must be able to look at the input from the TP sensor and compare it to other inputs, such as engine speed, MAP inputs, and airflow. If the PCM determines that the TP signal does not reflect a true value based on engine speed and load inputs, it will set a code.

DMM Checks. With the ignition on, connect a voltmeter from the sensor signal wire to ground. Slowly open the throttle and observe the meter. The reading should increase smoothly and gradually. If the TP sensor does not have the specified voltage at a particular throttle opening or if the voltage signal is erratic, replace the sensor.

Connect the meter between the reference wire and the ground. Normally, the voltage should be 5 volts. If the reference wire is not supplying the specified voltage, check the voltage on this wire at the computer terminal. If the voltage is within specifications at the computer, but low at the sensor, repair the reference wire. When the voltage is low at the computer, check the voltage supply wires and ground wires on the computer. If these wires are satisfactory, replace the computer.

A TP sensor can be checked with an ohmmeter or a voltmeter. If checked with an ohmmeter, you should be able to watch the resistance of the sensor change as the throttle is opened and closed. Often, there will be a resistance specification given in the service information. Compare your reading to this.

Lab Scope Checks. Testing a TP sensor with a lab scope is a good way to watch the sweep of the resistor. The waveform is a DC signal that moves up as the voltage increases. Most potentiometers in computer systems are fed a reference voltage of 5 volts. Therefore, the voltage output of these sensors will range from 0.5 to 4.5 volts. The change in voltage should be smooth. Connect the scope to the sensor's output and a good ground and watch the trace as the throttle is opened and closed. The resulting trace should look smooth and clean, without any sharp breaks or spikes in the signal (Figure 4-48). A bad sensor will typically have a glitch

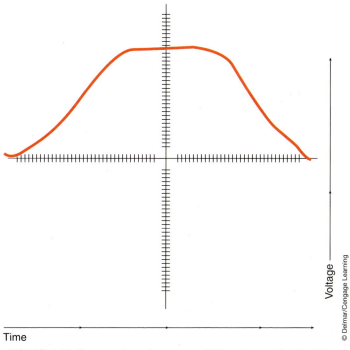

FIGURE 4-48 The waveform from a good TP sensor as the throttle is opened and closed.

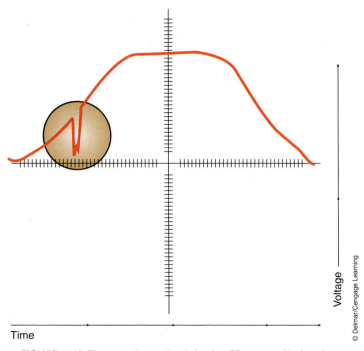

Voltage

Time

FIGURE 4-49 The waveform of a defective TP sensor. Notice the glitch while the throttle opens.

(a downward spike) somewhere in the trace (Figure 4-49) or will not have a smooth transition from high to low. These glitches are an indication of an open or short in the sensor. Also look for sudden changes in voltage. These can be caused by changes in resistance or an intermittent open in the circuit or sensor.

Mass-Airflow Sensor

A mass-airflow sensor is used to determine engine load by measuring the mass of the air being taken into the throttle body. When this sensor fails or sends faulty signals, the engine runs roughly and tends to stall as soon as you put the transmission into gear. The transmission may also not respond to changes in loads, which can lead to wrong gear operation.

The mass-airflow sensor is typically a wire, located in the intake air stream that receives a fixed voltage. The wire is designed so that it changes resistance in response to temperature changes. When the wire is hot, the resistance is high and less current flows through the wire. When the wire is cold, the resistance is low and larger amounts of current pass through it. The amount of air passing over the wire determines the amount of resistance the wire has. The computer monitors the amount of current flow and interprets the flow as the mass or volume of the air.

This sensor can be checked with a multimeter set to the Hz frequency range. Check the service information for specific values. Normally, at idle, 30 Hz is measured and the frequency will increase as the throttle opens.

A scan tool may also be used to test this sensor; most have a test mode that monitors MAF sensors. The output of some MAFs can be observed with a DMM; their output is variable DC voltage. While diagnosing these systems, keep in mind that cold air is denser than warm air.

DIAGNOSIS OF COMPUTER VOLTAGE SUPPLY AND GROUND WIRES

Basically, if the computer system's inputs and outputs and the engine and hydraulic and mechanical functions of the transmission are all good, but a problem exists, the computer or its circuit is bad. Replace the computer only as a last resort. Although there are DTCs that

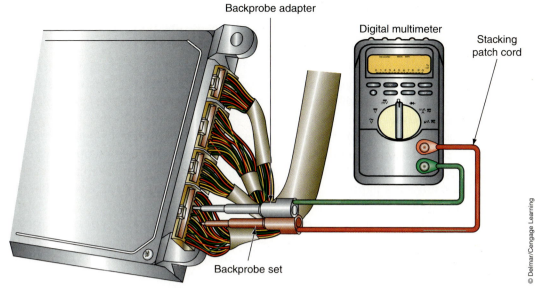

Backprobe adapter

Digital multimeter

Stacking
patch cord

© Delmar/Cengage Learning

Backprobe set

FIGURE 4-50 Using a digital multimeter to check the PCM's circuit.

relate directly to the PCM or TCM, it may be difficult to identify a faulty computer. Since the computer reacts to the inputs it receives, abnormal input signals may result in abnormal computer operation. Also, most computers are designed to react to out-of-range inputs by ignoring the signals and using a fixed value held in its memory. This results in less than efficient operation.

All PCMs cannot operate properly unless they have good ground connections and the correct voltage at the required terminals. A wiring diagram for the vehicle must be used for the following tests. Backprobe the battery terminal at the PCM and connect a digital voltmeter from this terminal to ground (Figure 4-50). Always ground the black meter lead.

The voltage at this terminal should be 12 volts with the ignition switch off. If 12 volts are not available at this terminal, check the computer's fuse and related circuit. Turn the ignition on and connect the red meter lead to the other battery terminals at the PCM with the black lead still grounded. The voltage measured at these terminals should also be 12 volts. When the specified voltage is not available, conduct a voltage drop test to determine the location of the unwanted resistance in the voltage supply wires to these terminals. These terminals may be connected through fuses, fuse links, or relays.

If the computer is receiving battery voltage, check the computer's ground. The best way of doing this is to place one probe of a voltmeter on the case or ground connection of the computer and the other at a known good ground. With the system powered, the meter should read less than 0.1 volt. If the reading is higher than that, conduct voltage drop tests on the rest of the ground circuit.

Ground Circuits

Ground wires usually extend from the computer to a ground connection on the engine or battery. With the ignition on, connect a voltmeter from the battery ground to the computer ground. The voltage drop across the ground wires should be 30 millivolts or less. If the voltage is greater than that, repair the ground wires or connection.

Not only should the computer ground be checked, but so should the ground (and positive) connection at the battery. Checking the condition of the battery and its cables should always be part of the initial visual inspection before diagnosing an engine control system.

A good ground is especially critical for all reference voltage sensors. The problem here is not obvious until it is thought about. A bad ground will cause the reference voltage (normally

SERVICE TIP:
Never replace a computer unless the ground wires and voltage supply wires are proven to be in satisfactory condition.

5 volts) to be higher than normal. As a result, the computer will be making decisions based on the wrong information. If the output signal is within the normal range for that sensor, the computer will not notice the wrong information and will not set a DTC.

EMI is something that occurs when voltage changes in a nearby wire or component and influences the voltage in the wire or component that it is close to.

Electrical Noise. Poor grounds can also allow EMI or noise to be present on the reference voltage signal. This noise causes small changes in the voltage going to the sensor. Therefore, the output signal from the sensor will also have these voltage changes. The best way to check for noise is to use a lab scope.

Connect the lab scope between the 5-volt reference signal wire at the sensor and its ground. The trace on the scope should be flat. If noise is present, move the scope's negative probe to a known good ground. If the noise disappears, the sensor's ground circuit is bad or has resistance. If the noise is still present, the voltage feed circuit is bad or there is EMI in the circuit from another source.

Circuit noise may be evident by a flickering MIL, a popping noise on the radio, or by an intermittent engine miss. The most common sources of noise are electric motors, relays and solenoids, AC generators, ignition systems, switches, and A/C compressor clutches. Typically, noise is the result of an electrical device being turned on and off.

Sometimes the source of the noise is a defective suppression device. Manufacturers include these devices to minimize or eliminate electrical noise. Some of the commonly used noise suppression devices are resistor-type secondary cables and spark plugs, shielded cables, capacitors, diodes, and resistors. Capacitors or chokes are used to control noise from a motor or generator. If the source of the noise is not a poor ground or a defective component, check the suppression devices.

DETAILED TESTING OF ACTUATORS

Once the TCM determines that a correction or adjustment must be made to the system, an output signal is sent to a control device or actuator. These actuators are as follows: solenoids, switches, relays, or motors; physically act or carry out the command sent by the TCM.

Actuators are electromechanical devices that convert an electrical current into mechanical action. This action can be used to open and close valves, engage or disengage gears, control vacuum to other components, or open and close switches. When the TCM receives an input signal indicating a change in one or more of the operating conditions, the TCM determines the best strategy for handling the conditions. The TCM then controls a set of actuators to achieve a desired effect or strategy goal. In order for the computer to control an actuator, it must rely on a component called an output driver.

The driver usually completes the ground circuit of the actuator. The ground can be applied steadily if the actuator must be activated for a selected amount of time. Or the ground can be pulsed to activate the actuator in pulses. Output drivers are transistors or groups of transistors that control the actuators. These drivers operate by the digital commands from the TCM. If an actuator can't be controlled digitally, the output signal must pass through an A/D converter before flowing to the actuator.

Most systems allow for testing an actuator through a scan tool. Actuators that are duty cycled are more accurately diagnosed this way. Serial data can be used to diagnose outputs. The data should be compared against specifications to determine the condition of an actuator. Also, when an actuator is suspected as being faulty, make sure the inputs related to the control of that actuator are within normal range. Faulty inputs will cause an actuator to appear faulty.

Many systems have operating modes that can be accessed with a scan tool to control the operation of an output. Common names for this mode are the output state control (OSC) and output test mode (OTM). In this mode, an actuator can be enabled or disabled or the duty cycle or the movement of the actuator can be increased or decreased. While the actuator is being controlled, related PIDs are observed as an indication of the how the system reacted to the changes. The actuators that can be controlled by this mode vary. Always refer to the service information to determine what can be checked and how it should be checked. Here is a summary of typical output controls and what happens on a late model vehicle:

TCC PC Solenoid

- The TCM commands the TCC PC solenoid pressure to apply and release the TCC.
- When the ignition is on and engine off, there are no limits to this control. The solenoid remains on until commanded OFF.
- When the engine is running, the solenoid cannot be commanded OFF and when the transmission is in Park, it cannot be commanded OFF for an extended time.

Shift Solenoids 1 and 2

- The TCM commands shift solenoids ON or OFF.
- When the ignition is on and the engine is off, the solenoid remains on until commanded OFF.
- When the engine is running, the transmission must be in Park or Neutral. Also there cannot be any active transmission DTCs.

Line PC Solenoid

- The scan tool can be used to request pressure in increments of 30 psi (207 kPa) from 0–270 psi (0–1862 kPa).
- When the ignition is on and the engine is off, there are no limits to the output control.
- When the engine is running and the transmission is in Park or Neutral, the engine's speed must be less than 1500 rpm. When the transmission is not in Park or Neutral, the TCM does not allow a pressure that may cause damage to the transmission. Also there cannot be any active transmission DTCs.

PC Solenoids 2, 3, 4, & 5

- The TCM commands the PC solenoids to apply or release the clutches.
- When the ignition is on and the engine is off, the solenoid remains on until commanded OFF.
- Also there cannot be any active transmission DTCs.

High Side Driver 2

- The TCM commands the driver to send voltage to the solenoid.
- When the ignition is on and the engine is off, there are no limits to this control.
- When the engine is running, the high side drivers cannot be commanded ON or OFF.

Testing with a DMM

Some actuators are easily tested with a voltmeter by checking input voltage at the actuator. If there is the correct amount of input voltage, check the condition of the ground. If both of these are good, then the actuator is faulty.

When checking anything with an ohmmeter, logic can dictate good and bad readings. If the meter reads infinite, this means there is an open. Based on what you are measuring across, an open could be good or bad. The same is true for very low resistance readings. Across some things, this would indicate a short. For example, you do not want an infinite reading across the windings of a solenoid. You want low resistance. However, you want an infinite reading from one winding terminal to the case of the solenoid. If you have low resistance, the winding is shorted to the case.

Testing with a Lab Scope

When actuators are faulty, it is because they are electrically faulty or mechanically faulty. By observing the action of an actuator on a lab scope, you will be able to watch its electrical

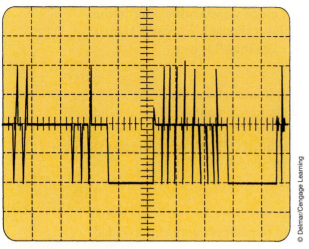

FIGURE 4-51 A typical control signal for a pulse width modulated solenoid.

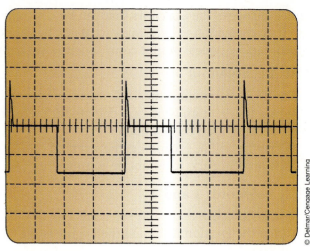

FIGURE 4-52 A typical control signal for a solenoid.

activity. Normally if there is a mechanical fault, this will affect its electrical activity as well. Therefore, you get a good sense of the actuator's condition by watching it on a lab scope.

Some actuators are controlled pulse-width modulated signals (Figure 4-51). These signals show a changing pulse width. These devices are controlled by varying the pulse width, signal frequency, and voltage levels. By watching the control signal, you can see the turning on and off of the solenoid (Figure 4-52). The voltage spikes are caused by the discharge of the coil in the solenoid.

Both waveforms should be checked for amplitude, time, and shape. You should also observe changes to the pulse width as operating conditions change. A bad waveform will have noise, glitches, or rounded corners. You should be able to see evidence that the actuator immediately turns off and on according to the commands of the computer.

Make sure you follow the instructions from the scope's manufacturer. If the scope is set wrong, the scope will not break. It just will not show you what you want to be shown. To help with understanding how to set the controls on a scope, keep the following things in mind: The vertical voltage scale must be adjusted in relation to the expected voltage signal; the horizontal time base or milliseconds per division must be adjusted so the waveform appears properly on the screen.

Minor adjustments of the trigger line may be necessary to position the waveform in the desired vertical position. Trigger slope indicates the direction in which the voltage signal is moving when it crosses the trigger line. A positive trigger slope means the voltage signal is moving upward as it crosses the trigger line, whereas a negative trigger slope indicates the voltage signal is moving downward when it crosses the trigger line.

Software packages, often programmed in a lab scope or GMM, are available to help you properly interpret scope patterns and set up the scope. These also contain an extensive waveform library that you can refer to and find what the normal waveform of a particular device should look like. The library also contains the waveforms caused by common problems. You can also add to the library by transferring waveforms to a PC from the lab scope. After the waveforms have been transferred, notes can be added to the file. The software may also include the theory of operation and diagnostic procedures for common inputs and outputs.

TESTING SOLENOIDS

Most scan tools have the capability of controlling shift points by manually controlling shift solenoids. If your scan tool does not have this feature, special testers are available to do this. This check operates the solenoids on command thereby bypassing the computer. Doing this, you can observe the operation of the solenoids and the transmission. If the customer's complaint is still evident when the computer is bypassed, you know the problem is not the computer. If the problem is not evident, you know the problem is the computer and/or computer circuit.

Before continuing, however, you must first determine if the solenoids are case grounded and fed voltage by the computer or if they always have power applied to them and the computer merely supplies the ground. While looking in the service information to find this, also find the section that tells you which solenoids are on and which are off for each of the different gears.

The special solenoid testing tools connect into the solenoid assembly and allow the technician to switch gears by depressing or flicking a switch. This check can be conducted on all transmission solenoids except those that have a duty cycle. Check the service information to identify this type of solenoid. Photo sequence 7 shows a typical procedure for checking solenoids with solenoid tester.

If the transmission shifted fine with the movement of the switches, you know that the transmission is fine. Any shifting problem must be caused by something electrical. If the transmission did not respond to the switch movements, the problem is probably in the transmission.

At times, a solenoid valve will work fine during light throttle operation but may not exhaust enough fluid when pressure increases. To verify that the valve is not exhausting, activate the solenoid then increase engine speed while pulling on the throttle cable. If the solenoid valve cannot exhaust, the transmission will downshift. Restricted solenoids should be suspect whenever the transmission shifts roughly or early under heavy loads or full throttle but shifts fine under light throttle.

Solenoids can be checked for circuit resistance and shorts-to-ground. This test can be conducted at the transmission case connector. By identifying the proper pins in the connector, individual solenoids can be checked with an ohmmeter (Figure 4-53). Remember, lower than normal resistance indicates a short, whereas higher than normal resistance indicates

SERVICE TIP:
This type of tester is easily made. Simply secure a harness for the transmission you want to test. Connect the leads from the harness to simple switches. Follow the solenoid/gear pattern when doing this. To change gears, all you will need to do is turn off one switch and turn on the next.

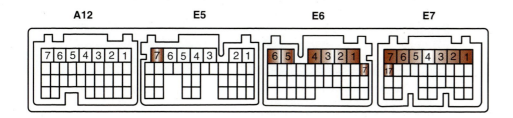

Terminal no. (symbol)	Color	Terminal description
E6-6 (SL1+) - E6-5 (SL1-)	Y - L	Shift solenoid valve SL1 signal
E6-4 (SL2+) - E6-3 (SL2-)	BR - Y	Shift solenoid valve SL2 signal
E6-2 (SL3+) - E6-1 (SL3-)	L - B	Shift solenoid valve SL3 signal
E7-7 (SL4+) - E7-6 (SL4-)	B - W	Shift solenoid valve SL4 signal
E7-5 (SL5+) - E7-4 (SL5-)	L - BR	Shift solenoid valve SL5 signal
E7-3 (SLT+) - E7-2 (SLT-)	B - G	Shift solenoid valve SLT signal
E7-1 (SLU+) - E7-8 (SLU-)	W - L/R	Shift solenoid valve SLU signal
E6-7 (SL) - E5-7 (E1-)	L/R - BR	Shift solenoid valve SL signal
E7-17 (SR) - E5-7 (E-)	G - GR	Shift solenoid valve SR signal

© Delmar/Cengage Learning

FIGURE 4-53 Solenoid-related pins and testing points at the TCM for a Toyota AA80E transmission.

199

CHECKING THE CONDITION OF SOLENOIDS WITH A SOLENOID TESTER

All photos in this sequence are © Delmar/Cengage Learning.

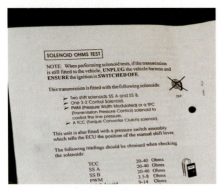

P7-1 In the service manual, locate the wiring diagram for the solenoids and determine if the solenoids are power controlled or ground controlled. While looking in the service manual, locate the section that tells you which solenoids are on and which are off for each of the different gears.

P7-2 The solenoid tester.

P7-3 Make sure the ignition switch is off and disconnect the car's wiring harness that leads to the solenoids.

P7-4 If the solenoids are controlled by their ground, connect a jumper wire with an in-line 20-amp fuse to the tester and a known good ground.

P7-5 Complete the connection of the tester to the transmission and then program it according to the tester's instructions.

P7-6 Start the engine and move the shift lever into "drive."

P7-7 Now turn on solenoid #1; the transmission should be in first gear. Pay attention to how it shifted.

P7-8 Increase your speed, and then turn on solenoid #2. The transmission should have immediately shifted into second gear. Pay attention to how it shifted.

P7-9 When third gear is selected, solenoid #1 will be turned off. Pay attention to how it shifted.

P7-10 Likewise, when you want fourth gear, solenoid #2 will be turned off. Pay attention to how it shifted.

P7-11 After checking all gears under light, half, and full throttle, return to the shop and summarize your findings.

P7-12 Disconnect the tester from the system and proceed to check solenoid circuits that seemed not to work correctly.

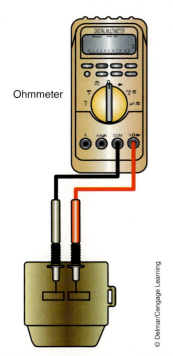

Ohmmeter

© Delmar/Cengage Learning

FIGURE 4-54 Checking a solenoid for a short to ground.

a problem of high resistance. If you get an infinite reading across the solenoid, the solenoid windings are open. The ohmmeter can also be used to check for shorts-to-ground. Simply connect one lead of the ohmmeter to one end of the solenoid windings and the other lead to a good ground (Figure 4-54). The reading should be infinite. If there is any measurable resistance, the winding is shorted to ground.

Solenoids can also be tested on a bench. Resistance values are typically given in the appropriate service information (Figure 4-55). A solenoid may be electrically fine but still may fail mechanically or hydraulically. A solenoid's check valve may fail to seat or the porting

Sump temperature		Resistance in Ohms	
°F	°C	TCC PWM	Shift and PC solenoids
32	0	9.5	20.0
68	20	10.5	22.0
104	40	11.5	24.5

FIGURE 4-55 The resistance specifications for the solenoids in an Allison series 1000 transmission.

can be plugged. This is not an electrical problem, rather it could be caused by the magnetic field collecting metal particles in the ATF and clogging the port or check valve. These would cause erratic shifting, no shift conditions, wrong gear starts, no or limited passing (kickdown) gear, or binding shifts. When a solenoid affected in this way is activated, it will make a slow dull thud. A good solenoid tends to snap when activated.

Chrysler Solenoids

The typical Chrysler solenoid assembly consists of four solenoids, four ball valves, three pressure switches, and three resistors. The ball valves are operated by the solenoids and together they cause the transaxle to shift. The solenoids should be checked for proper operation. This can be done through the eight-pin connector at the assembly (Figure 4-56).

Using an ohmmeter with the harness disconnected from the assembly, check the resistance across pins 1 and 4, 2 and 4, and 3 and 4. The resistance should be between 270 and 330 ohms. With this test, you have checked the resistance for each of the switches. To check the action of these switches, apply air pressure to the feed holes for the overdrive, low-reverse, and overdrive pressure switches. With air applied, the resistance reading across each of the pin combinations should now be 0 ohms.

Each of the four solenoids can also be checked with an ohmmeter. Connect the meter across pins 4 and 5, 4 and 6, 4 and 7, and 4 and 8. The resistance across any of these combinations should be 1.5 ohms.

Ford Solenoids

Ford's E4OD can be shifted without the computer by energizing the #1 and #2 shift solenoids in the correct sequence. Refer to Figure 4-57 for the location of the pins in the E4OD harness connector. Connect a 12-volt power source to pin #1, then connect a ground lead to the following pins to shift the transmission: First gear—Pin #3, Second gear—Pin #3 and #2, Third gear—Pin #2, and Fourth gear—no pins grounded. If the transmission does not shift, check the solenoids.

Using an ohmmeter, check the resistance of the #1 solenoid by measuring the resistance across pins #1 and #3. To check solenoid #2, measure across pins #1 and #2. And to check

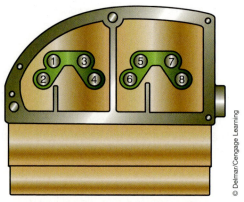

FIGURE 4-56 The connector at the solenoids in a typical Chrysler transmission.

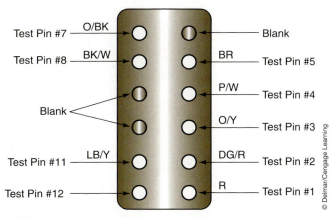

FIGURE 4-57 The transmission harness connector for a Ford E4OD. The pin numbers are needed for testing of the solenoids.

solenoid #3, measure the resistance across pins #1 and #4. The resistance across any of these should be 20–30 ohms.

To check the solenoids on an AX4S with an ohmmeter, disconnect the harness to the solenoid assembly. Measure the resistance across the appropriate pins and compare your readings to specifications. The shift solenoids should have a reading of 12–30 ohms; the EPC solenoid should have a reading of 2.5–6.5 ohms; the modulated converter clutch solenoid should have a reading of 0.75–22.0 ohms; and the regular converter clutch solenoid should have 16–40 ohms across its terminals.

GM Solenoids

The solenoids of 4T60-E and 4L80-E transmissions can be checked with an ohmmeter. The terminals for these solenoids are accessible at the transmission harness (Figure 4-58).

These solenoids are simple on/off devices. Because of this, they can be checked through normal diagnostic methods. However, some transmissions, such as the GM 4L80-E (Figure 4-59) controls line pressure according to engine load signals. A simple on/off shift solenoid cannot regulate the flow of oil in a metered manner. Therefore, the 4L80-E is equipped with a solenoid, which controls line pressure. This solenoid is a three port spool

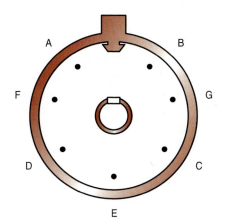

SOLENOID	ACROSS	the RESISTANCE should be:
SOLENOID A	E to F	20 to 40 ohms
SOLENOID B	E to G	20 to 40 ohms
PWM SOLENOID	E to B	10 to 15 ohms
ALL	A to D	20 to 40 ohms
ALL	A to D	2 to 15 ohms (reversed leads)

FIGURE 4-58 The transmission harness connector for a 4T60-E.

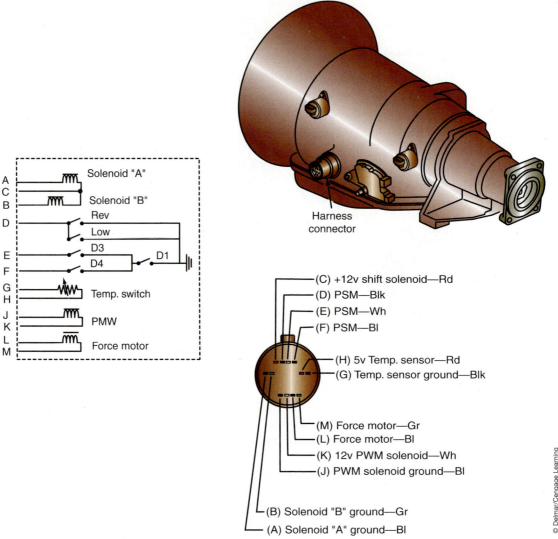

FIGURE 4-59 Testing the electrical components in a 4L80-E can be done through the transmission harness connector.

valve, electronic pressure regulator that controls pressure based on current flow through its coil winding. The amount of current flow through the solenoid is controlled by the PCM. As current through the pressure control solenoid increases, line pressure decreases.

Solenoids that are controlled by duty cycle or a constantly changing signal are more active than the simple on/off solenoids. Therefore, they are more likely to wear. Many transmission rebuilders recommend that this type of solenoid be automatically replaced during a transmission overhaul. Duty cycle solenoids are less prone to clogging because they are cycled full on every 10 seconds. This full on time helps to keep the regulator valve clean.

The action of a pressure control solenoid can be monitored on a DSO. Connect the DSO directly to the solenoid's circuit. The pattern should be compared to those given in the service information. A PWM solenoid is used to apply and release the TCC in such a manner that the TCC will apply and release smoothly under all conditions. The PWM solenoid in the 4L80-E uses a negative duty cycle and its activity can also be monitored on a scope.

Honda Solenoids

Honda's shift solenoids are normally closed. When the computer sends battery voltage to them, they open and allow fluid to flow through them. When the fluid flows through the solenoid, it releases pressure that was preventing a shift valve from moving. When the pressure is exhausted, the valve is able to move and the transaxle shifts.

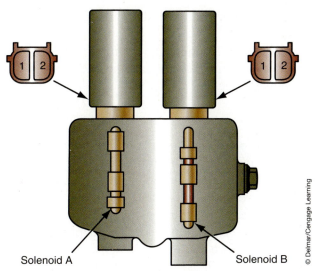

Solenoid A   Solenoid B

FIGURE 4-60 Identification of the terminals used to test Honda solenoids.

A Honda transmission can be shifted, without the computer, by jumping 12 volts to terminal 1 at the shift solenoids (Figure 4-60). Connect a good ground to terminal 2. Apply voltage to solenoid B to operate in first gear, apply voltage to both shift solenoids for second gear, to solenoid A for third, and to none of them for fourth gear. If the transaxle shifts normally when the solenoids are energized, the solenoids sensors or the controller are at fault and the mechanical and hydraulics of the transaxle are normal.

Disconnect the jumper wire, and check the valve at the fluid passage in the valve body. If the valve binds or moves slowly or if the solenoid does not work, replace solenoids A and B.

To observe the action of the solenoids, connect a pressure gauge to the transaxle taps for the solenoids. Observe the pressure readings when the transaxle upshifts. If the pressure is greater than 3 psi when the solenoid valve is open, the valve has a restriction or the solenoid is not opening fully and not allowing a full release of pressure. If the valving is restricted, the solenoids should be replaced.

Use an ohmmeter to electrically check the solenoids. The resistance across a shift solenoid should be 12–24 ohms and 14–30 ohms across the lockup solenoid. If the resistance of the solenoids is not within these figures, they should be replaced.

Repairing the System

After identifying the cause of the problem, repairs should be made. When servicing or repairing OBD II circuits, the following guidelines are important:

- Do not connect aftermarket accessories into an OBD II circuit.
- Do not move or alter grounds from their original locations.
- Always replace a relay with an exact replacement. Damaged relays should be thrown away, not repaired.
- Make sure all connector locks are in good condition and are in place.
- After repairing connectors or connector terminals, make sure the terminals are properly retained and the connector is sealed.
- When installing a fastener for an electrical ground, be sure to tighten it to the specified torque.

After repairs have been made, the system should be rechecked to verify that the repair took care of the problem. This may involve a road test in order to verify that the complaint has been resolved. Record the fail records or freeze frame data taken before the repair. Use the scan tool to erase any DTCs. Then operate the vehicle within the conditions noted in the fail records or the freeze frame data. After driving the vehicle through a variety of conditions,

SERVICE TIP:
There are small filters located in the valve body to protect the solenoids. If these filters are plugged, they will cause a restriction to oil flow. The filters are mounted in a preformed rubber gasket and can be cleaned, however the rubber gaskets should be replaced on every rebuild.

CAUTION:
Static electricity can destroy or render an electronic part useless. When handling any electronic part, do whatever is possible to reduce the chances of electrostatic buildup on your body and the inadvertent discharge to the electronic part.

recheck the DTCs. If the vehicle worked fine and no new codes were set, the repair has been verified.

Some control systems have ATF monitors that must be reset (sometimes with a scan tool) after service. Other systems may have relearn procedures that must be followed after the transmission has been overhauled.

Relearn Process

Vehicles equipped with engine or transmission computers may require a relearn procedure after the battery has been disconnected or after some repairs to the transmission and control system have been made. These include the replacement of the transmission, torque converter, clutches, and seals, as well as overhauling the valve body.

This is an important step because it tells the computer to look at new operating parameters. Based on these new values, the computer will readapt to a transmission that is functioning differently than the one before the repairs. When the vehicle's battery is disconnected, this memory is lost. The computer will use default data until new data from each key start is stored. As the computer memorizes vehicle operation for each new key start, driveability is restored. Vehicle computers may memorize vehicle's operation patterns for 40 or more key starts. Always refer to the service information as some transmissions have the ability to learn quickly when certain conditions are met.

Customers often complain of problems during the relearn stage because the vehicle acts differently than before it was serviced. Depending on the vehicle and how it is equipped, the following complaints may exist: harsh or poor shift quality, rough or unstable idle, hesitation or stumble, rich or lean running, and poor fuel economy. These complaints should disappear after a number of drive cycles. To reduce the possibility of complaints, after any service that requires that the battery be disconnected, the vehicle should be road tested. If a specific relearn procedure is not available, the following procedure may be used:

1) Set the parking brake and start the engine in P or N. Allow the engine to warm up to normal operating temperature or until the electric cooling fan cycles on.
2) Allow the vehicle to idle for about a minute in the N position, then move the gear selector into the D position and allow it to idle in gear for one minute.
3) Road test the vehicle. Accelerate at normal throttle openings (20–50 percent) until the vehicle shifts into to top gear.
4) Then maintain a cruising speed with a light to medium throttle.
5) Decelerate to a stop, make sure you allow the transmission to downshift, and use the brakes to bring the vehicle to a stop.
6) If a driveability problem still exists, repeat the sequence.

Some systems have a "quick learn" process. As a result of this procedure, all previously learned parameters are set to a default value. This may initially result in poor shift quality, however as the vehicle is driven the system will relearn and adjust the transmission's operation to provide good shift quality. Before initiated the quick learn, certain conditions must be met:

- There should be no stored DTCs.
- The transmission should be shifted through all gear ranges with the engine running.
- The fluid should be at its appropriate level.
- The brake pedal should be depressed.
- The engine should be idling.

Once these conditions are met, the scan tool will display instructions to complete the procedure. After completing the process, the transmission should be placed in Park. Then the vehicle can be driven to test it and to allow the computer to learn. After the test drive, the system should be scanned for DTCs and for abnormal serial data.

Some manufacturers recommend a specific relearn procedure that is designed to establish good driveability during the relearn process. These procedures are especially important for all vehicles equipped with an electronically controlled converter and/or transmission. Always complete the procedure before returning the vehicle to the customer.

Chryslers with 41TE and 42LE transaxles are relearned by first warming up the transaxle to normal operating temperatures by allowing the vehicle to idle. Then operating the vehicle and maintaining a constant throttle opening during upshifts. This sets the transaxle into the upshift relearn process.

Then accelerate the vehicle with a throttle opening of 10–50 degrees. Accelerating the vehicle from a stop to 45 mph with a moderate throttle is sufficient for this part of the procedure. Then operate the vehicle until the transaxle completes 1–2, 2–3, and 3–4 shifts at least 20 times.

Now operate the vehicle at a speed of less than 25 mph and force downshifts with a wide open throttle. Repeat this at least eight times.

Operate the vehicle above 25 mph in fourth gear and force downshifts with a wide open throttle. Repeat this at least eight times. The forced downshifts allow the computer to relearn kickdown operation.

Ford Motor Company also specifies a relearn procedure for some of their transmissions and transaxles. All of the specific procedures begin with a fluid check and warming the ATF to normal temperatures. The procedure has two segments, an idle relearn and a drive relearn. The idle relearn procedure begins with starting the vehicle in P with all accessories off and the parking brake set. Then move the gear selector to N and allow the engine to idle for one minute. Move the gear selector to D and again allow the engine to idle for one minute. After the idle relearn procedure is complete, the drive relearn procedure can begin. The drive relearn procedures are specific for the different transmissions.

If the vehicle is equipped with a 4R70W or AX4S transmission, road test the vehicle. With the gear selector in overdrive, moderately accelerate the vehicle to 50 mph for a minimum of 15 seconds. The transmission should be in fourth gear at the end of that time. Then hold the speed with a steady throttle and lightly apply and release the brake for about 5 seconds. Then stop and park the vehicle with the gear selector in the D position for a minimum of 20 seconds. Repeat this procedure five times.

If the vehicle is equipped with an E4OD transmission, put the gear selector in the D position and depress the O/D cancel button (the LED should light). Then moderately accelerate to 40 mph for a minimum of 15 seconds. Hold the throttle steady and depress the O/D cancel button (LED should go out). Accelerate to 50 mph. The transmission should be in fourth gear at this time. Hold that speed for 15 seconds, then lightly apply and release the brake. Maintain 50 mph while applying the brake. Then stop and park the vehicle with the gear selector in the D position for a minimum of 20 seconds. Repeat this procedure five times.

DIAGNOSING OBD I SYSTEMS

OBD I systems have limited self-diagnostic capabilities. By entering into a self-test mode, the system is able to evaluate its condition. If problems are found, they may be identified as either hard faults (on-demand) or intermittent failures. Each type of fault or failure is assigned a numerical trouble code that is stored in memory.

A hard fault is a problem found in the system at the time of the self-test. An intermittent fault, on the other hand, indicates that a malfunction occurred (for example, a poor connection causing an intermittent open or short) but was not present during the self-test. Nonvolatile RAM allows intermittent faults to be stored for up to a specific number of ignition key on/off cycles. If the trouble does not reoccur during that period, it is erased from the memory.

There are various ways to access the trouble codes stored in the computer. Most manufacturers have specific equipment designed to monitor and test the system's components. Aftermarket companies also manufacture scan tools that have the capability to read and record the system's input and output signals. Photo Sequence 8 covers the typical procedure for checking an early EAT with a scan tool and breakout box.

Typical Procedure for Diagnosing EAT Problems

All photos in this sequence are © Delmar/Cengage Learning.

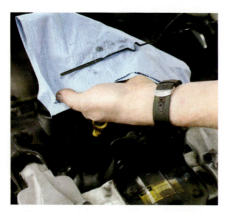

P8-1 Check fluid level and condition.

P8-2 Check all connections.

P8-3 Road test the vehicle to verify the complaint.

P8-4 Connect the STAR tester.

P8-5 Observe the codes.

P8-6 Connect the breakout box.

P8-7 Install the template over the breakout box.

P8-8 Perform pinpoint tests according to the service manual.

P8-9 After the fault has been corrected, rerun the Quicktest.

Chrysler

Chrysler has basically two electronic control systems. One was first used in 1988 and had been used in a variety of models. The other system is a multiplexed system, which is used on later models. The original system is quite basic and similar to those used by other manufacturers. The recommended way to retrieve trouble codes is through the use of Chrysler's scan tool, called a DRB-II.

The DRB-II is connected to the DLC, which is located by the left front shock tower. The scan tool has a variety of cartridges that contain the test parameters for the vehicle being tested. After the correct cartridge has been installed in the tester and the scan tool is connected to the vehicle, the tester's display will show a test pattern. After a few seconds, the display will change to read out the copyright date and revision level of the cartridge. After a few seconds, the display will ask for the vehicle's model year. The model year is then selected. Once the display shows the correct year, press the down arrow key.

The display will then ask for the system that will be tested. The system is selected by pressing either left or right arrow keys. Once the display indicates powertrain, press down arrow key. The display will then ask for the desired test selection. The diagnostic test mode is used to see if there are any fault codes stored in the system. The circuit Actuation Test Mode (ATM) is used to make the computer cycle a solenoid on and off. The sensor test mode is used to see output signals of certain sensors as the computer receives them when the engine is not running. The engine running test mode is used to see sensor output signals as received by the computer while the engine is running.

The other transmission control system is the Chrysler Collision Detection system, which has a serial data bus or network. The network interconnects the CCD modules. Each CCD module uses this network to communicate and exchange data with other modules on the network.

Diagnosis of this transaxle system is relatively simple. Nearly all of the important information is available through the use of a scan tool and the transaxle is equipped with many pressure taps. Both of these features give you an accurate look at the operation of the computer.

When the transaxle controller detects a serious problem, it will automatically go into the "limp-in" mode of operation. This mode defaults the transaxle to second gear only with second-gear starts, Reverse, and Park operations. To determine why the computer has defaulted to this mode, connect a scan tool to the blue six-pin CCD connector under the dash. Retrieve and record the trouble codes that are displayed. Then clear the memory and drive the vehicle until the complaint occurs again. Retrieve and write down the codes again. These codes are the most current and should be compared to the original codes.

If the scan tool is unable to display a code or is unable to recognize the computer or its circuits, the problem may be a faulty or dead bus. A dead bus is commonly caused by the body computer, transaxle controller, or any other module on the CCD bus. The exact bad module can be identified by disconnecting each module, one at a time, from the bus.

After you have one module disconnected, rerun the start-up of the scan tool. If the bus becomes active, the disconnected control module was the cause of the previous dead bus. Repeat this process, by reconnecting the disconnected module and unplugging another. The module that is disconnected when the bus becomes active is the dead module.

If the cause of the dead bus was not identified, voltage checks should be made. Measure the voltage drop across the groundside of the diagnostic connector by inserting the positive voltmeter lead to the ground lead of the connector and the negative lead to the negative terminal of the battery. The maximum allowable voltage drop across the ground is 0.1 volts. The source voltage to the computer should also be checked. If battery voltage is not present at the computer, check the voltage of the battery. If it is okay, check the voltage drop in the supply circuit. If there is more than 0.1 volts dropped, identify the source of the resistance and repair the problem. Also, check the bus bias voltage by connecting the voltmeter's positive lead to either bias terminal in the diagnostic connector and the negative lead to the ground terminal in the connector. The bias voltage should be between 2.1 and 2.6 volts. Any incorrect voltages could be the cause for a dead bus and the cause of the problems should be corrected. After the electrical problems are repaired, the scan tool test should be conducted.

Using the scan tool, the trouble codes are retrieved and compared to the trouble code chart given in the service manual. These codes will identify problems circuits. The service manual also gives step-by-step procedures for pinpoint diagnostics of each trouble code. Always follow these, they are efficient and exact.

Ford

Most early Ford EAT systems are based on their EEC IV or EEC V systems. Although the exact make-up of the systems varies according to the transmission used and the model vehicle being tested, all follow similar procedures. The following equipment is recommended to diagnose and test an EEC system:

- Self-Test Automatic Read-out (STAR) Diagnostic Tester—This tester is recommended but not required. The tester was designed for EEC systems and is used to display the service codes. There are also many aftermarket testers available for testing these systems. An Analog Volt/Ohmmeter with a 0–20 V DC range can be used as an alternate to a diagnostic tester.
- DMM—This multimeter must have a minimum impedance of 10 megohms.
- Breakout Box—This is a jumper wire assembly that connects between the vehicle harness and the PCM. The breakout box is required to perform certain tests on the system. During individual circuit tests, the procedures will call for probing particular pin numbers, these pin numbers relate to the pins of the breakout box.
- Vacuum gauge and Vacuum Pump—This can be one assembly or separate units.
- Tachometer—Must have a 0–6000 RPM range defined in 20-rpm increments.
- Spark tester—A modified side electrode spark plug with the electrode removed and an alligator clip attached may be used in place of the spark tester.
- MAP/BP Tester—This tester plugs into the MAP/BP sensor circuit and the DMM. It is used to check input and output signals from the sensor, which produces a frequency signal. This signal may also be monitored on a scope set to a milli-volt scale.
- Other equipment: Timing light, Fuel Injection Pressure Gauge, nonpowered 12-volt test light, and a jumper wire, about 15 inches long.

The EEC system offers a variety of test sequences. These steps are called the Quicktest. Failure to follow the recommended test sequence may result in misdiagnosis or the replacement of nonfaulty components.

After all tests, servicing, or repairs have been completed, the Quicktest should be repeated to ensure that all systems are working properly. The KOEO and KOER self-tests are designed to identify faults present at the time of the testing and not intermittent faults. Intermittent faults are detected by the Continuous self-test mode.

To retrieve codes with the analog VOM, turn the ignition key off. Then connect the jumper wire from Self-Test Input (STI) pigtail to pin #2 at the Self-Test connector. Set the VOM to the correct DC voltage scale. Connect the positive lead of VOM to the positive battery terminal. The negative lead of the meter is connected to the #4 pin of the Self-Test connector. Connect the timing light and proceed to the KOEO Self-Test. Codes are shown as voltage pulses (needle sweeps), pay attention to the length of the pauses in order to read the codes correctly.

To retrieve codes with the STAR tester, turn the ignition off. Then connect the color-coded adapter cable leads to the diagnostic tester. Connect the adapter cable's two service connectors to the vehicle's Self-Test connectors. Then connect the timing light and proceed to the KOEO Self-Test.

All service codes are two or three digit numbers that are generated one digit at a time. There will be a 2-second pause between each DIGIT in a code. There will be a 4-second pause between each CODE. The continuous memory codes are separated from the functional test service codes by a 6-second delay, a single 1/2-second sweep, and another 6-second delay.

Always record the codes in the order received. If a diagnostic tester is used, it will count the pulses and display them as a digital code. If the "Check Engine" light is used, numeric service codes will be displayed at the light.

The first set of codes that will be displayed are the KOEO self-test codes. These codes will be repeated twice. They are followed by the Separator Pulse signal and the continuous memory codes. These codes will also be repeated twice. A KOEO code 11 (or 111) indicates the system has passed the self-test. Whenever a repair is made, repeat the Quicktest. To clear codes, remove the jumper wire while the codes are being displayed and before turning off the ignition. Record all displayed codes.

General Motors

The 4T60-E and other GM transaxles use two electric solenoids to control transaxle upshifts and downshifts. The 4L80-E transmission functions in much the same way, however it is also fitted with a force motor, which controls hydraulic line pressure and a TCC solenoid. Some models have a TCM in addition to the PCM. In all systems, the PCM has self-diagnostic capabilities, which help identify which parts or circuits may need further testing.

The PCM constantly monitors all electrical circuits. If the PCM detects a problem or an out-of-range sensor input, it records a trouble code in its memory. If the problem continues for some time, the MIL will glow. It is possible for the PCM to have detected a trouble and not light the MIL. However, the appropriate code may be stored in its memory.

Many methods are commonly used to retrieve the trouble codes from the PCM's memory. The simplest method is to use the MIL. Other methods include the use of special scan tools.

To display the trouble codes with the MIL, locate the **Assembly Line Data Link (ALDL)** or DLC connector. Then, with the ignition on and engine off, jump across the A and B terminals of the connector. The MIL should begin to flash codes. Each code will be repeated three times. The first series of flashes is the first digit of the code and the second series is the second digit. Trouble codes are displayed starting with the lowest numbered code. Codes will continue until the jumper is disconnected from the ALDL.

A scan tool connected into the ALDL (Figure 4-61) can provide access to circuit voltage information, as well as trouble codes. By observing the various input voltages, you can

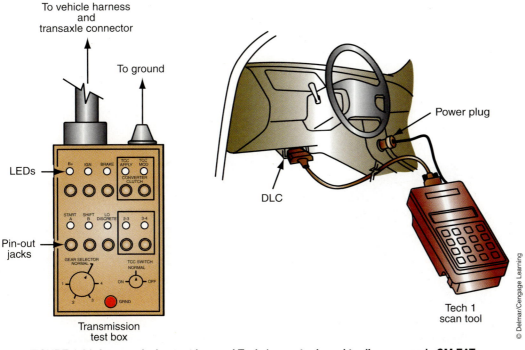

FIGURE 4-61 A transmission test box and Tech 1 scan tool used to diagnose early GM EATs.

identify out of specification input signals. Out of specification signals will not set a code unless they are out of the normal operating range. However, these signals will cause a drive-ability problem. To retrieve trouble codes with a scan tool, simply plug the tester lead into the ALDL. Then, program the scan tool for that particular vehicle according to the scan tools' instructions. Trouble codes will appear, digitally, on the tester.

To erase the codes after repairs have been made, turn the ignition off. With the jumper wire still connected, remove the control module fuse for 10 seconds. If the fuse cannot be located, disconnect the negative battery lead for 10 seconds. After the power has been disconnected from the PCM, the codes will be removed. So will the operating instructions on some models. For this reason, it is important that you follow the relearn procedures for the transmission.

Honda

Honda and Acura transaxle trouble codes are communicated by either flashing an LED on a side of the transaxle controller, or by flashing the S or D4 shifter status light on the instrument panel after connecting a jumper wire at the service connector (Figure 4-62). Always refer to the correct service manual when attempting to retrieve codes.

The connector is a two-pin, two-wire connector that is not connected to anything. Using a jumper wire, hook the two pins together, turn the ignition switch on, and watch either the S or D4 light on the instrument panel. The three possible locations for the connector are as follows: under the right side of the dash, behind the right side of the center console, or behind the left front edge of the passenger's carpet up against the firewall. If a trouble code is present, the light will either give short or long and short flashes. Short flashes are either a one or two-digit trouble code. Long flashes are the tens digit of a two-digit code. Short flashes are about a half second long and long flashes are about a second and a half long. There is about a 1-second pause between each code. All codes will be shown in sequence, lowest number first, up to the highest code. The sequence will repeat as long as the key is on and the connector jumped across.

Some models have the diagnostic LED on the side of the transaxle controller. Read the flashes of the LED to determine the trouble code. The LED blinks all short flashes. Whenever the key is on and a problem exists, the LED will blink a series of short flashes, pauses, and then either repeats the same number of flashes (if there is only one code), or advances to the next number of blinks. Just count each series of flashes, and you have the code. The computer for Honda cars is located either under the front edge of the front passenger's carpet, behind the center console, or under the driver's seat. Acuras locate the computer behind the left side of the instrument panel, next to the inner fender panel, or under the right or left front seat with the LED facing the rear of the car.

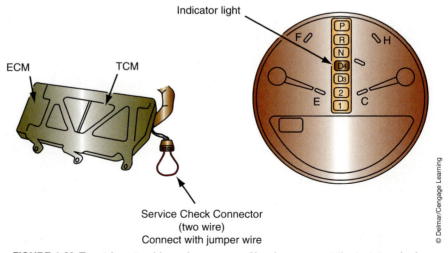

FIGURE 4-62 To retrieve trouble codes on some Hondas, connect the two terminals of the Service Check Connector and observe the D4 indicator.

The single- and double-digit codes should be compared to the appropriate trouble code chart provided in the service manual. Like other computer self-diagnostic systems, these systems do not determine the exact fault. Rather they identify the problem area. After repairs have been made, the codes should be erased.

To erase the codes, remove the appropriate fuse for at least 1 minute or pull the negative cable of the battery for a minute. Because Honda and Acura systems use different fuses for different models to keep the memory alive, make sure you check the service manual to identify the proper fuse.

Mazda

Trouble codes in Mazda EAT systems are displayed by either the Hold light or the Check Engine light. If no problem exists, the Hold or Check Engine light will come on for 3 seconds after turning the key on, and then it will go out. If a trouble does exist, the light will flash in a regular pattern until the trouble is no longer there. This is not a trouble code; rather, it is an indication that a problem exists.

To retrieve the codes, locate the Diagnostic Request lead. This lead may be a single pin and wire blue connector, a single pin and wire green connector, or one of the pins in an integrated Diagnostic connector. Connect the lead to a good ground with a jumper wire. When the diagnostic pin is grounded, the Hold or Check Engine light will display the code in long and short flashes. Long flashes, 1.2 seconds long, are the tens digit of a two-digit trouble code, short flashes, 0.4 seconds long, are the ones digit of a one- or two-digit code. The lowest number codes will be shown first, then the higher number codes. There is a 4-second pause between codes. The series of codes will repeat as long as the Diagnostic Request pin is grounded. If no codes are present when you ground this pin, the indicator light will remain off.

Nissan

Transmission trouble code retrieval on Nissan computer shifted transmissions and transaxles are easy and uniform throughout the different models of vehicles. Simply turn the ignition switch to the on position and move the shifter and the O/D off switch. Then an instrument panel light will flash codes. The codes are displayed by the O/D off light, Power, or A/T Check light.

To begin the procedure, make sure the Mode switch is in the Auto position. Then run the engine until it has reached normal operating temperature. Turn the key off, set the parking brake and do not touch the brake or throttle pedals until the procedure calls for you doing this. Move the shifter to D and turn the O/D switch off. Turn the ignition switch back to the on position. After waiting at least 2 seconds, move the shifter to "2" and turn the O/D switch on. Then, move the shifter to "1" and turn the O/D switch off. Now, slowly push the throttle pedal to the floor and slowly release it. The codes will now appear on the instrument panel.

The codes will appear in a sequence of 11 flashes. This sequence always begins with a long, 2 second, starter flash, followed by 10 shorter flashes. If a problem does not exist, the 10 following flashes will be very short flashes, 0.2 seconds. If a problem exists, one of the 10 short flashes will be a little longer, 0.8 seconds, than the rest in that sequence. To identify the code, count the short flashes and determine which one was the longer one. Begin counting after the longer initial flash. If the first flash was the longest short flash, the code is 1. After each sequence of 11 flashes, there is a 2.5-second pause, and then either the code or codes will repeat. All codes will automatically clear if the problem is fixed and the engine has been started two times after the system has been repaired.

Saturn

Saturn EATs use five electronic valves that are controlled by the PCM to control shift feel and timing. The PCM has a self-diagnostic function. If an electrical problem occurs, the SES or "Shift to D2" lamp will come on or will start flashing. The PCM stores trouble codes whenever it has detected a fault in the engine or transaxle circuit. Problems are stored as hard or intermittent problems. The PCM will also store other problems but may not turn on the SES light in response to the problem. These codes are stored in the memory to aid in diagnostics.

The PCM recognizes three different types of faults: hard, intermittent, and malfunction history/information flags. Hard codes cause the SES light to glow and remain on until the problem is repaired. These codes can be interpreted by looking up the code in the appropriate trouble code chart. Intermittent codes cause the SES to flicker or glow and go out approximately 10 seconds after the fault disappears. The corresponding trouble code will remain, however, in memory to be retrieved by a technician. If the intermittent fault does not occur for 50 engine starts, the code will be erased from the computer's memory. Engine information flags will not cause the SES light to glow. Unlike the other types of codes, information flags and malfunction history will not be erased from the PCM memory. These codes can only be retrieved and erased from memory by using a Saturn Portable Diagnostic Tool (PDT).

Trouble codes are read by counting the flashes of the SES lamp. The first series of flashes represents the first digit of the trouble code and the second series represents the second digit. The first code will always be code "12," followed by all other stored codes. Each code is repeated three times. When code "11" appears, that signals the end of the engine's self-diagnostic test and the beginning of the transaxle self-test. The "Shift to D2" lamp will start flashing any transaxle related codes stored in the transaxle controller. If the SES does not flash a code "11," there are no transaxle codes stored in memory.

To set the PCM into the self-diagnostic mode, turn the ignition on with the engine off. Connect a jumper wire from terminal B to terminal A at the PCM diagnostic connector. The SES should begin to flash codes. To end the self-diagnostic mode, turn the ignition off and remove the jumper wire.

To erase the trouble codes, turn the ignition on and make contact three times within 5 seconds with a jumper wire between from terminals A and B at the diagnostic connector.

The PCM has a relearn function and the proper learn process should be followed anytime the battery has been disconnected.

Toyota

A self-diagnostic feature is built into all Toyota EATs. A warning of a fault in indicated by the overdrive OFF indicator lamp. If a malfunction occurs in the speed sensor or solenoid circuits, the overdrive OFF lamp will blink to warn the driver of the fault. On some models, the diagnostic codes can be read by the number of blinks on the overdrive OFF lamp when terminals ECT and E2, in the diagnostic connector, are shorted together. Other models require that terminals TE and E be shorted together (Figure 4-63). Make sure you refer to

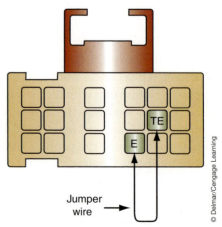

FIGURE 4-63 A Toyota diagnostic connector with a jumper wire inserted to activate the self-test mode.

the appropriate service manual to determine which terminals should be shorted. Once the computer is set into its diagnostic mode, codes will begin to be displayed.

If a malfunction is in memory, the overdrive OFF lamp will blink once every 0.5 seconds. The first number of blinks equals the first digit of a two-digit code. After a 1.5-second pause, the second number of blinks will equal the second digit. If there are two or more codes, there will be a 2.5-second pause between each code. All codes should be interpreted by using the trouble code chart provided in the service manual.

The diagnostic codes can be erased from the computer's memory removing the DOME, ECU B+, or EFI fuse for at least 10 seconds with the key off.

TERMS TO KNOW
(continued)

Digital multimeter

Digital storage oscilloscope (DSO)

Flashing

Freeze frame

Glitch

Impedance

Lab scope

Multimeter

On-Board Diagnostics II (OBD II)

Open

Parameter identification (PID)

Radio frequency interference (RFI)

Serial data

Short

Sine wave

Sinusoidal

Sweep rate

CASE STUDY

A customer came into the shop with a high-mileage Hyundai Excel with a Mitsubishi KM-175 transmission. A tow truck delivered the car because the transmission wouldn't shift gears. The technician assigned to the car pulled the transmission and overhauled it. After installing an overhaul kit and inspecting all other transmission parts, the transmission was reassembled and installed back into the car. The transmission still didn't shift.

Then the technician got serious and began to check the electronics involved with the transmission. After a little research in the service manual, he found that the transmission was in the fail-safe mode. With a multimeter, the technician then checked the solenoids, connectors, and wiring, only to find nothing wrong. Frustrated, he began to check complete circuits and found that the computer was doing nothing.

He located the computer under the passenger's seat and found heavy corrosion on the computer and the wiring harness terminals. The technician then cleaned the area up and replaced the computer. The transmission then shifted well.

ASE-STYLE REVIEW QUESTIONS

1. *Technician A* says a faulty TP sensor can cause delayed shifts.
   Technician B says delayed shifts can be caused by an open shift solenoid.
   Who is correct?
   A. A only
   B. B only
   C. Both A and B
   D. Neither A nor B

2. While checking a transmission range sensor:
   Technician A says typically the switch should be open in all positions except park and neutral.
   Technician B says a faulty TR sensor may allow the engine to start in other gear positions besides park and neutral.
   Who is correct?
   A. A only
   B. B only
   C. Both A and B
   D. Neither A nor B

3. *Technician A* says scan tools are needed to retrieve codes on all models of cars.
   Technician B says all scan tools provide a historical report of the computer system.
   Who is correct?
   A. A only
   B. B only
   C. Both A and B
   D. Neither A nor B

4. *Technician A* says an ammeter without an inductive pickup is always connected in series with the circuit it is testing.
   Technician B says an ammeter is only used when the power for the circuit is disconnected.
   Who is correct?
   A. A only
   B. B only
   C. Both A and B
   D. Neither A nor B

5. While checking voltage drops:

   Technician A connects the voltmeter across the load with the circuit activated.

   Technician B connects the negative meter lead to ground and the positive lead to the component being tested.

   Who is correct?

   A. A only
   B. B only
   C. Both A and B
   D. Neither A nor B

6. *Technician A* says some shift solenoids can be activated by providing a ground for the solenoid.

   Technician B says some shift solenoids can be activated by applying hydraulic pressure to their valve.

   Who is correct?

   A. A only
   B. B only
   C. Both A and B
   D. Neither A nor B

7. While checking the trouble codes on a Honda without OBD II:

   Technician A says the short flashes should be counted, as this is the trouble code.

   Technician B says there is about a one-second delay in the flashing between codes.

   Who is correct?

   A. A only
   B. B only
   C. Both A and B
   D. Neither A nor B

8. While checking a switch:

   Technician A uses a voltmeter at the output of the switch in the various switch positions.

   Technician B uses a test light at the output of the switch in the various switch positions.

   Who is correct?

   A. A only
   B. B only
   C. Both A and B
   D. Neither A nor B

9. *Technician A* says an ohmmeter can be used to test semiconductors.

   Technician B says a low-impedance ohmmeter should be used on electronic circuits.

   Who is correct?

   A. A only
   B. B only
   C. Both A and B
   D. Neither A nor B

10. While checking the codes on a Ford AX4S with an OBD II compliant computer system:

    Technician A says during the KOEO self-test, the codes are repeated only twice.

    Technician B says the continuous memory codes appear after the separator pulse signal.

    Who is correct?

    A. A only
    B. B only
    C. Both A and B
    D. Neither A nor B

ASE CHALLENGE QUESTIONS

1. An electronically controlled transmission has erratic shifting.

   Technician A says a poor PCM ground will cause this problem.

   Technician B says a bad AC generator-to-battery circuit will cause erratic performance of a transmission or transaxle.

   Who is correct?

   A. A only
   B. B only
   C. Both A and B
   D. Neither A nor B

2. A scope is displaying a nonsinusoidal wave at a rate of 10,000 times per second.

   Technician A says the device being checked may not be functioning correctly because the wave is nonsinusoidal.

   Technician B says the frequency cannot be calculated because the wave is nonsinusoidal.

   Who is correct?

   A. A only
   B. B only
   C. Both A and B
   D. Neither A nor B

3. A glitch appears in the waveform of a vehicle speed sensor. Which of the following is *not* a probable cause of the problem?

A. A loose connector

B. A damaged wire

C. A poorly mounted sensor

D. A damaged magnet in the sensor

4. A hot-wire MAF has low resistance across it when the wire is hot.

Technician A says this condition will cause the engine to stall as soon as you put the transmission into gear.

Technician B says this condition will cause the transmission to shift late.

Who is correct?

A. A only

B. B only

C. Both A and B

D. Neither A nor B

5. *Technician A* says the second character code in an OBD II DTC is a designated code for future use and has little value to a technician.

Technician B says that if the third character of a powertrain code is a 7 or 8, the problem is directly related to the transmission.

Who is correct?

A. A only

B. B only

C. Both A and B

D. Neither A nor B

Name _____ **Date** _____

USING A DSO ON SENSORS AND SWITCHES

Upon completion of this job sheet, you should be able to connect a DSO and observe the activity of various sensors and switches.

ASE Correlation

This job sheet is related to the ASE Automatic Transmission and Transaxle Test's Content Area *General Transmission and Transaxle Diagnosis*.
Task: Diagnose mechanical and vacuum control systems; determine necessary action.

Tools and Materials

A vehicle with accessible sensors and switches   A DSO
Service manual for the above vehicle   A DMM
Component locator manual for the above vehicle

Describe the vehicle being worked on:
Year _____ Make _____ VIN _____
Model _____

Procedure

Task Completed

_____

1. Connect the DSO across the battery. Make sure the scope is properly set. Observe the trace on the scope. Is there evidence of noise? Explain.

   _____

   _____

   _____

2. Locate the A/C compressor clutch control wires. Start and run the engine. Connect the DMM to read available voltage. Observe the meter, then turn the compressor on. What happened on the meter?

   _____

   _____

   Now connect the DSO to the same point with the compressor turned off. Observe the waveform, then turn the compressor on. What happened to the trace?

   _____

   _____

3. Turn off the engine but keep the ignition on. Locate the TP sensor. List each wire and describe the purpose of each.

   _____

   _____

   _____

   _____

4. Connect the DMM to read reference voltage at the TP sensor. What do you read?

_____

_____

Now move the leads to read the output of the sensor. Starting with the throttle closed, slowly open the throttle until it is wide open. Watch the voltmeter while doing this. Describe your readings.

_____

_____

5. Now connect the DSO to read reference voltage at the TP sensor. What do you see on the trace?

_____

_____

Now move the leads to read the output of the sensor. Starting with the throttle closed, slowly open the throttle until it is wide open. Watch the trace while doing this. Describe your readings.

_____

_____

6. Now run the engine. Locate the oxygen sensor and identify the purpose of each wire going to it. Connect the DMM to read voltage generated by the sensor. (To do this, you may use an electrical connector for the O_2 sensor positioned away from the hot exhaust manifold.) Watch the meter and describe what happened.

_____

_____

_____

_____

7. Now connect the DSO to read voltage output from the sensor. Watch the trace and describe what happened.

_____

_____

_____

_____

8. Explain what you observed as the differences between testing with a DMM and a DSO.

_____

_____

_____

_____

Instructor's Response _____

_____

_____

Name _____ Date _____

TEST AN ECT SENSOR

Upon completion of this job sheet, you should be able to check the operation of an engine coolant temperature sensor.

ASE Correlation

This job sheet is related to the ASE Automatic Transmission and Transaxle Test's Content Area *General Transmission and Transaxle Diagnosis.*
Task: Diagnose mechanical and vacuum control systems; determine necessary action.

Tools and Materials

DMM

Describe the vehicle being worked on:

Year _____ Make _____ VIN _____

Model _____

Procedure

_____

Task Completed

1. Describe the location of the ECT sensor.

   _____

   _____

2. What color wires are connected to the sensor?

   _____

   _____

3. Record the resistance specifications for a normal ECT sensor for this vehicle.

   _____

   _____

   _____

   _____

   _____

   _____

   _____

4. Disconnect the electrical connector to the sensor.

5. Measure the resistance of the sensor. It was _____ ohms at approximately
   _____ °F.

6. Conclusions from this test.

_____

_____

_____

_____

Instructor's Response _____

_____

_____

Name _____ Date _____

CHECK THE OPERATION OF A TP SENSOR

Upon completion of this job sheet, you should be able to test the operation of a throttle position sensor with a variety of test instruments.

ASE Correlation

This job sheet is related to the ASE Automatic Transmission and Transaxle Test's Content Area *General Transmission and Transaxle Diagnosis*.

Task: Diagnose mechanical and vacuum control systems; determine necessary action.

Tools and Materials

DMM   Lab scope

Describe the vehicle being worked on:

Year _____ Make _____ VIN _____

Model _____

Procedure

Task Completed

1. Connect the lab scope across the TP sensor.   ☐

2. With the ignition on, move the throttle from closed to fully open and then allow it to close slowly.   ☐

3. Observe the trace on the scope while moving the throttle. Describe what the trace looked like.

   _____

   _____

   _____

4. Based on the waveform of the TP sensor, what can you tell about the sensor?

   _____

   _____

5. With a voltmeter, measure the reference voltage to the TP sensor. The reading should be _____ volts. It was _____ volts.

6. What is the output voltage from the sensor when the throttle is closed? _____ volts

7. What is the output voltage from the sensor when the throttle is open? _____ volts

8. Move the throttle from closed to fully open and then allow it to close slowly. Describe the action of the voltmeter.

   _____

   _____

9. Conclusions from these tests.

_____

_____

_____

Instructor's Response _____

_____

_____

Name _____ **Date** _____

TESTING A MAP SENSOR

Upon completion of this job sheet, you should be able to test a manifold absolute pressure sensor in a variety of ways.

ASE Correlation

This job sheet is related to the ASE Automatic Transmission and Transaxle Test's Content Area *General Transmission and Transaxle Diagnosis*.
Task: Diagnose mechanical and vacuum control systems; determine necessary action.

Tools and Materials

Hand-operated vacuum pump   DMM
Lab scope

Describe the vehicle being worked on:

Year _____ Make _____ VIN _____
Model _____

Procedure

Task Completed

1. If the MAP sensor produces an analog voltage signal, follow this procedure. ☐

2. With the ignition switch on, backprobe the 5-volt reference wire. ☐

3. Connect a voltmeter from the reference wire to ground. The reading is _____ volts.

4. If the reference wire is not supplying the specified voltage, what should be checked next?

   _____

5. With the ignition switch on, connect the voltmeter from the sensor ground wire to the battery ground. ☐

6. What is the measured voltage drop? _____ volts

7. What does this indicate?

   _____

   _____

8. Backprobe the MAP sensor signal wire and connect a voltmeter from this wire to ground with the ignition switch on. ☐

9. What is the measured voltage? _____ volts

10. What does this indicate?

    _____

    _____

11. How do you determine the barometric pressure based on these voltage readings?

_____

_____

_____

_____

☐ 12. Turn the ignition switch on and connect a voltmeter to the MAP sensor signal wire.

13. Connect a vacuum hand pump to the MAP sensor vacuum connection and apply 5 inches of vacuum to the sensor. Record the voltage reading: _____ volts.

Note: On some MAP sensors, the sensor voltage signal should change 0.7–1.0 volt for every 5 inches of vacuum change applied to the sensor. Always use the vehicle manufacturer's specifications. If the barometric pressure voltage signal was 4.5 volts with 5 inches of vacuum applied to the MAP sensor, the voltage should be 3.5 V–3.8 V. When 10 inches of vacuum is applied to the sensor, the voltage signal should be 2.5 V–3.1 V. Check the MAP sensor voltage at 5-inch intervals from 0–25 inches.

If the MAP sensor voltage is not within specifications at any vacuum, replace the sensor.

14. Record the results of all vacuum checks.

_____

_____

15. What did these tests indicate?

_____

_____

☐ 16. Connect the scope to the MAP output and a good ground.

17. Accelerate the engine and allow it to return to idle. Observe and describe the trace.

_____

_____

18. What did the trace show about the sensor?

_____

_____

_____

_____

Note: If the MAP sensor produces a digital voltage signal of varying frequency, check the 5-volt reference wire and the ground wire with the same procedure used on other MAP sensors. This sensor diagnosis is based on the use of a MAP sensor tester that changes the MAP sensor varying frequency voltage to an analog voltage. Follow these steps to test the MAP sensor voltage signal:

☐ 1. Turn off the ignition switch, and disconnect the wiring connector from the MAP sensor.

☐ 2. Connect the connector on the MAP sensor tester to the MAP sensor.

☐ 3. Connect the MAP sensor tester battery leads to a 12-volt battery.

☐ 4. Connect a pair of digital voltmeter leads to the MAP tester signal wire and ground.

5. Turn on the ignition switch and observe the barometric pressure voltage signal on the meter. Observe and record the voltmeter readings.

_____

_____

_____

_____

6. Supply the specified vacuum to the MAP sensor with a hand vacuum pump. ☐

7. Observe the voltmeter reading at each specified vacuum. Record the readings.

_____

_____

_____

_____

8. What do these readings indicate?

_____

_____

Instructor's Response _____

_____

_____

Name _____ **Date** _____

CONDUCT A DIAGNOSTIC CHECK OF AN ENGINE EQUIPPED WITH OBD II

Upon completion of this job sheet, you should be able to conduct a system inspection and retrieve codes from the PCM of an OBD II–equipped engine.

ASE Correlation

This job sheet is related to the ASE Automatic Transmission and Transaxle Test's Content Area *General Transmission and Transaxle Diagnosis.*
Task: Diagnose mechanical and vacuum control systems; determine necessary action.

Tools and Materials

A vehicle equipped with OBD II   Service manual
Scan tool

Describe the vehicle being worked on:

Year _____ Make _____ VIN _____
Model _____
Engine size and type _____

Procedure

_____
Task Completed

1. Check all vehicle grounds, including the battery and computer ground, for clean and tight connections. Comments:

   _____

   _____

2. Perform a voltage drop test across all related ground circuits. State where you tested and what your findings were.

   _____

   _____

   _____

3. Check all vacuum lines and hoses, as well as the tightness of all attaching and mounting bolts in the induction system. Comments:

   _____

   _____

4. Check for damaged air ducts. Comments:

   _____

   _____

5. Check the ignition circuit, especially the secondary cables, for signs of deterioration, insulation cracks, corrosion, and looseness. Comments:

_____

_____

6. Are there any unusual noises or odors? If there are, describe them and tell what may be causing them.

_____

_____

7. Inspect all related wiring and connections at the PCM. Comments:

_____

_____

☐ 8. Gather all pertinent information about the vehicle and the customer's complaint. This should include detailed information about the symptom from the customer, a review of the vehicle's service history, published TSBs, and the information in the service manual.

9. Are there any vacuum leaks? _____ Yes _____ No

10. Is the engine's compression normal? _____ Yes _____ No

11. Is the ignition system operating normally? _____ Yes _____ No

12. Are there any obvious problems with the air/fuel system? _____ Yes _____ No

13. Note your conclusions from the preceding questions and answers.

_____

_____

14. Check the operation of the MIL by turning the ignition on. Describe what happened and what this means.

_____

_____

☐ 15. Connect the scan tool to the DLC.

☐ 16. Enter the vehicle identification information into the scan tool.

☐ 17. Retrieve the DTCs with the scan tool.

18. List all codes retrieved by the scan tool.

_____

_____

_____

19. Conclusions from these tests and checks.

_____

_____

_____

Instructor's Response _____

_____

Chapter 5

REBUILDING TRANSMISSIONS AND TRANSAXLES

BASIC TOOLS

Basic mechanic's tool set

Clean lint-free rags

Appropriate service manual

UPON COMPLETION AND REVIEW OF THIS CHAPTER, YOU SHOULD BE ABLE TO:

- Diagnose noise and vibration problems and determine needed repairs.

- Remove and install a transmission/transaxle assembly from a car or light truck.

- Disassemble, clean, and inspect a transmission/transaxle.

- Inspect, repair, and replace transmission cases.

- Inspect and repair the bores, passages, bushings, vents, and mating surfaces of a transmission case.

- Inspect, repair, and replace extension housing and extension housing bushings and seals.

- Inspect and replace speedometer drive gear, driven gear, and retainers.

- Inspect and replace external seals and gaskets.

- Reassemble a transmission/transaxle after servicing it.

DIAGNOSIS OF NOISE AND VIBRATION PROBLEMS

Abnormal noises and vibrations can be caused by faulty bearings, damaged gears, worn or damaged clutches and bands, or a bad oil pump, as well as by contaminated fluid or an improper fluid level. Torque converter and cracked flexplate problems can also be the cause of vibrations (Figure 5-1).

A customer will often complain of a transmission noise that in reality is caused by something else in the driveline and not the transmission or torque converter. Bad CV or U-joints, wheel bearings, brake calipers, and dragging brake pads can generate noises that customers, and unfortunately some technicians, mistakenly blame on the transmission or torque converter. The entire driveline should be checked before assuming the noise is transmission related.

Most vibration problems are caused by an unbalanced torque converter assembly, a poorly mounted torque converter, or a faulty engine or transmission mount, or a defective output shaft. The key to determining the cause of the vibration is to pay particular attention to the vibration in relation to engine and vehicle speed. If the vibration changes with a change in engine speed, the cause of the problem is most probably the torque converter. If the vibration changes with vehicle speed, the cause is probably the output shaft or the driveline connected to it. The latter type of problem can be a bad extension housing bushing or universal joint, which would become worse at higher speeds.

Begin your diagnosis by determining if the cause of the problem is the driveline or the transmission. To do this, put the transmission in gear and apply the foot brakes. If the noise is no longer evident, the problem must be in the driveline or the output of the transmission. If the noise is still present, the problem must be in the transmission or torque converter.

Noise problems are also best diagnosed by paying attention to the speed, operating gear, and the conditions at which the noise occurs. If the noise is engine speed related and is

This chapter covers the mechanical aspects of diagnosis, disassembly, inspection, service, and assembly of an automatic transmission. The hydraulic components are detailed in Chapter 8.

Refer to a good manual transmission textbook for procedures on diagnosing the entire driveline.

Classroom Manual

Chapter 5, page 156

Problem	Probable Cause(s)
Ratcheting noise	The return spring for the parking pawl is damaged, weak, or misassembled
Engine speed sensitive whine	Torque converter is faulty Faulty pump
Popping noise	Pump cavitation—bubbles in the ATF Damaged fluid filter or filter seal
Buzz or high-frequency rattle Whine or growl	Cooling system problem Stretched drive chain Broken teeth on drive and/or driven sprockets Nicked or scored drive and/or driven sprocket bearing surfaces Pitted or damaged bearing surfaces
Final drive hum	Worn final drive gear assembly Worn or pitted differential gears Damaged or worn differential gear thrust washers
Noise in forward gears	Worn or damaged final drive gears
Noise in specific gears	Worn or damaged components pertaining to that gear
Vibration	Torque converter is out of balance Torque converter is faulty Misaligned transmission or engine Output shaft bushing is worn or damaged Input shaft is out of balance The input shaft bushing is worn or damaged

FIGURE 5-1 A basic noise and vibration diagnostic chart.

present in all gears, including Park and Neutral, the oil pump is the most probable source because it rotates whenever the engine is running.

However, if the noise is engine speed related and is present in all gears except Park and Neutral, the most probable sources of the noise are those parts that rotate in all gears, such as the drive chain, the input shaft, and torque converter. The noise from a torque converter is typically a whining noise that is not present when the transmission is in Park or Neutral. It is important to note that a vibration will not normally accompany the noise if it is caused by an internal part of the transmission. This is because those parts do not have enough mass to set up a vibration.

Noises that only occur when a particular gear is operating must be related to those parts responsible for providing that gear, such as a brake or clutch. Often, the exact cause of noise and vibration problems can only be identified through a careful inspection of a disassembled transmission.

Regardless of the exact fault in the transmission that is causing a vibration or noise, the transmission will need to come out and the entire unit will have to be checked and carefully inspected. Proper diagnosis prior to disassembling the transmission will identify the specific areas that should be carefully looked at and can prevent unnecessary transmission removal and teardown.

Transmission/Transaxle Removal

 WARNING: Be sure to wear safety glasses or goggles when working under the vehicle and when handling ATF. ATF, as well as rust and dirt, can cause serious damage to your eyes.

RWD Vehicles

Removing the transmission from a rear-wheel-drive (RWD) car is generally more straightforward than removing one from a front-wheel-drive (FWD) model, as there is typically one crossmember, one driveshaft, and easy access to cables, wiring, cooler lines, and bell-housing bolts. Transmissions in FWD cars, because of their limited space, can be more difficult to remove because you may need to disassemble or remove large assemblies such as engine cradles, suspension components, brake components, splash shields, or other pieces that would not usually affect RWD transmission removal.

The following is a list of components typically removed or disconnected while removing an automatic transmission from a RWD vehicle. This list is arranged in a suggested order of events. Some vehicles require more than this, others require less.

Battery ground cable
Transmission oil pan
Torque converter access plate
Torque converter drain plug
Transmission dipstick tube
Transmission cooler lines
Speedometer cable
Vacuum hose to modulator
Electrical connectors to solenoids
Electrical connectors to sensors
Gear selector linkage
Throttle pressure linkage
Kickdown linkage or electrical connector to switch
Neutral safety switch
Reverse lamp switch
Starter motor
Exhaust heat shields
Electrical connectors to oxygen sensors
Exhaust pipes and catalytic converters
Driveshaft
Torque converter to flexplate bolts or nuts
Transmission mounts
Crossmember
Bell housing to engine bolts

The exact procedure for removing a transmission will vary with each year, make, and model of vehicle. Always refer to the service manual for the procedure you should follow. Normally the procedure begins with placing the vehicle on a hoist so that you can easily work under the vehicle and under the hood.

Once the vehicle is in position, disconnect the negative battery cable and place it away from the battery. Disconnect and remove any transmission linkages connected to the engine. Also remove the ATF dipstick.

Then raise the vehicle and disconnect the parts of the exhaust system that may interfere with transmission removal. Disconnect all electrical connections at the transmission (Figure 5-2).

SPECIAL TOOLS
Drain pan
Transmission jack
Engine support and/or heavy-duty chain

SERVICE TIP:
If you plug the cooler line fittings on the housing and the lines themselves, you will prevent the frustration and possible danger of having ATF drip down your neck or into your eyes while you are removing the transmission.

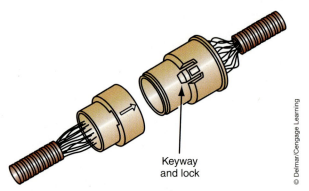

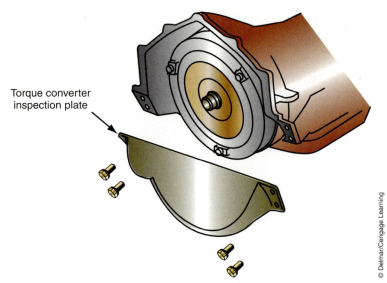

Torque converter
inspection plate

© Delmar/Cengage Learning

FIGURE 5-2 Be sure to examine the electrical connectors before attempting to separate them. Most have locks that must be unlocked.

FIGURE 5-3 Location of a typical torque converter (and flexplate) inspection cover and bolts.

Make sure you place them away from the transmission so that they are not damaged during transmission removal or installation. Disconnect and remove the starter motor.

Remove the torque converter inspection plate or dust cover (Figure 5-3). Place an index mark on the converter and the flexplate to ensure that the two will be properly mated during installation. Using a flywheel turning tool, rotate the flywheel until some of the converter-to-flexplate bolts are exposed. Loosen and remove the bolts. Then rotate the flywheel until more bolts are accessible. Remove them and continue the process until all the mounting bolts are removed. Once the bolts are removed, slide the converter back into the transmission.

Place a drain pan under the transmission and drain the fluid from the transmission. Once the fluid is out, place the oil pan back onto the transmission and keep it in place with three or four bolts. Move the drain pan under the rear of the transmission. Then remove the driveshaft.

Disconnect the shift and other linkages from the transmission. Then remove the speedometer cable (if there is one). Disconnect the cooler lines from the transmission (Figure 5-4). Cap the ends of the lines to prevent leakage and dirt from entering the cooling system.

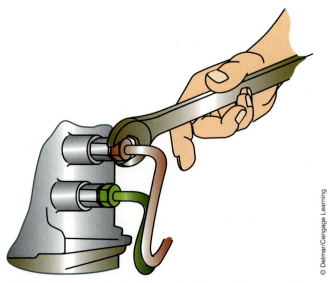

© Delmar/Cengage Learning

FIGURE 5-4 Disconnect the fluid cooler lines at the transmission.

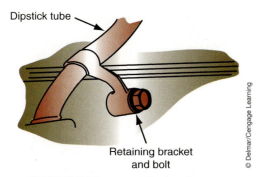

Dipstick tube

Retaining bracket and bolt

© Delmar/Cengage Learning

FIGURE 5-5 Disconnect the brackets for the transmission vent and dipstick tube and then remove the vent and dipstick tubes.

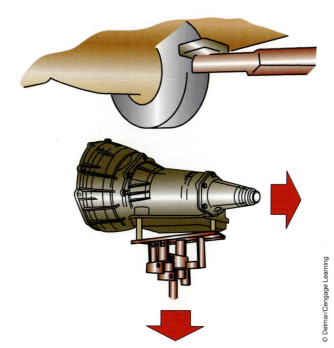

© Delmar/Cengage Learning

FIGURE 5-6 Move the transmission away from the engine until the unit is clear of the alignment dowels at the rear of the engine and the converter hub has a clear path while lowering the transmission.

Disconnect the brackets for the transmission vent and dipstick tube. Then remove the vent and dipstick tubes (Figure 5-5).

Place a transmission jack under the transmission and secure the transmission to it. The use of a transmission jack allows for easier access to parts hidden by crossmembers or hidden in the space between the transmission and the vehicle's floor pan. Using a chain or other holding fixture, secure the transmission to the jack's pad. Using two chains in an "X" pattern around the transmission works well to secure the transmission to the jack. If the transmission begins to slip on the jack while you are removing it, never try to catch it. Let it fall.

Remove the transmission mounting bolts. Then remove the crossmember at the transmission. After the mounts are free from the transmission, lower the transmission slightly so you can easily access the top transmission-to-engine bolts. Loosen and remove these bolts. Then remove the remaining transmission-to-engine bolts.

Move the transmission away from the engine until the unit is clear of the alignment dowels at the rear of the engine. Then slowly lower the transmission (Figure 5-6). Make sure the converter hub and any associated shafts have a clear path while you are lowering the transmission. Once the transmission is out of the vehicle, carefully move it to the work area and mount it to a stand.

FWD Vehicles

Obviously the procedure for removing a transmission from a FWD vehicle is different. Because all the components are confined to a very small area, the exact items that need to be removed or disconnected vary from vehicle to vehicle. The following is a list of components that are typically removed or disconnected. This list is arranged in a suggested order of events.

Battery ground cable
Underhood electrical connectors to transaxle
Front wheels
Electrical connectors at the wheel brake units
Brake calipers

CAUTION:
If the vehicle is a hybrid vehicle, make sure you follow all precautions and procedures for disconnecting the high-voltage circuit before proceeding to remove the transmission.

Steering knuckles
Drive axles
Transmission oil pan
Transaxle to engine brackets
Torque converter access plate
Torque converter drain plug
Transmission dipstick tube
Transmission cooler lines
Speedometer cable
Electrical connectors to solenoids
Electrical connectors to sensors
Gear selector linkage
Throttle pressure linkage
Kickdown linkage or electrical connector switch
Neutral safety switch
Reverse lamp switch
Starter motor
Exhaust heat shields
Electrical connectors to oxygen sensors
Exhaust pipes and catalytic converters
Torque converter to flexplate bolts or nuts
Transaxle to engine mount
Crossmember
Bell housing to engine bolts

On some vehicles, the recommended procedure may include removing the engine with the transaxle. Always refer to the appropriate service information before proceeding to remove the transaxle. You may waste much time and energy if you don't check the manual first.

> **CUSTOMER CARE:** Because much of your time will be spent working under the hood while you are removing an engine, make sure you use fender covers and take extra care not to damage the sheet metal or finish of the vehicle.

Begin removal by placing the vehicle on a lift. However, before raising the vehicle, take a look around the engine bay to see if any interference will occur between the firewall and engine components, such as distributors, fans and fan shrouds, fuel lines, exhaust systems or electrical components, when the transaxle is removed. If any causes for interference are found, these problems should be corrected before continuing. Also, any bell-housing bolts, wiring, or TV cables accessible from above should be removed. Before raising a FWD vehicle to begin transaxle removal, a support fixture should be attached to the engine (Figure 5-7).

Raise the vehicle to a comfortable height and remove all but the three or four corner bolts of the transmission pan, depending on the shape of the pan. Place a drain pan large enough to catch all the oil under the transaxle. Carefully remove bolts from one side of the pan. Back off the bolt or bolts on the other side just enough to allow the pan to drop slightly as you pry it loose. Be careful, as some pans will come loose without being pried and if you loosen the pan too much, you may have a large mess to clean up! When the fluid stops draining, replace the pan with a minimum of bolts, taking care not to lose the remaining bolts.

To remove FWD driveshafts, you must first loosen the large nut that retains the outer CV joint, which is splined on the shaft to the hub.

Now raise the vehicle and remove the front wheels. Tap the splined CV joint shaft with a soft-faced hammer to see if it is loose. Most will come loose with a few taps. Many Ford FWD

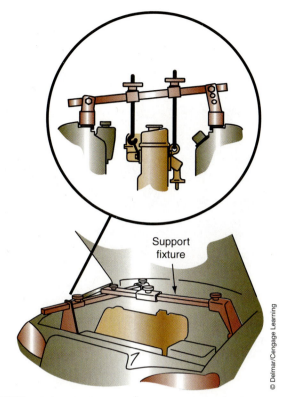

FIGURE 5-7 A typical engine support fixture for a FWD vehicle.

© Delmar/Cengage Learning

cars use an interference fit spline at the hub. You will need a special puller for this type of CV joint. The tool pushes the shaft out and on installation pulls the shaft back into the hub.

The lower ball joint must now be separated from the steering knuckle. The ball joint will be either bolted to the lower control arm or held into the knuckle with a pinch bolt. Once the ball joint is loose, the control arm can be pulled down and the knuckle can be pushed outward to allow the splined CV joint shaft to slide out of the hub (Figure 5-8). The inboard joint can then be pried out or it will slide out. Some transaxles have retaining clips that must be removed before the inner joint can be removed (Figure 5-9).

The speedometer drive gears may need to be removed before pulling out the driveshafts, as they may be damaged when the shafts are removed. On some cars, the inner CV joints have flange-type mountings. These must be unbolted for removal of the shafts. In some cases, the flange-mounted driveshafts may be left attached to the wheel and hub assembly and unbolted only at the transmission flange. The free end of the shafts should be supported and placed out of the way. Doing this will greatly decrease the amount of time needed to remove and install the transmission.

At this time, you should study the underbody layout to aid reinstallation. This is another time when use of an instant camera will help your memory when it comes time to reinstall the transmission. In many shops, the technician repairing the transmission is not the same person who removes and reinstalls it. If the installer does not remember the exact location of all the underbody and engine bay components, or the correct adjustments of control cables, the transmission may not operate properly when it is reinstalled. Improper installation can ruin a transmission or cause driveability, noise, and vibration problems.

Now the shift linkages, vacuum hoses, electrical connections, speedometer drives, and control cables should be disconnected (Figure 5-9). The exhaust system may also need to be lowered or partially removed. The inspection cover between the transaxle and engine should be removed to allow access to the torque converter bolts. There will be three to six bolts or nuts securing the converter to the flexplate, depending on the application.

SERVICE TIP:

To control the mess, some technicians disconnect a cooler line at the radiator and place the end of the line in a drain pan. The engine is then cranked and fluid pumped into the drain pan. This can be done to move most of the fluid out of the transmission. Do not crank the engine for more than 30 seconds at one time.

CAUTION:

It is recommended that the large nut that retains the outer CV joint be loosened with the vehicle on the floor and the brakes applied to reduce possible damage to the CV joints and wheel bearings.

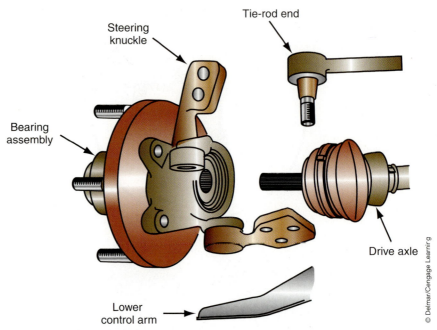

FIGURE 5-8 The knuckle assembly is moved out of the way to allow the stub shaft to be pulled out of the hub.

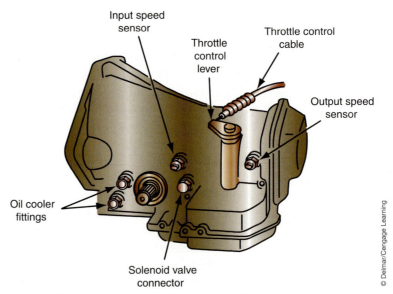

FIGURE 5-9 Location of the various switches, connectors, and levers on a typical transaxle.

Mark the position of the converter to the flexplate to help maintain balance or runout. It will be necessary to rotate the crankshaft to remove the converter bolts. This can be done by using a long ratchet and socket on the front crankshaft bolt or by using a flywheel-turning tool, if space permits.

Carefully remove the cooler lines by holding the case fitting with one wrench and loosening the line nut with a line wrench. Doing this assures that you will not twist the steel lines, which would damage them and restrict their flow.

With a transmission jack supporting the transmission, remove the transaxle mounts. The mounts may be at the top, side, or bottom of the unit. Some mounts consist of two parts (Figure 5-10). Make sure you remove both parts as well as all transaxle mounts.

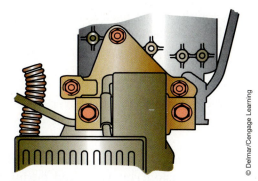

FIGURE 5-10 Remove the bolts from the transaxle insulator, then lower the transaxle and remove the nuts and bolts to remove the insulator bracket.

Now remove the starter. The starter wiring may be left connected but you will need to hang the starter with heavy mechanic's wire to avoid damage to the cables. You can also remove the starter from the vehicle to get it totally out of the way.

Now pull the transaxle away from the engine. It may be necessary to use a pry bar between the transaxle and engine block to separate the two units. Make sure the converter comes out with the transaxle. This prevents bending the input shaft, damaging the oil pump, or distorting the drive hub. After separating the transaxle from the engine, retain the torque converter in the bell housing. This can be done simply by bolting a small combination wrench to a bell-housing bolt hole across the outer edge of the converter.

Installation

Transmission installation is generally a reverse of the removal procedure. Care must be taken to avoid destroying the new or rebuilt transmission during installation. A quick check of the following list will greatly simplify your installation and reduce the chances of destroying something during installation.

- Make sure the block alignment pins (**dowels**) are in the appropriate bores and are in good shape and that the alignment holes in the bell housing are not damaged (Figure 5-11).
- Make sure the pilot hole in the crankshaft is smooth and not out-of-round. This will allow the converter to move in and out on the flexplate.
- Make sure the pilot hub of the converter is smooth and cover it with a light coating of chassis lubricant to prevent chafing or rust.
- Make sure the converter's drive hub is smooth and coat it with trans gel or petroleum jelly.
- Secure all wiring harnesses out of the way to prevent their being pinched between the bell housing and engine block. If the wires get pinched, not only will there be a large electrical short, but you may destroy the car's computer.

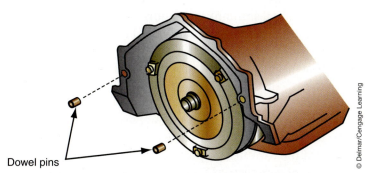

Dowel pins

FIGURE 5-11 The dowel pins may remain in the engine or the transmission case when the transmission is removed. In either case, they should be inspected.

CAUTION:
Never allow the starter to hang by the wires attached to it. The weight of the starter can damage the wires or, worse, break the wire and allow the motor to fall, possibly on you or someone else. Always securely support the starter and position it out of the way after you have unbolted it from the engine.

CAUTION:
Never force the torque converter back into the oil pump if it has slipped out.

Alignment pins are commonly referred to as dowels.

CAUTION:
Dowels or locating pins are used to keep the engine and transmission in perfect alignment. Torque converter housing bolts alone cannot maintain perfect alignment. Perfect alignment is necessary to prevent flexplate breakage.

- Flush out the oil cooler lines and the cooler itself to remove any material that could damage the transmission. Likewise, the converter should be flushed. It is recommended that clutch-type converters be replaced, as it is not possible to tell how much the unit has been damaged.
- Always perform an endplay check and check the overall height before reinstalling a torque converter or installing a fresh unit out of the box.
- Pour one quart of the recommended fluid into the converter before mounting the converter to the transmission. This will assure that all parts in the converter have some lubrication before start-up.

Slide the converter into the transmission, making sure that all drives are engaged. Double-check this by using the height dimension you measured during transmission removal. In older transmissions, the torque converter must engage into the input shaft stator splines and the oil pump, later models also have those or a direct driveshaft or an oil pump driveshaft.

Secure the converter in the transmission as you did during removal. Then transfer the transmission to the jack and move it under the car. Raise the transmission to get close alignment to the engine block. If the transmission has a full-circle bell housing, you will need to align the converter drive studs or bolt pads before you push the transmission up against the block, as this is difficult to do once the transmission is against the engine block. Make sure you set the torque converter onto the flexplate in alignment with the marks made during removal. These will provide proper balance for the unit. Once the converter is aligned, be sure the block dowels line up with the bell housing, then push the transmission against the engine block. Check to be sure that nothing has been caught between the block and the bell housing. Start two bell housing bolts across from each other and slowly tighten them. Check for free rotation of the torque converter while doing this. Never use the bolts to pull the transmission against the engine block. Then, install the rest of the bolts and torque them to specifications.

On FWD cars, if the engine is held in place with a support bar, the transmission jack may be removed at this time. If the car has a split cradle-type subframe, it should be installed now.

On RWD vehicles, it may be necessary to leave the transmission jack in place while components such as exhaust crossovers or frame crossmembers are installed. Do not connect the gearshift linkage until the transmission is mounted to the crossmember and the transmission jack is removed. Once the rear transmission mount is attached to the crossmember, the jack may be removed.

Install the converter drive bolts (or nuts). You should notice while installing these fasteners that the converter has some noticeable fore and aft movement. This amount varies with different transmissions, but it is generally between $\frac{1}{8}$ and $\frac{1}{4}$ inch. This is normal and necessary, as it allows the converter to move on the flexplate and also allows for a noninterference fit at the oil pump drive gear. If there were no movement, premature pump failure would result.

The cooler lines should now be installed. Make sure the lines are connected to the bores they were originally in. Then, tighten the line fittings, making sure you do not twist or distort the line. The remaining components—starter, throttle cables, electrical connections, dipstick tube, and so on—can now be installed.

While connecting the manual shift linkage (Figure 5-12), pay attention to the condition of the plastic bushings. If these are worn or missing, hard or inaccurate linkage movement will result. Replace the bushings if needed.

Some other areas that require special attention during installation are the grounding straps and any rubber tubes. These are often overlooked during transmission removal and installation and can cause problems if not checked. All rubber tubes are suspect, as they are exposed to heat, which causes them to crack. New parts should be used if necessary. The ground straps must be in good condition and free of corrosion, as these provide an electrical ground path to the body of the car during operation. Failure to clean or attach these straps or cables can also result in poor signals to the PCM, voltage spikes that can damage the PCM, electrolysis through the fluid that welds internal transmission components, or even fused manual linkage cables as the current looks for a path to flow through.

Classroom Manual

Chapter 5, page 163

CAUTION:

Never use an impact wrench on torque converter bolts. Impact wrenches can drive the bolts through the cover, which will warp the inside surface and prevent proper clutch apply or may damage the clutch pressure plate.

CAUTION:

After the transmission has been installed, make sure everything is properly realigned. Repositioning the transmission an inch or less can have a big effect on the manual shift linkage adjustment.

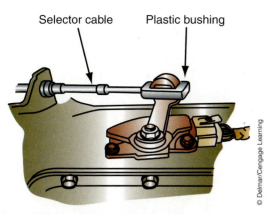

Selector cable   Plastic bushing

© Delmar/Cengage Learning

FIGURE 5-12 Connect the gear selector rod or cable. Make sure the plastic bushings are in good shape and that they are properly seated.

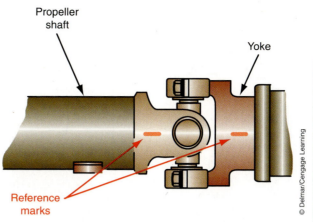

Propeller shaft

Yoke

Reference marks

© Delmar/Cengage Learning

FIGURE 5-13 The index marks on the driveshaft flange yoke and the pinion flange ensure the driveshaft is installed at the correct phase.

SERVICE TIP:

On governor-equipped transmissions, put the transmission into reverse as soon as the engine starts. Many transmissions do not send fluid to the governor in reverse; putting the transmission in reverse prevents the possibility of moving debris from the rebuild into the governor, which would cause it to stick or seize.

SERVICE TIP:

If the transmission is computer-controlled, check the service manual before taking it on its initial road test. Some computer-controlled automatic transmissions require that a "Learning Procedure" be followed, which includes various driving conditions. Since you need to road test the transmission and need to teach the transmission, you may as well do both at the same time.

This sequence covers the general disassembly of an automatic transmission. Complete overhaul of common transmissions is covered in Chapter 10 of this manual.

On RWD vehicles, the driveshaft must be installed using the marks you made during removal (Figure 5-13). Be sure to coat the slip yoke with ATF before sliding it into the extension housing. The driveshafts of FWD cars are installed in the reverse manner in which they were removed and the related components (speedometer gears, strut lower ball joint bolts, and so on) should be installed just as they were before they were removed. Use a new nut on the outer CV joint stud, if the nut is the self-locking type. The torque of this nut is critical. Torquing the nut should be done with the car's tires on the ground and with the brakes held. Air-type impact tools should not be used, as they can damage the wheel bearing or hub.

Connect the battery cables. Then add about half of the total quantity of the proper ATF to the transmission. This amount varies, but an average amount is four quarts. Some technicians will also connect a pressure gauge to the transmission and take pressure readings during the initial operation of the fresh transmission. Some imported-transmission manufacturers do require that this be done because line pressures must be set after start-up. Connecting the pressure gauges will also give you an indication of whether there is sufficient oil pressure to road test the car.

Put the vehicle on a lift, then start the engine and apply the brakes. Move the shift selector through all of the gear ranges. Then place the selector into park. Check the fluid level and add ATF until you reach the "add" or "cold" mark. Then, inspect the transmission for any signs of leakage or loose bolts. If a pressure gauge was attached to the transmission and the pressure reading was fine, disconnect the gauge after you shut down the engine.

Road test the vehicle to check the operation of the transmission and anything else that may have been affected by your work. Make any necessary adjustments. The road test should cover at least 20 miles in order to completely warm up the transmission. Recheck and adjust the fluid level. It should be between the "add" and "full" marks on the dipstick. DO NOT OVERFILL! After the road test, visually inspect the transmission again for signs of leakage. Also carefully look around the engine and transmission to see if any wires, hoses, or cables are disconnected or positioned in an undesirable spot.

DISASSEMBLY

The disassembly of a transmission and a transaxle are similar; therefore, the disassembly, inspection, and reassembly guidelines can be grouped together.

Once the transmission has been removed from the vehicle, certain things should be checked before beginning to disassemble the transmission. Measure the depth at which the converter fits into the transmission case (Figure 5-14). To do this, hold a straightedge across the bell housing and measure into the pilot hub or a drive pad on the converter. Record this dimension and save it for use when installing the converter.

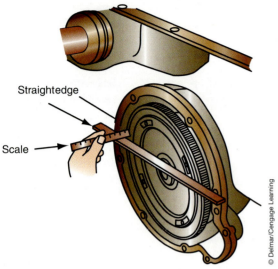

Straightedge

Scale

FIGURE 5-14 After the transmission and engine have been separated, measure and record the depth of the converter into the housing. This measurement will be used during reassembly to ensure the torque converter and oil pump are properly assembled.

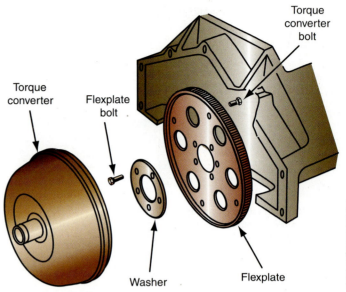

Torque converter bolt

Torque converter

Flexplate bolt

Washer

Flexplate

FIGURE 5-15 Carefully inspect the flexplate and torque converter mounting hardware.

A transmission overhaul normally includes replacing all rubber and paper gaskets, frictional materials, and worn bushings; cleaning the planetary gearsets and the valve body; and making correct endplay adjustments.

The term **teardown** is often used to describe the process of disassembling a transmission.

Classroom Manual

Chapter 5, page 166

Also inspect the flexplate for warpage and cracks (Figure 5-15). Pay attention to the condition of the teeth on the starter ring gear. Also check the flexplate for excessive runout and elongated bolt holes. Replace the flexplate if there is evidence of damage. Also inspect the converter-attaching bolts. Replace any worn bolts with suitable equivalents. While inspecting the flexplate, check the starter ring gear for looseness and damage. Pay special attention to the welds that secure the ring gear to the plate. It is common for these and the areas around them to crack.

Check the flexplate for cracks at the crankshaft mounting bolt holes. The best way to do this is to remove the plate and hold it up toward a bright light. If there are cracks, the light will shine through the cracks. If there is any flexplate damage, the flexplate should be replaced. Before reinstalling the flexplate, check the service manual to make sure the attaching bolts are either reusable or should be replaced. Also make sure the flexplate is installed in the correct direction and that all spacers are in place.

Check the drive hub of the torque converter. It should be smooth and show no signs of wear. Pay particular attention to the area the seal rides on. If the hub is worn, the torque converter should be replaced and the oil pump drive should be carefully inspected for scoring or other damage.

Before disassembling the transmission, check out the causes of any leakage. If there is evidence of leakage or leakage is the reason for the **teardown**, determine the path of the leakage before cleaning the area around the seals. At times, the leakage may be from sources other than the seal. Leakage could be from worn gaskets, loose bolts, cracked housings, or loose line connections.

Inspect the outside sealing area of the seal to see if it is wet or dry. If it is wet, see whether the oil is running out or if it is merely a lubricating film. Check both the inner and outer parts of the seals for wet oil, which means leakage.

When removing a seal, inspect the sealing surface, or lips, before cleaning it. Look for signs of unusual wear, warping, cuts and gouges, or particles embedded in the seal. On spring-loaded lip seals, make sure the spring is seated around the lip, and that the lip was not damaged when first installed. If the seal's lip is hardened, this was probably caused by heat from either the shaft or the fluid.

If the seal is damaged, check all shafts for roughness, especially at seal contact areas. Look for deep scratches or nicks that could have damaged the seal. Determine if a shaft spline, keyway, or burred end could have caused a nick or cut in the seal lip during installation. Inspect the bore into which the seal was fitted. Look for nicks and gouges that could create a path of oil leakage. A coarsely machined bore can allow oil to seep out through a spiral path. Sharp corners at the bore edges could have scored the metal case of the seal when it was installed. These scores can make a path for oil leakage.

Cleaning and Inspection

Before disassembling the automatic transmission, care should be taken to clean away any dirt, undercoating, grease, or road grime on the outside of the case. This ensures that dirt will not enter the transmission during disassembly. Once the transmission is clean outside, you may begin the disassembly.

When cleaning automatic transmission parts, avoid the use of solvents, degreasers, and detergents that can decompose the friction composites used in a transmission. It is best not to attempt to clean the friction members, as this will damage the parts. Use compressed air to dry components; don't wipe down parts with a rag. The lint from a rag can easily destroy a transmission after it has been rebuilt.

There are many different ways to clean automatic transmission parts. Some rebuilding shops use a parts washing machine, which does an excellent job in a small amount of time, to thoroughly clean a transmission case, converter housing, and extension housing. These parts washers use hot water and a special detergent that are sprayed onto the parts as they rotate inside the cleaner. The key to good cleaning with these machines is to use a very small amount of soap with very hot water. Many rebuilders simply clean the parts in a mineral spirits tank, where the parts are brushed and hand cleaned. No matter what type of cleaning procedure is followed, the transmission and parts should be rinsed with water, then thoroughly dried with compressed air before reassembly.

 WARNING: **Always wear safety goggles when using compressed air to dry something. The air pressure can easily move dirt, metal, or other debris around the work area. If these get into your eye, they can cause permanent damage.**

After the case is clean, remove the torque converter and carefully inspect it for damage. Check the converter hub for grooves caused by hardened seals. Also check the bushing contact area. To remove the converter, slowly rotate it as you pull it from the transmission, and have a drain pan handy to catch the fluid. It should come right out without binding. This is a good time to check the input shaft splines, stator support splines, and the converter's pump

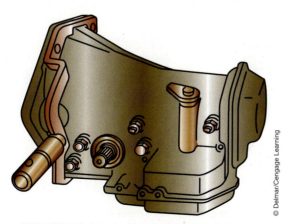

© Delmar/Cengage Learning

FIGURE 5-16 **A typical transaxle holding fixture.**

The 4R70W and 4T60 are examples of transmissions that use a direct driveshaft.

SERVICE TIP:

Record endplay measurements before disassembly and during reassembly.

SPECIAL TOOLS

Dial indicator mounting fixture

Dial indicator

Transmission stand

drive hub for any wear or damage. Converters with direct driveshafts should be checked to be sure that no excessive play is present at the drive splines of the shaft or the converter. If any play is found in the converter, the converter or the shaft must be replaced.

Position the transmission to perform an endplay check. The transmission endplay checks can provide the technician with information about the condition of internal bearings and seals, as well as clues to possible causes of improper operation found during the road test. These measurements will also determine the thickness of the thrust washer(s) during reassembly. Thrust-washer thickness sets the endplay of various components. Excessive endplay allows clutch drums to move back and forth too much, causing the transmission case to wear. If there is insufficient clearance, rapid wear may occur due to inadequate oil clearances. Assembled endplay measurements should be between minimum and maximum specifications, but preferably at the low end of the specifications. Photo Sequence 9 covers a typical procedure for checking endplay.

Position the transmission so that the shaft centerline is vertical. This allows the weight of the internal components to load the shafts toward the rear of the transmission. Most GM transmissions can be checked by mounting a dial indicator to read the input shaft movement. Zero the dial indicator and lift upward on the input shaft. Most Chrysler and some Ford transmissions require removing the oil sump pan and valve body and prying the input shell upward. Chrysler and Ford transmissions may be measured horizontally (Figure 5-17), but you may want to center the output shaft with a slip yoke for more accurate readings.

Input shaft endplay is measured with a dial indicator. The dial indicator should be solidly clamped to the bell housing and the plunger positioned so that it is centered on the end of the input shaft. Move the input shaft in and out of the case and note the reading on the indicator. Compare this reading with the specifications for that transmission. If the endplay is incorrect, it should be corrected during assembly by installing a thrust washer, snap ring, or spacer of the correct thickness between the input shaft and the output shaft. Some transmissions require additional endplay checks during disassembly. The manufacturer's recommended procedure for checking endplay should always be followed.

Some transaxles may require more than one check. As most transaxles have their shaft centerlines to one side, the transaxle may need to be partially disassembled in order to gain access to the shafts.

It is possible that an output or final drive endplay check should be made on either a transmission or transaxle (Figure 5-18). The purpose of these settings is to provide long gear life

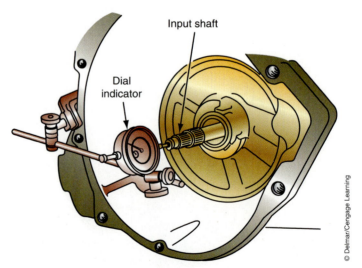

© Delmar/Cengage Learning

FIGURE 5-17 The endplay of the input shaft should be checked during disassembly and reassembly.

MEASURING INPUT SHAFT THRUST PLAY (ENDPLAY)

All photos in this sequence are © Delmar/Cengage Learning.

P9-1 Position the transmission so the input shaft is facing up.

P9-2 Mount a dial indicator to the transmission case so the indicator's plunger can move with the input shaft as the shaft is pulled up and pushed down.

P9-3 Lightly pull up on the shaft, then push it down until it stops. Then zero the indicator.

P9-4 Pull the shaft up until it stops and then read the indicator. This is the total amount of endplay.

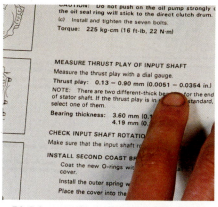

P9-5 Locate the endplay specifications in the service manual.

P9-6 Compare your measurements to the specifications.

and to eliminate engagement "clunk." All measurements taken during this phase of teardown should be kept during overhaul and used as a guide during the rebuild. Having this information gives the technician places to look for worn parts during the teardown and helps obtain the necessary shims to correct undesired endplay. If the endplay is excessive, this may indicate that the thrust washers or bearings are worn or the sealing rings and grooves are worn. During rebuild, endplay settings should be kept to the minimum allowable amount. Once the endplay checks have been made and recorded, disassembly can begin.

A clean work area, clean tools, and an orderly sequence during teardown will help you avoid lost time and frustration during reassembly. As you remove the various assemblies from the transmission, place them on the workbench in the order that they were removed. The correct position of thrust washers, snap rings, and even bolts and nuts is of great importance and should be noted. Unless the transmission suffered a major breakdown or was overheated, the internal components of the transmission will be fairly clean due to the nature of the fluid. A light cleaning with mineral spirits or a simple blowing off with low-pressure air is all that is necessary for cleaning most parts for inspection.

SERVICE TIP:
When working on automatic transmissions, there is no such thing as being too clean. All dirt, grease, and other materials should be cleaned off any parts that are going to be reused.

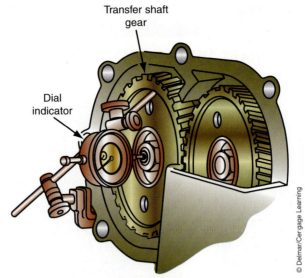

Transfer shaft gear

Dial indicator

© Delmar/Cengage Learning

FIGURE 5-18 **Some transaxles require additional end play checks during assembly and disassembly.**

Each subassembly should be completely disassembled and the parts thoroughly cleaned and checked for signs of excessive wear, scoring, overheating, chipping, or cracking. If there is any question as to the condition of a part, replace it.

Basic Disassembly

The following is a general procedure for disassembling a transmission. Always refer to the manufacturer's recommended procedure for the specific transmission.

With the transmission mounted on a fixture, remove the oil pan. Carefully inspect the oil pan for types of foreign matter. An analysis of the type of matter may help you determine what parts need to be carefully examined.

Remove the bell housing from the transmission case, if it is not part of the case. Unscrew and remove all externally mounted solenoids and switches, such as the electrical kickdown solenoid. Remove the O-rings and seals for each after the part is removed.

If the transmission has a vacuum modulator, unscrew and remove it. Remove the speedometer drive assembly, with its gear and O-ring.

Many transaxles have a valve body cover at the side of the unit (Figure 5-19). Remove the retaining bolts and then remove the cover. Be careful not to bend the cover's flange while removing it. With the cover removed, a wiring harness to various solenoids and sensors may be exposed. Disconnect the connectors of the harness before attempting to remove the valve body.

Unbolt and remove the solenoids from the valve body (Figure 5-20). In some transmissions the solenoids are part of a single unit, along with the TCM, remove this unit. Whenever disconnecting any electrical part, make sure to inspect the wiring and connectors.

Unbolt the valve body. Most valve bodies are assemblies of two or more valve units with separators and gaskets (Figure 5-21). Carefully separate the total assembly. Make sure you note the location of all check balls before removing them. Also, the length of the bolts may vary; therefore, keep track of where each bolt came from.

Remove the manual valve from the valve body to prevent the valve from dropping out. Back off the servo piston stem locknut and snugly tighten the piston stem to prevent the front clutch drum from dropping out when removing the front pump.

Using the correct puller, remove the front pump from the case (Figure 5-22). Lay this assembly to the side for further inspection.

SPECIAL TOOLS

Transmission mount

Mineral spirits

Snapring pliers

Dial indicator

Assortment of special tools as needed for particular transmission

SERVICE TIP:

The magnets inside electronic shift control solenoids will attract any ferrous metal that is floating around the inside of the transmission. Thoroughly clean or replace these solenoids during a transmission rebuild.

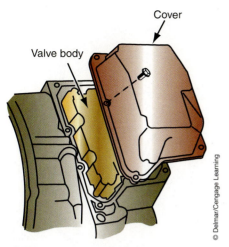

FIGURE 5-19 If the transaxle has a side valve body cover, remove it to gain access to the valve body.

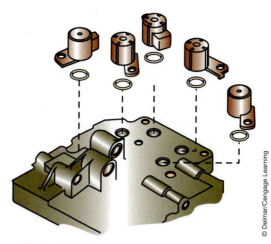

FIGURE 5-20 Remove the solenoids from the valve body. Keep track of their exact location.

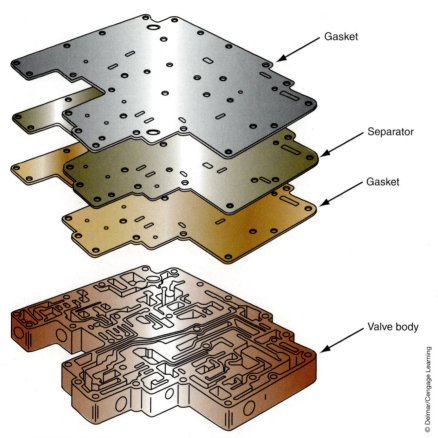

Gasket

Separator

Gasket

Valve body

FIGURE 5-21 This valve body is comprised of several separate assemblies.

Remove the front clutch thrust washer and bearing race. Now back off the front brake band servo piston stem to release the band. Remove the brake band strut and front brake band. The drum and band may be removed together.

Remove the front and rear clutch assemblies (Figure 5-23). Note the positions of the front pump thrust washers and rear clutch thrust washer, if the transmission has them.

Remove the rear clutch hub, front planetary carrier, and connecting shell. Note the positions of the thrust bearings and the front planetary carrier and thrust washer.

CAUTION:
When removing the valve body, the steel check balls may fall out. Be prepared for this to happen, and take steps so that you don't lose them. Better yet, try to remove the valve body so the balls will remain in place while the valve body is separated from the case.

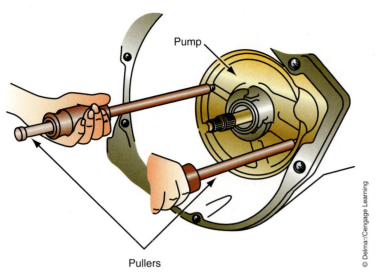

FIGURE 5-22 The proper tool should be used to pull the oil pump from the transmission housing.

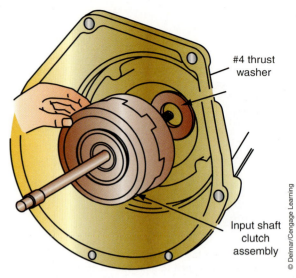

FIGURE 5-23 When removing clutch assemblies, be careful not to lose track of the placement of the thrust washers.

Remove the output shaft snap ring. It will often be easier to remove the snap ring if the carrier is removed first. Remove the carrier snap ring and remove the carrier. Now remove the output shaft snap ring.

Remove the rear connecting drum from the housing. Then using a screwdriver, remove the large retaining snap ring of the rear brake assembly (Figure 5-24). Tilt the extension housing upward and remove the rear brake assembly.

Then unbolt and remove the extension housing. Be careful not to lose the parking pawl, spring, and retaining washer. Pull out the output shaft, without the governor. Now remove the governor with its attachments (such as the oil distributor, thrust washer, and needle bearing assembly).

Remove the inner race of the one-way clutch, the thrust washer, piston return spring, and thrust ring. Using compressed air, remove the rear brake piston and front servo.

 WARNING: Cover the servo with a rag to prevent ATF from blowing into your face and the servo piston from popping into your face.

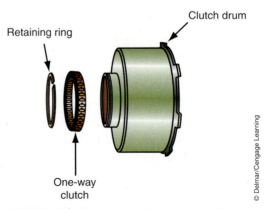

FIGURE 5-24 One-way clutches are normally retained by a snap ring.

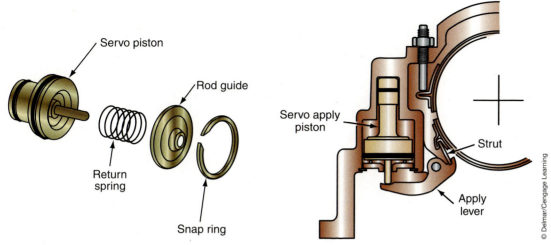

FIGURE 5-25 All servos and accumulators should be removed from their bores in the case and the bores carefully inspected.

Band servos and accumulators are pistons with seals in a bore held in position by springs and retaining rings. Remove the retaining rings and pull the assembly from its bore for cleaning (Figure 5-25). Check the condition of the piston and springs. Cast iron seal rings may not need replacement, but elastomer seals should always be replaced.

Once the transmission has been disassembled into its various subassemblies, each subassembly should be disassembled, cleaned, inspected, and reassembled.

TRANSMISSION CASE SERVICE

The transmission case should be thoroughly cleaned and all passages blown out. Make sure all electrical components have been removed from the case before cleaning it. After the case has been cleaned, all bushings, fluid passages, bolt threads, clutch plate splines, and the governor bore should be checked. The passages can be checked for restrictions and leaks by applying compressed air to each one. If the air comes out the other end, there is no restriction. To check for leaks, plug off one end of the passage and apply air to the other. If pressure builds in that passage, there are probably no leaks in it.

Modern transmission cases are made of aluminum, primarily to save weight. Aluminum is a soft material that is susceptible to corrosion and can be deformed, scratched, cracked, or scored much more easily than cast iron. Special attention should be given to the following areas: clutch, oil pump, servo, and accumulator bores. All bores should be smooth to avoid scratching or tearing the seals. The servo piston could also hang up in a bore that is deeply scored. Check the fit of the servo piston in the bore without the seal, if possible, to be sure it has free travel. There should be no tight spots or binding over the whole range of travel. Any deep scratches or gouges that cause binding of the piston will require case replacement.

Case-mounted accumulator bores are checked the same as servo bores. The oil pump bore at the front of the case should be free of any scratches that would keep the O-ring from sealing the outer diameter of the pump to the front of the case. Case-mounted hydraulic clutch bores are prone to the same problems as servo bores. Look for any scratches or gouges in the sealing area that would affect the rubber seals. It is possible to damage these areas during disassembly, so be careful with tools used during overhaul.

Sealing surfaces of the case should be inspected for surface roughness, nicks, or scratches where the seals ride (Figure 5-26). Any problems found in servo bores, clutch drum bores, or governor support bores can cause pressure leakage in the affected circuit. Imperfections in steel or cast iron parts can usually be polished out with **crocus cloth**. Care should be taken not to disturb the original shape of the bore. Under no circumstances should sandpaper be

The inspection, service, and testing of all hydraulic components are detailed in Chapter 7 of this manual.

Classroom Manual
Chapter 5, page 153

Crocus cloth is a very fine polishing paper. It is designed to remove very little metal, so it is safe to use on critical surfaces. Never use regular sandpaper, as it will remove too much metal.

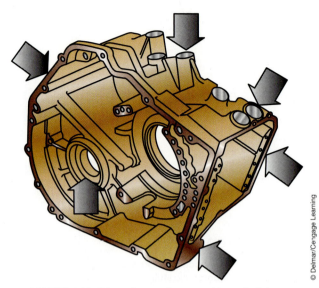

FIGURE 5-26 All sealing surfaces and bored of the transmission case should be carefully inspected for cracks, grooves, and scratches.

© Delmar/Cengage Learning

The oil passages in the case are commonly called worm tracks.

SERVICE TIP:
The Ford E4OD has shown this to be a problem with as much as 0.0625-inch warpage across the lower surface. Since at the time of writing the E4OD case costs approximately $625, replacing the case is rather expensive. It is possible to push the case back into the correct shape and then, with the center support bolted in, flat-file the case to obtain the desired flatness. This type of repair is used in the aftermarket to save an otherwise good case and to reduce costs. This is not a Ford-authorized repair.

used. Sandpaper will leave too deep a scratch in the surface. Use the crocus cloth inside clutch drums to remove the polished marks left by the cast iron sealing rings. This will help the new rings rotate with the drum as designed. As a rule, all sealing rings, either cast iron or Teflon, are replaced during overhaul, as this gives the desired sealing surface required for proper operation.

Passages in the case guide the flow of fluid through the case. Although not common, porosity in this area can cause cross-tracking of one circuit to another. This can cause bind-up (two gears at once) or a slow bleed-off of pressure in the affected circuit, which can lead to slow burnout of a clutch or band. If this is suspected, try filling the circuit with solvent and watching to see if the solvent disappears or leaks away. If the solvent goes down, you will have to check each part of the circuit to find where the leak is. Be sure to check that all necessary check balls were in position during disassembly.

Check the valve body mounting area for warpage with a straightedge and feeler gauge. This should be done in several locations so that any crossover from one worm track to another is evident. If there is a slight burr or high spot, it can be removed by flat-filing the surface.

A long straightedge should be laid across the lower flange of the case to check for distortion. Any warpage found here may result in circuit leakage, causing any number of hydraulically related problems. Case warpage should be less than 0.002 inch. Cases with bolted-in center supports, such as 3L80, 4L80, and E4OD models, should be checked with the support bolted in place. Cases are being made lighter and often will distort during service. This causes a pressure loss to some circuits, causing band or clutch burnouts.

Be sure to check all bell-housing bolt holes and dowel pins. Cracks around the bolt holes indicate that the case bolts were tightened with the case out of alignment with the engine block. The case should be replaced if the following problems are present: broken worm tracking, cracked case at the oil pump to case flange, cracked case at clutch housing pressure cavity, ears broken off the bell housing, or oil pan flange broken off the case. Although it is possible to weld the aluminum case, it is not possible to determine if the repair will hold. A transmission case is very thin and welding may distort the case.

If any of the bolts that were removed during disassembly have aluminum on the threads, the thread bore is damaged and should be repaired. Thread repair entails installing a thread insert, which serves as new threads for the bolt, or retapping the bore. After the threads have been repaired, make sure you thoroughly clean the case to remove all metal filings.

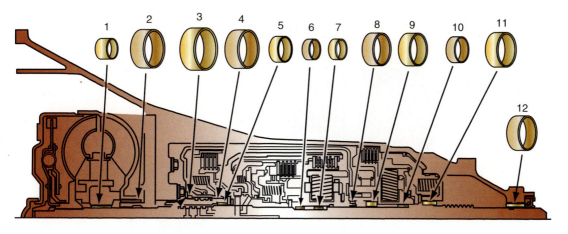

LEGEND

1. Bushing, stator shaft (front)
2. Bushing, oil pump body
3. Bushing, reverse input clutch (front)
4. Bushing, reverse input clutch (rear)
5. Bushing, stator shaft (rear)
6. Bushing, input sun gear (front)
7. Bushing, input sun gear (rear)
8. Bushing, reaction carrier shaft (front)
9. Bushing, reaction gear
10. Bushing, reaction carrier shaft (rear)
11. Bushing, case
12. Bushing, case extension

FIGURE 5-27 **Bushings are used throughout a transmission.**

The small screens found during teardown should be inspected for foreign material. These screens are used to prevent valve hang-up at the pressure regulator and governor and they must be in place. Most screens can be removed easily. Care should be taken when cleaning, because some cleaning solvents will destroy the plastic screens. Low air pressure (approximately 30 psi) can be used to blow the screens out in a reverse direction.

Bushings (Figure 5-27) in a transmission case are normally found in the rear of the case and require the same inspection and replacement techniques as other bushings in the transmission. Always be sure that the oil passage to a pressure-fed bushing or bearing is open and free of dirt and foreign material. It does no good to replace a bushing without checking to be sure it has good oil flow.

Vents are located in the pump body or transmission case and provide for equalization of pressures in the transmission. These vents can be checked by blowing low-pressure air through them, squirting solvent or brake cleaning spray through them, or by pushing a small-diameter wire through the vent passage. A clean, open passage is all you need to verify proper operation.

EXTENSION HOUSING

Check the extension housing (Figure 5-28) for cracks, especially around the case mounting surface and the pad that attaches to the transmission mount. Using a straightedge and feeler gauge set, check the flatness of the mating surface. Lay the straightedge across the surface and attempt to insert feeler gauges of various sizes between the bottom of the straightedge and the surface. If a 0.002-inch gauge fits under the straightedge, the surface may need to be resurfaced or the housing replaced. Minor problems may be corrected by filing down the surface. However, filing should only be done when it is necessary. To correctly file the surface, select the largest single-cut file available. Place the file across one end of the surface and pull or draw the file across the surface to the opposite end. Lift the file off the surface and place it back at its original position, then draw the file to the other end. Repeat this process until the surface is corrected. Never move the file from side to side.

Carefully inspect all threaded and nonthreaded bores in the housing. All damaged threaded bores should be repaired by running a tap through the bore or by installing threaded inserts. If any condition exists that cannot be adequately repaired, the case should be replaced.

SERVICE TIP:

The oil pump gears may seize to the pump plate in a 5R55E if the bell housing is warped around the bushing bore. Be extra careful when inspecting this bell housing as this fault can destroy a transmission rebuild.

Classroom Manual

Chapter 5, page 163

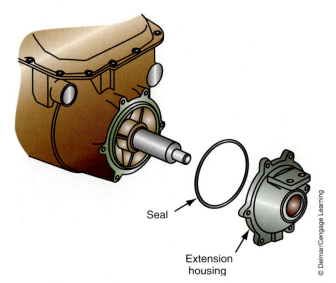

Seal

Extension
housing

© Delmar/Cengage Learning

FIGURE 5-28 The extension housing should be checked for defects and flatness. Always replace its seals and gaskets during reassembly.

At the rear of the extension housing is the slip yoke bushing. This bushing will normally wear to one side due to loads imparted on it during operation. Oil-feed holes at this bushing must be checked to make sure oil can get to this bushing. The speedometer drive gear is often responsible for throwing oil back to the rear bushing. A sheared or otherwise inoperative speedometer gear could cause the extension housing bushing to fail. Always make sure this bushing is aligned correctly during replacement or premature failure can result.

Before installing the rear extension housing, assemble the parking pawl pin, washer, spring, pawl, and any other assembly that is enclosed by the extension housing. Be sure they are assembled properly. Install the housing and tighten the bolts to specifications.

GASKETS

Classroom Manual

Chapter 5, page 166

Gaskets are used to create a seal between two flat surfaces. This seal must be able to prevent fluid leaks while undergoing changes in pressure and temperature. Whenever a transmission is overhauled, new gaskets should be used throughout the transmission. The only exception to this is the transmission oil pan gasket, which can be reused if it is in good condition. If the oil pan gasket is damaged or deformed in any way, do not reuse it. Most rebuilders don't reuse oil pan gaskets; they always replace them. The risk of the gasket not sealing is not worth the money saved.

While disassembling a transmission, keep all of the old gaskets. These will not be reinstalled, but will be used to select the correct gaskets for reassembly. Often transmission rebuilding kits come with a few different gaskets for the same part. To help select the correct gasket for the transmission and the part, compare the old gasket with the new ones. Obviously, the new gasket that matches the old one is the gasket that should be used.

When installing a gasket, make sure both surfaces are clean and flat. Any imperfections in a sealing surface should be corrected before installing a new gasket. These imperfections will prevent the gasket from providing a good seal. To remove minor scratches from a sealing surface, use a fine flat file. Make sure you don't remove more metal than is necessary. Also make sure the surface is flat after you have filed away the scratches.

To ensure flatness, spread a piece of medium-grade (approximately 300 grit) sandpaper on a flat surface. Then set the sealing surface of the object on the paper. Move the object in a figure 8 pattern around the surface of the sandpaper. Apply an even amount of pressure on the object while doing this. When the sandpaper leaves a mark on all parts of the surface, the sealing surface is flat. Stop sanding it and clean it off.

Oil pans are typically made of stamped steel. The thin steel tends to become distorted around the attaching bolt holes. These distortions can prevent the pan from fitting tightly against the transmission case. To flatten the pan, place the mounting flange of the pan on a block of wood and flatten one area at a time with a ball-peen hammer.

Transmission gaskets should not be installed with any type of liquid adhesive or sealant, unless specifically noted by the manufacturer. If any sealer gets into the valve body, severe damage can result. Also, sealant can clog the oil filter. If a gasket is difficult to install, a thin coating of transmission assembly lube can be used to hold the gasket in place while assembling the parts.

One type of gasket that presents unique installation problems is the cork-type gasket. These gaskets tend to change shape and size with changes in humidity. If a cork-type gasket is slightly larger than it should be, soak it in water and lay it flat on a warm surface. Allow it to dry before installing it. If the gasket is slightly smaller than required, soak it in warm water prior to installation.

Whenever you are using a cork-type gasket to create a seal between two parts, make sure you properly tighten the two surfaces together. Tighten the attaching bolts or nuts, in a staggered pattern, to the specified torque so that the gasket material is evenly squeezed between the two surfaces. If too much torque is applied, the gasket may split.

SEALS

Four types of seals are used in automatic transmissions: O-ring, square-cut, lip, and sealing rings (Figure 5-29). These seals are designed to stop fluid from leaking out of the transmission and from moving into another circuit of the hydraulic circuit. The latter ensures proper shifting of the transmission.

O-ring and square-cut seals are used to seal nonrotating parts (Figure 5-30). These seals can provide a good seal at high pressure points within the transmission. Common points that are sealed with these seals are oil pumps, servos, clutch pistons, speedometer drives, and vacuum modulators.

When installing a new O-ring or square-cut seal, coat the entire surface of the seal with assembly lube. Make sure you don't stretch or distort the seal while you work it into its holding groove. Some stretching may be necessary to work the seal over a shaft or fitting, but do not stretch it more than is needed. After a square-cut seal is installed, double-check to make sure it is not twisted. The flat surface of the seal should be parallel with the bore. If it is not, fluid will easily leak past the seal.

Lip seals are commonly used around rotating shafts and apply pistons. Lip seals that are used to seal a shaft typically have a metal flange around their outside diameter. The shaft rides on the lip seal at the inside diameter of the seal assembly. The rigid outer diameter provides a mounting point for the lip seal and is pressed into a bore. Once pressed into the bore, the outer diameter of the seal prevents fluid from leaking into the bore while the inner lip seal prevents leakage past the shaft.

Piston lip seals are set into a machined groove on the piston. These lip seals are not housed in a rigid metal flange. They are designed to be flexible and provide a seal while the piston moves up and down. While the piston moves, the lip flexes up and down. Any distortion or damage to the seal will allow fluid to escape. If fluid escapes from around a piston, the piston will not move with the force and speed that it should.

The most important thing when installing a lip seal is to make sure the lip is facing in the correct direction. The lip should always be aimed toward the source of pressurized fluid. If installed backward, fluid under pressure will easily leak past the seal. Also remember to make sure the surfaces to be sealed are clean and not damaged.

Teflon or metal sealing rings are commonly used to seal servo pistons, oil pump covers, and shafts. These rings may be designed to provide for a seal, but they may also be designed

SERVICE TIP: Another way to make a cork-type gasket grow is to lay it on a flat, clean, and hard surface, and then, with a hammer, strike it all the way around until it is the correct size.

The square-cut seal is commonly called a lathe-cut seal.

Most seal problems after a rebuild are caused by incorrect installation. Never rush through the installation process and always use the correct tool for the job.

Classroom Manual Chapter 5, page 166

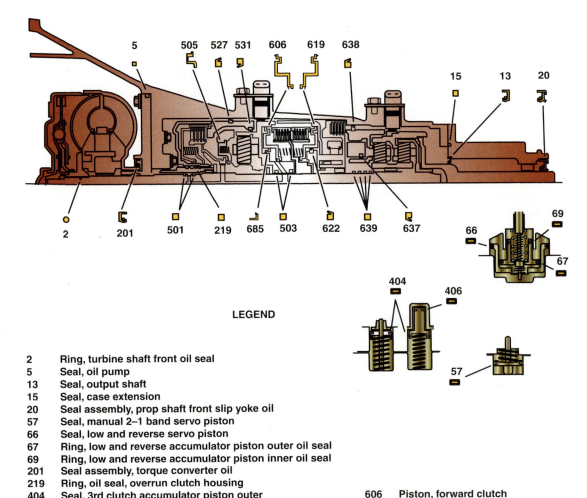

LEGEND

2	Ring, turbine shaft front oil seal
5	Seal, oil pump
13	Seal, output shaft
15	Seal, case extension
20	Seal assembly, prop shaft front slip yoke oil
57	Seal, manual 2–1 band servo piston
66	Seal, low and reverse servo piston
67	Ring, low and reverse accumulator piston outer oil seal
69	Ring, low and reverse accumulator piston inner oil seal
201	Seal assembly, torque converter oil
219	Ring, oil seal, overrun clutch housing
404	Seal, 3rd clutch accumulator piston outer
406	Seal, 3rd clutch accumulator piston inner
501	Ring, turbine shaft rear oil seal
503	Ring, turbine shaft intermediate oil seal
505	Piston assembly, overrun clutch
527	Seal, 4th clutch piston inner
531	Seal, 4th clutch piston outer

606	Piston, forward clutch
619	Piston, direct clutch
622	Seal, direct clutch piston intermediate
637	Seal, intermediate clutch piston inner
638	Seal, intermediate clutch piston outer
639	Ring, direct clutch housing oil seal
685	Seal assembly, forward clutch piston

FIGURE 5-29 Location of the various gaskets and seals used in today's transmissions.

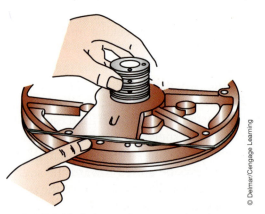

FIGURE 5-30 O-ring and square-cut seals are used to seal nonrotating parts, such as the front pump.

to allow a controlled amount of fluid leakage. Leakage may be allowed to lubricate some shaft bushings in the transmission. Sealing rings are either solid or cut rings. Cut sealing rings are one of three designs: open-end, butt-end, or locking-end.

Solid sealing rings are commonly used in late-model transmissions. These rings are made of a Teflon-based material and are never reused. To remove them, pry them out of their groove and carefully cut through the seal. Installing a new seal requires two tools: an installation tool and a resizing tool. It is important to note that the seals used in a particular transmission or in different transmissions may be a different size. This means two tools are required for each of the different sizes unless you have a universal tool set.

To install a Teflon seal, place the installation tool over the hub or shaft. Warm the new seal in hot water to soften the material, this will make installation easier. Lubricate the new seal and tool. Slide the seal over the tool and seat it into its groove on the hub or shaft. The installation tool will stretch the seal uniformly and just enough to install it. Due to the stretch, the resizing tool must be used to ensure a proper fit. The resizing tool is tube or cylinder with a long chamfer or bevel. The tool is designed to compress the seal into its original size after installation. To resize the seal, lubricate it and the resizing tool. Then carefully work the tool over the seal. While doing this, make sure the seal is sitting in its groove.

Open-end sealing rings fit loosely into a machined groove. The ends of the rings do not touch when they are installed. This type of ring is typically removed and installed with a pair of snap ring pliers. The ring should be expanded just enough to move it off or onto the shaft.

Butt-end sealing rings are designed so that their ends butt or touch each other once the seal is in place. This type of seal can be removed with a small screwdriver. The blade of the screwdriver is used to work the ring out of its groove. To install this type of ring, use a pair of snap ring pliers and expand the ring to move it into position.

Locking-end rings may have hooked ends that connect or ends that are cut at an angle (these are called scarf cut seals) to hold the ends together. These seals are removed and installed in the same way as butt-end rings. After these rings are installed, make sure the ends are properly positioned and touching.

REASSEMBLY AND TESTING

Before proceeding with the final assembly of all components, it is important to verify that the case, housing, and parts are clean and free from dust, dirt, and foreign matter (use an air gun). Have a tray available with clean ATF for lubricating parts. Also have ready a jar or tube of Vaseline for securing washers during installation.

Coat all parts with the proper type of ATF. Soak bands and clutches in the fluid for at least 15 minutes before installing them. All new seals and rings should have been installed before beginning final assembly.

Seal Installation

All seals should be checked in their own bores. They should be slightly smaller or larger (± 3 percent) than their groove or bore. If a seal is not the proper size, find one that is. Do not assume that because a particular seal came with the overhaul kit, it is the correct one.

Never install a seal when it is dry. The seal should slide into position and allow the part it seals to slide into it. A dry seal is easily damaged during installation. Coat all seals with transmission assembly lube. Following are some guidelines for installing a seal in an automatic transmission:

1. Install only genuine seals recommended by the manufacturer of the transmission.
2. Use only the proper fluids as stated in the appropriate service manual.
3. Keep the seals and fluids clean and free of dirt.
4. Before installing seals, clean the shaft or bore area. Carefully inspect these areas for damage. File or stone away any burrs or bad nicks and polish the surfaces with a fine crocus

Locking-end sealing rings may be called locking rings if they are made of metal, or scarf-cut rings if they are made of Teflon.

SERVICE TIP:
When tightening any fastener that directly or indirectly involves a rotating shaft or other part, rotate that part during and after tightening to ensure that the part does not bind.

Automatic transmission parts distributors package part kits to fit the level of service to be performed. Two common ones are the resealing and reconditioning kits. The resealing kit includes all the necessary parts, such as gaskets and cast iron, Teflon, rubber, and metal rings and seals, to reseal a transaxle or transmission. A transmission reconditioning kit includes all the necessary seals, gaskets, filters, bushings, clutch friction and steel discs, and brake bands.

cloth. Then clean the area to remove the metal particles. In dynamic applications, the sliding surface for the seal should have a mirror finish for best operation.

5. Lubricate the seal, especially any lips, with transmission assembly lube to ease installation.

6. All metal sealing rings should be checked for proper fit. Since these rings seal on their outer diameter, the seal should be inserted in its bore and should feel tight there. If the seal has some form of locking ends, these should be interlocked prior to trying the seal in its bore (Figures 5-31 and 5-32). The fit of the sealing rings in their shaft groove should also be checked. If the ring can move laterally in the groove, the groove is worn and this will cause internal fluid leaks. To check the side clearance of the ring, place the ring into its groove and measure the clearance between the ring and the groove with a feeler gauge. Typically, the clearance should not exceed 0.003 inch.

7. While checking the clearance, look for nicks in the grooves and for evidence of groove taper or stepping. If the grooves are tapered or stepped, the shaft will need to be replaced. If there are burrs or nicks in the grooves, they can be filed away.

8. Any distorted or undersized sealing rings should be replaced.

9. Always use the correct driver when installing a seal (Figure 5-33) and be careful not to damage the seal during installation. When possible, press the seal into position. This usually prevents the garter spring from moving out of position during installation.

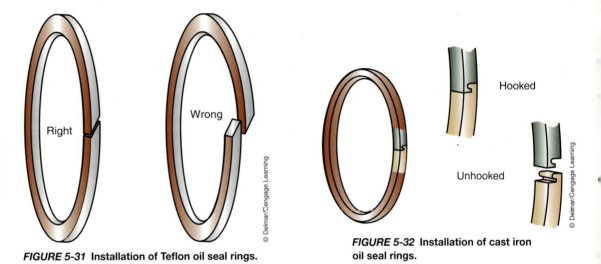

FIGURE 5-31 Installation of Teflon oil seal rings.

FIGURE 5-32 Installation of cast iron oil seal rings.

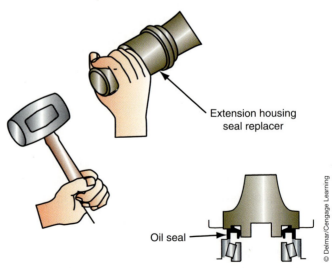

FIGURE 5-33 Using a special driver to install an oil seal. Note how the tool prevents damage to the seal.

Transmission Reassembly

Prior to reassembly, check the service manual for the order and procedure for installing the different components and units. Also be sure to examine all thrust washers carefully and coat them with petroleum jelly before placing them in the housing. Install the thrust ring, piston return spring, thrust washer, and one-way clutch inner race into the case. Align and start the bolts into the inner race from the rear of the case. Torque the bolts to specifications.

Lubricate and install the rear piston into the case. After determining the correct number of friction and steel plates, install the steel dished plate first. Make sure the steel disc is facing in the correct direction, then install the steel and friction plates, the retaining plate, and snap ring (Figure 5-34).

Using a suitable blowgun with a rubber tapered tip, air check the rear brake operation. After the rear brake has been completely assembled, measure the clearance between the snap ring and the retainer plate. Select the proper thickness of selective snap ring or retaining plate that will give the correct ring-to-plate clearance if the measurement does not meet the specified limits.

Slide the governor distributor assembly onto the output shaft from the front of the shaft. Install the shaft and governor distributor into the case, using care not to damage the distributor rings.

On some models, the output shaft, bearing, and appropriate gauging shims are placed into the transmission housing. The output shaft washer and bolt are then installed. While holding the output shaft and gear assembly, torque the output shaft nut to specifications. Then install a dial indicator and check the travel of the output shaft as it is pushed and pulled. Remove the gauging shims and install the correct sizes of service shims (Figure 5-35) output shaft gear, washer, and nut. Torque the output shaft nut to specifications. Using a pound-inch torque wrench, check the turning torque of the output shaft and compare this reading to specifications.

CAUTION:
Check and verify that the return spring is centered onto the race before tightening.

SERVICE TIP:
The ends of snap rings are slightly tapered. When installing snap rings, make sure the taper faces up or toward the outside of the shaft. This will provide a good gripping surface for the snap ring pliers during installation and for removal the next time the transmission is taken apart.

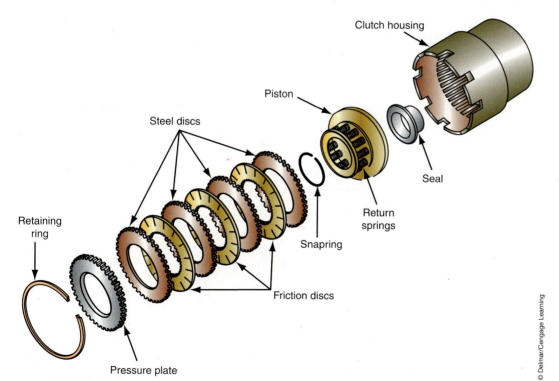

© Delmar/Cengage Learning

FIGURE 5-34 Be sure the steel disc is facing the correct direction, then install the steel and friction plates, the retaining plate, and snap-ring when assembling a clutch pack.

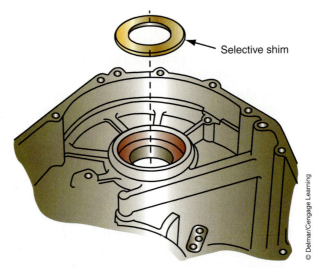

Selective shim

FIGURE 5-35 A differential bearing race selective shim. Measuring rotational torque will determine the correct size shim to use. If the measurement was less than the specification, a thicker selective shim should be used, likewise if the measurement was greater than the specification, use a thinner shim.

Place the small thrust washer on the pilot end of the transaxle output shaft. Then place the rear clutch assembly, front clutch drum, turbine shaft, and thrust washer into the housing. Locate and align the rear clutch over its hub. Gently move the rear clutch and turbine shaft around, rotating the assembly to engage the teeth of the friction discs with the rear clutch hub. Align the direct clutch assembly over the front clutch hub. Move the input shaft back and forth, rotating it so the front clutch friction disks engage with the front clutch hub.

Position the thrust washer to the back of the rear planetary carrier. Install the rear planetary carrier and thrust washer into the housing to engage the rear planetary ring gear. Install the front thrust washer and the drive shell assembly, engaging the common sun gear with the planetary pinions in the rear planetary carrier. Assemble the front planetary gear assembly into the front planetary ring gear. Make sure the planetary pinion gear shafts are securely locked to the planetary carrier.

Install the one-way sprag into the one-way clutch outer race with the arrow on the sprag facing the front of the transmission. Some overrunning clutches do not have an arrow to use for guidance. For these, the manufacturers recommend that an index mark be made on the clutch and the transmission case (Figure 5-36). This index mark should be made at the

SERVICE TIP:

This snap ring may be thinner than the clutch drum snap ring, so be sure you are using the correct size. Should you have insufficient space to install the snap ring into the drum groove, pull the connecting drum forward as far as possible. This will give you sufficient groove clearance to install the drum snap ring.

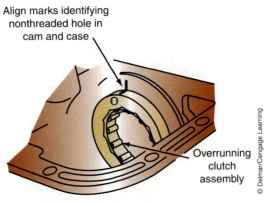

Align marks identifying nonthreaded hole in cam and case

Overrunning clutch assembly

FIGURE 5-36 For proper installation, an overrunning clutch should be indexed to the case prior to removal.

nonthreaded bore in the clutch. The bores for the mounting of the cam are slightly counter-sunk on one side. The clutch needs to be installed with this side facing rearward or toward the piston retainer. Install the connecting drum and sprag by rotating the drum clockwise, using a slight pressure and wobbling to align the plates with hub and sprag assembly. The connecting drum should now be free to rotate clockwise only. This check will verify that the sprag is correctly installed and operative. Now install the rear internal gear and the shaft's snap ring.

Secure the thrust bearing with petroleum jelly and install the rear planet carrier and the snap ring.

Assemble the front and rear clutch drum assemblies together and lay them flat on the bench. Be sure the rear hub thrust bearing is properly seated, then measure from the face of the front clutch drum to the top of the thrust bearing. Install the thrust washer and pump front bearing race to the pump. Then measure from the pump shaft (bearing race included) to the race of the thrust washer. Normally, the difference in measurements should be about 0.02 inch. If the thrust washer is not within the limits, replace it with one of the correct thickness.

Total endplay should now be checked. Set the transmission case on end, front end up. Be sure the thrust bearings are secure with assembly lube. Pick up the complete front clutch assembly and install it into the case. Be sure all parts are seated before proceeding with the measurement. Using a dial indicator or caliper, measure the distance from the rear hub thrust bearing to the case. Next measure the pump with the front bearing race and gasket installed. Tolerance should fall within specifications. If the difference between the measurements is not within tolerance, select the proper size front bearing race. If it is necessary to change the front bearing race, be sure to change the front clutch thrust washers the same amount.

Install the brake band servo. Use extreme care not to damage the O-rings. Lubricate around the seals. Then install and torque the retainer bolts to specifications. Loosen the piston stem. Install the brake band and strut (Figure 5-37) and finger-tighten the band servo piston stem just enough to prevent the band and strut from falling out. Do not adjust the band at this time. Air check for proper performance.

Place some assembly lube in two or three spots around the oil pump gasket and position it on the transaxle housing. Next align the pump and install it with care (Figure 5-38). Tighten the pump, attaching bolts to specifications in the specified order. After the bolts are tight, check the rotation of the input shaft. If the shaft does not rotate, disassemble the transmission to locate the misplaced thrust washer.

Before installing the bell housing, check the bolt hole alignment. Install the bell housing and torque the retaining bolts to specifications.

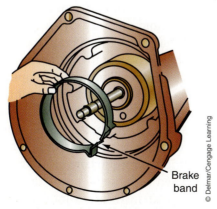

© Delmar/Cengage Learning

Brake band

FIGURE 5-37 Install the brake band and strut and finger tighten the band servo piston stem just enough to keep the band and strut snug from falling out.

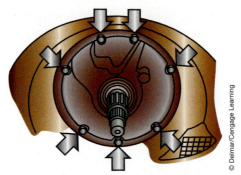

FIGURE 5-38 When installing the oil pump, make sure it is in the proper position. Also, tighten the bolts in the correct order and to the correct specification.

Adjust the band after you have checked to make sure that the brake band strut is correctly installed. Torque the adjuster bolt to specifications. Back off two (or the number specified by the manufacturer) full turns and secure with a locknut. Tighten the locknut to specifications.

Before proceeding with the installation of the valve body assembly, it is good practice to perform a final air check of all assembled components. This will ensure that you have not overlooked the tightening of any bolts or damaged any seals during assembly.

Assemble the parking pawl assembly. Place the assembly into its position and install the extension housing with a new gasket, then tighten the attaching bolts to the proper specifications.

On transaxles, the differential assembly should be disassembled, cleaned, inspected, and reassembled. After it has been reassembled, measure its endplay with gauging shims. Then select a shim thick enough to correct the endplay. After installing the proper shims, measure the differential turning torque with a pound-inch torque wrench. Increase or decrease the shim thickness to provide the correct turning torque.

Install the valve body. Be sure the manual valve is in alignment with the selector pin. Tighten the valve body attaching bolts to the specified torque.

Before installing the vacuum modulator valve, it is good practice to measure the depth of the hole in which it is inserted. This measurement determines the correct rod length to ensure proper performance. Refer to the service manual to determine the correct rod length based on your measurements. You should note that the actual rod size is slightly longer than the measurement taken. If you do not have the correct chart, it is fairly safe to simply add 0.07 inch to the measurement taken.

Before installing the solenoids, check to verify that they are operating properly. Connect a solenoid checker to each solenoid. When the solenoids are activated, you should hear the solenoid click on. You can also check the integrity of a solenoid with an ohmmeter. If the solenoid is good, install it. If the solenoid does not check out, replace it.

Install the kickdown switch. Again check the operation of the switch. This is best done by connecting an ohmmeter across the switch and moving the switch through its different settings.

Before installing the oil pan, check the alignment and operation of the control lever and parking pawl engagement. Make a final check to be sure all bolts are installed in the valve body. Install the oil pan with a new gasket. Torque the bolts to specifications.

Lubricate the oil pump's lip seal and the converter neck before installing the converter. Also put a coat of petroleum jelly on the seal's garter spring. This will prevent the spring from being knocked loose while installing the torque converter. Install the converter, making sure that the converter is properly meshed with the oil pump drive gear.

The transmission is now ready for installation into the vehicle. Use the reverse of the removal procedures. Never bolt the converter onto the flexplate and then try to install the transmission onto the converter. Remember to follow the correct fluid filling procedure and check the service manual for transmission relearn instructions.

CAUTION:
Be certain that the proper length bolts are installed in the related depth holes. The lengths of the bolts vary.

SERVICE TIP:
It is a good idea to partially prefill all torque converters before installing them in the vehicle.

CASE STUDY

A customer brought a late-model GM vehicle with a 2.8L engine into a transmission rebuilding shop to have the transmission rebuilt. This was the second time in three years that the transmission had gone out.

The technician diagnosed the problem as low pump pressure in all gears. A faulty pump was suspected. While pulling out the transmission, she noticed that the flexplate was cracked. This somewhat verified the diagnosis, since a faulty or excessively worn oil pump body bushing would cause this problem. She assumed that by replacing the oil pump and the flexplate, the customer would be happy and have many years of troublefree transmission service.

After completing the repairs and during the installation of the transmission, the technician noticed the dowel pins. There was nothing wrong with them, but they made her remember something she had read in a technical bulletin. She researched the TSBs, and found the one she was looking for. The two 15.0-mm dowel pins at the rear of the engine needed to be replaced with two 19.0-mm dowel pins. This extra length gave more rigidity to the mating of the transmission and the engine.

The slight shifting of the transmission against the engine caused both the oil pump bushings and the flexplate to go bad. Had the technician not read the correct TSB, she would have replaced the flexplate and the oil pump, only to have the customer come back again for the same problem.

TERMS TO KNOW

Crocus cloth

Dowels

Teardown

ASE–STYLE REVIEW QUESTIONS

1. *Technician A* says before tearing down a transmission, you should inspect the material trapped in the fluid filter.

   Technician B says installing an overhaul kit will take care of all the problems with a transmission.

   Who is correct?

   A. A only   C. Both A and B

   B. B only   D. Neither A nor B

2. While discussing endplay checks,

   Technician A says these checks should be taken before the transmission is disassembled.

   Technician B says these checks should be taken after the transmission is reassembled.

   Who is correct?

   A. A only   C. Both A and B

   B. B only   D. Neither A nor B

3. *Technician A* says abnormal noises from a transmission will never be caused by faulty clutches or bands.

   Technician B says abnormal noises from a transmission are typically caused by a faulty torque converter.

   Who is correct?

   A. A only   C. Both A and B

   B. B only   D. Neither A nor B

4. An output shaft or final drive endplay check should be made on either a transmission or transaxle.

   Technician A says this is done to provide long gear life and to eliminate engagement "clunk."

   Technician B says endplay should be measured during teardown, which helps you obtain the correct shims necessary to correct undesired endplay.

   Who is correct?

   A. A only   C. Both A and B

   B. B only   D. Neither A nor B

5. *Technician A* says abnormal noises and vibrations can be caused by damaged or worn gears, and by damaged clutches or bands.

   Technician B says abnormal noises can be caused by a bad oil pump or contaminated fluid.

   Who is correct?

   A. A only   C. Both A and B

   B. B only   D. Neither A nor B

6. *Technician A* says vibration problems can be caused by an unbalanced torque converter.

   Technician B says vibration problems can be caused by a faulty output shaft.

   Who is correct?

   A. A only   C. Both A and B

   B. B only   D. Neither A nor B

7. *Technician A* says once the transmission is removed and before it is disassembled, it should be cleaned in a solvent tank.

   Technician B says no matter what type of cleaning procedure is followed, the transmission and parts should be rinsed with water and then thoroughly dried with compressed air before reassembly.

   Who is correct?

   A. A only   C. Both A and B

   B. B only   D. Neither A nor B

8. While checking a transmission's vent,

   Technician A applies a vacuum to it and says that if the vent leaks, it should be replaced.

   Technician B runs ATF through the vent and says that if the vent cannot hold fluid, it must be replaced.

   Who is correct?

   A. A only   C. Both A and B

   B. B only   D. Neither A nor B

9. *Technician A* says seals should be kept clean and free of dirt, before and during installation.

   Technician B says most seals should be installed dry.

   Who is correct?

   A. A only   C. Both A and B

   B. B only   D. Neither A nor B

10. While discussing computer relearn procedures:

    Technician A says this type of computer strategy is designed to optimize driveability.

    Technician B says the computer on some cars needs to learn about the vehicle before it can control the systems properly.

    Who is correct?

    A. A only   C. Both A and B

    B. B only   D. Neither A nor B

ASE CHALLENGE QUESTIONS

1. The customer complains of a buzzing noise that increases with an increase in engine speed.

   Technician A says cavitation in the oil pump may be the noise source.

   Technician B says a leaking band apply servo may be the source.

   Who is correct?

   A. A only   C. Both A and B

   B. B only   D. Neither A nor B

2. *Technician A* says all shafts that extend out of the transmission should have the endplay measured before transmission disassembly.

   Technician B says to use a straightedge and ruler to measure shaft endplay during reassembly.

   Who is correct?

   A. A only   C. Both A and B

   B. B only   D. Neither A nor B

3. *Technician A* says compressed air may be used to remove some servos.

   Technician B says compressed air is used to check clutch and servo action during transmission reassembly.

   Who is correct?

   A. A only   C. Both A and B

   B. B only   D. Neither A nor B

4. *Technician A* says the best way to diagnose noise problems is to take a road test and pay attention to the operating gear, speed, and the conditions at which the noise occurs.

Technician B says noise diagnosis should begin with putting the vehicle in gear and apply the brake. If the noise or vibration is not evident, the problem is undoubtedly in the driveline or output of the transmission.

Who is correct?

A. A only   C. Both A and B
B. B only   D. Neither A nor B

5. After completing a transmission overhaul:

Technician A checks the service manual for the relearn procedures before taking the vehicle out on a road test.

Technician B connects a scan tool to the electronic control system and checks for DTCs before operating the transmission.

Who is correct?

A. A only   C. Both A and B
B. B only   D. Neither A nor B

Name _____ Date _____

PREPARE TO REMOVE A TRANSAXLE OR TRANSMISSION

Upon completion of this job sheet, you should be able to describe the procedures that must be followed in order to remove a transaxle or transmission from a vehicle.

ASE Correlation

This job sheet is related to the ASE Automatic Transmission and Transaxle Test's Content Area *Off-Vehicle Transmission and Transaxle Repair, Removal, Disassembly, and Assembly.*

Task: Remove and replace transmission/transaxle; inspect engine core plugs, transmission dowel pins, and dowel pin holes.

Tools and Materials

Vehicle with automatic transmission

Hoist

Transmission jack

Engine support fixture

Drain pan

Droplight or good flashlight

Service manual

Describe the vehicle being worked on:

Year _____ Make _____ VIN _____

Model _____

Model and type of transmission _____

Procedure

This job sheet is designed to allow you to take a good look at what would be involved in removing a transmission from an assigned vehicle.

_____
Task Completed

1. Refer to the service manual and look for any precautions or special instructions that relate to the removal of a transmission from this vehicle. Describe them here.

   _____

   _____

2. Before beginning to remove the transmission from this vehicle, what is the first thing that should be disconnected? Why?

   _____

   _____

3. Look under the hood and identify everything that should be removed or disconnected from above before raising the vehicle on the hoist. List those items here.

   _____

   _____

   _____

   _____

   _____

4. Raise the vehicle to a comfortable working height.

5. Carefully examine the area around the transmission and identify everything that should be disconnected or removed before unbolting the transmission from the engine. List those items here.

_____

_____

_____

_____

_____

_____

_____

_____

_____

_____

Instructor's Response _____

_____

_____

Name _____ Date _____

DISASSEMBLE AND INSPECT A TRANSMISSION

Upon completion of this job sheet, you should be able to disassemble and inspect the major components of a transmission.

ASE Correlation

This job sheet is related to the ASE Automatic Transmission and Transaxle Test's Content Area *Off-Vehicle Transmission and Transaxle Repair, Removal, Disassembly, and Assembly.* Task: Disassemble, clean, and inspect.

Tools and Materials

Automatic transmission or transaxle on a bench

Compressed air and air nozzle

Supply of clean solvent

Lint-free shop towels

Service manual

Describe the transmission being worked on:

Model and type of transmission _____

Year _____ Make _____ VIN _____

Model _____

Procedure

Task Completed

1. Disassemble the transmission into major units. Set each unit aside until this job sheet refers to it. Describe any problems you encountered while disassembling the transmission. Be sure to follow the procedures given in the appropriate service manual while taking the transmission apart.

   _____

   _____

   _____

   _____

2. Clean the planetary gearset in clean solvent and allow it to air dry. If compressed air is used to help the drying process, firmly hold the pinion gears to prevent the bearings from moving. ☐

3. Clean all thrust washers, thrust bearings, and bushings in clean solvent and allow them to air dry. ☐

4. Place one member of the planetary gearset on the sun gear. Rotate the gearset slowly. Do the same for the other parts of the gearset. Describe the feel of the gears' rotation.

   _____

   _____

   _____

5. Inspect each member of the gearset for damaged or worn gear teeth. Describe their condition.

_____

_____

6. Inspect the splines of each gearset member and describe their condition.

_____

_____

_____

☐

7. Remove any buildup of material or dirt that may be present between the teeth of the gears.

8. Look at the carrier assemblies and check for cracks or other damage. Describe your findings.

_____

_____

_____

9. Inspect the entire gearset for signs of discoloration. Describe your findings and explain what is indicated by them.

_____

_____

_____

10. Check the output shaft and its splines for wear, cracking, or other damage. Describe their condition.

_____

_____

_____

11. Carefully inspect the driving shells and drive lugs for wear and damage. Describe their condition.

_____

_____

12. Summarize the condition of the planetary gearsets and shaft.

_____

_____

_____

13. Check the thrust washers and bearings for damage, excessive wear, and distortion. Describe their condition.

_____

_____

_____

14. Check all of the bushings for signs of wear, scoring, and other damage. Describe their condition.

_____

_____

_____

_____

15. Check the shafts that ride in the bushings. Describe their condition.

_____

_____

_____

16. Summarize the condition of the thrust washers, bearings, and their associated shafts.

_____

_____

_____

Instructor's Response _____

_____

_____

Name: _____ **Date:** _____

REASSEMBLY OF A TRANSMISSION/TRANSAXLE

Upon completion of this job sheet, you should be able to properly measure and set endplay and preload during the assembly of a transmission or transaxle.

ASE Correlation

This job sheet is related to the ASE Automatic and Transaxle Test's Content Areas *Off-Vehicle Transmission Repair; Removal, Disassembly, and Assembly and Gear Train, Shafts, Bearings, and Case.*

Tasks: Assemble transmission/transaxle and measure endplay or preload; determine necessary action.

Tools and Materials

Basic hand tools   Air nozzle
Special tools for the transmission   Micrometer
Clean ATF in a tray   Dial indicator
Petroleum jelly

Describe the transmission that was assigned to you and the vehicle it came out of:

Year _____ Make _____ Model _____

VIN _____ Engine type and size _____

Model and type of transmission _____

Procedure

Task Completed

1. Before proceeding with the final assembly of all components, it is important to verify that the case, housing, and parts are clean and free from dust, dirt, and foreign matter. All new seals and rings should have been installed before beginning final assembly.   ☐

2. Coat all parts with the proper type of ATF. Soak bands and clutches in the fluid for at least 15 minutes before installing them.   ☐

3. Examine all thrust washers carefully and coat them with petroleum jelly before placing them in the housing.   ☐

4. Install the thrust ring, piston return spring, thrust washer, and one-way clutch inner race into the case. Align and start the bolts into the inner race from the rear of the case. Torque the bolts to specifications. What are the specifications?

   _____

   _____

5. Lubricate and install the rear piston into the case.   ☐

6. After determining the correct number of friction and steel plates, install the steel dished plate first, then the steel and friction plates, and finally the retaining plate and snap ring. How many steel plates did you have?

_____

7. Using a suitable blowgun with a rubber tapered tip, air check the rear brake operation. What were the results?

_____

_____

8. After the rear brake has been completely assembled, measure the clearance between the snap ring and the retainer plate. Select the proper thickness of retaining plate that will give the correct ring to plate clearance if the measurement does not meet the specified limits. What were the results?

_____

_____

9. Slide the governor distributor assembly onto the output shaft from the front of the shaft; install the shaft and governor distributor into the case, using care not to damage the distributor rings.

☐

10. On some models, the output shaft, bearing, and appropriate gauging shims are placed into the transmission housing. The output shaft washer and bolt are then installed. While holding the output shaft and gear assembly, torque the output shaft nut to specifications. What are the specifications?

_____

11. Install a dial indicator and check the travel of the output shaft as it is pushed and pulled. What were the results?

_____

_____

12. Remove the gauging shims and install the correct sizes of service shims, output shaft gear, washer, and nut.

☐

13. Torque the output shaft nut to specifications. Using a pound-inch torque wrench, check the turning torque of the output shaft and compare this reading to specifications. What were the results?

_____

_____

☐ **14.** Place the small thrust washer on the pilot end of the transaxle output shaft.

☐ **15.** Place the rear clutch assembly, front clutch drum, turbine shaft, and thrust washer into the housing.

16. Locate and align the rear clutch over its hub. Gently move the rear clutch and turbine shaft around, rotating the assembly to engage the teeth of the friction discs with the rear clutch hub. Align the direct clutch assembly over the front clutch hub. Move the input shaft back and forth, rotating it so the front clutch friction discs engage with the front

☐ clutch hub.

17. Position the thrust washer to the back of the rear planetary carrier.

☐

18. Install the rear planetary carrier and thrust washer into the housing to engage the rear planetary ring gear.

☐

19. Install the front thrust washer and the drive shell assembly, engaging the common sun gear with the planetary pinions in the rear planetary carrier.

☐

20. Assemble the front planetary gear assembly into the front planetary ring gear. Make sure the planetary pinion gear shafts are securely locked to the planetary carrier.

☐

21. Install the one-way sprag into the one-way clutch outer race with the arrow on the sprag facing the front of the transmission.

☐

22. Install the connecting drum with sprag by rotating the drum clockwise using a slight pressure and wobbling to align the plates with hub and sprag assembly. The connecting drum should now be free to rotate clockwise only. This check will verify that the sprag is correctly installed and operative. What were the results?

_____

_____

23. Install the rear internal gear and the shaft's snap ring.

☐

24. Secure the thrust bearing with petroleum jelly and install the rear planet carrier and the snap ring.

☐

25. Assemble the front and rear clutch drum assemblies together and lay them flat on the bench.

☐

26. Make sure the rear hub thrust bearing is properly seated, then measure from the face of the front clutch drum to the top of the thrust bearing. What did you measure, and how does it compare to the specifications?

_____

_____

_____

27. Install the thrust washer and pump front bearing race to the pump.

☐

28. Measure from the pump shaft (bearing race included) to the race of the thrust washer. If the thrust washer is not within the limits, replace it with one of the correct thickness. What were the results?

_____

_____

29. Total endplay should now be checked. Set the transmission case on end, front end up.

☐

30. Make sure the thrust bearings are secure with petroleum jelly. Pick up the complete front clutch assembly and install it into the case. Be sure all parts are seated before proceeding with the measurement. Using a dial indicator or caliper, measure the distance from the rear hub thrust bearing to the case. What were the results?

_____

_____

31. Measure the pump with the front bearing race and gasket installed. Tolerance should fall within specifications. If the difference between the measurements is not within tolerance, select the proper size front bearing race. If it is necessary to change the front bearing race, be sure to change the front clutch thrust washers the same amount. What were the results?

_____

_____

_____

☐ 32. Install the brake band servo. Use extreme care so as not to damage the O-rings. Lubricate around the seals.

33. Install and torque the retainer bolts to specifications. What are the specifications?

_____

☐ 34. Loosen the piston stem.

☐ 35. Install the brake band strut and finger tighten the band servo piston stem just enough to keep the band and strut snug and prevent them from falling out. Do not adjust the band at this time.

☐ 36. Air check for proper performance. What were the results?

☐ 37. Place some petroleum jelly in two or three spots around the oil pump gasket and position it on the transaxle housing.

☐ 38. Align the pump and install the pump with care.

39. Tighten the pump attaching bolts to specifications in the specified order. What are the specifications and the specified order?

_____

40. Check the rotation of the input shaft. If the shaft does not rotate, disassemble the transmission to locate the misplaced thrust washer. What were the results?

_____

_____

41. Install the bell housing and torque the retaining bolts to specifications. What are the specifications?

_____

☐ 42. Adjust the band after you check to make sure that the brake band strut is correctly installed.

43. Torque the piston stem to specifications. What are the specifications?

_____

44. Back off two (or the number specified by the manufacturer) full turns and secure with the locknut. What are the specifications?

_____

45. Tighten the locknut to specifications. What are the specifications?

_____

46. Before proceeding with the installation of the valve body assembly, it is good practice to perform a final air check of all assembled components. This will ensure that you have not overlooked the tightening of any bolts or damaged any seals during assembly. What are the specifications?

_____

47. Assemble the parking pawl assembly. Place the assembly into its position and install the extension housing with a new gasket, then tighten the attaching bolts to the proper specifications. What are the specifications?

_____

48. On transaxles, the differential assembly should be disassembled, cleaned, inspected, and reassembled. After it has been reassembled, measure its endplay with gauging shims. Then select a shim thick enough to correct the endplay. What are the specifications?

_____

49. After installing the proper shims, measure the differential turning torque with a pound-inch torque wrench. What were your results?

_____

_____

50. Install the valve body. Be sure the manual valve is in alignment with the selector pin. Tighten the valve body attaching bolts to the specified torque. What are the specifications?

_____

51. Before installing the vacuum modulator valve, it is good practice to measure the depth of the hole in which it is inserted. This measurement determines the correct rod length to ensure proper performance. Refer to the service manual to determine the correct rod length based on your measurements. ☐

52. Before installing the kickdown solenoid or other solenoids, check to verify that they are operating properly. Connect the solenoid to a 12-volt source and ground the other terminal. What happened?

_____

_____

_____

53. Install the kickdown switch. ☐

54. Before installing the oil pan, check the alignment and operation of the control lever and parking pawl engagement. Make a final check to be sure all bolts are installed in the valve body. ☐

55. Install the oil pan with a new gasket. Torque the bolts to specifications. What are the specifications?

_____

56. Lubricate the oil pump's lip seal and the converter neck before installing the converter. ☐

☐

57. Install the converter, making sure that the converter is properly meshed with the oil pump drive gear.

☐

58. The transmission is now ready for installation in the vehicle. Use the reverse of the removal procedures. Remember to follow proper fluid filling procedures.

Instructor's Response _____

_____

_____

Chapter 6

TORQUE CONVERTER AND OIL PUMP SERVICE

BASIC TOOLS

Basic mechanic's tool set

DMM

Appropriate service manual

UPON COMPLETION AND REVIEW OF THIS CHAPTER, YOU SHOULD BE ABLE TO:

- Diagnose torque converter problems and determine needed repairs.
- Perform a stall test and determine needed repairs.
- Perform converter clutch system tests and determine needed repairs.
- Diagnose hydraulically and electrically controlled torque converter clutches.
- Inspect converter flexplate, converter attaching bolts, converter pilot, and converter pump drive surfaces.

- Measure torque converter endplay and check for interference.
- Check the stator clutch.
- Check a torque converter and transmission cooling system for contamination.
- Inspect, leak test, flush, and replace cooler, lines, and fittings.
- Inspect, measure, and replace an oil pump assembly and related components.

GENERAL DIAGNOSTICS

Many transmission problems are related to the operation of the torque converter. Normally torque converter problems will cause abnormal noises, poor acceleration in all gears, normal acceleration but poor high-speed performance, or transmission overheating.

To test the operation of the torque converter, many technicians perform a stall test. The stall test checks the holding capacity of the converter's stator overrunning clutch assembly, as well as the clutches and bands in the transmission.

However, torque converter problems can often be identified by the symptoms, and therefore the need for conducting a stall test is minimized. If the vehicle lacks power when it is pulling away from a stop or when passing, it has a restricted exhaust or the torque converter's one-way stator clutch is slipping. To determine which of these problems is causing the power loss, test for a restricted exhaust first.

Exhaust Restriction Test

The easiest way to test for an exhaust restriction is to use a vacuum gauge. Connect the vacuum gauge to a source of engine manifold vacuum. Observe the vacuum reading with the engine at idle. Quickly open the throttle and observe the vacuum reading. Then quickly release the throttle to allow it to close. The vacuum reading should increase about 5 in./Hg at the initial closing of the throttle (Figure 6-1). If the vacuum did not increase when the throttle was closed, the exhaust is restricted. Common causes for exhaust restrictions are a plugged catalytic converter or a collapsed exhaust pipe.

Many engine problems can cause a vehicle to behave as if the torque converter is malfunctioning, or actually cause the converter clutch to either lockup early or not at all. Clogged fuel injectors or bad spark plug wires can be misinterpreted as torque converter complaints. Bad vacuum lines, EGR valves, or engine speed sensors can prevent the clutch from locking up at the proper time.

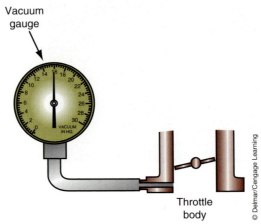

Vacuum gauge

Throttle body

© Delmar/Cengage Learning

FIGURE 6-1 To check for an exhaust restriction, connect a vacuum gauge to the engine and observe the readings while the engine speed is raised and quickly lowered.

SPECIAL TOOLS

Vacuum gauge
Various tee fittings
Various lengths of hoses

Common exhaust restrictions are a plugged catalytic converter or a collapsed exhaust pipe.

Sources for engine manifold vacuum are found in the intake manifold, below the throttle plate.

Classroom Manual

Chapter 6, page 188

Another way to determine if the exhaust is restricted is with a pressure gauge. Insert the gauge into the exhaust manifold's bore for the oxygen sensor. Then, start the engine. Bring the engine to 2000 rpm and observe the gauge. If the exhaust is not restricted, the reading will be less than 1.25 psi. A very restricted exhaust will have a reading of over 2.75 psi.

If there is no evidence of a restricted exhaust from these tests, it can be assumed that the torque converter's stator clutch is slipping and not allowing any torque multiplication to take place in the converter. To repair this problem, the torque converter should be replaced.

Other Common Problems. If the engine's speed flares up during acceleration in Drive and does not have normal acceleration, the clutches or bands in the transmission are slipping. This symptom is similar to the slipping of a clutch in a manual transmission. Often, this problem is mistakenly blamed on the torque converter.

The torque converter is often blamed as the cause for problems simply based on the customer's complaint. Complaints of thumping or grinding noises are often blamed on the converter when they are really caused bad thrust washers or damaged gears and bearings in the transmission. This type of noise can also be caused by nontransmission parts, such as bad CV joints and wheel bearings.

Also, many engine problems can cause a vehicle to act as if it has a torque converter problem. This is especially true of converter clutches, which may engage early or not at all. Engine problems can cause the converter clutch to lockup early or not at all. Common engine problems that can cause torque converter like symptoms include the following: clogged fuel injectors and faulty spark-plug wires, vacuum lines, EGR valves, or speed sensors.

A faulty engine dampner can also feel like a bad T/C, allowing vibrations that seem to be related to the torque converter. These vibrations can be caused by an out of balance torque converter. If the T/C is not properly balanced, the vibration will increase at higher speeds and while the transmission is in gear.

TESTING CONVERTER CLUTCHES

Late-model transmissions are equipped with a torque converter clutch (TCC). Most converter clutches are controlled by the powertrain control module (Figure 6-2). The computer turns on the converter clutch solenoid, which opens a valve and allows fluid pressure to engage the clutch. When the computer turns the solenoid off, the clutch disengages.

One of the trickiest parts of diagnosing a converter clutch problem is differentiating a normal-acting converter clutch from an abnormal-acting one. You should pay attention to the action of all converter clutches, whether or not they are suspected of having a problem.

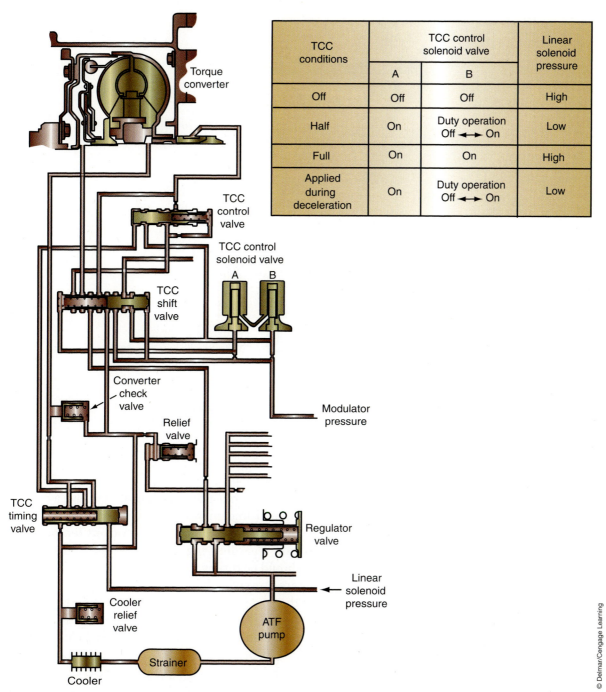

TCC conditions	TCC control solenoid valve		Linear solenoid pressure
	A	B	
Off	Off	Off	High
Half	On	Duty operation Off ←→ On	Low
Full	On	On	High
Applied during deceleration	On	Duty operation Off ←→ On	Low

© Delmar/Cengage Learning

FIGURE 6-2 **Typical electrical and hydraulic circuit for the clutch in a lockup torque converter.**

By knowing what a normal clutch feels like, it is easier to feel abnormal clutch activity. A malfunctioning converter clutch can cause a wide variety of driveability problems. Normally, the application of the clutch should feel like a smooth engagement into another gear. It should not feel harsh, nor should there be any noises related to the application of the clutch.

To properly diagnose converter clutch problems, you must know when they should engage and disengage and understand the function of the various controls involved with the system. Although the actual controls for a converter clutch vary with the different manufacturers and models of transmissions, they all will have certain operating conditions that must be met before the clutch can be engaged.

Care should be taken during diagnostics because abnormal clutch action can be caused by engine, electrical, clutch, or torque converter problems.

Classroom Manual

Chapter 6, page 180

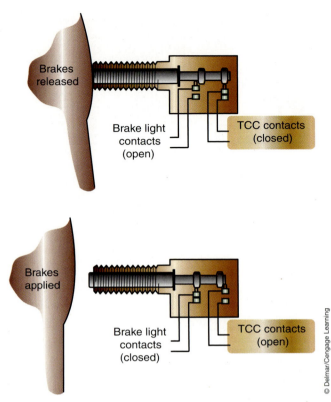

FIGURE 6-3 Proper adjustment of the brake light switch is essential for proper operation of a torque converter clutch.

Before the converter clutch is applied, the vehicle must be traveling at or above a certain speed. The vehicle speed sensor sends this speed information to the computer. Also, the converter clutch will not engage when the engine is cold; therefore, a coolant temperature sensor provides the computer with information regarding engine temperature. During sudden deceleration or acceleration, the clutch should be disengaged. One of the sensors used to tell the computer when these driving modes are present is the TP sensor. Some transmissions use a third or fourth gear switch to inform the computer when the transmission is in those gears. The computer will then allow clutch engagement if other operating conditions are satisfactory. A brake switch is also used in some clutch circuits to disengage the clutch when the brakes are applied (Figure 6-3). These key sensors—the VSS, ECT, TP, third/fourth gear switch, and brake switch—should be visually checked as part of your diagnosis of converter problems.

Diagnosis of a converter clutch circuit should be conducted in the same way as any other computer system. The computer will recognize problems within the system and store trouble codes that reflect the problem area of the circuit. The codes can be retrieved and displayed by an instrument panel light or a handheld scan tool.

Engagement Quality

The vehicle should be taken on a road test to verify the customer's concern and to gather more information about the vehicle. Prior to operating the vehicle, make sure you have as much detail about the concern as possible. This should include the driving conditions at which the concern is most noticeable.

During the road test, determine if the TCC engages and disengages. If it does neither, detailed testing of the control system is required. The engagement of the clutch should be

SPECIAL TOOLS
Jumper wire
Paper clip
Diagnostic key
Scan tool

Torque converter clutch shudder is sometimes called chatter or a stick-slip condition.

Step	Findings	Remedy
1. Does the TCC engage and disengage?	Yes	Go to step #2
	No	Go to symptoms chart—diagnose the system
2. Describe the vibration or shudder during a 3–4 or 4–3 shift.	Light	Go to step #3
	Medium	Go to step #3
	Heavy	Go to symptoms chart—is not TCC related
2. Is the vibration or shudder vehicle speed related and not gear related?	Yes	Go to symptoms chart—is not TCC related
	No	Go to step #4
4. Is the vibration or shudder engine speed related and not gear related?	Yes	Go to symptoms chart—is not TCC related
	No	Go to step #5
5. Is the vibration or shudder occur in coast, cruise, or reverse gear?	Yes	Go to symptoms chart—is not TCC related
	No	Go to step #6
6. Does the Vibration or shudder occur during long periods of light braking?	Yes	Go to symptoms chart—is not TCC related
	No	Go to step #7
7. Did the Vibration or shudder only occur in step #2?	Yes	There is a probable TCC problem—diagnose the system
	No	Go to symptoms chart—is not TCC related

© Delmar/Cengage Learning

FIGURE 6-4 An evaluation form to use while doing a road test to check the torque converter.

smooth. Also pay attention to any noise or vibration that may take place during shifting. If the vibration only occurs during the gear changes that should cause the TCC to engage or disengage, there is evidence of T/C shudder. If the vibration is present at other times, the cause is probably something other than the TCC (Figure 6-4).

If the clutch prematurely engages or is not being applied by full pressure, a shudder or vibration results from the rapid grabbing and slipping of the clutch. The clutch begins to engage, then slips, because it cannot hold the engine's torque and become fully engaged. The torque capacity of the clutch is determined by the oil pressure applied to the clutch and the condition of the frictional surfaces of the clutch assembly.

If the shudder, sometimes called a stick-slip condition, is only noticeable during the engagement of the clutch, the problem is typically the converter. When the shudder is evident after the engagement of the clutch, the engine, transmission, or another component of the driveline may be the cause. *Tip-in shudder* is a commonly used term used to describe the condition where the clutch releases and reapplies under a light load. This is a clutch control problem.

When clutch apply pressure is low and the clutch cannot firmly engage fully, shudder will occur. A faulty clutch solenoid or its return spring may cause this. The valve controlled by the solenoid is normally held in position by a coil-type return spring. If the spring loses tension, the clutch will be able to prematurely engage. Because insufficient pressure is available to hold the clutch, shudder occurs as the clutch begins to grab and then slips. If the solenoid valve and/or return spring is faulty, they should be replaced, as should the torque converter.

An out-of-round torque converter prevents full clutch engagement, which will cause shudder, as will contaminated clutch frictional material. The frictional material can become contaminated by metal particles circulating through the torque converter and collecting on the clutch. Broken or worn clutch dampener springs will also cause shudder.

Besides replacing the torque converter and/or replacing other components of the system, one possible correction for shudder is the use of an ATF with friction modifiers. Some rebuilders may recommend that an oil additive be added to the ATF. The additive is designed to improve or alter the friction capabilities of regular ATF.

The ability of a converter clutch to hold torque is its torque capacity.

Classroom Manual

Chapter 6, page 187

Nearly all torque converters are welded-together units and therefore cannot be disassembled for inspection or repair. The primary objective when servicing a torque converter should be to check the serviceability of the unit.

SPECIAL TOOLS

Short piece of hose

Graduated container

OSHA-approved air nozzle

Classroom Manual

Chapter 6, page 186

TCC is a commonly used acronym for torque converter clutch.

TC-Related Cooler Problems

Vehicles equipped with a converter clutch may stall when the transmission is shifted into reverse gear. The cause of this problem may be plugged transmission cooler lines, or the cooler itself may be plugged. Fluid normally flows from the torque converter through the transmission cooler. If the cooler passages are blocked, fluid is unable to exhaust from the torque converter and the converter clutch piston remains engaged. When the clutch is engaged, there is no vortex flow in the converter and therefore little torque multiplication is taking place in the converter.

To verify that the transmission cooler is plugged, disconnect the cooler return line from the radiator or cooler (Figure 6-5). Connect a short piece of hose to the outlet of the cooler and allow the other end of the hose to rest inside an empty container. Start the engine and measure the amount of fluid that flows into the container after 20 seconds. Normally one quart of fluid should flow into the container. If less than that filled the container, a plugged cooler is indicated.

To correct a plugged transmission cooler, disconnect the cooler lines at the transmission and the radiator. Blow air through the cooler, one end at a time, then through the cooler lines. The air will clear large pieces of debris from the transmission cooler. Always use low air pressure of no more than 50 psi. Higher pressures may damage the cooler. If

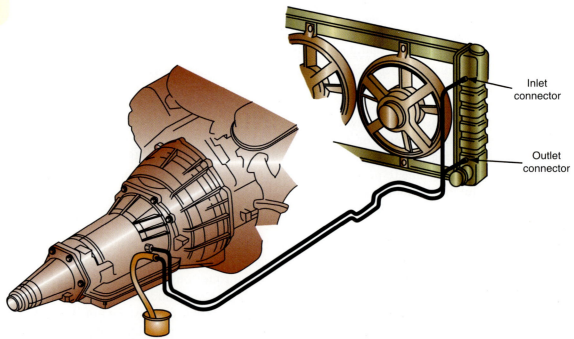

Inlet connector

Outlet connector

© Delmar/Cengage Learning

FIGURE 6-5 To verify if a fluid cooler is plugged, disconnect the return fluid line and observe the flow of fluid in and out of the cooler.

there is little air flow through the cooler, the radiator or external cooler must be removed and flushed or replaced.

GENERAL CONVERTER CONTROL DIAGNOSTICS

All testing of TCC controls should begin with a basic inspection of the engine and transmission. Too often, technicians skip this basic inspection and become frustrated during diagnostics because of conflicting test results. The basic inspection should include the following:

1. A road test to verify the complaint and further define the problem.
2. A careful inspection of the engine and transmission.
3. A check of the PCM for codes.
4. A check of the mechanical condition of the engine, the output of ignition system, and the operation of the fuel system.
5. An idle speed and ignition timing check. If the timing is nonadjustable, check the operation of the electronic spark control system.
6. A check of the entire intake system for vacuum leaks.

When inspecting wires, look for burnt spots, bare wires, and damaged or pinched wires. Make sure the wiring harness to the electronic control unit has a tight and clean connection. Also, check the source voltage at the battery before beginning any detailed tests. If the voltage is too low or too high, the system cannot function properly.

> **AUTHOR'S NOTE:** Engine problems can cause transmission and torque converter problems. The opposite is also true; transmission problems can cause engine problems. For example, if the TCC does not disengage, the engine will stall when the vehicle is at a stop.

Hydraulic Systems

In early TCC systems, clutch engagement was controlled hydraulically. A switch valve was controlled by two other valves, the lockup and fail-safe valves, in the clutch control assembly (Figure 6-6). The lockup valve responds to governor pressure and prevents lockup at speeds below 40 mph. The fail-safe valve responds to throttle pressure and permitted clutch engagement in high gear only. Problems with this system are diagnosed in the same way as other hydraulic circuits.

The clutch in most hydraulic systems is applied when oil flow through the torque converter is reversed. This change can be observed on a pressure gauge. Connect the gauge to the hydraulic line, with a "tee" fitting, from the transmission to the cooler. Position the gauge so that it is easily seen from the driver's seat. Then raise the vehicle on a hoist with the drive wheels off the ground and able to spin freely. Operate the vehicle until the transmission shifts into high gear. Then maintain a speed of approximately 55 mph. Once the speed is maintained, watch the pressure gauge.

If the pressure decreases 5–10 psi, the converter clutch was applied. With this action, you should feel the engagement of the clutch, as well as a drop in engine speed. If the pressure was changed but the clutch did not engage, the problem may be inside the converter or at the end of the input shaft. If the input shaft end is worn or the O-ring at the end is cut or worn, there will be a pressure loss at the converter clutch. This loss in pressure will prevent full engagement of the clutch. If the pressure did not change and the clutch did not engage, suspect a faulty clutch valve or control solenoid, or a fault in the solenoid control circuit.

Classroom Manual
Chapter 6, page 177

SERVICE TIP:
It is important to realize that some oil coolers cannot be flushed effectively. Check the service information for any notations regarding this.

SERVICE TIP:
Nearly all electronic converter and transmission controls have a self-test mode and are capable of displaying DTCs. However, the basic inspection is important because the computer will not display codes for low compression or other common engine problems.

SPECIAL TOOLS
Hydraulic pressure gauge
Various T-fittings
Various lengths of hoses

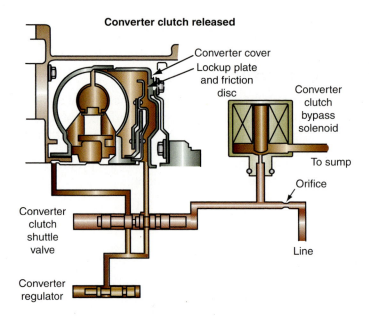

Converter clutch released

Converter cover

Lockup plate and friction disc

Converter clutch bypass solenoid

To sump

Orifice

Converter clutch shuttle valve

Line

Converter regulator

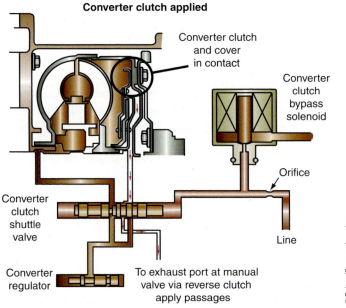

Converter clutch applied

Converter clutch and cover in contact

Converter clutch bypass solenoid

Orifice

Converter clutch shuttle valve

Line

Converter regulator

To exhaust port at manual valve via reverse clutch apply passages

© Delmar/Cengage Learning

FIGURE 6-6 The action of the hydraulic fluid from the lockup valve to the torque converter.

Classroom Manual

Chapter 6, page 191

ELECTRONIC CONVERTER CLUTCH CONTROL DIAGNOSTICS

The TCC in late-model transmissions is controlled by the PCM or TCM. The computer turns on the converter clutch solenoid (Figure 6-7), which opens a valve and allows fluid pressure to engage the clutch. When the computer turns the solenoid off, the clutch disengages.

A faulty clutch solenoid or its return spring may cause low apply pressure. The valve controlled by the solenoid is normally held in position by a coil-type return spring. If the spring loses tension, the clutch will engage too soon. Because there is insufficient pressure to hold the clutch, shudder occurs as the clutch begins to grab and then slip. If the solenoid valve and/ or return spring are faulty, they should be replaced, as should the torque converter. If the TCC fails to release, it can cause the engine to jerk and stall when the vehicle is stopping.

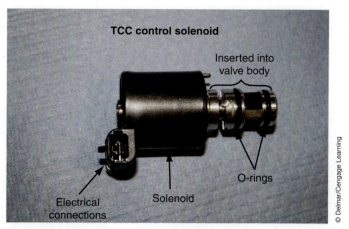

FIGURE 6-7 A typical torque converter clutch solenoid.

A malfunctioning converter clutch can cause a wide variety of driveability problems. Normally, the application of the clutch should feel like a smooth engagement into another gear. It should not feel harsh, nor should there be any noises related to the application of the clutch.

Electrical Checks

Each manufacturer has specific test sequences for testing an electronic converter clutch. In all cases, the system is first taken on a road test and then checked for DTCs. Normally vehicles equipped with an OBD-II system, will display TCC-related DTCs along with the DTCs of other systems. Early systems required special test procedures. Photo Sequence 10 shows a typical procedure for testing early systems.

To properly diagnose converter clutch problems, you must know when the TCC should engage and disengage and understand the function of the various components involved with the system (Figure 6-8). Although the actual controls for a TCC vary with the different models of transmissions, they all have certain operating conditions that must be met before the clutch can be engaged. Figure 6-9 shows the status of a typical TCC solenoid during different gear ranges.

The TCM monitors vehicle speed, gear range, and throttle position to determine when to apply the TCC. Specific conditions must be present (these depend on the model of vehicle and transmission) before the TCM will apply the TCC solenoid. If one of those conditions changes is no longer within the specified parameters, the TCC will be disengaged. Normally all of the following conditions must be met:

- The transmission is operating in the specified gear(s).
- The vehicle is traveling above the minimum speed.
- The TP sensor senses a throttle opening that is greater than idle.
- The brake light switch is not on (brakes are not applied).
- The temperature of the engine coolant or ATF is above a predetermined level.
- The vehicle's speed does not drop 6 or more mph below the set cruise control speed.
- No knocking is detected by the knock sensors.

Checking TCC Operation

The following test verifies that the TCC and T/C are working correctly.

1. Check for DTCs. Keep the scan tool connected.
2. Connect a tachometer to the engine (if the vehicle does not have one).

TYPICAL PROCEDURE FOR DIAGNOSING A COMPUTER-CONTROLLED TCC SYSTEM

All photos in this sequence are © Delmar/Cengage Learning.

P10-1 Turn the ignition on with the engine off Verify that the "CHECH ENGINE" or "*Service Engine Soon*" lamp is lit.

P10-2 Locate the DLC.

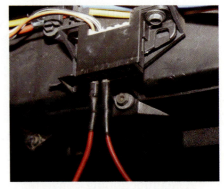

P10-3 Ground the diagnostic test terminal by connecting a jumper wire from terminal A to terminal B at the DLC connector.

P10-4 The MIL should flash a code 12. The code 12 is indicated by a flash, a pause, and two flashes.

P10-5 Connect scan tool to the DLC.

P10-6 Connect the scan tool power lead to the cigarette lighter receptacle.

P10-7 The scan tool must be programmed with the correct identification of the vehicle.

P10-8 The scan tool will display the trouble codes numerically.

P10-9 A scan tool not only will display the trouble codes held in the memory of the PCM, but also can be used to display the data or signals that the PCM is receiving from inputs or is sending to outputs.

P10-11 Compare the trouble codes that are received with the appropriate trouble code chart.

P10-12 After the codes have been retrieved, remove the jumper wire or scan tool.

P10-13 Than remove the PCM fuse to erase the codes from memory.

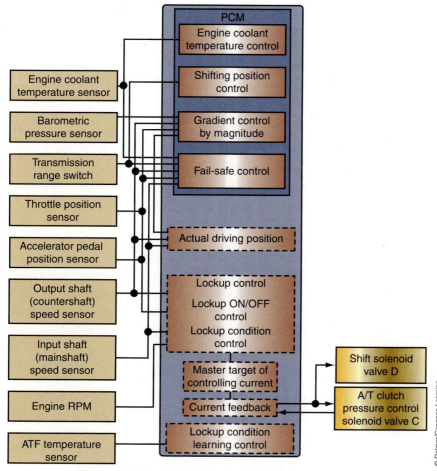

FIGURE 6-8 A typical electronic control system for a TCC.

© Delmar/Cengage Learning

Gear selector position	PCM commanded gear	TCC solenoid status
P/R/N	1	Hydraulically disabled
D	1	Hydraulically disabled
D	2	Electronically controlled
D	3	Electronically controlled
D	4	Electronically controlled
D w/OD OFF	1	Hydraulically disabled
D w/OD OFF	2	Electronically controlled
D w/OD OFF	3	Electronically controlled
MANUAL 2	2	Electronically controlled
MANUAL 1	1	Hydraulically disabled
MANUAL 1	2	Electronically controlled

© Delmar/Cengage Learning

FIGURE 6-9 The status of a typical TCC solenoid while the transmission is operating in different gears.

3. Drive the vehicle, with the gear selector in the D position, at highway speeds for about 15 minutes to bring the engine to normal operating temperature.

4. Using the scan tool, make sure that the ATF temperature is within specification. Normally this is about 110°F (43°C).

5. Once normal operating temperature is reached, drive at a constant speed of about 50 mph (80 km/h).

6. Lightly depress the throttle and pay attention to changes in engine speed. If there is a sudden increase in engine speed, the TCC was not locked.

7. If the TCC was locked, resume the cruising speed.

8. Tap brake pedal with your left foot and observe the speed of the engine. Engine speed should increase when brake pedal is tapped and decrease about 5 seconds after pedal is released.

9. Reduce speed to about 30 mph (48 km/h) and maintain that speed.

10. Place the scan tool at position where it can be safely monitored. Set the tool to read the PID "TCCAMP_MES."

11. Fully release the throttle pedal and then depress the throttle to partially open it again. Observe the scan tool. The PID will read 0 when the TCC is released, and 1 when the TCC is applied. Likewise, engine speed will increase when the TCC is released and decrease when the TCC is applied. If this not occur, further diagnosis is necessary.

TCC Diagnostics

Diagnosis of a TCC circuit should be conducted in the same way as any other computer system. Make sure you understand the customer's concern and verify the problem by operating the vehicle. Before conducting any test on the system, make sure the fluid in the transmission is at the correct level and is in good condition. Also conduct a good inspection of the vehicle. This should include a gear selector linkage check and a look at the vehicle to see if any modifications were made to it that may affect the operation of the TCC. Also make sure to check for a TSB that may relate to the concern. If there is such a bulletin, follow the given test and repair procedure before proceeding.

With the engine and transmission at normal operating temperature, connect the scan tool to the vehicle. Retrieve and record all DTCs. The DTCs are set when the computer recognizes problems within the system. The stored DTCs reflect the problem area of the circuit.

It is important that you retrieve all codes and deal with them in the proper sequence. Use the DTC chart in the service information to define the code and identify the possible causes. All engine-related DTCs should be examined first and repairs made to the faulty system. Then attention should be paid to the transmission- and torque converter-related codes (Figure 6-10). Pinpoint tests are conducted to diagnose the exact cause of the DTCs. It is also important that the system be checked for DTCs after repairs have been made.

DTC	Description	Condition	Symptom
P0741	TCC slippage detected	The PCM noticed an excessive amount of slippage during normal operation.	TCC slippage, erratic operation, or no TCC operation. Flashing transmission MIL.
P0743	TCC solenoid circuit failure during on-board diagnostic	The TCC solenoid circuit does not provide the specified voltage drop across the solenoid. The circuit is open or shorted or the PCM driver failed during the diagnostic test.	With a short circuit, the engine will stall in second gear at low idle speeds with brake applied. With an open circuit, the TCC will not engage. This may cause the MIL to flash.
P0740	TCC electrical failure	The TCC solenoid circuit does not provide the specified voltage drop across the solenoid. The circuit is open or shorted or the PCM driver failed during the diagnostic test.	With a short circuit, the engine will stall in second gear at low idle speeds with brake applied. With an open circuit, the TCC will not engage. This may cause the MIL to flash.
P1740	TCC malfunction	A mechanical failure of the solenoid was detected.	If the solenoid is stuck in the ON position, the engine will stall in second gear at low idle speeds with brake applied. If the solenoid stays in the OFF position, the TCC will not engage. This may cause the MIL to flash.
P1742	TCC solenoid failed on	The TCC solenoid has failed, which was caused by an electric, mechanical, or hydraulic problem.	The transmission will exhibit harsh shifts.
P2758	TCC solenoid circuit failure, stuck ON.	The TCC solenoid circuit does not provide the specified voltage drop across the solenoid. The circuit is open or shorted or the PCM driver failed during the diagnostic test.	The TCC will never engage and there will be no adaptive or self learning strategies.

FIGURE 6-10 A sample of TCC-related DTCs.

It is important to note that before the TCM will set a DTC, a transmission fault must occur during four consecutive driving cycles. A torque converter DTC will be set when the fault occurs during five consecutive driving cycles. In some transmissions, the TCC system has its own monitor, similar to those found with emission-related systems. The monitor is continuously run and will set a DTC if a fault occurs during two driving cycles. Make sure the vehicle has completed its necessary driving cycles before responding to a no-code situation. Before driving the vehicle through its required drive cycle, record and erase all DTCs. A typical driving cycle includes the following steps:

1. Warm engine to its normal operating temperature.
2. Check the ATF level and correct it as necessary.
3. With transmission in its normal Drive position, moderately accelerate to 50 mph (80 km/h).
4. Maintain that speed for a minimum of 15 seconds.
5. Lightly depress and release the brake pedal.
6. Resume the speed for maintain it for at least 5 seconds.
7. Bring the vehicle to a stop and keep it there for a minimum of 20 seconds.
8. Repeat Steps 3 through 7 at least five times.

TCC Monitors

The TCM determines the lockup status of the torque converter by comparing the engine speed to the turbine speed. The TCM then calculates the actual gear the transmission is in by comparing the turbine speed to the counter or output gear speed. When the TCM sees the conditions are suitable, it will do a quick check of the operation of the TCC, along with other transmission components. It checks the operation of the TCC by applying voltage to the appropriate solenoid to cause a lockup condition. When there is a difference of 70 rpm or more between engine speed and turbine speed and the TCM commands lockup, the TCM will set a code. In most cases, it will also turn on the MIL.

© Delmar/Cengage Learning

Although the monitor runs continuously, it will not run if certain codes are set. These codes typically relate to CAN communications, the electronic throttle control system, certain engine sensors, and the circuits for the transmission's solenoids. The monitor will also not run if the engine is not running, the vehicle is not moving more than 15 mph (25 km/h), the transmission is in low, the engine's temperature is lower than 104°F (40°C), or the temperature of the ATF is lower than 14°F (−10°C).

Other Scan Tool Checks

Most systems have the provision for operating outputs from the scan tool. This allows you to control the operation and main components of the transmission. The scan tool also displays PID information, along with the output control functions. With the output control, you can engage and disengage the TCC. The output control allows a technician to command three separate modes:

- ON—Commands the TCC solenoid On (100 percent duty cycle) to lock the TCC
- OFF—Commands the TCC solenoid Off (0 percent duty cycle) to unlock the TCC
- XXX—Cancels control of the TCC (returns to normal operation)

Most output control systems have a drive mode that allows you to control the TCC during a road test. During the drive mode, certain conditions must be met before the TCC solenoid can be commanded ON or OFF. When these conditions are not met, the scan tool will display an error message and the control function is ended. Examples of these conditions are the following:

To turn OFF the solenoid the following should be checked:

- The engine must be running
- The gear selector is in its highest range
- The vehicle must be traveling at least 2 mph (3 km/h)

To command the solenoid ON the following should be checked:

- The engine must be running
- The gear selector is in its highest range
- The vehicle must be traveling at least 2 mph (3 km/h)
- The ATF temperature must be between 60 and 240°F (6–116°C)
- There are no excessive loads on the engine
- The engine's speed is higher than 1500 rpm

Intermittent Problems

When there is an intermittent problem, using pending or soft codes may allow you to isolate the fault. These codes along with the customer's description of the problem and the road test results can lead you to the most probable problem area.

One way to identify the cause of an intermittent problem is by conducting a wiggle test with the scan tool connected to the vehicle. On the scan tool, select the TCC and TCCF PIDs. These will display the activity of the solenoid. Then with the scan tool, command the TCC solenoid ON and OFF, while monitoring the voltage and state of the solenoid on the scan tool (ON and OFF). When the solenoid changes from ON or OFF, it should click. Now wiggle the wiring and connectors at the transmission and observe the scan tool. If the state of voltage changed, there is a short or an open in the wiring or that connector.

Solenoid Checks

A common way to assess the condition of a solenoid is to check its resistance. Normally this is done by connecting an ohmmeter across the terminals of the solenoid (Figure 6-11). The specified resistance is typically listed for room temperature. This means the solenoid should not be checked if it is cold or hot. Most solenoids should have low resistance. The solenoid

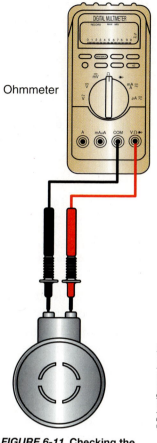

FIGURE 6-11 Checking the resistance of a TCC solenoid.

should also be checked for shorts-to-ground. This is done by connecting one of the meter's leads to a terminal at the solenoid and the other to a known good ground. Observe the meter. Repeat this test at the other solenoid terminal. If there is no short-to-ground, the readings will be close to infinite. If the solenoid is shorted, it should be replaced.

No Code Diagnosis

While testing any system, if no codes or codes indicating the system is operating normally are retrieved, conduct a thorough inspection of the system. Visually inspect all of the wires and connectors in the circuit. Make sure they are not loose, disconnected, or corroded. Also, refer to the symptom diagnostic chart given by the manufacturer.

If the clutch does not engage, make sure the converter clutch circuit has power. Check the circuit's fuse, make sure it is not blown. Check for voltage at the solenoid. If it is available, make sure the circuit's ground is good. If there is power available and the ground is good, check the voltage drop across the solenoid. The solenoids should drop very close to source voltage. If less than that is measured, check the voltage drop across the power and ground sides of the circuit. If the voltage drop testing results in good results, remove the solenoid and test it with an ohmmeter. If the solenoid checks out fine with the ohmmeter, suspect clutch material, dirt, or other material plugging up the solenoid valve passages. If blockage is found, attempt to flush the valve with clean ATF. If the solenoid has a filter assembly (Figure 6-12), replace the filter after cleaning the fluid passages. If the blockage cannot be removed, replace the solenoid.

If the clutch engages at the wrong time, a sensor or switch in the circuit is probably the cause. If clutch engagement occurs at the wrong speed, check all speed related sensors. A faulty temperature sensor may cause the clutch not to engage. If the sensor is not reading the

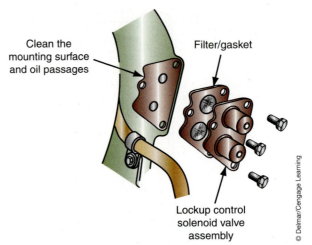

Clean the mounting surface and oil passages

Filter/gasket

Lockup control solenoid valve assembly

© Delmar/Cengage Learning

FIGURE 6-12 A torque converter clutch solenoid with a replaceable filter.

correct temperature, the PCM may never realize the temperature is suitable for engagement. Checking the appropriate sensors can be done with a scan tool, DMM, and/or lab scope. A check of the sensors is normally part of the manufacturer's system check.

To verify that a component is faulty, test it with the appropriate meter. All faulty wires and connectors should be repaired or replaced. TCC solenoids are not rebuildable and are replaced when they are faulty.

Mechanical Problems

Mechanical problems can cause the TCC not to operate properly. The causes of these problems typically result in the replacement of the torque converter with a new or rebuilt unit. It is important to keep these causes in mind when all electrical tests prove that system to be fine.

If the TCC does not apply, the cause can be fluid leaking from the converter or past its seals. This problem can also be caused by a severely worn TCC disc or friction material has fallen off the clutch disc. These problems can be verified by a careful inspection of the T/C.

If the TCC stays applied and tends to stall the engine at a stop, the problem may be a damaged piston plate or the plate is stuck to the T/C cover. These problems can be caused by overheating the converter. Look at the converter for signs of heat to verify this problem. This problem can also be caused by the lack of end clearance inside the converter. In either case, the T/C should be replaced.

STALL TESTING

Classroom Manual

Chapter 6, page 184

Two methods are commonly used to check the operation of the stator's one-way clutch: the stall test and the bench testing. Bench testing a converter will be covered later in this chapter. For now, we are still trying to determine if the converter is the cause of the customer's complaint. To bench test a converter, it must be removed from the vehicle and there is no need to do that unless we know the converter is at fault.

 WARNING: Make sure no one is around the engine or the front of the vehicle while a stall test is being conducted. A lot of stress is put on the engine, transmission, and brakes during the test. If something lets go, somebody can be seriously hurt.

To conduct a stall test, connect a tachometer to the engine and position it so that it can be easily read from the driver's seat (Figure 6-13). Set the parking brake, raise the hood, and place

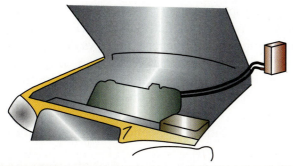

Trouble	Probable cause
Stall rpm high in D₄, 2, 1 & R	Low fluid level or oil pump output Clogged oil strainer Pressure regulator valve stuck closed Slipping clutch
Stall rpm high in R	Slippage of 4th clutch
Stall rpm high in 2	Slippage of 2nd clutch
Stall rpm high in D₄	Slippage of 1st clutch or 1st gear one-way clutch
Stall rpm low in D₄, 2, 1 & R	Engine output low Torque converter one-way clutch slipping

© Delmar/Cengage Learning

FIGURE 6-13 Before conducting a stall test, chock the wheels and place the tachometer in a position where it can be easily seen from the driver's seat.

blocks in front of the vehicle's nondriving tires. Conduct the test outdoors if possible, especially if it is a cold day. If the test is conducted indoors, place a large fan in front of the vehicle to keep the engine cool. With the engine running, press and hold the brake pedal. Then move the gear selector to the "drive" position and press the throttle pedal to the floor. Hold the throttle down for two seconds, note the tachometer reading, and immediately let off the throttle pedal and allow the engine to idle. Compare the measured stall speed to specifications (Figure 6-14).

If the torque converter and transmission are functioning properly, the engine will reach a specific speed. If the tachometer indicates a speed above or below specifications, a possible problem exists in the transmission or torque converter. If a torque converter is suspected of being faulty, it should be removed and the one-way clutch checked on the bench.

1. Never conduct a stall test if there is an engine problem.
2. Check the fluid levels in the engine and transmission before conducting the test.
3. The engine should be at normal operating temperature during the test.
4. Never hold the throttle wide open for more than five seconds during the test.
5. Do not perform the test in more than two gear ranges without driving the vehicle a few miles to allow the engine and transmission to cool down.
6. After the test, allow the engine to idle for a few minutes to cool the transmission fluid before shutting off the ignition.

TORQUEFLITE TRANSMISSION STALL SPEED CHART

Engine liter	Transaxle type	Converter diameter	Stall rpm
1.7	A-404	9 ½ in. (241 mm)	2300–2500
2.2	A-413	9 ½ in. (241 mm)	2200–2410
2.6	A-470	9 ½ in. (241 mm)	2400–2630

© Delmar/Cengage Learning

FIGURE 6-14 A typical torque converter stall speed chart.

SERVICE TIP:
Stall testing is not recommended on many late-model transmissions. This test places extreme stress on the transmission and should only be conducted if recommended by the manufacturer.

CAUTION:
If a stall test is not correctly conducted, the converter and/or transmission can be damaged.

CAUTION:
To prevent serious damage to the transmission, follow these guidelines while conducting a stall test.

SPECIAL TOOLS
Tachometer
Large fan
Wheel chocks

If the stall speed is below the specifications, a restricted exhaust or slipping stator clutch is indicated. If the stator's one-way clutch is not holding, ATF leaving the turbine of the converter works against the rotation of the impeller and slows down the engine. With both of these problems, the vehicle would exhibit poor acceleration. This is caused by either a lack of power from the engine or no torque multiplication occurring in the converter. If the stall speed is only slightly below normal, the engine is probably not producing enough power and should be diagnosed and repaired.

If the stall speed is above specifications, the bands or clutches in the transmission may be slipping and not holding properly.

If the vehicle has poor acceleration but had good results from the stall test, suspect a seized one-way clutch. Excessively hot ATF in the transmission is a good indication that the clutch is seized. However, other problems can cause these same symptoms; therefore, be careful during your diagnosis.

A normal stall test will generate a lot of noise, most of which is normal. However, if you hear any metallic noises during the test, diagnose the source of those noises. Operate the vehicle, at low speeds, on a hoist with the drive wheels free to rotate. If the noises are still present, the source of the noise is probably the torque converter.

Visual Inspection

If there was a noise coming from the torque converter during the stall test, visually inspect the converter before pulling the transmission out and testing or replacing the converter. Remove the torque converter access cover on the transmission (Figure 6-15) and rotate the engine with a remote starter button. While the torque converter is rotating with the engine, check to make sure the torque converter bolts are not loose and are not contacting the bell housing. Also, observe the action of the converter as it is spinning. If the converter wobbles, it may be due to a damaged flexplate or converter.

 WARNING: Before rotating the engine with a remote starter button, make sure the engine's ignition or fuel injection system is disabled. Also keep all parts of your body away from the rotating flexplate. Serious injury can result from being careless during this check.

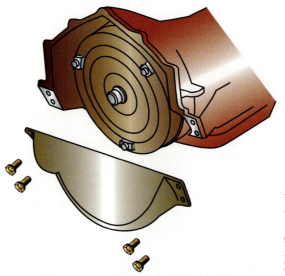

© Delmar/Cengage Learning

FIGURE 6-15 To inspect a torque converter while it is still in the vehicle, remove the access cover.

Check the converter's balance weights to make sure they are still firmly attached to the unit. Carefully inspect the flexplate for evidence of cracking or other damage. Also check the condition of the starter ring gear, the teeth of which should not be damaged. The gear should be firmly attached to the flexplate.

Check the torque converter for **ballooning**. If excessive pressure was able to build up inside the converter, the converter will expand, or balloon. A stuck converter check valve typically causes this. If the converter is ballooned, it should be replaced and the cause of the problem also repaired.

A ballooned torque converter will also damage other driveline components. If the converter is ballooned toward the rear of the unit, the transmission's oil pump is most likely bad. If it is ballooned toward the front, the crankshaft's thrust bearings are worn or damaged.

Converter Fluid. A stall test works well to circulate fluid inside the converter, therefore a small sample of the fluid can give a good indication of the condition of the converter. Pour a small amount of fluid from the converter onto a white absorbent cloth or paper towel or pour it through a fine filter. Examine and smell the fluid. Now examine the paper towel. Look for any signs of residue. If particles were found in the fluid or if there is any indication of engine coolant in the fluid, the converter should be replaced. If the fluid was not contaminated, empty the converter of all fluid.

> **Ballooning** makes the torque converter look like it has been blown up like a balloon; it is caused by excessive pressure in the converter.

External Measurements

Certain parts of the torque converter should be measured for wear. These also should be carefully inspected for signs of damage. The measurements are taken with the appropriately sized micrometer. Begin by measuring the distance between the two flats on the hub. Then measure the outside diameter (OD) of the hub (Figure 6-16). The latter measurement determines if the hub is worn where it rides in its bushing. Most specifications list a minimum dimension for these diameters. If either measurement of the hub is less than the minimum, the converter should be replaced.

The OD of the pilot on the cover of the torque converter should also be measured (Figure 6-17). The pilot is also a potential wear area. If the diameter is less than the minimum given in the specifications, replace the converter.

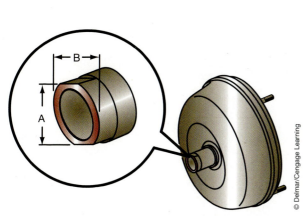

FIGURE 6-16 Measure the torque converter's hub and compare it to specifications. (A) the OD between the flats of the torque converter hub, and (B) the OD of the hub at the bushing wear area.

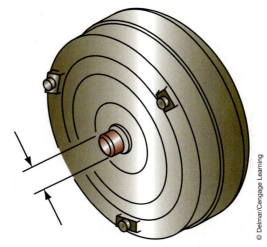

FIGURE 6-17 Measure the OD of the torque converter's pilot and compare it to specifications.

© Delmar/Cengage Learning

BENCH TESTING

Further inspection of the torque converter requires that it be removed from the vehicle. These inspections (Figure 6-18) should also be done anytime you have the torque converter out of a vehicle, especially if the oil pump is damaged or if the customer's complaint appears to be related to the torque converter.

When removing the transmission, pay particular attention to the condition of the flexplate and converter mounting bolts. Problems that appear to be torque converter problems may be caused by a damaged flexplate or a loose torque converter.

Once the converter is separated from the transmission, inspect the drive studs or lugs used to attach the converter to the flexplate. These are designed to hold the converter firmly to the flexplate and to keep the converter in line with the engine. Damaged studs or lugs will cause runout, misalignment, and vibration problems, and can result in damage to the bushings of the oil pump. If the threads are damaged slightly, they can be cleaned up with a thread file, tap, or die set. However, if they are badly damaged, the converter should be replaced. The converter should also be replaced if the studs or lugs are loose or damaged. Also check the shoulder area around the lugs and studs for cracked welds or other damage. If any damage is found, the converter should be replaced. An exception to this is when the internal threads of a drive lug are damaged. These can often be repaired by tapping the threads or by installing a threaded insert. Also inspect the converter attaching bolts or nuts and replace them if they are damaged.

Check the pilot of the converter for wear and other damage. Also check the area around the pilot for cracks. In either case, the converter should be replaced. If the area around the pilot has dimples and looks like it has contacted the flywheel bolts, the converter has ballooned. If the converter has ballooned, it should be replaced and the cause of the ballooning corrected.

Inspect the flexplate for signs of damage, warpage, and cracks. Check the condition of the teeth on the starter ring gear. Replace the flexplate if there is any evidence of damage.

Check the drive hub of the torque converter; it should be smooth and not show any signs of wear. The easiest way to check the smoothness of the hub is to run a fingernail across the surface. If any irregularities are felt, the hub should be smoothed with crocus cloth or the converter replaced. If the hub is worn, carefully inspect the oil pump drive and replace the torque converter. Light scratches, burrs, nicks, or scoring marks on the hub surface can be polished with fine crocus cloth. Be careful not to allow dirt to enter into the converter while polishing the hub. Use a rag to cover the opening in the hub. Dirt and dust

SPECIAL TOOLS

Vee-blocks

Dial indicator and mounting bracket

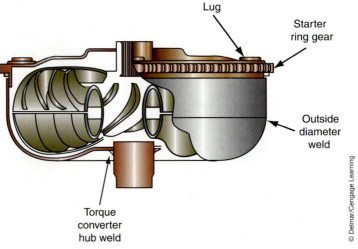

FIGURE 6-18 Some of the areas of a torque converter that need to be carefully checked.

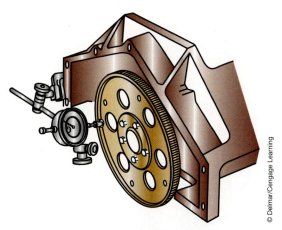

FIGURE 6-19 Setup for checking flexplate runout.

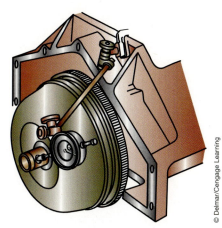

FIGURE 6-20 **Setup for checking converter hub runout.**

that enter the converter through this opening can cause the converter to wear rapidly. After polishing, clean the hub with solvent and a clean, lint-free rag. If the hub has deep scratches or other major imperfections, the converter should be replaced.

Since misalignment can cause many different problems, it is important that you check for excessive runout of the flexplate and the converter's hub. Check flexplate runout with a dial indicator. Mount the indicator onto the engine block and set its plunger on the flexplate (Figure 6-19). Rotate the flexplate and observe the readings on the indicator. Typically, the maximum allowable runout is 0.03 inches.

To check the runout of the converter's hub, mount the converter onto the flexplate. Tighten the retaining bolts to the specified torque. Position a dial indicator on the engine block. Set the plunger of the indicator on the hub (Figure 6-20). Rotate the torque converter and observe the readings on the indicator. Again, the typical maximum allowable runout is about 0.03 inches. If the runout is excessive and there is a normal amount of flexplate runout, remove the torque converter and set it at a new position on the flexplate. Check the runout again. If the runout is now within specifications, mark its position on the flexplate. If repositioning the converter on the flexplate does not correct hub runout, the converter should be replaced.

Because of the weight of torque converters, the above check is not always reliable. To get an accurate check of torque converter runout, mount the converter on two vee-blocks. Set the dial indicator so it can read lateral runout of the converter housing. To do this, place the indicator's plunger against the top of the converter housing's flywheel mounting surface. Rotate the housing and observe the runout. To check radial runout, place the dial indicator on the outside end of the pump hub. Rotate the housing and observe the runout. Then move the indicator on the hub and toward the housing. Again rotate and measure the runout.

In general, a torque converter should be replaced if there is an excessive amount of metal in the transmission fluid. The debris is very difficult to remove from inside the torque converter. The entire unit should also be replaced if has signs of overheating, fluid leakage around its seams or welds, worn drive stud shoulders, loose or damaged drive studs, a heavily grooved hub, or excessive hub runout. Of course, the torque converter should be bench tested before you decide to reuse it.

Remote Driveshafts. Transaxles that don't have their oil pump driven directly by the torque converter use a driveshaft (Figure 6-21) that fits into a support bushing inside the converter's hub (Figure 6-22). This bushing should be checked for wear. To do this, measure the inside diameter of the bushing and the outside diameter of the driveshaft. The difference between the two is the amount of clearance. This measurement should be compared to factory specifications. Normally the maximum allowable clearance is 0.004 inches. Excessive clearance can cause oil leaks. If the clearance is excessive, the bushing should be replaced.

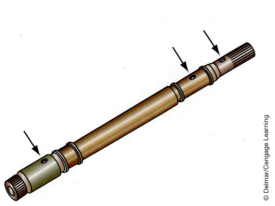

FIGURE 6-21 The pump driveshaft should be inspected in the areas that ride on the bushings and bearings. The shaft seals should always be replaced.

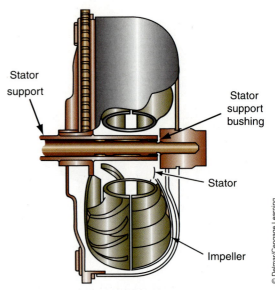

FIGURE 6-22 Typical stator support and oil flow in a torque converter.

Classroom Manual

Chapter 6, page 179

Stator One-Way Clutch Diagnosis

The operation of the stator one-way clutch inside the torque converter is critical to overall effectiveness of the torque converter. If a problem occurs in this clutch assembly, the clutch will either fail to lock when rotated in either direction or fail to unlock when rotated in either direction. Although these problems are similar, they affect efficiency at opposite ends of the engine's operating speeds. However, in either case, fuel economy will be affected.

When the stator clutch does not lock, there is a disruption in vortex flow and a loss of torque multiplication in the torque converter. A vehicle with this problem will have sluggish low-speed performance, but will perform normally at higher speeds when the stator is supposed to freewheel.

A vehicle with a constantly locked stator will have good low-speed and poor high-speed performance. Torque multiplication will always occur, as will speed reduction. A vehicle with a constantly locked stator will show signs of overheating. If you suspect a locked stator, check for a bluish tint on the hub of the converter. This discoloration typically results from overheating. It is normal for some blue to be evident at the spot where the hub was welded to the housing. If the most of the hub is blue, the converter has overheated.

To check the stator's one-way clutch with the converter on a bench, insert a finger into the splined inner race of the clutch. Attempt to turn the inner race in both directions. You should be able to turn the race freely in one direction and feel lockup in the opposite direction. If the clutch rotates freely in both directions or if the clutch is locked in both directions, the converter should be replaced.

Because this check does not put a load on the clutch assembly, it does not totally check the unit. Therefore, some manufacturers, such as Ford, recommend the use of a special tool set that holds the inner race and exerts a measurable amount of torque on the outer race, thereby allowing the technician to observe the action of the clutch while under load (Figure 6-23).

Internal Interference Checks

Internal converter parts hitting each other or hitting the housing may also cause noises. To check for any interference between the stator and turbine, place the converter face down on a bench (Figure 6-24). Then install the oil pump assembly. Make sure the oil pump drive engages with the oil pump. Insert the input shaft into the hub of the turbine. Hold the oil pump and converter stationary, and then rotate the turbine shaft in both directions. If the shaft does not move freely or makes noise, the converter must be replaced.

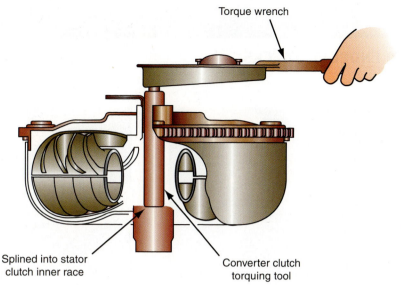

Splined into stator
clutch inner race

Converter clutch
torquing tool

FIGURE 6-23 Checking one-way clutch with special tool fixture.

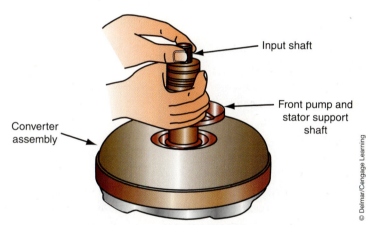

Input shaft

Front pump and
stator support
shaft

Converter
assembly

FIGURE 6-24 Checking stator-to-turbine interference.

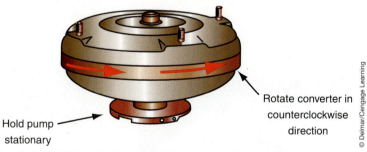

Rotate converter in
counterclockwise
direction

Hold pump
stationary

FIGURE 6-25 Checking stator-to-impeller interference.

To check for any interference between the stator and the impeller, place the transmission's oil pump on a bench and fit the converter over the stator support splines (Figure 6-25). Rotate the converter until the hub engages with the oil pump drive. Then hold the pump stationary and rotate the converter in a counterclockwise direction. If the converter does not freely rotate or makes a scraping noise during rotation, the converter must be replaced. Photo Sequence 11 covers the procedure for checking internal interference and endplay.

The input shaft may be referred to as the turbine shaft on some transaxles.

INTERNAL INTERFERENCE AND ENDPLAY CHECKS

All photos in this sequence are © Delmar/Cengage Learning.

P11-1 To check for any interference between the stator and turbine, place the converter, face down, on a bench.

P11-2 Then install the oil pump assembly. Make sure the oil pump drive engages with the oil pump.

P11-3 Insert the input shaft into the hub of the turbine.

P11-4 Hold the oil pump and converter stationary, then rotate the turbine shaft in both directions. If the shaft does not move freely and/or makes noise, the converter must be replaced

P11-5 To check for interference between the stator and the impeller, place the transmission's oil pump on a bench and fit the converter over the stator support splines.

P11-6 Rotate the converter until the hub engages with the oil pump drive. Then hold the pump stationary and rotate the converter in a counterclockwise direction. If the converter does not rotate freely or makes a scraping noise during rotation, the converter must be replaced.

P11-7 Special tools are required to check the internal endplay of a torque converter. These include a holding tool and a dial indicator with a holding fixture.

P11-8 Insert the holding tool into the hub of the converter and once bottomed, tighten it in place. This locks the tool into the splines of the turbine.

P11-9 Fasten the dial indicator to the hub.

P11-10 Set the plunger of the dial indicator so it can read the movement of the tool.

P11-11 Zero the dial on the indicator.

P11-12 Lift up on the tool and observe the reading on the indicator. This is the amount of endplay inside the converter. Compare your reading to specifications.

Endplay Check

The special tools required to check the internal endplay of a torque converter are typically part of the essential tool kit recommended by each manufacturer. However, these specialty tools can be individually purchased through specialty tool companies. Basically, the special tools are a holding tool and a dial indicator with a holding fixture. The holding tool is inserted into the hub of the converter and, once bottomed, is tightened in place. This locks the tool into the splines of the turbine. The dial indicator is fixed onto the hub. The amount indicated on the dial indicator, as the tool is lifted up, is the amount of endplay inside the converter. If this amount exceeds specifications, replace the converter.

SPECIAL TOOLS

Manufacturer-specified holding tool

Dial indicator and mounting bracket

Converter Leakage Tests

If the initial visual inspection suggested that the converter has a leak, special test equipment can be used to determine if the converter is leaking. This equipment uses compressed air to pressurize the converter. Leaks are found in much the same way as tire leaks are; that is, the converter is submerged in water and the trail of air bubbles lead the technician to the source of leakage.

Oil Pump Seal

If the backside of the torque converter is wet, it is very likely that the transmission pump seal is bad. There are many possible causes for leaks at the seal. These must also be looked at and corrected when the new seal is installed (Figure 6-26). These are some of the many causes of pump seal leakage:

- The seal has a missing or damaged spring.
- The lip of the seal is cut or damaged.
- The pump bore is worn, scratched, or damaged.
- The pump seal bore is off the centerline of the crankshaft.
- The pump drain hole is too small or is restricted.
- The pump bushing is worn, loose, or missing.
- The hub seal area is worn, rusted, or pitted.
- The hub has excessive runout.

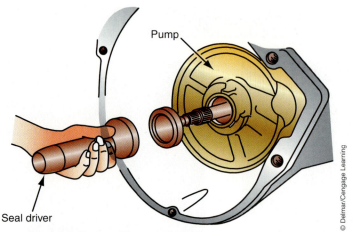

FIGURE 6-26 The converter/pump seal must be installed with the proper driver in order to seal.

Pump

Seal driver

© Delmar/Cengage Learning

- The hub is cracked or machined roughly.
- The hub is the incorrect length.
- The hub or converter is out of balance.
- There is excessive crankshaft end clearance (should be no more than 0.006 inches).
- The crankshaft pilot bore is worn.
- The crankshaft pilot sleeve is missing.
- The flexplate spacer is broken or missing.
- The flexplate is broken or has excessive runout.
- The converter housing is loose.
- The dowels in the converter housing are worn or missing.
- The replacement seal has incorrect elastomer or is missing an auxiliary lip.
- The inside or outside diameter of the seal is incorrect.
- The elastomer of the seal is too thick and is restricting the drain hole.
- The helix direction of the seal is incorrect.
- The seal was installed upside down.
- The seal was damaged during installation by the use of a defective or wrong driver.
- The seal was damaged by careless installation of the torque converter.

Summary of Checks

If certain conditions are found while inspecting a torque converter, the converter must be replaced. Specifically, the converter should be replaced:

1. If there is a stator clutch failure.
2. If there is internal interference.
3. If the transmission's front pump is badly damaged.
4. If the converter hub is severely damaged or scored.
5. If there are signs of external fluid leaks.
6. If the drive studs or lugs are damaged or loose.
7. If there are signs of overheating.
8. If heavy amounts of metal were found in the fluid.
9. If any damage is evidence that the converter is no longer balanced.

Although some specialty shops will rebuild a converter, nearly all technicians replace the converter when any of the above exists. This is especially true of converters with a clutch. Rebuilding a converter is for the automatic transmission specialist and is not a normal task for an automobile technician. Because this procedure requires special equipment and knowledge, do not attempt to repair a faulty converter.

Starter Ring Gear Replacement

The starter ring gear is most often part of the flexplate. Therefore, whenever the teeth of the ring gear are damaged, the entire flexplate is replaced. There are some transmissions that are equipped with a torque converter fitted with a ring gear around the outer circumference of the torque converter cover. The ring gear is welded to the front of the converter cover and gear replacement involves breaking or cutting the welds to remove the old gear, then replacing the gear and welding it back onto the converter. This procedure is typically not recommended for torque converters fitted with a clutch, as the heat from welding can destroy the frictional surfaces inside the converter. In these cases, if the ring gear is damaged, the entire converter should be replaced.

CLEANING

A torque converter without a clutch should be flushed any time it will be reused and there was contaminated fluid in the transmission or if the transmission overheated. There are many small corners inside a converter for dirt to become trapped in. Leftover debris in the converter can lead to converter and transmission damage. Remember, the fluid that circulates through the converter circulates through the transmission next. Converters are typically cleaned by flushing the inside of the housing with solvent.

Flushing with solvent is not recommended for most converters with a clutch. It is recommended that a clutch-type converter be replaced, rather than flushed or cleaned. If the converter will be used again and had no signs of being contaminated, pour about 2 quarts (1.9 liters) of clean transmission fluid into the converter. Swirl the fluid around the inside of the converter, then drain the fluid out.

 WARNING: Never use water-based cleaners or mineral spirits to clean or flush the torque converter. Use only clean automatic transmission fluid designated for that transmission.

Flushing removes all debris and sludge from the inside of the converter. Flushing is either done by a machine or by hand. Some transmission shops use a torque converter flushing machine that pumps a solvent through the converter as it is rotated by the machine. This method keeps the fluid moving inside the converter. The moving fluid is quite efficient at picking up any dirt present. The dirty solvent is pumped out of the converter and new solvent added during the flushing procedure.

In shops without a flushing machine, it is possible to clean the inside of converters by hand. But this method is risky and not very effective, so it is not generally recommended. Basically, the procedure involves pouring about two quarts of clean solvent into the converter. Then, forcefully rotate and vibrate the converter. This action should dislodge any trapped debris. It is also helpful to use the input shaft and spin the turbine while the fluid is inside. The solvent is drained and the process repeated until the drained solvent is clean.

Since any amount of dirt can destroy a transmission, most rebuilders replace the torque converter during a transmission overhaul. There is no accurate way of knowing if dirt remains in the converter after cleaning and flushing it. The cleaning process may loosen up the debris, which will break down and contaminate the fluid once the torque converter starts spinning and gets hot.

Some transmission shops may cut the torque converter shell in half, then clean the parts, examine them for wear, and replace any that are worn or broken. The shell is then welded back together. This is a job only for shops that are equipped to do this job right.

After the converter has been flushed, disconnect it from the machine. If the converter has a drain plug, invert and drain the complete assembly. Converters without drain holes or plugs

can be drilled with a $\frac{1}{8}$-inch drill bit between the top end of the impeller fin dimples. This hole will act as an air bleed to maximize flushing. After flushing and draining the converter, the bleeder hole is sealed with a closed-end pop rivet covered with sealant. Some technicians use a MIG welder to seal the hole rather than use a rivet. However, this is not recommended because the heat can destroy the T/C clutch.

TORQUE CONVERTER REPLACEMENT

Extra care should always be taken when replacing a torque converter. Size and fit are not the only important variables. Nor does size alone determine the stall speed of the converter. Even if the converter has exactly the same stall speed, it may have a different torque ratio and should not be used. Always check and double-check the part or model number of the torque converter you are removing and compare it to the one you are going to install. Converters are typically identified by a sticker or a number code stamped into the converter housing.

When replacing a torque converter, never use an impact wrench on the torque converter bolts. Impact wrenches can drive the bolts through the cover, which will warp the inside surface and prevent proper clutch engagement, or it will damage the clutch's pressure plate.

Always perform an endplay check and check the depth of the torque converter in the bell housing (Figure 6-27) before reinstalling a torque converter or installing a new unit.

Correct installation of a torque converter requires that the converter's hub be fully engaged with the transmission. To help in seating the torque converter into the transmission, push on the converter while rotating it. Care should be taken not to damage the pump seal while installing the converter. To verify full engagement, place a straightedge across the bell housing or transaxle flange. Then measure from the straightedge to the pilot (Figure 6-28). Compare your readings to those specified by the manufacturer. If your reading is less than the specifications, the torque converter is not properly engaged in the transmission.

Converter Balance

When replacing a torque converter, make sure the manufacturer or rebuilder balanced it. If the converter is not balanced, noise and vibrations can result, as well as damage to the transmission.

When reusing a torque converter, make sure it is positioned in its original position on the flexplate and make sure the mounting bolts or nuts are tightened to specifications. Sometimes manufacturers install a slightly unbalanced torque converter in a precise position on the flexplate. This is done to offset or counter engine vibrations, and you have no way of knowing if this is what they did.

When installing a known balanced converter, it can be positioned anywhere on the flexplate as long as it is fully seated against it and the mounting bolts or nuts are properly torqued.

Classroom Manual

Chapter 6, page 185

SERVICE TIP:
Many OBD II systems have an operational mode that allows for converter clutch break-in. The purpose of this mode is to allow the system to readjust to a new clutch in the torque converter. Whenever a torque converter is replaced in a vehicle equipped with OBD II, set the system into the clutch break-in mode. Doing this will allow the system to realize things have changed and will allow the converter clutch to work more efficiently.

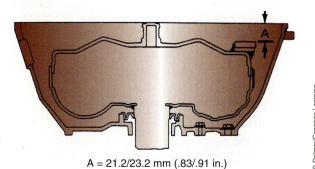

A = 21.2/23.2 mm (.83/.91 in.)

FIGURE 6-27 **Before removing a torque converter, check its installed depth in the bell housing.**

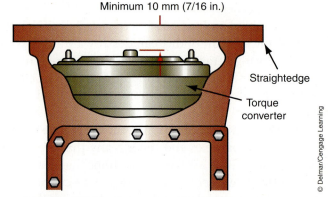

FIGURE 6-28 **Checking proper installation of a torque converter in a transmission.**

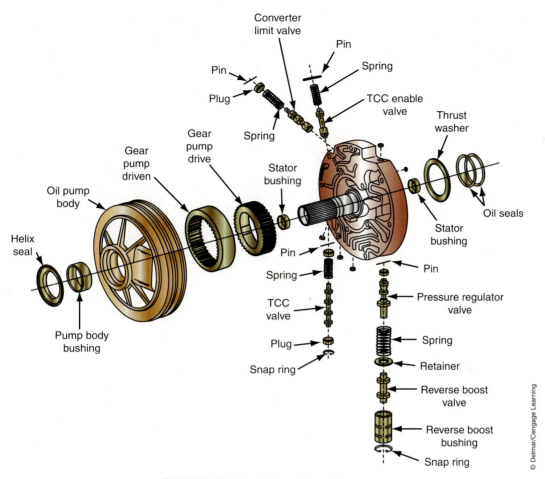

Figure labels: Converter limit valve, Pin, Pin, Spring, Plug, Spring, TCC enable valve, Gear pump drive, Thrust washer, Gear pump driven, Stator bushing, Oil pump body, Oil seals, Helix seal, Stator bushing, Pin, Pin, Spring, Pressure regulator valve, TCC valve, Pump body bushing, Plug, Spring, Retainer, Snap ring, Reverse boost valve, Reverse boost bushing, Snap ring

© Delmar/Cengage Learning

FIGURE 6-29 A typical gear-type oil pump.

OIL PUMP SERVICE

The oil pump (Figure 6-29) of an automatic transmission should be carefully inspected during any transmission overhaul and especially when low line pressure was measured during a pressure test. Carefully remove the oil pump assembly (Figure 6-30). Some transmissions require the use of a special puller to remove the oil pump from the transmission case (Figure 6-31). Never pry the pump out; it is easy to damage the case if this is done.

 WARNING: Some transmissions are fitted with oil control floats located between the oil pump and the gearsets. These floats can be easily damaged during pump removal and installation. The floats close the opening between the oil pan and the gearsets to prevent the chamber that contains the gears from overfilling with fluid, which can cause fluid aeration. The float can be easily dislodged and become stuck in one position.

Disassembly

Before disassembling the pump, mark the alignment of the gears. If acceptably worn gears are reinstalled in a position other than their wear pattern, excessive noise will result. When the pump halves have been separated, look at the relationship between the inner and outer gears. Most gears will have a mark on them indicating the top side of the gear (Figure 6-32).

If no mark is present, you should use a nondestructive type of marker to be sure you install the gears in the same position they were originally in (Figure 6-33). This ensures that the converter drive hub will mate correctly with the inner gear.

Classroom Manual
Chapter 6, page 206

SPECIAL TOOLS
Machinist bluing
Paint stick
Feeler gauge set
RTV

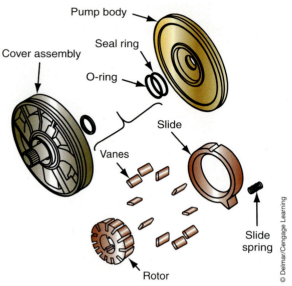

FIGURE 6-30 While removing a pump assembly, make sure to keep track of the position of all of the parts.

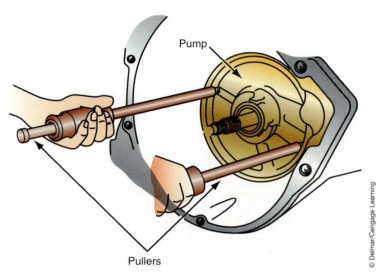

FIGURE 6-31 Using a puller to remove an oil pump from a transmission case.

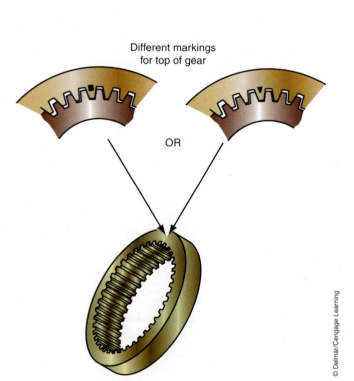

FIGURE 6-32 Identification marks on oil pump gearsets.

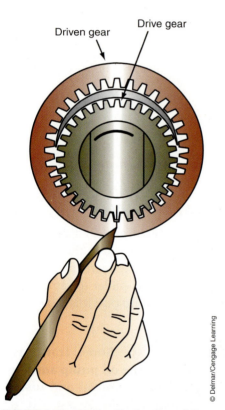

FIGURE 6-33 Marking the location of the outer and inner gears so that they can be properly meshed during reassembly.

Another way of ensuring the proper position of the gears is to observe the existing wear pattern on the gears while they are being removed and make a note of this for use during reassembly. If the inner gear is installed upside down, there will not be enough free movement, which can result in one or more of the following problems: broken inner gear, broken flexplate, pump cover scoring, or broken transmission case ears. The proper procedure for disassembling an oil pump is shown in Photo Sequence 12. This procedure is given as an example; always refer to the procedures for the specific transmission you are working on.

PROPER PROCEDURE FOR DISASSEMBLING AN OIL PUMP

All photos in this sequence are © Delmar/Cengage Learning.

P12-1 Remove the front pump bearing race, front clutch thrust washer, gasket, and O-ring. Inspect the pump bodies, pump shaft, and ring groove areas.

P12-2 Unbolt and separate the pump bodies.

P12-3 Mark the gears with machinist bluing ink or paint before removing them, so that the gears will remain in the same relationship during reassembly.

P12-4 Inspect the gears and all internal surfaces for defects and visible wear.

P12-5 Measure between the outer gear and the pump housing. Clearance should be 0.003–0.006 in. or 0.08–0.15 mm. Replace the pump if the measurements exceed specifications.

P12-6 Measure between the outer gear teeth and the crescent. Clearance should be 0.006–0.007 in. or 0.015–0.18 mm. Replace the pump if clearance exceeds 0.008 in.

P12-7 Place the pump flat on the bench and, with a feeler gauge and straight edge, measure between the gears and the pump cover. The clearance should be no more than 0.002 in. (0.05 mm). Replace the pump if the measurement exceeds specifications.

P12-8 Put the halves back together.

P12-9 Torque the securing bolts to specifications. Replace all O-rings and gaskets.

Inspection

Inspect the pump bore for scoring on both its bottom and sides. A converter that has a tight fit at the pilot hub may hold the converter drive hub and inner gear too far into the pump, causing cover scoring. A front pump bushing that has too much clearance may allow the gears to run off center, causing them to wear into the crescent or the sides of the pump body.

The stator shaft should be inspected for looseness in the pump cover (Figure 6-34). The shaft's splines and bushing should also be carefully looked at. If the splines are distorted, the shaft and the pump cover should be replaced. Because the bushings control oil flow through the converter and cooler, their fit must be checked (Figure 6-35). Bushings must be tight inside the shaft and provide the input shaft with a 0.0005- to 0.003-inch clearance.

Inspect the gears and pump parts for deep nicks, burrs, or scratches. Examine the pump housing for abnormal wear patterns. The fit of each gear into the pump body, as well as the centering effect of the front bushing, controls oil pressure loss from the high-pressure side of the pump to the low-pressure input side. Scoring or body wear will greatly reduce this sealing capability.

On positive-displacement pumps, use a feeler gauge to measure the clearance between the outer gear and the pump pocket in the pump housing (Figure 6-36). If possible, also check the clearance between the outer pump gear teeth and the crescent (Figure 6-37), and between the inner gear teeth and the crescent. Compare these measurements to the specifications. Use a straightedge and feeler gauge to check gear side clearance (Figure 6-38) and

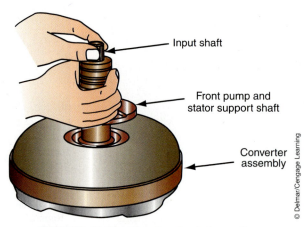

FIGURE 6-34 Checking the play between the splines of the oil pump and the input shaft.

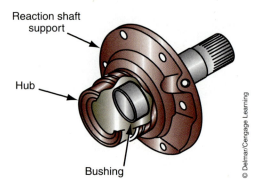

FIGURE 6-35 Location of a stator shaft bushing.

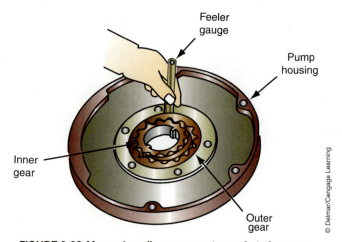

FIGURE 6-36 Measuring oil pump gear-to-pocket clearance.

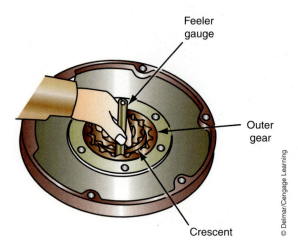

FIGURE 6-37 Measuring the clearance between the outer gear and the crescent.

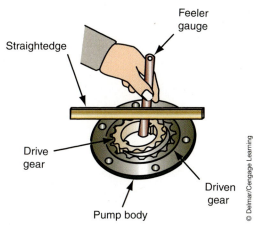

FIGURE 6-38 Measuring the side clearance of the pump's gears.

compare the clearance to the specifications. If the clearance is excessive, replace the pump. Normally, the maximum clearance is 0.0015 in (0.038 mm)

Variable-displacement vane-type pumps require different measuring procedures. However, the inner pump rotor to converter drive hub fit is checked in the same way as described for the other pumps. The pump rotor, vanes, and slide are originally selected for size during assembly at the factory (Figure 6-39). Changing any of these parts during overhaul can destroy this sizing and possibly the body of the pump. You must maintain the original sizing if any parts are found to need replacement. These parts are available in select sizes for just this reason.

The vanes are subject to edge wear, as well as cracking and subsequent breakage. The outer edge of the vanes should be rounded with no flattening (Figure 6-40). These pumps have an aluminum body and cover halves; therefore, any scoring indicates that they should be replaced.

Inspect the reaction shaft's seal rings. If the rings are made of cast iron, check them for nicks, burrs, scuffing, or uneven wear patterns; replace them if they are damaged (Figure 6-41). Make sure the rings are able to rotate in their grooves. Check the clearance between the reaction shaft support ring groove and the seal ring. If the seal rings are the Teflon full-circle type, cut them out, and use the required special tool to replace them.

The outer area of most pumps utilizes a rubber seal (Figure 6-42). Check the fit of the new seal by making sure the seal sticks out a bit from the groove in the pump. If it does not, it will leak. The seal at the front of the pump is always replaced during overhaul. Most of these seals are the metal-clad lip seal type. Care must be taken to avoid damage to the seating area when removing the old seal.

Check the area behind the seal to be sure the drainback hole is open to the sump. If this hole is clogged, the new seal may blow out. The drainback hole relieves pressure behind the seal. A loose-fitting converter drive hub bushing can also cause the front pump seal to blow out.

Wear Plate Inspection. Some oil pumps have a wear plate. These are common in vane-type pumps. The plate should be carefully inspected for damage and distortion. The plate should not be scored, nicked, or have grooves cut into it. The thickness of the plate at the wear area should be measured and compared to specifications (Figure 6-43). If the plate is damaged or excessively worn, it should be replaced.

Oil Pump Body Bushing. The bushing of the oil pump should be carefully checked. It should be replaced if it is damaged or worn. The best way to determine if it is worn is to measure its inside diameter (Figure 6-44). If the measurement exceeds specifications, the bushing should be replaced.

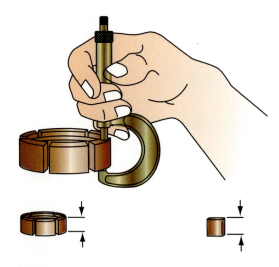

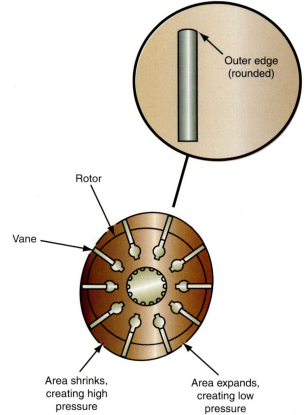

Rotor selection

Thickness (mm)	Thickness (in.)
17.593–17.963	0.7068–0.7072
17.963–17.973	0.7072–0.7076
17.973–17.983	0.7076–0.7080
17.983–17.993	0.7080–0.7084
17.993–18.003	0.7084–0.7088

Vane selection

Thickness (mm)	Thickness (in.)
17.943–17.961	0.7064–0.7071
17.961–17.979	0.7071–0.7078
17.979–17.997	0.7078–0.7085

Slide selection

Thickness (mm)	Thickness (in.)
17.983–17.993	0.7080–0.7084
17.993–18.003	0.7084–0.7088
18.003–18.013	0.7088–0.7092
18.013–18.023	0.7092–0.7096

FIGURE 6-39 Vane-type oil pump measurements and selection chart.

© Delmar/Cengage Learning

FIGURE 6-40 Placement of vanes in the pump's rotor.

© Delmar/Cengage Learning

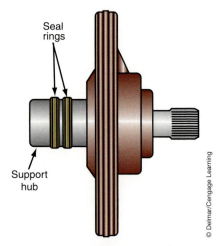

FIGURE 6-41 Location of oil seal rings on a typical oil pump assembly.

© Delmar/Cengage Learning

FIGURE 6-42 Outer oil pump seal.

© Delmar/Cengage Learning

FIGURE 6-43 The wear plate's thickness should be measured.

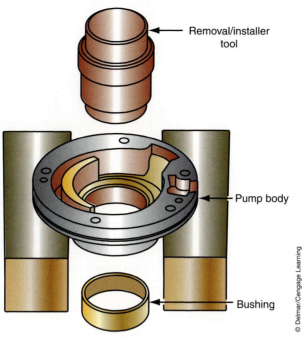

FIGURE 6-45 Special tools are required to remove and install the oil pump's bushing.

Removal/installer tool

Pump body

Bushing

© Delmar/Cengage Learning

FIGURE 6-44 Checking the oil pump's bushing for wear.

© Delmar/Cengage Learning

Special tools are required to remove the old bushing and install a new one. To replace the bushing, remove all gears from the pump's body. Support the body so both ends of the bushing's bore are away from the bench (Figure 6-45). Install the removal tool into the bore and press or drive the bushing out of the bore. Clean the bore. Before installing the new bushing, apply a coat of Loctite® #620, or equivalent, on the inside of the bore. Then press or drive the bushing into the bore.

Reassembly

The pump seal can be installed with a hammer and seal driver. Apply a very thin coating of RTV sealant around the outside surface of the seal case, when installing the seal. Place some transmission fluid in the pocket of the pump housing and install the gears into the housing according to their alignment marks. Align and install the reaction shaft support and tighten the bolts to the specified torque. Make sure the pump is not binding after you have tightened it by using the torque converter to rotate the pump.

Place a layer of clean ATF in two or three spots around the oil pump gasket and position it on the transmission case. Tighten the pump, attaching bolts to specifications in the specified order (Figure 6-46). After the bolts are tight, check the rotation of the input shaft. If the shaft does not rotate, disassemble the transmission to locate the misplaced thrust washer or misaligned friction plate.

Transmission Cooler Service

Vehicles equipped with an automatic transmission can have an internal or external transmission fluid cooler, or both (Figure 6-47). The basic operation of either type of cooler is that of a **heat exchanger**, meaning that heat from the fluid is transferred to something else, such as a liquid or air. Hot ATF is sent from the transmission to the cooler, where it has some of its heat removed, then the cooled ATF returns to the transmission.

Internal-type coolers are located inside the engine's radiator. Heated ATF travels from the torque converter to a connection at the radiator. Inside the radiator is a small internal cooler,

A **heat exchanger** may also be called an intercooler.

Heat is transferred because of a law of nature: the heat of an object will always attempt to heat a cooler object. Heat is exchanged.

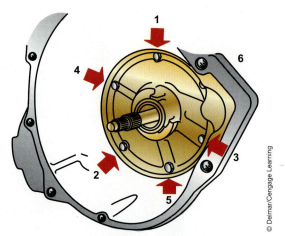

FIGURE 6-46 Oil pump bolts must be torqued in the sequence recommended by the manufacturer.

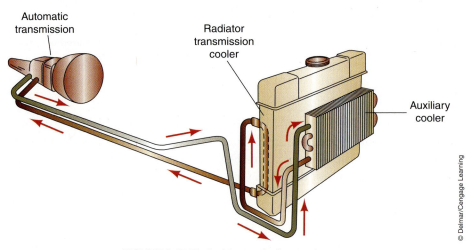

FIGURE 6-47 Typical transmission coolers.

which is sealed from the liquid in the radiator. ATF flows through this cooler and its heat is transferred to the liquid in the radiator. The ATF then flows out of a radiator connection, back to the transmission.

> **CUSTOMER CARE:** This type of cooler is less efficient as engine temperatures increase; therefore, you should recommend that an external cooler be installed in addition to the internal one if the vehicle is used in ways that require the engine to operate at high temperatures, such as carrying heavy loads or pulling trailers.

The process of cooling is actually the process of removing heat.

External coolers are mounted outside the engine's radiator, normally just in front of it. Air flowing through the cooler removes heat from the fluid before it is returned to the transmission.

The engine's cooling system is the key to efficient transmission fluid cooling. If anything affects engine cooling, it will also affect ATF cooling. The engine's cooling system should be carefully inspected whenever there is evidence of ATF overheating or a transmission cooling problem.

If the problem is the transmission cooler, examine it for signs of leakage. Check the dipstick for evidence of coolant mixing with the ATF. Milky fluid indicates that engine coolant is leaking

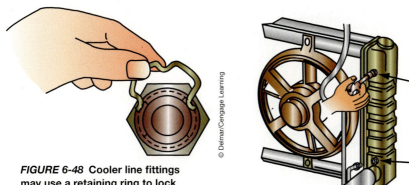

FIGURE 6-48 Cooler line fittings may use a retaining ring to lock the fitting to the line. These rings should never be reused and must be properly installed.

Inlet connector

Outlet connector (plug)

FIGURE 6-49 Before testing the cooling system, identify which line is the in and which is the out line.

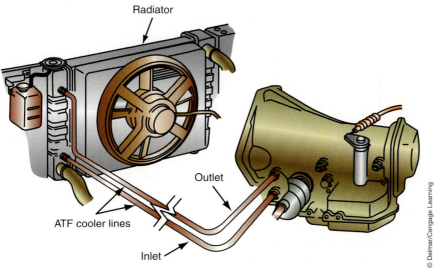

Radiator

Outlet

ATF cooler lines

Inlet

FIGURE 6-50 Typical routing of transmission oil cooler lines for a transaxle.

into and mixing with the ATF because of a leak in the cooler. At times, the presence of ATF in the radiator will be noticeable when the radiator cap is removed, as ATF will tend to float to the top of the coolant. A leaking transmission cooler core can be verified with a leak test.

External cooler leaks are evident by the traces or film buildup of ATF around the source of the leak and typically requires little time to determine the source of the leakage. However, internal coolers present a little bit more difficulty and a leak test should be conducted. To do this, place a catch pan under the fittings that connect the cooler lines to the radiator (Figure 6-48). Then, disconnect (Figure 6-49) and plug both cooler lines from the transmission (Figure 6-50). Tightly plug one end of the cooler and apply compressed air (50–75 psi) into the open end of the cooler.

 WARNING: Be sure the radiator is cool to the touch. Hot coolant may spray on you if the system is hot. The hot coolant can cause serious injuries. Wearing safety goggles is a good precaution when working around hot liquids.

 WARNING: Wear safety goggles when working with compressed air.

SPECIAL TOOLS
Catch pan
Various lengths
of hoses
Graduated container
OSHA-approved
air nozzle

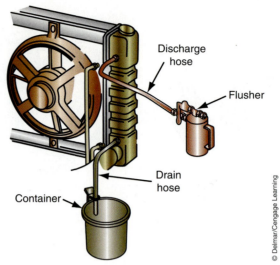

© Delmar/Cengage Learning

FIGURE 6-51 Setup for flushing a transmission's cooling system.

Carefully check the coolant through the top of the radiator; if there is a leak in the transmission cooler, bubbles will be apparent in the coolant. If it leaks, the cooler must be replaced.

The inability of the transmission fluid to cool can also be caused by a plugged or restricted fluid cooling system. If the fluid cannot circulate, it cannot cool. To verify that the transmission cooler is plugged, disconnect the cooler return line from the radiator or cooler (Figure 6-51). Connect a short piece of hose to the outlet of the cooler and allow the other end of the hose to rest inside an empty container. Start the engine and measure the amount of fluid that flows into the container after 20 seconds. Normally, one quart of fluid should flow into the container. If less than that filled the container, a plugged cooler is indicated.

A tube-type transmission cooler can be cleaned by using cleaning solvent or mineral spirits and compressed air. A fin-type cooler, however, cannot be cleaned in this way. Therefore, the normal procedure includes replacing the radiator (which includes the cooler). In both cases the cooler lines should be flushed to remove any debris.

If the cooler is plugged, disconnect the cooler lines at the transmission. Apply compressed air through one port of the cooler, then to the other port. The air should blow any debris out of the cooler. Always use low air pressure, no more than 50 psi. Higher pressures may damage the cooler. If little air passes through the cooler, the cooler is severely plugged and may need to be replaced.

If the vehicle has two coolers, pull the inlet and outlet lines from the side of the transmission and, using a hand pump, pump **mineral spirits** into the inlet hose until clear fluid pours out of the outlet hose. To remove the mineral spirits from the cooler, release some compressed air into the inlet port, then pump some ATF into the cooler.

The transmission's cooling system can be flushed. This will remove nearly all of the debris, leaving a clean, dry system. To do this, a transmission cooler flusher (Figure 6-44) is required. The flusher's container is filled with biodegradable flushing fluid. Compressed air pushes the fluid through the entire cooling system of the transmission. The system is then cleaned with forced hot water. When the discharge from the system is cleared, compressed air is expelled through the system until no moisture is evident in the discharge.

Check the condition of the cooler lines from their beginning to their ends. A line that has been accidentally damaged while the transmission has been serviced will reduce oil flow through the cooler and shorten the life of the transmission. If the steel cooler lines need to be replaced, use only double-wrapped and brazed steel tubing. Never use copper or aluminum tubing to replace steel tubing. The steel tubing should be double-flared and installed with the correct fittings.

CASE STUDY

A customer brought her late-model Chevy Silverado to a shop. Her concern was the truck sometimes stalled when it came to a stop. The technician verified the concern and began diagnosis with a check of the engine. He found nothing unusual. He then checked the TSBs available for the truck and found several. Most were related to a stuck torque converter clutch solenoid; however, there were other possible causes. He began by unplugging the TCC solenoid to see if the problem was still there. If the solenoid is stuck and disconnected, the concern should disappear which is what it did. The technician knew the solenoid could be the cause but the concern could also be caused by dirt in the valve body or a bad signal from the PCM.

He raised the vehicle on a lift with the driving wheels off the ground. He checked the ground and power feed to the solenoid and found it to be within specifications. He then connected a test light across terminals A and D at the transmission. With the engine running while the transmission is Drive, he observed the test light. The vehicle was then slowly accelerated to about 60 mph (97 km/hr). The test light should have illuminated if the PCM was sending a signal to the solenoid. In this case, the test light did light.

While the vehicle was moving at this speed, he applied the brake pedal. This should cause the TCC to disengage, it did not. He therefore moved to check the brake pedal switch and its circuit. During the initial inspection, he found a potential source of the concern. A connector in the wiring harness near the brake switch was damaged. The connection was very loose. To verify that this was the cause of the problem, he taped the connector tightly together. He then repeated his tests and found the TCC disengaging when the brake pedal was pressed down. He then replaced the connector and reconnected everything he had disconnected for his tests. During a road test, he accelerated and stopped several times. Never get the truck stall. His conclusion was the poor connection caused an intermittent problem that would sometimes not turn off the TCC solenoid.

TERMS TO KNOW

Ballooning

Heat exchanger

Mineral spirits

ASE-STYLE REVIEW QUESTIONS

1. *Technician A* says a ballooned torque converter can cause damage to the oil pump.
   Technician B says a ballooned torque converter is caused by excessive pressure in the torque converter.
   Who is correct?
   A. A only
   B. B only
   C. Both A and B
   D. Neither A nor B

2. While checking torque converter endplay:
   Technician A says the torque converter must be installed in the transmission.
   Technician B says the torque converter endplay is corrected by installing a thrust washer between the oil pump and the direct or front clutch.
   Who is correct?
   A. A only
   B. B only
   C. Both A and B
   D. Neither A nor B

3. While servicing a variable displacement vane-type pump:
   Technician A says the pump rotor, vanes, and slide have selective sizes and may destroy the pump if the correct ones are not used.
   Technician B says the outer edge of the vanes should be flat.
   Who is correct?
   A. A only
   B. B only
   C. Both A and B
   D. Neither A nor B

4. *Technician A* says the gears in a gear-type oil pump should be replaced if the outer edges of the teeth are worn flat.
   Technician B says all parts of gear-type pumps are selectively sized.
   Who is correct?
   A. A only
   B. B only
   C. Both A and B
   D. Neither A nor B

5. *Technician A* says a fin-type cooler can be cleaned by using cleaning solvent or mineral spirits and compressed air.

Technician B says a tube-type transmission cooler can be cleaned by using cleaning solvent or mineral spirits and compressed air.

Who is correct?

A. A only
C. Both A and B
B. B only
D. Neither A nor B

6. *Technician A* says that if the stall speed of a torque converter is below specifications, a restricted exhaust or slipping stator clutch is indicated.

Technician B says that if the stall speed is above specifications, the bands or clutches in the transmission may not be holding properly.

Who is correct?

A. A only
C. Both A and B
B. B only
D. Neither A nor B

7. *Technician A* says torque converter clutch control problems are always caused by electrical malfunctions.

Technician B says nearly all converter clutches are engaged through the application of hydraulic pressure on the clutch.

Who is correct?

A. A only
C. Both A and B
B. B only
D. Neither A nor B

8. While checking the stall speed of a torque converter:

Technician A maintains full-throttle power until the engine stalls.

Technician B manually engages the TCC before running the test.

Who is correct?

A. A only
C. Both A and B
B. B only
D. Neither A nor B

9. *Technician A* says a converter that has a tight fit at the pilot hub could hold the converter drive hub and inner gear too far into the pump, causing cover scoring.

Technician B says a front pump bushing that has too much clearance may allow the gears to run off center, causing them to wear into the crescent or the sides of the pump body.

Who is correct?

A. A only
C. Both A and B
B. B only
D. Neither A nor B

10. *Technician A* says a seized one-way stator clutch will cause the vehicle to have good low-speed operation but poor high-speed performance.

Technician B says a freewheeling or nonlocking one-way stator clutch will cause the vehicle to have poor acceleration.

Who is correct?

A. A only
C. Both A and B
B. B only
D. Neither A nor B

ASE Challenge Questions

1. *Technician A* says a lower-than-specified stall speed may be caused by a faulty stator one-way clutch.

   Technician B says a higher-than-specified stall speed may be caused by faulty clutch packs.

   Who is correct?

   A. A only   C. Both A and B

   B. B only   D. Neither A nor B

2. *Technician A* says a shudder after the converter clutch engages could be caused by a damaged or missing clutch check ball.

   Technician B says driveline or converter shudder can be isolated by disconnecting the converter's clutch solenoid.

   Who is correct?

   A. A only   C. Both A and B

   B. B only   D. Neither A nor B

3. The backside of a torque converter is found to be wet.

   Technician A says it may be caused by excessive torque converter hub runout.

   Technician B says it can be caused by insufficient input shaft endplay.

   Who is correct?

   A. A only   C. Both A and B

   B. B only   D. Neither A nor B

4. A customer complains that her vehicle's engine seems to surge when she is driving at about 48 mph.

   Technician A says there may be a problem with the torque converter clutch.

   Technician B says a faulty vehicle speed sensor may cause this problem.

   Who is correct?

   A. A only   C. Both A and B

   B. B only   D. Neither A nor B

5. *Technician A* says a plugged transmission cooler or cooler lines may cause the vehicle to stall when the transmission is shifted into a forward gear.

   Technician B says a plugged transmission cooler or cooler lines may overheat the converter.

   Who is correct?

   A. A only   C. Both A and B

   B. B only   D. Neither A nor B

Name _____ Date _____

Road Test a Vehicle to Check the Operation of a Torque Converter Clutch

Upon completion of this job sheet, you should be able to road test a vehicle with an automatic transmission to verify and analyze the operation of a torque converter clutch.

ASE Correlation

This job sheet is related to the ASE Automatic Transmission and Transaxle Test's Content Area *General Transmission and Transaxle Diagnosis.*
Task: Perform lockup converter mechanical/hydraulic system tests; determine necessary action.

Tools and Materials

A vehicle with an automatic transmission and lockup converter
Service manual
Pad of paper and pencil

Describe the vehicle being worked on:

Year _____ Make _____ Model _____
VIN _____ Engine type and size _____
Model and type of transmission _____

Procedure

Task Completed

1. Park the vehicle on a level surface. ☐

2. Describe the condition of the fluid (color, condition, smell).

   _____

   _____

3. What is indicated by the fluid's condition?

   _____

   _____

4. In the appropriate service manual, find the chart that shows the conditions that must be present in order for the converter clutch to be engaged. Describe these conditions.

   _____

   _____

   _____

   _____

5. Allow the engine and transmission to cool by letting the vehicle sit.

6. Take the vehicle for a test drive. Bring the vehicle to normal lockup speed in the appropriate gear and observe the action of the clutch. Record the speed and feel of the clutch engagement.

_____

_____

_____

7. Was the engagement of the clutch normal for the engine temperature and vehicle speed? Why or why not?

_____

_____

8. Now drive the car through the following conditions and note the action of the torque converter clutch.

Manual upshifting of the transmission during acceleration

Clutch action: _____

Part-throttle downshifting

Clutch action: _____

Wide-open-throttle downshifting

Clutch action: _____

Cruising at highway speed, then lightly applying the brakes

Clutch action: _____

Cruising at highway speed, then letting completely off the throttle but not applying the brakes

Clutch action: _____

9. Summary of clutch operation:

_____

_____

_____

_____

Instructor's Response _____

_____

_____

Name _____ Date _____

VISUAL INSPECTION OF A TORQUE CONVERTER

Upon completion of this job sheet, you should be able to conduct a visual inspection of a torque converter.

ASE Correlation

This job sheet is related to the ASE Automatic Transmission and Transaxle Test's Content Area *General Transmission and Transaxle Diagnosis*.

Task: Listen to driver's concern and road test vehicle to verify mechanical/hydraulic system problems; determine necessary action.

Tools and Materials

Vehicle with automatic transmission	Remote starter switch
Clean white paper towels	Droplight or good flashlight
Hoist	Service manual

Describe the vehicle being worked on:

Year _____ Make _____ Model _____

VIN _____ Engine type and size _____

Model and type of transmission _____

Procedure

Task Completed

1. Park the vehicle on a level surface. ☐

2. Describe the condition of the fluid (color, condition, smell).

   _____

   _____

3 . What is indicated by the fluid's condition?

   _____

   _____

4. Connect a remote starter switch to the vehicle and disable the ignition. Explain how you did this.

   _____

   _____

5. Raise the vehicle on a hoist to a comfortable working height. ☐

6. Put on eye protection. ☐

7. Remove the torque converter access cover or shield.

8. Inspect the converter through the access hole. Use the remote starter switch to observe all of the converter. Describe your findings.

_____

_____

_____

_____

Instructor's Response _____

_____

_____

Name _____ Date _____

TESTING A LOCKUP CONVERTER

Upon completion of this job sheet, you will be able to test the operation of a lockup converter.

ASE Correlation

This job sheet is related to the ASE Automatic Transmission and Transaxle Test's Content Area *In-vehicle Transmission and Transaxle Repair*.

Task: Perform lockup converter system tests; determine necessary action.

Tools and Materials

Service manual
Scan tool
DMM
Goggles or safety glasses with side shields

Describe the vehicle being worked on.

Year _____ Make _____ Model _____
VIN _____ Engine type and size _____

Procedure

Task Completed

1. Check with the service or owner's manual to identify the lamp that warns the driver of a transmission concern. What lamp is it?

   _____

   _____

2. Start the engine and observe the above lamp and the engine MIL. What did you observe?

   _____

   _____

3. Connect the scan tool and retrieve all DTCs. What are they?

   _____

   _____

4. If an engine-related problem is suspected, it should be corrected before moving on to the torque converter tests. ☐

5. Operate the vehicle with the scan tool connected. Drive it at 40 mph. ☐

6. With the scan tool, engage the lockup converter's solenoid. Watch the tachometer while making the change. What did you observe? Engine speed should drop when the converter locks.

_____

_____

7. Disengage the clutch and observe the engine's speed. What did you observe?

_____

_____

8. The action of the lockup converter can also be checked with the vehicle stopped and the engine running. Place the shifter into the "drive" position.

9. With the scan tool, engage the lockup solenoid. Describe what happens.

_____

_____

10. If there was no change in engine speed, what is evident?

_____

_____

11. The lockup solenoid can also be checked with the ignition on but the engine off. Engage the solenoid with the scan tool and listen. Did you hear a click?

_____

_____

12. If a problem with the solenoid is suspected, remove it from the transmission for testing.

13. Refer to the service manual, and locate the resistance checks for the solenoid. What are the specifications?

_____

_____

14. Measure the resistance of the solenoid across the test points specified. What were your results?

_____

_____

15. What can you conclude from these tests?

_____

_____

Instructor's Response _____

_____

_____

Name _____ Date _____

CONDUCT A STALL TEST

Upon completion of this job sheet, you should be able to conduct a stall test.

ASE Correlation

This job sheet is related to the ASE Automatic Transmission and Transaxle Test's Content Area *General Transmission and Transaxle Diagnosis.*
Task: Perform stall tests; determine necessary action.

Tools and Materials

Vehicle with an automatic transmission   Stethoscope
Tachometer   Droplight or good flashlight
Hoist   Service manual

Describe the vehicle being worked on:

Year _____ Make _____ Model _____
VIN _____ Engine type and size _____
Model and type of transmission _____

_____

Procedure   Task Completed

1. Park the vehicle on a level surface.   ☐

2. Describe the condition of the fluid (color, condition, smell).

   _____

   _____

3. What is indicated by the fluid's condition?

   _____

   _____

4. Connect a tachometer to the engine. If the vehicle has one in the instrument panel, there is no need to connect another one. Explain how you connected the tachometer.

   _____

   _____

5. Place the tachometer on the inside of the vehicle so that it can be easily read.   ☐

6. Mark the face of the tachometer with a grease pencil at the recommended maximum rpm for this test. What is the maximum rpm? _____

7. Check the engine's coolant level.   ☐

8. Block the front wheels and set the parking brake.   ☐

9. Place the transmission in park and allow the engine and transmission to warm.   ☐

10. Place the gear selector into the gear that is recommended for this test. What is the recommended gear? _____

☐ 11. Press the throttle pedal to the floor with your right foot and firmly press the brake pedal with your left. Hold the brake pedal down.

12. When the tachometer needle stops rising, note the engine speed and let off the throttle. What was the highest rpm during the test? _____

13. Place the gearshift into neutral and allow the engine to run at 1000 rpm for at least one minute. What is the purpose of doing this?

_____

_____

☐ 14. If noise is heard from the transmission during the stall test, raise the vehicle on a hoist.

☐ 15. Use the stethoscope to determine if the noise is from the transmission or the torque converter.

16. What are your conclusions from this test?

_____

_____

_____

_____

Instructor's Response _____

_____

_____

Name _____ Date _____

SERVICING AN AUTOMATIC TRANSMISSION OIL PUMP

Upon completion of this job sheet, you should be able to inspect, measure, and replace the components of an oil pump assembly.

ASE Correlation

This job sheet is related to the ASE Automatic Transmission and Transaxle Test's Content Area *Off-Vehicle Transmission and Transaxle Repair; Oil Pump and Converter.*
Task: Inspect, measure, and replace oil pump assembly and components.

Tools and Materials

Machinist dye   Feeler gauge

Describe the transmission being worked on and the vehicle the transmission is from:

Year _____ Make _____ Model _____

VIN _____ Engine type and size _____

Model and type of transmission _____

Procedure

Task Completed

1. With the oil pump removed from the transmission, remove the front pump bearing race, front clutch thrust washer, gasket, and O-ring. Inspect the pump bodies, pump shaft, and ring groove areas. Record your findings.

    _____

    _____

    _____

2. Mount the pump on a stand and unbolt and separate the pump bodies.

    ☐

3. Mark the gears with machinist bluing ink or paint before removing them so that the gears will remain in the same relationship during reassembly. What did you use to mark them and where did you mark them?

    _____

    _____

4. Inspect the gears and all internal surfaces for defects and visible wear. Record your findings.

    _____

    _____

5. With the pump mounted on the stand, use a feeler gauge to measure between the outer gear and the crest in the pump housing crest. What tolerance do the specifications call for? What did you measure?

_____

_____

6. Measure between the outer gear teeth and the crescent. What tolerance do the specifications call for? What did you measure?

_____

_____

7. Place the pump flat on the bench and, using a feeler gauge and straightedge, measure between the gears and the pump cover. What tolerance do the specifications call for? What did you measure?

_____

_____

8. Measure the clearance between the C-ring and the ring groove. What tolerance do the specifications call for? What did you measure?

_____

_____

9. Using the stand to center the pump, torque the securing bolts to specifications. What is the specified torque?

_____

_____

☐   10. Replace all O-rings and gaskets.

Instructor's Response _____

_____

_____

Chapter 7

GENERAL HYDRAULIC SYSTEM SERVICE

UPON COMPLETION AND REVIEW OF THIS CHAPTER, YOU SHOULD BE ABLE TO:

- Remove and install the valve body from common transmissions.
- Check valve body mating surfaces.
- Inspect and measure valve body bores, springs, sleeves, retainers, brackets, check balls, and screens.

- Inspect and replace valve body spacers and gaskets.
- Check and/or adjust valve body bolt torque.
- Inspect and replace the governor cover and seals.
- Inspect, adjust, repair, and replace the governor sleeve, valve, weights, springs, retainers, and gear.

The hydraulic control center for a transmission is the valve body. All EATs have electronic controls that modify the action of clutches. These too have a central control center, called the TCM or PCM. But, the actual control of the hydraulics takes place in the valve body. Therefore, the valve body is an extremely important part of the transmission and faults in it can cause a wide variety of problems.

A valve body can be a single unit or an assembly of many subassemblies. When it is comprised of several parts, the assembly may be made up of a main valve body, regulator valve body, secondary valve body, and accumulator body. The valve body can be mounted in several different locations, the common of which is to the side of the torque converter housing on below the gears.

The oil pump is driven by the torque converter. Fluid flows through the regulator valve to maintain a specified pressure through the main valve body to the manual valve, which directs pressure to each of the clutches. The actual application of the clutches is controlled by the action of the shift solenoids.

The main valve body typically contains the manual valve, modulator valve, torque converter check valve, some shift valves, servo control valves, oil relief valve, lockup shift valve, lockup timing valve, and lubrication control and check valves. The main valve body basically controls the hydraulic pressure going to the rest of the hydraulic control system.

A typical regulator valve body is made up of the regulator valve, cooler check valve, lockup control valve, and some accumulators. The regulator valve keeps the hydraulic pressure from the ATF pump to the hydraulic control system constant. While doing this, the regulator valve is maintaining the correct level of line pressure. Excess pressure exhausts through the relief valve. Normal pressure moves through the transmission and through the manual shift valve to the various control valves. It also sends fluid to the torque converter. Line pressure will change according to need. This may be accomplished by the regulator valve. Some systems have a stator shaft arm that can work against a spring cap in the regulator valve. During acceleration or heavy loads, the torque demand causes the arm to move against the regulator's spring cap. This action allows line pressure to increase.

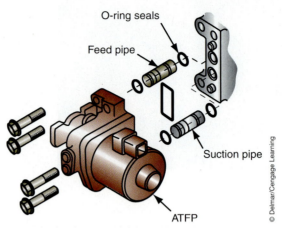

O-ring seals

Feed pipe

Suction pipe

ATFP

© Delmar/Cengage Learning

FIGURE 7-1 An auxiliary transmission fluid pump.

Auxiliary Transmission Fluid Pump

Some transmissions, especially those made for hybrid vehicles, have an auxiliary fluid pump for the transmission. With hybrids, the auxiliary pump ensures that the transmission has fluid flow when the system is in stop/start. Since the engine is off during those times the transmission's pump is not providing fluid flow. Therefore, the transmission will not be ready to move the vehicle once the engine is restarted. The auxiliary pump provides fluid flow during those conditions.

The **auxiliary transmission fluid pump (ATFP)** is normally mounted on the outside of the torque converter housing (Figure 7-1). When the vehicle comes to a stop and the engine turns off while the ignition is still on, the pump immediately begins to supply fluid to the regulator valve. The ATFP runs until the engine restarts. A typical ATFP is a three-phase brushless DC motor. The TCM has total control of the pump. If the transmission cannot move the vehicle after the engine restarts, check the ATFP motor.

ATFP Motor Test. To check the motor, disconnect the wiring connector at the ATFP. Place the positive lead of an ohmmeter to one terminal at the connector on the pump. Connect the other meter lead to a good ground. You should receive an infinite reading. Do this at each terminal in the connector. If there is continuity between any of the terminals and ground, the motor should be replaced.

Replacement of ATFP. Normally, the replacement of the motor is straightforward. Disconnect electrical connector at the pump. Carefully inspect the connector, make sure it is free of dirt, oil, and rust. Remove the pump's retaining bolts and lift off the pump. After the pump has been removed, the feed and suction pipes, and the dowel pins can be removed. Inspect all parts and replace them if they show signs of damage or wear. Inside the feed pipe there may be a check ball, shake the pipe to see if the ball can move freely. If it cannot, attempt to clean the inside of the pipe. Then, clean the mounting surface for the pump.

To install a new pump, fit new O-rings on the suction pipe and install the pipe into the converter housing with the filter end toward the outside of the housing. Then fit new O-rings on the feed pipe and install the large end of the pipe into the housing. Carefully install the dowel pins into the housing. Then fit a new O-ring into its groove in the pump assembly and install the pump onto the housing. Tighten the retaining bolts to specifications and securely reattach the electrical connector.

Feed Pipes

Some transmissions have fluid feed pipes that move fluid from one point of the valve body to another. At times, it is necessary to remove the pipes to gain access to the valve body.

The pipes can be retained in the valve body in a number of different ways. Some are retained by plate that is held in the housing by a snap ring. Others have flanges that are bolted

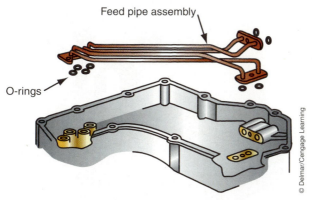

FIGURE 7-2 A feed pipe assembly.

to various locations on the valve body (Figure 7-2). Some others are merely pressed into bores in the valve body. To remove the pipes, the snap rings or bolts must be removed. Then the pipes are lifted away. If the pipes are pressed into a bore, they should be carefully pried out of the bore. It is important to be careful when removing them, make sure you do not bend or damage them. Also, make sure you do not damage their sealing surface on the valve body.

In most cases, the pipe's seals or O-rings should be replaced when the pipes are reinstalled. Some feed pipes have reusable seals. These are glued to the pipes by the manufacturer. If these seals are damaged, the entire pipe should be replaced.

Wires and Control Module

Often, there is a wiring harness or control module that must be removed before the valve body can be removed. Some of these harnesses are molded together and are removed as a unit. Others are individual wires that must be disconnected from the various components.

Molded leads are secured by their fit into the various connections (Figure 7-3). While removing the harness, inspect it for damage. Make sure all terminals are secure with their molded connector shell. Also when reinstalling it, make sure all connections are secure and tight.

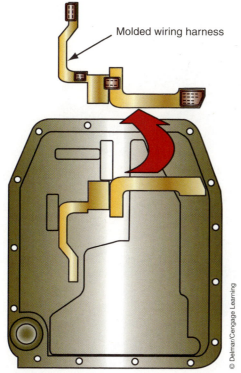

FIGURE 7-3 A molded wiring harness that is secured to the valve body by its connections.

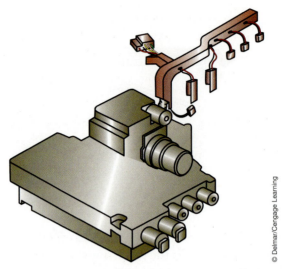

FIGURE 7-4 A wiring harness with connectors for the individual outputs and inputs located on the valve body.

Some wiring harnesses are comprised of several groups of wires and connectors that go to individual components, such as the EPC solenoid, shift solenoids, TCC solenoid, and switches (Figure 7-4). To remove this type of harness, it must be disconnected for component. In some cases, the harness is secured to the valve body by retaining bolts or clips. Some have plastic locks at the point where the harness passes into the valve body that must be released. It is important that the harness and the various connectors are not damaged during removal or installation.

Control Module. Some transmissions have a single assembly that contains all of the solenoids, the TCM, and a lead frame assembly. The lead frame connects to the individual components, therefore no wires are required. The assembly bolts directly to the lower and upper valve body assemblies (Figure 7-5). The assembly connects to an engine harness with

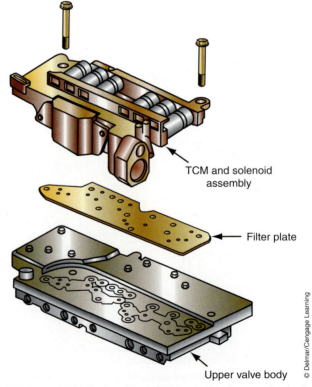

TCM and solenoid assembly

Filter plate

Upper valve body

FIGURE 7-5 A TCM/solenoid assembly mounted on a valve body.

a 16-pin connector through a pass-through sleeve in the valve body cover. Often, to remove the assembly, some part of the shift linkage must be removed. This is then followed by disconnecting the electrical connectors.

 WARNING: Be careful when handling the assembly. If it is jarred or dropped, damage to its internal components can result.

Always replace the filter when reinstalling the assembly. Once the filter has been removed and reinstalled, it may be contaminated which could cause damage to the transmission.

VALVE BODY REMOVAL

If the pressure test indicated that there is a problem associated with the valves in the valve body, a thorough disassembly, cleaning in fresh solvent, careful inspection, and the freeing up and polishing of the valves may correct the problem. Sticking valves and sluggish valve movements are caused by poor maintenance, the use of the wrong type of fluid, and/or over-heating the transmission. The valve body of most transmissions can be serviced when the transmission is in the vehicle; however, it is typically serviced when the transmission has been removed for other repairs.

Typically, to remove a valve body from a transmission while it is still in the vehicle, begin by draining the fluid and removing the oil pan. Then disconnect the manual and throttle lever assemblies. Carefully remove the detent spring and screw assemblies. Loosen and remove the valve body screws (Figure 7-6). Before lowering the valve body and separating the assembly, hold the assembly with the valve body on the bottom and the transfer and separator plates on top. Holding the assembly in this way will reduce the chances of dropping the steel balls that are located in the valve body (Figure 7-7). Lower the valve body and note where these steel balls are located in the valve body; remove them and set them aside, along with the various screws.

BASIC TOOLS

Basic hand tools
Cleaning pan
OSHA-approved
air nozzle
Cleaning solution

Classroom Manual

Chapter 7, page 220

SERVICE TIP:

To avoid having to spend hours crawling on the floor of the shop looking for lost parts, place your hand or fingers over spring-loaded valves or plugs when removing them from the valve body.

Valve body

Attachment bolts

© Delmar/Cengage Learning

FIGURE 7-6 When removing a valve body from the transmission, remove only the bolts that are necessary to lower the valve body. Also be careful not to lose any springs and check balls.

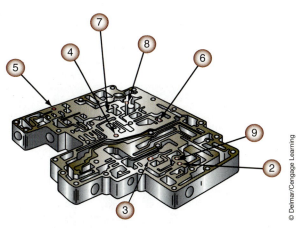

© Delmar/Cengage Learning

FIGURE 7-7 Location of the check balls in a typical valve body.

On transmissions with a horizontal valve body, you can leave the separator plate in the transmission to keep the check balls above the spacer from falling out. The check balls in the valve body will not fall out if the valve body is lowered carefully. Then the spacer plate can be carefully removed, keeping it level as it is lowered, and the check balls on the plate will remain in their proper position.

Removal of a valve body from a transmission on a bench is similar. Make sure you have the transmission positioned so that the valve body can be lowered in such a way that you can keep the check balls in place.

There are many different designs and configurations used in today's transmissions. Each of them requires specific steps for removal and installation. It is extremely important that you refer to the service manual for these specifics. To illustrate these variables, valve body removal and installation procedures for some common transmissions follow, as well as a description of the components that are included in the valve body assembly.

Chrysler Transaxles

Chrysler's 41TE transaxle is found in many of their cars and minivans. The valve body assembly of these units is comprised of the valve body, a transfer plate, and a separator plate. The valve body contains many different valves and check balls (Figure 7-8). All of these control fluid flow to the TCC, the **solenoid/pressure switch** assembly, and the various hydraulic apply devices.

To remove the valve body, move the manual valve lever clockwise into low gear. Loosen and remove the valve body retaining bolts. Then, using a screwdriver, push the park rod rollers out of the guide bracket. Next, remove the valve body from the housing. Carefully pull the valve body assembly from the case. The **manual shaft** is attached to the valve body and must be carefully pulled from its bore.

Once removed, the valve body can be disassembled. This begins with the removal of the TR sensor assembly (Figure 7-9). The accumulator retaining plate and accumulator are then removed.

The transfer plate can now be unbolted and separated from the valve body. While separating these units, pay attention to the check balls. It is very possible that a loose check ball will drop out.

Remove the oil screen and overdrive clutch check valve from the separator plate. Then remove the separator plate (Figure 7-10). Then remove the thermal valve and check balls from the valve body. Keep track of their location and count the balls as they are removed. Disassemble, inspect, and clean the valve body components as needed.

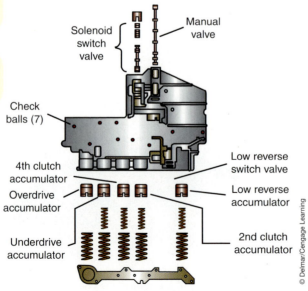

FIGURE 7-8 The various components in the valve body of a Chrysler 45RFE.

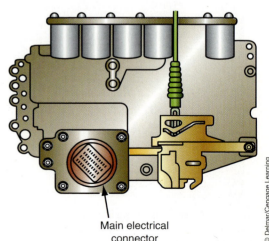

FIGURE 7-9 The main connector, TR sensor, and other associated parts in this Chrysler valve body.

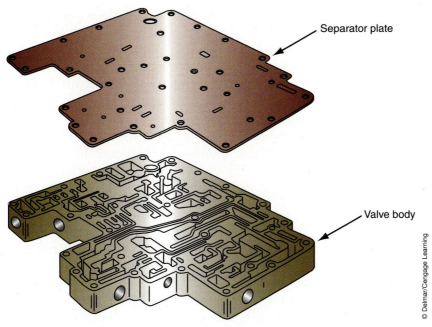

FIGURE 7-10 A valve body with its separator plate.

To install the valve body, guide the park rod rollers into the guide bracket while positioning the valve body. Install the valve body-to-case mounting bolts and tighten them to specifications.

Ford Transmissions

The 4R70E transmission is commonly found in Ford pickups and SUVs. Removing the valve body from this transmission is a rather simple process. The valve body is bolted directly to the transmission case and few items need to be removed to remove it. The electrical connectors to the various solenoids mounted to or around the valve need to be disconnected. Once they are, the wiring harness can be removed by disconnecting it from its main connector. After the harness is removed, the shift and TCC solenoids can be unbolted and removed. Then remove the retaining bolt for the manual valve detent lever spring and then the spring (Figure 7-11). Now unbolt the valve body from the transmission case. Remove and discard the pump outlet screen.

Remove the separator plate and discard the gaskets. Note the location of the check balls (Figure 7-12) and remove them before cleaning the valve body. Once the valve body is

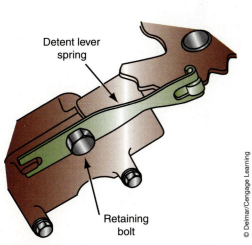

FIGURE 7-11 The detent lever spring for the manual valve.

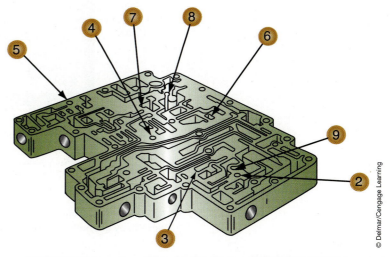

FIGURE 7-12 Location of the check balls in a 4R70W transmission.

serviced, install the check balls and the separator plate. Install a new pump outlet screen and gaskets during reassembly.

To ensure proper alignment of the valve body, this transmission has two alignment bolts. Position the valve body gasket and valve body onto the case by using these alignment bolts as a guide. Loosely install all of the valve body retaining bolts. Then install the manual valve detent lever spring and tighten it in place. Before doing this, make sure the drive pin of the detent lever is correctly positioned in the manual valve.

Now tighten the valve body retaining bolts according to the specified sequence and to the correct torque. Install the shift and TCC solenoids and reconnect the wiring harness to the main connector and to the solenoids.

Six-Speed Transaxle. This transaxle is used in the Fusion-based FWD vehicles from Ford Motor Company. The main valve body can be serviced with the transaxle in the vehicle, as well as with the unit removed from the vehicle. To gain access to the valve body, the bolts retaining the oil pan should be removed and discarded. With the pan removed, clean the sealing surfaces of the pan and housing, being careful not to damage the surfaces. Then disconnect the wiring harness from the solenoids and speed sensors (Figure 7-13) and remove the harness. Then unbolt the lock plate for the temperature sensor and remove the sensor. Mounted in a corner of the valve body is the suction cover. This must be removed before removing the attaching bolts for the valve body. The valve body bolts should be loosened in a staggered sequence. Once those bolts are removed, the valve body can be removed. While removing the valve body, the manual control lever may fall out of its bore, so remove the valve body carefully.

Installation of the valve follows the opposite sequence from removal. Take extra care to keep the harness away from the mounting surface for the valve body. If the harness is caught between the valve body and housing, it can be damaged. All retaining bolts must be torqued to specifications and all harnesses and connectors must be securely in place. Now, the oil pan must be sealed with a thin application of sealer and new bolts tightened to specifications.

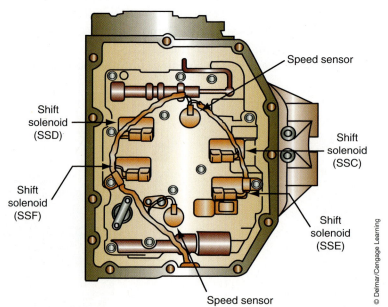

FIGURE 7-13 The various solenoids and switches located on the main valve body of Ford's six-speed transaxle.

General Motors Transmissions

GM's 4L60-E transmission is used in a wide variety of RWD vehicles. Like the valve body in a Ford transmission, the valve body of a 4L60 is rather straightforward. Although the procedures are similar, the 4L60 requires a different sequence and, due to its design, additional steps. Photo Sequence 13 covers the procedure for removing and installing a valve body when the 4L60 is still in the vehicle.

The removal process begins with disconnecting the electrical connections to the various switches and solenoids mounted to or around the valve body (Figure 7-14). The wiring harness cannot be removed until the TCC solenoid has been removed. The order in which the solenoids are removed is important. Some retaining bolts and brackets are covered by the solenoids. Remove the PWM solenoid to gain access to the retaining bolts for the TCC solenoid. Once the TCC solenoid is removed, the electrical harness retaining bolts can be removed.

Remove the pressure switch retaining bolts and switch. Then unbolt and remove the **transmission fluid pressure switch** assembly (Figure 7-15). This unit needs to be carefully inspected and checked for any damage or debris.

The manual detent spring can now be unbolted and removed. Once this is off, the remaining valve body mounting bolts can be removed. Once loosened, the valve body should be slightly lowered and the manual valve's linkage disconnected. This will allow for removal of the valve body.

Installation is done in the reverse order of removal. Again, it is important that the solenoids be installed according to the sequence. All of the valve body mounting bolts should be loosely installed as components are installed, until all of the bolts are in their proper location. Then the bolts should be tightened in the specified order and to the specified torque.

> The **transmission fluid pressure switch** assembly contains five different pressure switches and is connected to five different hydraulic circuits.

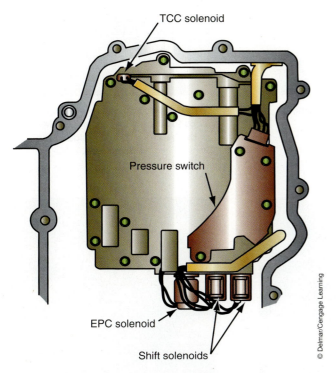

FIGURE 7-14 The various solenoids and switches located on or near the valve body of a 4T45E transaxle.

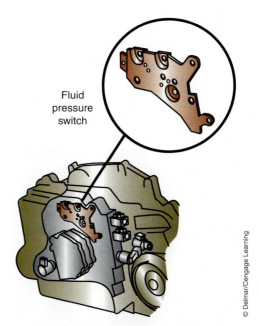

FIGURE 7-15 A typical transmission fluid pressure (TFP) switch assembly.

© Delmar/Cengage Learning

REMOVING AND INSTALLING A VALVE BODY

All photos in this sequence are © Delmar/Cengage Learning.

P13-1 Position the vehicle on a lift and loosen the transmission pan to drain the fluid.

P13-2 After the fluid has drained, remove the pan and the transmission filter.

P13-3 Allow the fluid to drip off the transmission, then disconnect the electrical connections to the various switches and solenoids mounted to or around the valve body.

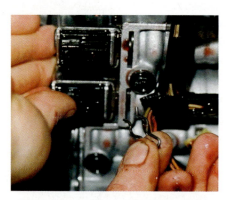

P13-4 Remove the retaining clip for the PWM solenoid, and then do the same for the shift solenoids.

P13-5 Remove the TCC solenoid.

P13-6 Remove the retaining bolts or clips for the electrical harness. Move the harness off to the side of the transmission.

P13-7 Now remove the pressure switch retaining bolts and switch.

P13-8 Now unbolt and remove the transmission fluid pressure switch assembly. Carefully inspect this assembly for any damage or debris.

P13-9 Unbolt and remove the manual detent spring.

P13-10 Loosen and remove the remaining valve body mounting bolts.

P13-11 Lower the valve body just enough to disconnect the manual valve's linkage, then lower the valve body away from the transmission and remove it.

P13-12 Unbolt and remove the accumulator assembly.

P13-13 Before reinstalling the valve body, check the fit of the new gasket to the transmission.

P13-14 Reinstall the accumulator.

P13-15 Position the valve body on the case and loosely install some of the mounting bolts. Then, install the manual detent spring.

P13-16 Install the transmission fluid pressure switch assembly. Do not tighten the mounting bolts.

P13-17 Position the electrical harness and loosely install the retaining bolts.

P13-18 Install the TCC solenoid but do not tighten the mounting bolts.

P13-19 Now tighten all of the valve body mounting bolts according to the specified order and to the specified torque.

P13-20 Install the various solenoids and switches to the valve body. Then tighten all retaining bolts to the specified torque.

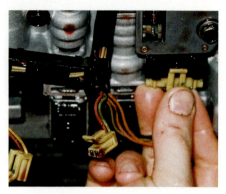

P13-21 Reconnect the electrical connectors to the various solenoids and switches.

P13-22 Install a new fluid filter.

P13-23 Install the transmission pan with a new gasket and tighten the retaining bolts to specifications.

P13-24 Fill the transmission with fluid, following the correct procedure and using the correct fluid.

Honda Transaxles

The valve body assembly in this unique transaxle is comprised of a main valve body, the regulator valve body, the servo valve body, and the accumulator body (Figure 7-16). The main valve body contains the pump gears and the manual, modulator, shift, servo control, TCC control, and cooler check valves. The shift valves are controlled by fluid flow directed by one of the three shift solenoids.

The regulator valve body contains the regulator, TCC timing, and relief valves. The servo valve body contains additional shift valves, two forward speed accumulators, and the servo valve. The accumulator valve body contains two forward speed accumulators and a lubrication check valve.

Transmission fluid from the regulator passes through the manual valve to the different shift and control valves. Oil feed pipes and internal passages are used to move the fluid to the various clutches in the transmission.

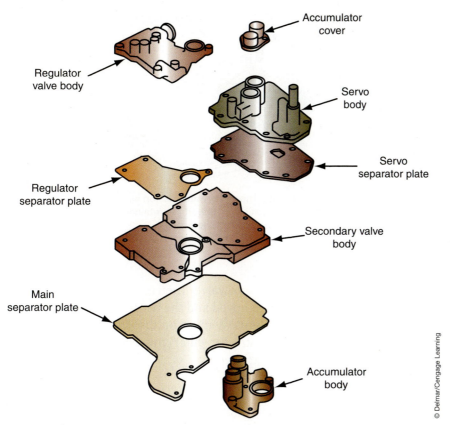

FIGURE 7-16 The valve body assembly for a late-model Honda transaxle.

© Delmar/Cengage Learning

To remove this valve body, the ATF feed pipes must be removed first. Then the ATF strainer and servo detent plate should be unbolted and removed. Now press down on the accumulator cover while loosening its mounting bolts. Loosen the bolts in a staggered pattern. The cover is spring-loaded and the threads in the servo valve body will strip if the cover is not evenly removed.

Now the servo valve body, servo separator plate, accumulator valve body, and regulator valve body can be removed. After these are removed, the stator shaft and shaft stop should be removed. Then unhook the detent spring from the detent arm and remove the detent arm shaft, detent arm, and control shaft.

Note the location of the cooler check ball and spring and remove them. Once these are removed, the main valve body can be unbolted and removed. While removing the valve body, be prepared to catch the TCC control valve and spring. These are held in position by the valve body.

Remove the pump gears, noting which side of the gears faces up. Now the main separator plate and its dowels can be removed. Clean and inspect all parts. Installation of the valve body assembly follows the reverse order of the removal procedure.

Toyota Transmissions

The valve body in most Toyota transmissions is comprised of an upper valve body, a lower valve body, a manual valve body, various covers, and separator plates. The individual hydraulic circuits extend to the housing of the transmission and may be connected by oil tubes.

To remove the valve body from an AA80E, the electrical connections to the various solenoids must be disconnected. The manual shaft detent assembly must be unbolted and removed, along with its spring and cover. The valve body can then be unbolted. Once the valve body is

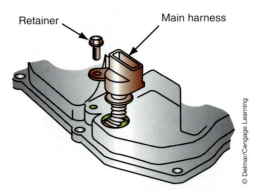

FIGURE 7-17 To remove the transmission wire in an AA80E transmission, unbolt the retainer and pull the harness out of the case.

free, disconnect the shift lever from the manual valve. Then remove the transmission wire from the case. To do this, the retainer's bolt is removed and the harness pulled out (Figure 7-17).

In an U250E transaxle, the wiring harness must be disconnected and removed, along with the ATF temperature sensor. The oil strainer (filter) must be unbolted and removed with its O-ring. Then the valve body can be removed.

VALVE BODY SERVICE

> **CUSTOMER CARE:** If while you are disassembling and inspecting the valve body you discover signs of poor transmission maintenance or abuse, make note of it. When you next talk to the customer, explain what you found and what the resulting problem was. Make sure you explain this in an understanding and nonoffensive way.

If previous tests suggest a problem with only one or two valves, start your inspection at those valves. Doing this will not only save you time, but will also reduce the chance of something being misplaced or ruined during a total disassembly. If the transmission had heavily contaminated fluid, the entire valve body should be inspected and cleaned, or replaced.

Disassembly

A valve body contains many valves. These valves are typically held in their bores by a plug or cover plate. These must be removed to gain access to the valves. The cover plates are bolted or screwed to the valve body (Figure 7-18). Plugs can be held in place in a number of different ways. All of them can be released by pressing the plug slightly into the bore.

Begin disassembly by removing the manual shift valve from the valve body (Figure 7-19). Then, remove the pressure regulator retaining screws while keeping one hand around the spring retainer and adjusting screw bracket. Remove the pressure regulator valve. Then remove all of the valves and springs from the valve body. Make sure that you keep all springs and other parts with their associated valves.

It is important to keep track of the placement and position of the springs on each valve. Doing this will make reassembly easy and will ensure you are doing things correctly. One way of doing this is to draw a diagram indicating where the valve was and the position of the spring on the valve. Many technicians use a digital camera or the camera on a cell phone to capture the spring–valve relationship. It is also wise to photograph the placement of all check balls during disassembly.

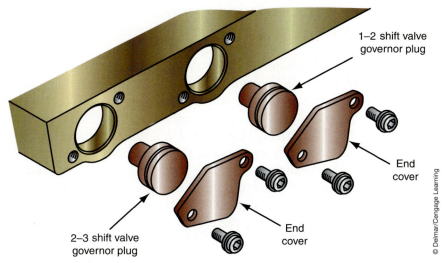

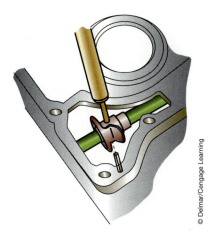

FIGURE 7-19 Remove the manual shift valve. Often, there is a retaining pin that must be driven out to release the valve.

2–3 shift valve governor plug

1–2 shift valve governor plug

End cover

End cover

FIGURE 7-18 Examples of the covers used to retain valves and their springs in the bores of a valve body.

Cleaning and Inspection

Remove the check balls and springs. Note their exact location and count them as they are removed. Compare your count to the number given in the service information. This will ensure that all have been removed and some will not be lost during cleaning. Never use a magnet to remove the check balls. The balls may become magnetized and tend to collect particles in the fluid. Be sure to do something that will help you remember where the balls were, such as photographing or drawing their location.

Valve Body, Plates, and Gaskets. Check the separator plate for scratches or other damage (Figure 7-20). Scratches or score marks can cause oil to bypass correct oil passages and result in system malfunction. If the plate is defective in any way, it must be replaced. Check the oil passages in the upper and lower valve bodies for varnish deposits, scratches, or other damage that could restrict the movement of the valves. Make sure all fluid drain back openings are clear and have no varnish or dirt buildup. Check all of the threaded holes and related bolts and screws for damaged threads, replace as needed.

SERVICE TIP:
Be sure to inspect the shift solenoid O-rings and the TCC solenoid O-rings for damage during disassembly. If they are damaged, attempt to identify the cause of the damage.

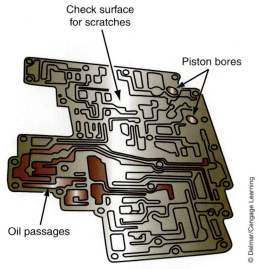

Check surface for scratches

Piston bores

Oil passages

FIGURE 7-20 Carefully check the bores, passages, and surface of the separator plate. It must be replaced if it is damaged.

If the valve body has an oil strainer, make sure it is clean. This can be checked by pouring clean ATF through the strainer and observing the fluid flow out of the strainer. If there is poor flow, the strainer should be replaced.

With a straightedge laid across the sealing surface of the valve body, use a feeler gauge to check its flatness. If it is slightly warped, it can be flat filed. Be very careful when doing this. Keep the file flat and always file in one direction. If the surface is warped beyond repair, the valve body must be replaced. Warpage can produce a cross leak; which is a leak from one track to another. These can cause unwanted, partial engagement of a clutch or band.

After the valves and springs have been removed, soak the valve body and separator and transfer plates in mineral spirits or lacquer thinner, then wash the parts off with water. Thoroughly clean all parts and make sure all passages within the valve body are clear and free of debris. Carefully blow-dry each part with compressed air. Never wipe the parts with a rag or paper towel. Lint from either will collect in the valve body passages and cause shifting problems. Make sure to clean the worm tracks in the valve body and case.

⚠ **WARNING:** **Wear safety glasses when drying parts with compressed air. The overspray of solvent and other cleaners can damage your eyes and lead to blindness.**

Valves. Continue disassembly by removing the valves and springs (Figure 7-21). The valves and springs can be retained in many different ways. Some have a cover plate that is bolted or screwed to the valve body. These may retain one or more valves. Others have a plug or sleeve retained by a pin, plate, or key at the outside of each bore. The plug can be removed by pushing it in slightly and removing the pin, plate, or key.

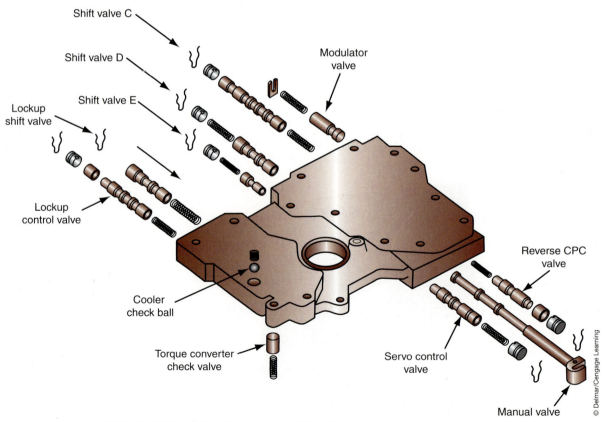

FIGURE 7-21 **Typical location of some of the valves and springs in a valve body.**

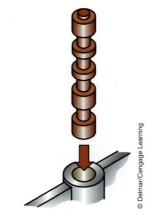

FIGURE 7-22 Check the fit of the valve in its bore by dropping it into its bore.

Examine each valve for nicks, burrs, and scratches. If the valves are coated with varnish, they must be soaked in an appropriate cleaner. If the valve lands have worn areas, this means the valve has been rubbing in its bore. The edges of the spools should be damage-free but should not be sharp. To identify a bent valve, install it into its bore and rotate it. If it binds or drags while it is being rotated, it is bent.

Make sure each valve properly fits into its respective bore. To do this, hold the valve body vertically and install an unlubricated valve into its bore. Let the valve fall of its own weight into the valve body until the valve stops (Figure 7-22). Then, place your finger over the valve bore and turn the valve body over. The valve should again drop by its own weight. If the valve moves freely under these conditions, it will operate freely with fluid pressure. Repeat this test on all valves.

If the valves do not move freely, the problem may be corrected by polishing the valve lands if the valves are made of steel. If the valves are aluminum, the valve body should be replaced. To polish a valve, use a polishing (Arkansas) stone or crocus cloth. It is helpful to let the cloth soak in ATF before using it to polish a valve. Evenly rotate the valve on the polishing material, make sure this does not round the edges of the valve. Polish the valves only enough to ensure free movement in the bore.

After the valve is polished, it must be thoroughly cleaned to remove all of the cleaning and abrasive materials. After the valve has been recleaned, it should be tested in its bore again.

Valve Bores. If a valve is smooth but sticks in its bore, carefully check the bore for dirt or nicks. Minor imperfections can be addressed; however, serious damage requires the replacement of the valve body.

To remove minor imperfections, do the following steps:

1. Soak a sheet of #600-grit abrasive paper in ATF for about 30 minutes.
2. Roll up half of the abrasive paper and insert it into the bore of the sticking valve (Figure 7-23).
3. Twist the roll of paper so it unrolls and expands to the size of the bore.
4. Then polish the bore by twisting it while moving it in and out of the bore.
5. Remove the paper.
6. Then clean the entire valve body in solvent and dry it with compressed air.

Place the valve into its bore and check its free movement. If the valve still cannot move freely in its bore, the valve body should be replaced. Individual valve body parts are usually not available.

SERVICE TIP:
A good aluminum valve in an unworn valve body bore may not drop in its bore if it is wet, the fluid may cause the valve to stick in the bore. Always check these without a coating of fluid.

SERVICE TIP:
Nearly all valve bodies are aluminum and therefore little polishing is required to remove the burrs or other defects.

FIGURE 7-23 Roll up some fine sandpaper and move it in bore to remove any minor imperfections.

Valve to Bore Tolerances. One of the most common problems today involves excessive valve body bore wear due to the constant shuffling of the valves by a duty-cycle solenoid. This wear causes excessive valve to bore clearance. The correct clearance allows the valves to slide back and forth while preventing fluid leakage. The clearance also allows for maintaining a thin film of oil for the valve to ride on. The typical valve to bore clearance is 0.0005 inch to 0.0016 inch (0.0127mm–0.040mm). If the bore is worn, or if there is excessive clearance, the entire valve body should be replaced.

Springs. Check each spring for signs of distortion. If any spring is damaged, the valve body should be replaced. A good way to check the spring is to lay them on their side and roll them. If they roll true, they are not distorted. If they wobble, they are distorted.

Reassembly

After the valve body has been cleaned, it should be allowed to air dry. Once dried, it should be dipped into a pan of clean ATF. While the valve body is soaking in ATF, locate the installation specifications in the service information.

Lubricate all parts with clean ATF. Then, install the valves and associated springs into their bores. It is important that you place the valve retaining plugs or caps in the correct bore and in the correct direction (Figure 7-24). Once the cap is in position, carefully depress it with a small screwdriver and install the retaining clip (Figure 7-25).

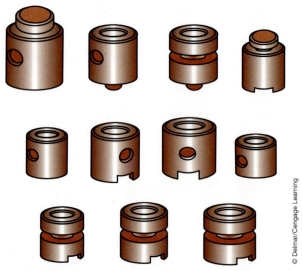

FIGURE 7-24 The plugs used to retain the valves in a valve body have designated bores and must face the correct direction.

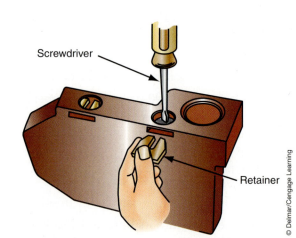

Screwdriver

Retainer

FIGURE 7-25 Once the cap is in position, carefully depress it with a small screwdriver and install the retaining clip.

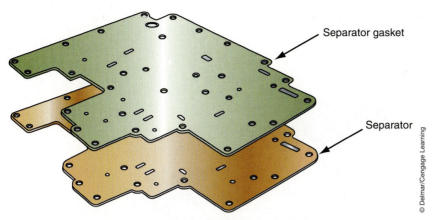

FIGURE 7-26 Always check a new valve body gasket by laying it over the separator plate and making sure it does not block any holes. If the gasket is correct, use clean ATF to hold it in place during assembly.

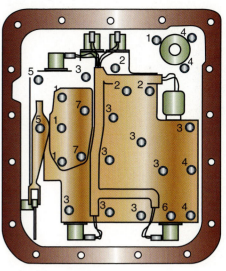

Each numbered bolt location corresponds to a specific bolt size and length, as indicated here:

1. M6 × 1.0 × 65.0   5. M8 × 1.25 × 20.0
2. M6 × 1.0 × 54.4   6. M6 × 1.0 × 12.0
3. M6 × 1.0 × 47.5   7. M6 × 1.0 × 18.0
4. M6 × 1.0 × 35.0

FIGURE 7-27 When installing the valve body mounting bolts, make sure you use the correct bolt size and length in the correct location. Refer to the service manual for guidance.

Install all check balls and springs in their correct location. If you have any doubts as to where they should be placed, refer to the service information. Count the check balls as you install them to make sure you have inserted all of them.

Before beginning to install the valve body, check the new valve body gasket to make sure it is the correct one by comparing it to the old gasket. If the gasket appears to be the correct one, lay it over the separator plate and hold it up to a light, make sure no oil holes are blocked (Figure 7-26). Also, check to make sure the gasket seals off the worm tracks and will not allow the fluid to go where it should not go. Then install the bolts to hold valve body sections together and the valve body to the case. Tighten the bolts to the torque specifications to prevent valve body warpage and possible leakover. Overtorquing can also cause the bores to distort which would not allow the valves to move freely once the valve body is tightened to the transmission case.

Many transmissions use bolts of various lengths to secure the valve body to the case. It is important that the correct length bolt is used in each bore. It is so important that service manuals list the size and location of the mounting bolts (Figure 7-27).

GOVERNOR SERVICE

If the transmission has a governor and tests suggest it may have problem, it should be removed, disassembled, cleaned, and inspected. Some governors are mounted internally and the transmission must be removed to service the governor. Others can be serviced by removing the extension housing or oil pan, or by detaching an external retaining clamp and then removing the unit (Figure 7-28).

Improper shift points can be caused by a faulty governor or governor drive gear system. However, many transmissions do not rely on the hydraulic signals from a governor; rather, they rely on the electrical signals from sensors, such as speed and load sensors. Faulty electrical components and/or loose connections can cause improper shift points.

CAUTION:
The primary cause of valve sticking is the overtightening of the valve body bolts. Always be careful when handling a valve body, they are very precise components.

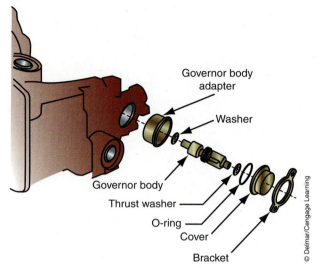

FIGURE 7-28 Some governor assemblies are contained in a separate housing and retained by a bolted retainer or cover.

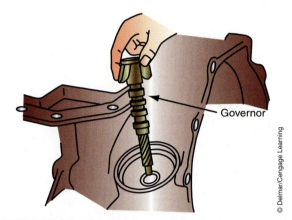

FIGURE 7-29 After the retaining cover has been removed, the governor assembly can be pulled out of its bore.

If the transmission has a shaft-mounted governor, it is driven by the output shaft and can be accessed by removing the extension housing (Figure 7-29). Some transaxles require complete disassembly of the transaxle to access the governor. Other transmissions may have a protrusion off the side of the extension housing that contains the governor. These governors are typically driven by a gear and are accessible by removing the governor cover from the protrusion.

Disassembly

To disassemble a typical gear-driven governor, remove the governor cover by carefully prying it out of its bore. Once the cover has been removed, remove the primary governor valve from its bore in the governor housing. Then remove the secondary valve retaining pin, secondary valve spring, and valve.

Check the action of the governor valve by moving the weights. With the weights held to the shaft, the exhaust port of the valve should be open. With the weights held in their fully extended position, the exhaust port should be closed and the inlet port opened. The amounts the port opened can be measured with a feeler gauge. Compare your measurement to specifications. While moving the weights, pay attention to their movement. They should move freely and return to their rest position without much effort.

Thoroughly clean and dry all parts. Test the valve in its bore in the governor housing, it should move freely in the bore without sticking or binding. Also, check the valve for any signs of burning or scoring and replace it, if necessary. Inspect the springs for a loss of tension and

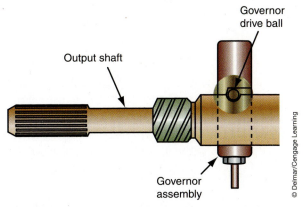

Output shift

Governor
drive ball

Governor
assembly

© Delmar/Cengage Learning

FIGURE 7-30 Governor drive ball in output shaft.

burning marks and replace if necessary. Make sure to check the ports of the governor for any buildups that may restrict fluid flow.

Reassembly

To reassemble a shaft-mounted governor, place the spring around the secondary valve and insert them into the bore. Then insert the retaining pin. Now, install the primary valve into the housing. The governor cover should then be driven in place with a new seal. Make sure you lubricate the seal with ATF before driving the cover into the bore.

To reassemble a shaft-driven governor, use a press and install the drive gear. Then install the weights, springs, valve, and thrust cap. Insert new retaining pins through the thrust cap and weights and crimp both ends of the pins to prevent them from working out.

After assembly, install the governor and torque all bolts to specifications. Overtightening can cause the valve to stick. Some transmissions use a drive ball on the output shaft, which locks the governor to the shaft (Figure 7-30). Make sure it is in place when installing the governor.

⚠️ **CAUTION:**
Never interchange components of the primary and secondary governors. Also, note that the flat faces of the primary valve must face outward when it is installed.

CASE STUDY

A customer with a late-model Honda complains that his automatic transmission seems to start out in the wrong gear when drive is selected. The technician takes the car for a road test to verify the complaint. She observed and verified the complaint. It did seem like the vehicle started out in second or third gear and had very poor and unsafe acceleration. But once the car got going, the transmission seemed to work fine.

Following the normal diagnostic routine, she found nothing all that unusual. So she took the car out again on a road test, this time with pressure gauges installed. She jotted down the test results and returned to the shop to search published information about this particular transmission.

She found that on Honda AS, AK, CA, and F4 model transmissions that are governor controlled, a complaint of no reverse and/or wrong gear starts can be caused by either stuck valves in the valve body or high governor pressure at a stop. The information

further stated that if the pressure gauge reads less than 2 psi when the vehicle is at a stop, the problem is sticky valves in the valve body. If the gauge reads 2 psi or more, then high governor pressure is the problem. Either the governor valve is sticking or something else is allowing a higher pressure to leak into the governor circuit.

Since she experienced more than 2 psi, she knew the problem was high governor pressure. She then removed the governor and used 800-grit sandpaper to polish the valve. Then she reassembled and installed the governor. She kept the pressure gauges connected and took a road test. She found that the car accelerated as it should and found the pressure at a stop to be normal.

There is a small tube that must be installed into the governor shaft. This tube separates line pressure from governor pressure, and if it is left out, line pressure will flow directly into the governor circuit, causing high governor pressure and the same problem.

TERMS TO KNOW

Auxiliary transmission fluid pump (ATFP)

Manual shaft

Solenoid/pressure switch

Transmission fluid pressure switch

ASE-STYLE REVIEW QUESTIONS

1. *Technician A* says all parts of the valve body should be soaked in mineral spirits before reassembling.
   Technician B says a lint-free rag is a must when wiping down valves.
   Who is correct?
   A. A only
   B. B only
   C. Both A and B
   D. Neither A nor B

2. While removing scratches in a valve:
   Technician A uses a flat file to remove the scratch.
   Technician B uses a sand blaster or glass bead machine to polish the surface of the valve.
   Who is correct?
   A. A only
   B. B only
   C. Both A and B
   D. Neither A nor B

3. *Technician A* says overtorquing the hold-down bolts of the valve body can cause the valves to stick in their bore.
   Technician B says flat-filing the surface of the valve body will allow the valve body to seal properly and will therefore allow the valves to move more freely in their bores.
   Who is correct?
   A. A only
   B. B only
   C. Both A and B
   D. Neither A nor B

4. While assembling a valve body after cleaning it:
   Technician A lubricates all parts with clean ATF.
   Technician B says it is important that the valve retaining plugs or caps be placed in the correct bore and in the correct direction.
   Who is correct?
   A. A only
   B. B only
   C. Both A and B
   D. Neither A nor B

5. After soaking valve body parts in mineral spirits:
   Technician A wipes the parts off with a paper towel.
   Technician B blow-dries each part individually with compressed air.
   Who is correct?
   A. A only
   B. B only
   C. Both A and B
   D. Neither A nor B

6. While inspecting a valve body:
   Technician A says scratches and score marks on the separator plate can cause oil to bypass correct oil passages and result in system malfunction.
   Technician B says scratches in the oil passages of the upper and lower valve bodies could restrict the movement of the valves.
   Who is correct?
   A. A only
   B. B only
   C. Both A and B
   D. Neither A nor B

7. While inspecting a valve body:
   Technician A says deep scratches in a valve's bore can be removed with #600-grit abrasive paper.
   Technician B says one of the most common transmission problems is excessive valve body bore wear due to the constant shuffling of the valves by a duty-cycle solenoid.
   Who is correct?
   A. A only
   B. B only
   C. Both A and B
   D. Neither A nor B

8. *Technician A* says that if diagnosis suggests a problem with only one or two valves, visual inspection of the valve body should start at those valves.
   Technician B says that if the transmission has heavily contaminated fluid, the entire valve body should be inspected and cleaned, or replaced.
   Who is correct?
   A. A only
   B. B only
   C. Both A and B
   D. Neither A nor B

9. While reassembling a valve body:
   Technician A says the valve springs and caps can be of different sizes and should be installed in their appropriate bores.
   Technician B makes sure all parts are clean, dry, and absent of any fluid before installing them.
   Who is correct?
   A. A only
   B. B only
   C. Both A and B
   D. Neither A nor B

10. *Technician A* says that if a valve cannot be cleaned well enough to move freely in its bore, the valve body should be replaced.

Technician B says that if there is even the slightest bit of damage or varnish buildup in a valve's bore, the entire valve body must be replaced.

Who is correct?

A. A only
B. B only
C. Both A and B
D. Neither A nor B

ASE Challenge Questions

1. *Technician A* says that if the governor pressure is slow to build, early upshifts will result.

Technician B says that if line pressure is higher than normal, early shifts will take place.

Who is correct?

A. A only
B. B only
C. Both A and B
D. Neither A nor B

2. Which of the following is *not* a common cause for sticking valves and sluggish valve movements?

A. Overtorqued valve body bolts.
B. A faulty pressure control solenoid.
C. The use of the wrong type of fluid.
D. Overheating the transmission.

3. *Technician A* says the valve body of all transmissions and transaxles can be serviced with the transmission still in the vehicle, providing that surrounding components are removed first.

Technician B says the governor in all transmissions and transaxles can be serviced with the transmission still in the vehicle, providing that surrounding components are removed first.

Who is correct?

A. A only
B. B only
C. Both A and B
D. Neither A nor B

4. *Technician A* says oil feed pipes are used to move the fluid to the various clutches and apply devices in a transmission.

Technician B says shift and TCC solenoids are often mounted directly to the valve body.

Who is correct?

A. A only
B. B only
C. Both A and B
D. Neither A nor B

5. *Technician A* says that if aluminum valves are scored or otherwise damaged, the individual valve or entire valve body should be replaced.

Technician B says problems rarely result from excessive clearance between the valve and its bore.

Who is correct?

A. A only
B. B only
C. Both A and B
D. Neither A nor B

Name _____ Date _____

SERVICE A VALVE BODY

Upon completion of this job sheet, you should be able to disassemble, clean, inspect, and reassemble a valve body.

ASE Correlation

This job sheet is related to the ASE Automatic Transmission and Transaxle Test's Content Area *In-Vehicle Transmission and Transaxle Repair.*

Tasks: Inspect valve body mating surfaces, bores, valves, springs, sleeves, retainers, brackets, check balls, screens, spacers, and gaskets; replace as necessary.

Check and adjust valve body bolt torque.

Tools and Materials

Automatic transmission or transaxle on a bench   Supply of clean solvent

Compressed air and air nozzle   Lint-free shop towels

Inch-pound torque wrench   Service manual

Measuring calipers

Describe the transmission being worked on:

Model and type of transmission _____

Year _____ Make _____ VIN _____

Model _____

Procedure

Task Completed

1. Remove the valve body attaching screws from the transmission. Start at the outside bolts and work toward the center if the service manual does not give specific instructions.   ☐

2. Remove the valve body assembly and place it in a container of clean solvent.   ☐

3. Remove the end plates and covers from the assembly.   ☐

4. In the space below, draw a simplified view of the valve body. Note the location of all the valves, check balls, and springs.   ☐

5. Begin to disassemble the unit. Lay all the parts on a clean surface in the order in which they were disassembled. This will help during the reassembly of the valve body. Be careful with the check balls; they may not be all the same size. Some transmissions use slightly larger or smaller check balls in circuits. How many check balls are in the valve body? Are they all the same size?

_____

_____

_____

6. Remove all the gaskets from the assembly. Place them aside. Do not throw them away. They will be needed for comparison when installing new gaskets.

7. Thoroughly clean the main body and plates of the assembly. Do not dry the valve body with anything that would leave lint. Allow the unit to air dry or dry it off with compressed air.

8. Inspect the valve body for damage and cracks. Also check the flatness of the body and the plates. Describe your findings.

_____

_____

_____

_____

9. What are the valves made of? _____

10. Check all the valves for free movement in their bores. Also check the valves for wear, scoring, and signs of sticking. Describe your findings and recommendations.

_____

_____

_____

_____

☐ 11. Correctly reinstall the valves and springs in their bores.

12. Replace the end plates or covers. Install the retaining screws by hand, then tighten them to the specific torque. The specified torque is _____.

☐ 13. Install the check balls into their proper location.

☐ 14. Compare the new gaskets with the old and install the correct new ones on the valve body.

☐ 15. Align the spring-loaded check balls and all other parts as needed.

16. Place the gasket on top of the transfer plate. Align the gasket holes with the transfer plate and the bores in the valve body.

☐ 17. Position the valve body on the transmission. Align the parking and other internal linkages. Then install the retaining screws by hand.

18. Tighten the screws to the specified torque. Tighten the screws in the center first, then work to the outside of the valve body. The specified torque is _____.

19. Summarize this job sheet.

_____

_____

_____

_____

Instructor's Response _____

_____

_____

Name _____ Date _____

SERVICING GOVERNORS

Upon completion of this job sheet, you should be able to inspect, repair, and replace a governor assembly.

ASE Correlation

This job sheet is related to the ASE Automatic Transmission and Transaxle Test's Content Area *In Vehicle Transmission and Transaxle Repair.*

Task: Inspect, repair, and replace governor assembly.

Tools and Materials

Torque wrench

Describe the vehicle that was assigned to you:

Year _____ Make _____ Model _____

VIN _____ Engine type and size _____

Transmission type and model _____

Procedure

_____

Task Completed

1. If the pressure tests suggested that there was a governor problem, it should be removed, disassembled, cleaned, and inspected. Some governors are mounted internally and the transmission must be removed to service the governor. Others can be serviced by removing the extension housing or oil pan, or by detaching an external retaining clamp and then removing the unit. How did you remove yours?

   _____

   _____

   _____

2. To disassemble a typical governor, remove the primary governor valve from its bore in the governor housing. ☐

3. Remove the secondary valve retaining pin, secondary valve spring, and valve. ☐

4. Thoroughly clean and dry these parts.

5. Test each valve in its bore in the governor housing; they should move freely in their bores without sticking or binding. Record your findings. ☐

   _____

   _____

6. Check the valves for any signs of burning or scoring and replace the valve body, if necessary. Record your findings.

   _____

   _____

Task Completed

7. Inspect the springs for a loss of tension and burning marks and replace, if necessary. Record your findings.

_____

_____

☐ 8. To reassemble the governor, place the spring around the secondary valve, and insert it into the secondary valve bore.

☐ 9. Insert the retaining pin into the governor housing pinholes.

☐ 10. Install the primary valve into the governor housing.

11. If the governor assembly was removed from the governor support and parking gear, be sure to tighten the bolts to specifications with a torque wrench. After assembly, install the governor and torque the bolts to specifications. What are the specifications?

_____

_____

Instructor's Response _____

_____

_____

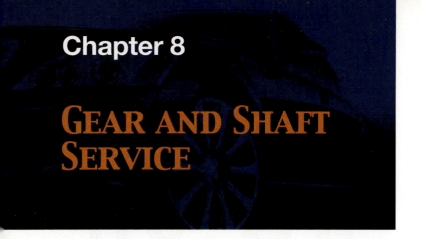

Chapter 8

GEAR AND SHAFT SERVICE

BASIC TOOLS
Basic mechanic's tool set
Clean lint-free rags
Appropriate service manual
Clean ATF

UPON COMPLETION AND REVIEW OF THIS CHAPTER, YOU SHOULD BE ABLE TO:

- Inspect, measure, and replace thrust washers and bearings.

- Inspect and replace bushings.

- Inspect and measure a planetary gear assembly and replace parts as necessary.

- Inspect and replace shafts.

- Remove and install the various types of shaft seals.

- Inspect, repair, or replace transaxle drive chains, sprockets, gears, bearings, and bushings.

- Inspect and replace parking pawl, shaft, spring, and retainer.

- Inspect, measure, repair, adjust, or replace transaxle final drive components.

- Inspect and service the transmission housing and cases.

The overall reliability of a transmission depends on the structural integrity of the shafts and gears. Therefore these parts, as well as all thrust washers, bearings, and bushings (Figure 8-1), should be carefully inspected and replaced if damaged or worn.

THRUST WASHERS, BUSHINGS, AND BEARINGS

A thrust washer is designed to support a **thrust load** and keep parts from rubbing together, preventing premature wear on parts such as planetary gearsets (Figure 8-2) and transfer and final drive assemblies. **Selective thrust washers** come in various thicknesses to take up clearances and adjust endplay at many different locations. Some transmissions have a selective thrust washer between the front and the rear multiple-disc assemblies (Figure 8-3). Thrust washers and thrust bearings are used wherever rotating parts must have their endplay maintained. To control the endplay of nonrotating parts, selective shims or spacers are used.

Many transmissions are fitted with thrust bearings, often called Torrington bearings (Figure 8-4). These thrust washers have small roller or needle bearings placed into them. The bearings reduce the rotational friction between the two surfaces they separate. Thrust bearings should be carefully inspected for distortion, cracks, and wear. If there is any evidence of damage, the bearing should be replaced.

Most flat thrust washers and bearings are not selective and have a fixed thickness. Typically, the selective types are numbered or colored by the manufacturer for easy identification (Figure 8-5). Thrust washers should be inspected for scoring, flaking, and wear. Flat thrust washers should also be checked for broken or weak tabs (Figure 8-6). These tabs are critical for holding the washer in place. On metal thrust washers, the tabs may appear cracked at the bend; however, this is a normal appearance due to the characteristics of the materials used to manufacture them.

Thrust load is a load placed in parallel with the center of an axis.

Thrust washers are often referred to as thrust plates.

The use of the name **selective thrust washer** means the thrust washer for this application is available in different thicknesses. The correct thrust washer must be selected to provide the correct endplay or clearance.

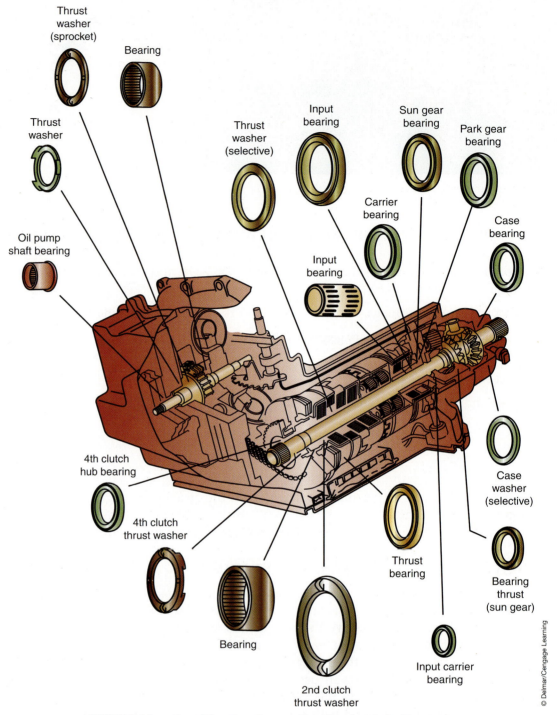

Thrust washer (sprocket)

Bearing

Thrust washer

Oil pump shaft bearing

Thrust washer (selective)

Input bearing

Sun gear bearing

Park gear bearing

Carrier bearing

Case bearing

Input bearing

4th clutch hub bearing

4th clutch thrust washer

Bearing

2nd clutch thrust washer

Thrust bearing

Input carrier bearing

Case washer (selective)

Bearing thrust (sun gear)

© Delmar/Cengage Learning

FIGURE 8-1 **Location of thrust washers and bearings in a typical transaxle.**

Thrust washers are not included in an overhaul kit because they are select fit. Even if the existing thrust washers are the correct size, they should be replaced with new ones if they are pitted or scored.

Plastic thrust washers are typically color-coded to denote their size. These washers will not show wear unless they are damaged. The only way to check their wear is to measure the thickness with a micrometer and compare them to a new part. All damaged and worn thrust washers and bearings should be replaced.

Proper thrust washer thicknesses are important to the operation of an automatic transmission. After following the recommended procedures for checking the endplay of various components, refer to the manufacturer's chart for the proper thrust plate thickness for each application.

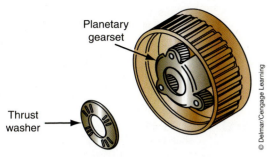

FIGURE 8-2 The purpose of a thrust washer is to support a thrust load and keep parts from rubbing together, such as planetary gearsets.

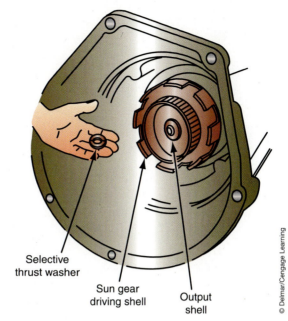

FIGURE 8-3 A selective thrust washer between the front and rear multiple-friction disc assemblies.

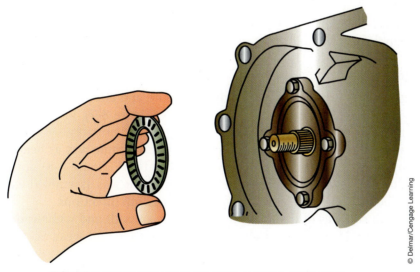

FIGURE 8-4 Thrust bearings are often used in transfer shaft or countershaft assemblies.

Colorcode	Depth	Thickness
Green	1.485–1.503 in. (37.706–38.184 mm)	0.050–0.054 in. (1.270–1.372 mm)
Yellow	1.504–1.521 in. (38.185–38.641 mm)	0.068–0.072 in. (1.727–1.829 mm)
Natural	1.522–1.538 in. (38.642–39.073 mm)	0.085–0.089 in. (2.159–2.261 mm)
Red	1.539–1.555 in. (39.074–39.505 mm)	0.102–0.106 in. (2.591–2.692 mm)
Blue	1.556–1.581 in. (39.506–40.165 mm)	0.119–0.123 in. (3.023–3.124 mm)

FIGURE 8-5 An example of the color code used to identify selective thrust washers.

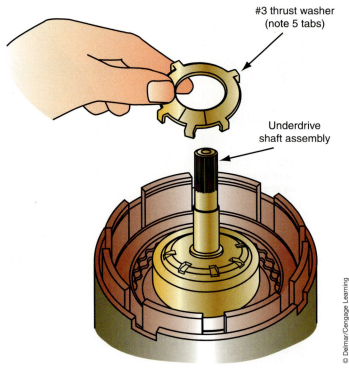

#3 thrust washer
(note 5 tabs)

Underdrive
shaft assembly

© Delmar/Cengage Learning

FIGURE 8-6 A thrust washer with tabs. These tabs must be carefully inspected for cracks or other damage.

Etching is a discoloration or removal of some material caused by corrosion or some other chemical reaction.

Use transjel or similar type lubricant to hold thrust washers in place during assembly. This will keep them from falling out of place, which will affect endplay.

Bearings

All bearings should be checked for roughness before and after cleaning. Carefully examine the inner and outer races, and the rollers, needles (Figure 8-7), or balls for cracks, pitting, **etching**, or signs of overheating.

Sprag and roller clutches should be inspected in the same way as bearings. Check their operation by attempting to rotate them in both directions (Figure 8-8). If working properly, they will allow rotation in one direction only. If the unit was assembled with the sprags or rollers installed backward, the unit will spin freely in both directions.

The smooth surfaces of the race should be very smooth and without defects. The roller clutch springs should be checked for signs of overheating and wear. Sprag units should be checked for loose sprags.

Bushings/Sleeves

Bushings and Sleeves should be inspected for pitting and scoring. Always check the depth that bushings are installed to and the direction of the oil groove, if so equipped, before you remove them. Many bushings that are used in the planetary gearing and output shaft areas have oiling holes in them. Be sure to line these up correctly during installation or you may block off oil delivery and destroy the geartrain. If any damage is evident on the bushing, it should be replaced.

Bushing wear can be checked visually as well as by observing the lateral movement of the shaft that fits into the bushing. Any noticeable lateral movement indicates wear and the bushing should be replaced. The amount of clearance between the shaft and the bushing can be checked with a wire-type feeler gauge. Insert the wire between the shaft and the bushing; if the gap is greater than the maximum allowable, the bushing should be replaced. Normally,

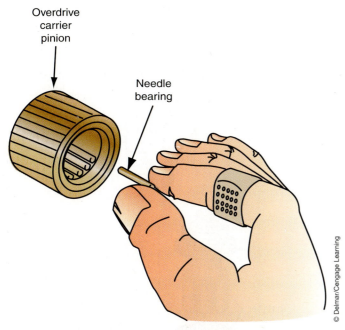

Overdrive carrier pinion

Needle bearing

© Delmar/Cengage Learning

FIGURE 8-7 Needle bearings are often located inside the small pinion gears. Coating the inside of the gear with transgel before the needle bearings are positioned will help to keep them in place.

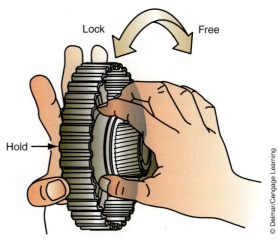

Lock   Free

Hold

© Delmar/Cengage Learning

FIGURE 8-8 To check the action of a one-way clutch, hold the inner race and rotate the clutch in both directions. The clutch should rotate smoothly in one direction and lock in the other.

bushings must fit the shafts they ride on with about a 0.0015- to 0.003-inch clearance. You can check this fit by measuring the inside diameter of the bushing and the outside diameter of the shaft with a dial caliper or micrometer (Figure 8-9). This is a critical fit throughout the transmission and especially at the converter drive hub, where 0.004 inches is the desired fit. Excessive clearance will allow TIC fluid to drain back.

Most bushings or sleeves are press-fit into a bore or on a shaft. To remove one, drive it out of the bore with a properly sized bushing tool. Some bushings can be removed with a slide hammer fitted with an expanding or threaded fixture that grips the bushing. Another way to remove bushings is to carefully cut one side of the bushing and collapse it. Once collapsed, the bushing can be easily removed with a pair of pliers. Small-bore bushings that are located in areas where it is difficult to use a bushing tool can be removed by tapping the inside bore of the bushing with threads that match a selected bolt that fits into the bushing. After the bushing has been tapped, insert the bolt and use a slide hammer to pull the bolt and bushing out of its bore.

Whenever possible, all new bushings should be installed with the proper bushing driver. The use of these tools prevents damage to the bushing and allows for proper seating of the bushing into its bore (Figure 8-10).

PLANETARY GEAR ASSEMBLIES

The purpose of inspection at this point is to try to eliminate the possibility of putting noise into a newly rebuilt unit. This would result in a costly setback, so close inspection of the planetary gearset is a must. All planetary gear teeth should be inspected for chips or stripped teeth. Any gear that is mounted to a splined shaft needs the splines checked for mutilated or shifted splines. Helical gears have many advantages over straight-cut gears, such as providing low operating noise, but you must check the endplay of the individual gears during your inspection. The helical cut of the gears makes them thrust to one side during operation. This can put a lot of load on the thrust washers and may wear them beyond specification. Checking these was discussed earlier, but particular attention should be given to the planetary carriers.

SPECIAL TOOLS

Wire-type feeler gauge set
Micrometer
Bushing pullers and drivers
Hydraulic press

Classroom Manual

Chapter 8, pages 248

361

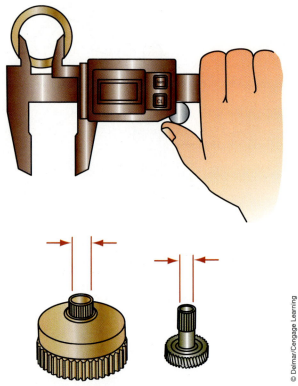

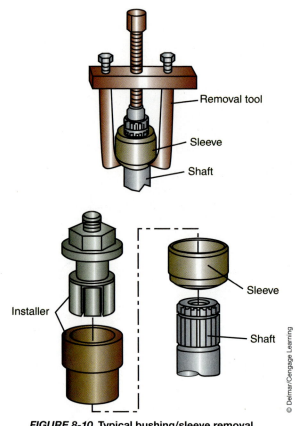

FIGURE 8-9 The inside diameter of bushings should be measured for wear with a telescoping gauge and a micrometer.

© Delmar/Cengage Learning

FIGURE 8-10 Typical bushing/sleeve removal and installation tools.

Removal tool
Sleeve
Shaft

Installer
Sleeve
Shaft

© Delmar/Cengage Learning

SPECIAL TOOLS

Feeler gauge set
Micrometer

Look first for obvious problems like blackened gears or pinion shafts. These conditions indicate severe overloading and require that the carrier be replaced. Occasionally the pinion gear and shaft assembly can be replaced individually. When looking at the gears themselves, a bluish condition can be a normal condition, as this is part of a heat-treating process used during manufacture. Check the planetary pinion gears for loose bearings. Check each gear individually by rolling it on its shaft to feel for roughness or binding of the needle bearings. Wiggle the gear to be sure it is not loose on the shaft. Looseness will cause the gear to whine when it is loaded. Also, inspect the gear teeth for chips or imperfections, as these will also cause whine.

Check the gear teeth around the inside of the front planetary ring gear. Check the fit between the front planetary carrier to the output shaft splines. Remove the snap ring and thrust washer from the front planetary ring gear (Figure 8-11). Examine the thrust washer and the outer splines of the front drum for burrs and distortion. The rear clutch friction discs must be able to slide on these splines during engagement and disengagement. With the snap ring removed, the front planetary carrier can be removed from the ring gear. Check the planetary carrier gears for endplay by placing a feeler gauge between the planetary carrier and the planetary pinion gear (Figure 8-12). Compare the endplay to specifications. On some Ravigneaux units, the clearance at both ends of the long pinion gears must also be checked and compared to specifications.

Check the splines of the sun gear. Sun gears should have their inner bushings inspected for looseness on their respective shafts. Also check the fit of the sun shell to the sun gear (Figure 8-13). The shell can crack where the gear mates with the shell. The sun shell should also be checked for a bell-mouthed condition where it is tabbed to the clutch drum. Any variation from a true round should be considered junk and should not be used. Look at the tabs and check for the best fit into the clutch drum slots. This involves trial fitting the shell

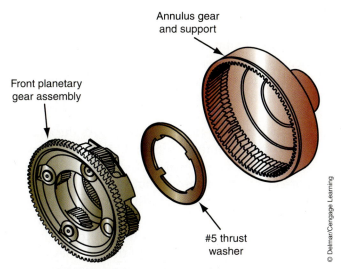

Front planetary gear assembly

Annulus gear and support

#5 thrust washer

© Delmar/Cengage Learning

FIGURE 8-11 The planetary assembly, its thrust washer, and annulus gear should be inspected for signs of abnormal wear.

© Delmar/Cengage Learning

FIGURE 8-12 The clearance between the pinion gears and the planetary carrier should be checked and compared to specifications.

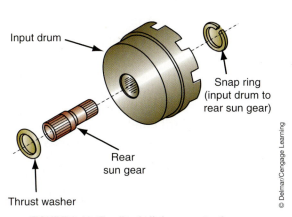

Input drum

Snap ring (input drum to rear sun gear)

Rear sun gear

Thrust washer

© Delmar/Cengage Learning

FIGURE 8-13 The fit of all drums onto the splines of their mating shafts should be checked.

© Delmar/Cengage Learning

FIGURE 8-14 Measure the inside diameter of all bushings to determine wear.

and drum at all the possible combinations and marking the point where they fit the tightest. A snug fit here will eliminate bell mouthing due to excess play at the tabs. It can also reduce engagement noise in reverse, second, and fourth gears. This excess play allows the sun shell tabs to strike the clutch drum tabs as the transmission shifts from first to second or when the transmission is shifted into reverse.

The gear carrier should have no cracks or other defects. Replace any abnormal or worn parts. Check the thrust bearings for excessive wear, and if required, correct the input shaft thrust clearance by using a washer with the correct thickness. Determine the correct thickness by measuring the thickness of the existing thrust washer and comparing it to the measured endplay. Now move the gear back and forth to check its endplay. Some shop manuals will give a range for this check, but if none can be found you can figure about 0.007 to 0.025 inch as an average amount. All the pinions should have about the same endplay.

Also inspect all bushings that may be inserted into the planetary gearsets. These are commonly found in sun gears (Figure 8-14). Measure the inside diameter and compare that dimension to specifications. If the diameter exceeds those specifications, replace the gear assembly.

SHAFTS

Carefully examine the area on all shafts that rides in a bushing, bearing, or seal. Check the entire length of the shaft for signs of overheating and other damage. If the shaft has gears, the teeth of the gears should be inspected for damage and breaks. Also inspect the splines for wear, cracks, or other damage (Figure 8-15). A quick way to determine spline wear is to fit the mating splines and check for lateral movement.

Shafts are checked for scoring in the areas where they ride in bushings. As the shaft is a much harder material than the bushings, any scoring on the shaft would indicate a lack of lubrication at that point. The affected bushing should appear worn into the backing metal. Because shaft-to-bearing fit is critical to correct oil travel throughout the transmission, a scored shaft should be replaced. Lubricating oil is carried through most shafts; therefore, an internal inspection for debris is necessary. A blocked oil delivery hole can starve a bushing, resulting in a scored shaft. The internal oil passage of a shaft may not be able to be visually inspected and only observation during cleaning will give an indication of the openness of the passage. Washing the shaft passage out with a solvent and possibly running a piece of small-diameter wire through the passage will dislodge most particles. Be sure to check that the ball that closes off the end of the shaft, if the shaft is so equipped, is securely in place. A missing ball could be the cause of burned planetary gears and scored shafts due to a loss of oil pressure. Any shaft that has an internal bushing should be inspected, as described earlier. Replace all defective parts as necessary.

Shafts should be checked for wear in the ring groove area. This is especially critical if the groove accommodates a metal ring. Make sure there is no step wear in the groove and that the sides and bottom are square. Also make sure the groove is not too wide for the ring. If a 0.005-inch feeler gauge will fit in the groove with the ring in place, the groove is worn and the shaft should be replaced.

Input and output shafts can be solid, drilled, or tubular. The solid and drilled shafts are supported by bushings, so the bushing journals of the shafts should be free of noticeable wear at these points. Small scratches can be removed with 320-grit emery paper. Grooved or scored shafts require replacement. The splines should not show any sign of waviness along their length. Check drilled shafts to be sure the drilled portion is open and free of any foreign material. Wash out the shaft with solvent and run a small-diameter wire through the shaft to dislodge any particles. After running the wire through the opening, wash out the shaft once more and blow it out with compressed air.

If the shaft has a check ball, such as the 4L60 turbine shaft (Figure 8-16), be certain the ball seats in the correct direction. This particular check ball controls oil flow direction to the converter. Some shafts have a ball pressed into one end to block off one end of the shaft. This is used to hold oil in the shaft so the oil is diverted through holes in the side of the shaft. These

SPECIAL TOOLS

Feeler gauge set
320-grit sandpaper
Cleaning solvent
OSHA-approved air nozzle

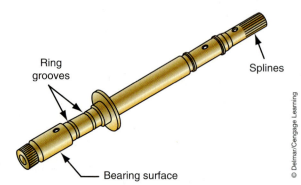

Ring grooves

Splines

Bearing surface

© Delmar/Cengage Learning

FIGURE 8-15 All shafts, including their splines and ring grooves, should be carefully inspected for wear or other damage.

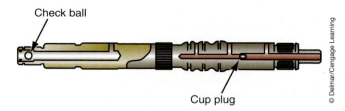

Check ball

Cup plug

© Delmar/Cengage Learning

FIGURE 8-16 If the turbine shaft is fitted with a check ball, make sure it is able to seat and unseat.

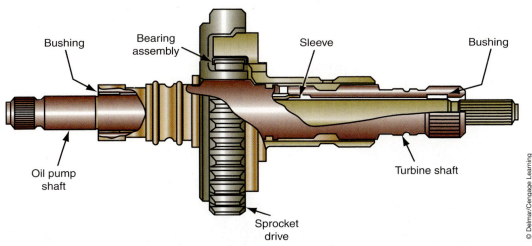

FIGURE 8-17 Some transmission shafts support another shaft through the bushings fitted to the inside diameter of one shaft. These bushings should be carefully inspected.

holes supply oil to bushings, one-way clutches, and planetary gears. If the ball does not fully block the end of the shaft, oil pressure can be lost, causing failure of these components. Some shafts may be used to support another shaft (Figure 8-17), as in the GM 4L30. The output shaft uses the rear of the input shaft to center and support itself. The small bushing found in the front end of the output shaft should always be replaced on this transmission during rebuild. If the input shaft pilot is worn or scored, a replacement shaft will be necessary.

All hubs, drums, and shells should be carefully examined for wear and damage. Especially look for nicked or scored band application surfaces on drums, worn or damaged lug grooves in clutch drums, worn splines, and burned or scored thrust surfaces. Minor scoring or burrs on band application surfaces can be removed by lightly polishing the surface with a 600-grit crocus cloth. Any part that is heavily scored or scratched should be replaced.

PARKING PAWL

> **CUSTOMER CARE:** Always remind your customers that they should not rely totally on the park gear selector position when parking the vehicle. The parking brake should be set in addition to fully placing the selector into park. This is especially of concern when the engine is running. Many accidents have happened because the transmission slipped out of park while the engine was running.

The parking pawl assembly (Figure 8-18) can be inspected after the transmission is disassembled or, on some transmissions, while the transmission is still in the vehicle. Examine the engagement lug on the pawl; make sure it is not rounded off. If the lug is worn, it may allow the pawl to slip out or not fully engage in the parking gear. Most parking pawls pivot on a pin; this also needs to be checked to make sure there is no excessive looseness at this point. The spring that pulls the pawl away from the parking gear must also be checked to make sure it can hold the pawl firmly in place. Also check the position and seating of the spring to make sure it will remain in that position during operation.

The pushrod or operating shaft (Figure 8-19) must provide the correct amount of travel to engage the pawl to the gear. Make sure the shaft is not bent or that the pivot hole in the internal shift linkage is not worn oblong. Also, make sure the bushing or sleeve that supports the manual shaft is in good condition.

Any components found unsuitable should be replaced. It should be noted that the components that make up the parking lock system are the only parts holding the vehicle in

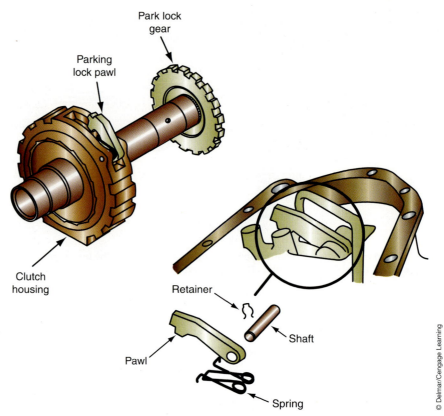

FIGURE 8-18 Each part of the parking pawl assembly should be carefully inspected.

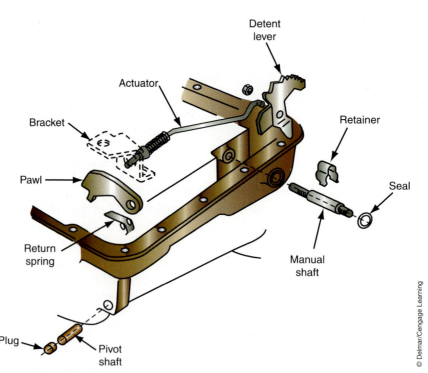

FIGURE 8-19 The push rod or operating shaft for the parking pawl must provide the correct amount of travel to engage the pawl to the gear. Make sure the shaft is not bent or the pivot hole in the internal shift linkage is not worn oblong.

place when parked. If they do not function correctly, the car may roll or even drop into reverse when the engine is running, causing an accident or injury. Replace any questionable or damaged parts.

 WARNING: A careful inspection of the parking pawl assembly is essential to avoid possible injury, death, and/or lawsuits.

Before installing the rear extension housing, assemble the parking pin, washer, spring, and pawl. Be sure they are assembled properly. Then install the housing and tighten the bolts to specifications.

DRIVE CHAINS

Classroom Manual
Chapter 8, pages 271

The drive chains used in some transaxles should be inspected for side play and stretch. These checks are made during disassembly and should be repeated as a double-check during reassembly. Chain deflection is measured between the centers of the two sprockets. Typically very little deflection is allowed.

Deflect the chain inward on one side until it is tight (Figure 8-20). Mark the housing at the point of maximum deflection. Then deflect the chain outward on the same side until it is tight (Figure 8-21). Again mark the housing in line with the outside edge of the chain at the point of maximum deflection. Measure the distance between the two marks. If this distance exceeds specifications, replace the drive chain.

Check each link of the chain by pushing and pulling the links away from the pin that holds them together. All of the links should move very little and each move the same amount. It is important to realize that a chain is only as strong as its weakest link. Check each link carefully.

Be sure to check for an identification mark on the chain during disassembly. These can be painted or dark-colored links, and may indicate either the top or the bottom of the chain, so be sure you remember which side was up.

The sprockets should be inspected for tooth wear and for wear at the point where they ride. If the chain is found to be too slack, it may have worn the sprockets in the same manner that engine timing gears wear when the timing chain stretches. A slightly polished appearance on the face of the gears is normal.

SPECIAL TOOLS
Paint stick
Machinist's rule
Bushing pullers
and drivers
Hydraulic press

Chain deflection is commonly referred to as chain slack.

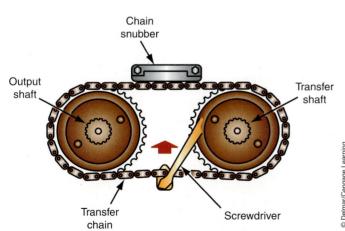

FIGURE 8-20 While measuring the slack of the drive chain, outwardly deflect the chain, and make a mark to that point of deflection.

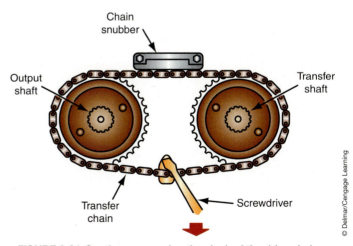

FIGURE 8-21 Continue measuring the slack of the drive chain by deflecting the chain inwardly. Make the point of deflection. The distance between the outward and inward marks is the amount of chain slack.

Bearings and Bushings

The bearings and bushings used on the sprockets need to be checked for damage. The radial needle thrust bearings must be checked for any deterioration of the needles and cage. The running surface in the sprocket must also be checked, as the needles may pound into the gear's surface during abusive operation. The bushings should be checked for any signs of scoring, flaking, or wear. Replace any defective parts.

Typically, the bearings and bushings are removed with a puller (Figure 8-22) and installed with a driver and a press (Figure 8-23).

The removal and installation of the chain drive assembly of some transaxles requires that the sprockets be spread slightly apart (Figure 8-24). The key to doing this is to spread the

FIGURE 8-22 The bearings and bushings in a chain drive assembly should be removed with a puller.

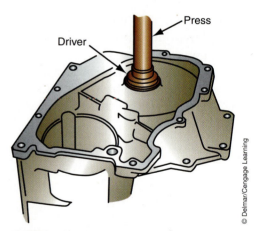

FIGURE 8-23 The bearings and bushings in a chain drive assembly should be installed with a driver and a hydraulic press.

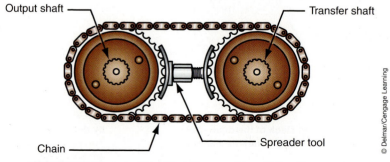

FIGURE 8-24 To remove and install some drive chains, they must be slightly spread apart with a special tool.

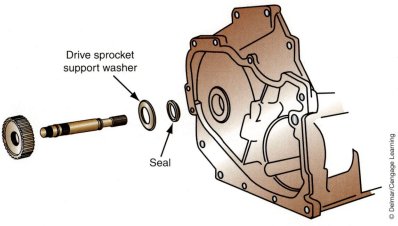

FIGURE 8-25 The shafts and gears of a chain drive assembly have numerous seals that must be replaced whenever the unit is disassembled.

sprockets just the right amount. If they are spread too far, they will not be easy to install or remove.

The shafts and gears of the assembly have numerous seals and thrust washers (Figure 8-25). The seals must be replaced whenever the unit is disassembled.

TRANSFER GEARS

Some transaxles use gears, instead of a drive chain, to move or transfer the output of the transmission to the final drive unit (Figure 8-26). The shafts and gears in these transaxles must be carefully inspected and replaced if they are damaged.

To remove and install the transfer shaft gear, a holding tool must be used to stop the transfer gear from turning while loosening or tightening the retaining nut (Figure 8-27). The nut is typically tightened to 200 ft.-lbs. To remove the transfer shaft from the transaxle case, the retaining snap ring must be removed (Figure 8-28). Then the shaft can be pulled out with its bearing (Figure 8-29). The shaft's bearing must be pressed on and off the shaft.

Classroom Manual
Chapter 8, pages 270

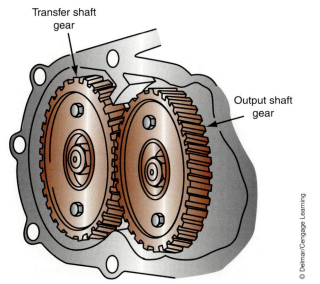

FIGURE 8-26 The transfer shafts and gears must be carefully inspected and replaced if they are damaged.

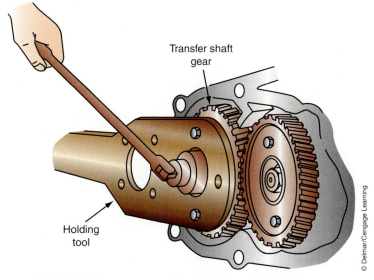

FIGURE 8-27 To remove and install the transfer shaft gear, a holding tool must be used to stop the transfer gear from turning while loosening or tightening the retaining nut.

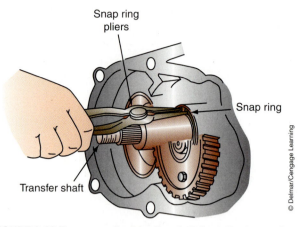

FIGURE 8-28 To remove the transfer shaft from the transaxle case, the retaining snap ring must be first removed.

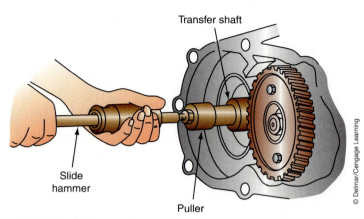

FIGURE 8-29 The transfer shaft can be pulled out with its bearing with a suitable puller tool.

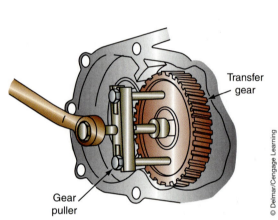

FIGURE 8-30 A puller must be used to remove the transfer gear from the output shaft, after its retaining bolt has been removed.

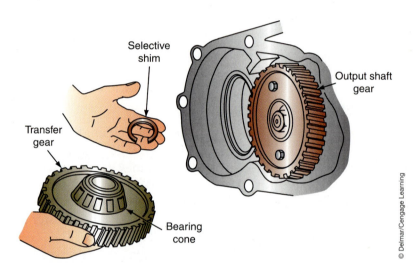

FIGURE 8-31 Behind the transfer shaft gear is a selective shim used to provide correct meshing of the teeth of the transfer shaft gear and the output shaft transfer gear and to control transfer shaft endplay.

SPECIAL TOOLS

Bushing pullers and drivers

Hydraulic press

Dial indicator

A puller must be used to remove the transfer gear from the output shaft (Figure 8-30) after its retaining bolt has been removed.

Behind the transfer shaft gear is a selective shim used to provide correct meshing of the teeth of the transfer shaft gear and the output shaft transfer gear (Figure 8-31) and to control transfer shaft endplay. Endplay is measured with a dial indicator. To ensure good contact with the indicator's plunger and the end of the transfer shaft, Chrysler recommends that a steel check ball be placed between the plunger tip and the shaft (Figure 8-32). To help hold the ball in place during the check, coat the ball in heavy grease. If the endplay is not within specifications, select a thrust washer with the correct thickness and install it behind the transfer gear.

The output shaft should also be checked for endplay. If the endplay is incorrect, a different size shim should be installed behind the output shaft transfer gear (Figure 8-33). Since the gear is splined to the shaft, a thrust bearing is not needed here. After endplay has been corrected and after the output shaft gear has been reinstalled, the turning torque of the shaft should be measured. If the turning torque is too high, a slightly thicker shim should be

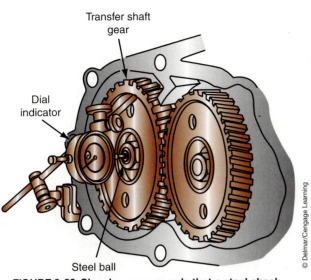

FIGURE 8-32 Chrysler recommends that a steel check ball coated in heavy grease be placed between the plunger tip and the shaft to ensure good contact with the indicator's plunger and the end of the transfer shaft.

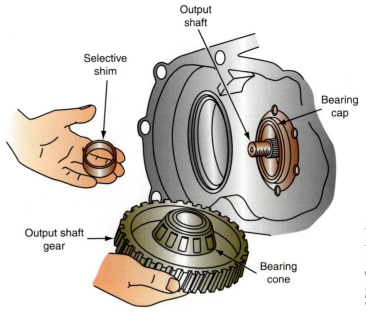

FIGURE 8-33 A selective shim is installed behind the output shaft transfer gear to control output shaft endplay.

installed. If the turning torque is too low, a slightly thinner shim should be installed. The bearings for the transfer and output shafts are pressed on and off the shafts.

It is very important that the transfer gears be tightened to specifications. The torque setting not only ensures that the gears will remain on the shafts, but it also maintains the correct bearing adjustments.

FINAL DRIVE UNITS

Transaxle final drive units should be carefully inspected. Examine each gear, thrust washer, and shaft for signs of damage. If the gears are chipped or broken, they should be replaced. Also inspect the gears for signs of overheating or scoring on the bearing surface of the gears.

Carefully check the differential pinion shaft and gear for wear. Also check to make sure the gear is not seized to the shaft. Seized pinion gears are a common cause of shaft failure and can result in the total destruction of the final drive unit and transaxle case.

Helical Gear Drives

Helical type final drive units should be checked for worn or chipped teeth, overloaded tapered roller bearings, and excessive differential side gear wear. Excessive play in the differential is a cause of engagement clunk (Figure 8-34). Be sure to measure the clearance between the side gears and the differential case and to check the fit of the gears on the gear shaft. Proper clearances can be found in the appropriate shop information.

Carefully check the condition of the bearings. Replace them if there are signs of wear or damage. Normally the bearings are pressed on and off the case using special tools (Figure 8-35). The side bearings of some final drive units are preloaded with shims (Figure 8-36). Select the correct size shim to bring the unit into specifications. With a torque wrench, measure the amount of rotating torque. Compare your readings against specifications (Figure 8-37).

If the bearing preload and end play is fine, as is the condition of the bearings, the parts can be reused. However, always install new seals during assembly. New bearings require that preload be set to the specifications for a new bearing. Used bearings should be set to the amount found during teardown or about one-half the preload of a new bearing.

Classroom Manual
Chapter 8, pages 269

SPECIAL TOOLS
Inch-pound
torque wrench

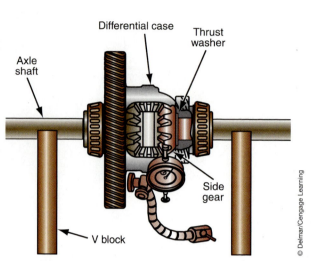

FIGURE 8-34 The backlash of the side gears in a final drive unit should be checked prior to assembling the transaxle.

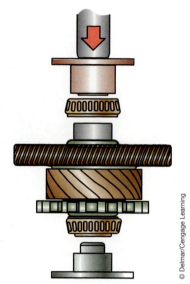

FIGURE 8-35 The side bearings are normally pressed on and off the unit.

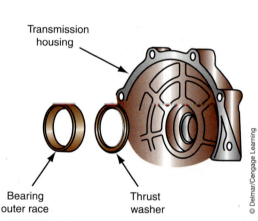

FIGURE 8-36 The side bearings of some final drive units are preloaded with selective shims.

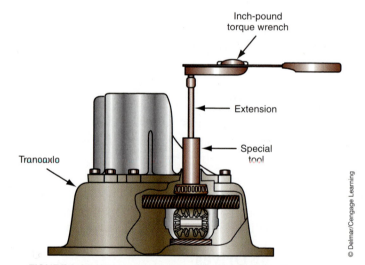

FIGURE 8-37 Using an inch-pound torque wrench, the turning torque of the transaxle assembly should be checked after assembly.

Planetary Gear Assemblies

Planetary type final drives (Figure 8-38) are also checked for the same differential case problems that the helical type would encounter. All planetary gear teeth should be inspected for chips and damage. Pay particular attention to the planetary carriers. Look obvious problems like blackened gears or pinion shafts. These conditions indicate severe overloading and require that the carrier be replaced. When looking at the gears, a bluish condition can be a normal condition as this could be from the heat-treating process used during manufacture. Check the planetary pinion gears for loose bearings. Check each gear by rolling it on its shaft to feel for roughness or binding of the needle bearings. Wiggle the gear to be sure it is not loose on the shaft, which can cause the gear to whine when it is loaded. Also, inspect the gears' teeth for chips or imperfections, as these will also cause whine.

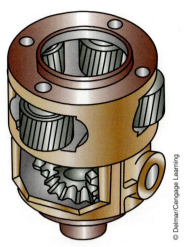

FIGURE 8-38 Planetary type final drives should be checked in the same basic way as helical type final drives.

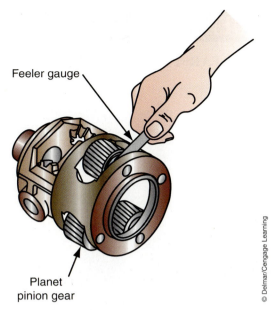

Feeler gauge

Planet pinion gear

© Delmar/Cengage Learning

FIGURE 8-39 The endplay of the planetary pinion gears should be checked with a feeler gauge.

Check the gear teeth around the inside of the planetary ring gear. Check the carrier assembly for cracks or other defects. Check the endplay of the pinion gears by placing a feeler gauge between the planetary carrier and each pinion gear (Figure 8-39). Compare the endplay to specifications. If the endplay is too low, the pinion assembly must be removed and the correct thrust washer installed. If the endplay is too much, the differential assembly must be replaced.

Any problems normally result in the replacement of the carrier as a unit as most pinion bearings and shafts are not sold as separate parts. Photo Sequence 14 covers a typical procedure for servicing a planetary gear-type final drive unit.

Adjusting Side Gear Endplay

Most of the procedures for servicing FWD final drive units are the same as for RWD units. Because FWD units typically use the end of the output shaft as the drive pinion gear, all normal pinion shaft adjustments are not necessary. However, ring gear and side bearing adjustments are still necessary. These are typically made with the differential case assembled and out of the transaxle case. Always refer to your service information before proceeding to make these adjustments on a transaxle. What follows is a typical procedure for adjusting side gear endplay:

1. Install the correct adapter into the differential bearings (Figure 8-40).
2. Mount the dial indicator to the ring gear with the plunger resting against the adapter.
3. With your fingers or a screwdriver, move the ring gear up and down (Figure 8-41).
4. Record the measured endplay.
5. Measure the old thrust washer with the micrometer.
6. Install the correct size of thrust washer (Figure 8-42).
7. Repeat the procedure for the other side.

Adjusting Bearing Preload

The following procedure is typical for the measurement and adjustment of the differential bearing preload in a transaxle. Always refer to the applicable service manual before proceeding to make these adjustments on a particular transaxle.

SPECIAL TOOLS

Adapter kit for transaxle
Dial indicator
0- to 1-inch micrometer

SPECIAL TOOLS

Set of gauging shims
Dial indicator
Fresh lubricant
Inch-pound torque wrench

SERVICING PLANETARY GEAR-TYPE FINAL DRIVE UNITS

All photos in this sequence are © Delmar/Cengage Learning.

P14-1 Place the final drive unit into the transaxle's oil pan. Doing this will lessen the chances of losing bearings while disassembling the unit. Make sure the pan is clean.

P14-2 With a pin punch and hammer, remove the differential pinion shaft retaining ring.

P14-3 Remove the retaining ring from the end of the output shaft. This ring must not be reused!

P14-4 Pull the output shaft from the differential carrier.

P14-5 Remove the final drive sun gear.

P14-6 Remove the differential pinion shaft retaining pin, the differential pinion shaft, pinion gears, and thrust washers and place them aside. Keep these orientation and in the order and orientation in which they were assembled.

P14-7 Remove the differential side gears and thrust washers.

P14-8 Using a screwdriver, carefully remove the final drive carrier retaining ring.

P14-9 Remove the planet pinion pins, then the planet pinion gears, thrust washers, needle bearings, and the bearing spacers.

P14-10 Inspect the needle bearings, thrust washers, pinion gears, and planet pinion pins.

P14-11 Remove the final drive sun gear to carrier thrust bearing.

P14-12 To begin reassembly, install the pinion gear needle bearing spacer, thrust washer, and pinion gear onto the planet pinion gear pin.

P14-13 Install the needle bearings, one at a time, into the top and bottom of the planet pinion gear. Use transjel to help hold the bearings in position.

P14-14 Coat the sun gear to final drive carrier thrust washer with transjel. Then install the thrust bearing onto the final drive carrier.

P14-15 Separate the pinion pin from the gear. Assemble the planet pinion gear thrust washers and gears into the final drive carrier. Make sure the gears face in the same direction as they faced when they were removed. Also be careful not to let the bearings fall out.

P14-16 Install the planet pinion gear pins into the carrier. Then, install the final drive carrier retaining ring.

P14-17 With a feeler gauge, check the endplay of the pinion gears. If the endplay is too low, the planet pinion assembly must be removed and the correct thickness thrust washer installed. If the endplay is too much, the differential assembly must be replaced.

P14-18 Install the differential side gears and thrust washers into the carrier.

SERVICING PLANETARY GEAR-TYPE FINAL DRIVE UNITS

P14-19 Apply transjel onto the thrust washers for the pinion gears, and then install them into the carrier.

P14-20 Align the bores of the pinion gears with the bore in the carrier for the pinion pin, and then install the pin. Make sure the hole in the pivot pin, for its retaining pin, is aligned with the retaining pin bore.

P14-21 Install the pinion pin's retaining pin.

P14-22 Insert the sun gear into the carrier, making sure the gear is placed in the correct direction.

P14-23 Install the output shaft into the final drive assembly.

P14-24 Install a new retaining snap ring onto the end of the output shaft.

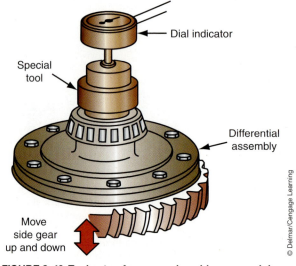

Dial indicator

Special tool

Differential assembly

Move side gear up and down

© Delmar/Cengage Learning

FIGURE 8-40 Tool setup for measuring side gear endplay.

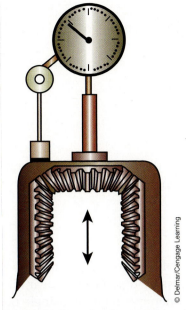

FIGURE 8-41 Move the ring gear up and down and observe the dial indicator.

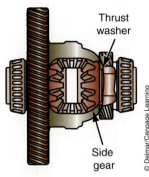

Thrust washer

Side gear

FIGURE 8-42 Location of thrust washers.

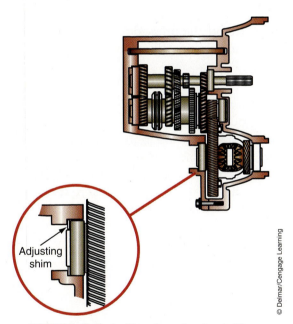

Adjusting shim

FIGURE 8-43 Typical location of preload shim.

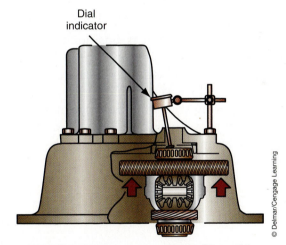

Dial indicator

FIGURE 8-44 Setup for measuring differential bearing preload.

1. Remove the bearing cup and existing shim from the differential bearing retainer.
2. Select a gauging shim that will allow for 0.001- to 0.010-inch endplay.
3. Install the gauging shim into the differential bearing retainer (Figure 8-43).
4. Press in the bearing cup.
5. Lubricate the bearings and install them into the case.
6. Install the bearing retainer.
7. Tighten the retaining bolts.
8. Mount the dial indicator with its plunger touching the differential case (Figure 8-44).
9. Apply medium pressure in a downward direction while rolling the differential assembly back and forth several times.

10. Zero the dial indicator.
11. Apply medium pressure in an upward direction while rotating the differential assembly back and forth several times.
12. The required shim-to-set preload is the thickness of the gauging shim plus the recorded endplay.
13. Remove the bearing retainer, cup, and gauging shim.
14. Install the required shim.
15. Press the bearing cup into the bearing retainer.
16. Install the bearing retainer and tighten the bolts.
17. Check the rotating torque of the transaxle. If this is less than specifications, install a thicker shim. If the torque is too great, install a slightly thinner shim.
18. Repeat the procedure until the desired torque is reached.

PART REPLACEMENT

Planetary-type final drives, like helical final drives, are available in more than one possible ratio for a given type of transaxle, so care should be taken to assure that the same gear ratios are used during assembly. This is not normally a problem when overhauling a single unit; however, in a shop where many transmissions are being repaired, it is possible to mix up parts causing problems during the rebuild.

Side Bearings

The side bearings of most final drive units must be pulled off and pressed onto the differential case. Always be sure to use the correct tools for removing and installing the bearings.

Final Drive Bearing Replacement

The final drive unit is positioned either in the transaxle case or in a separate housing mounted to the transaxle case. Regardless of its location, it is supported by bearings. Normally, tapered roller bearings are used. These bearings are removed with pullers (Figure 8-45) and installed with a press and driver. When replacing the bearings, make sure the bearing seats are free from nicks and burrs. These defects will not allow the bearing to seat properly and will give false endplay measurements. If the bearings were replaced, endplay and gear clearances must be checked and corrected before installing the final drive unit into the housing.

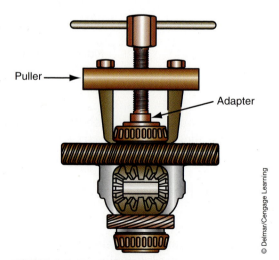

Puller

Adapter

© Delmar/Cengage Learning

FIGURE 8-45 Typical setup for removing side bearings from a differential case.

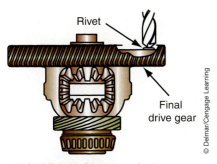

FIGURE 8-46 Ring gear rivets must be drilled and driven out to separate the ring gear from the differential case.

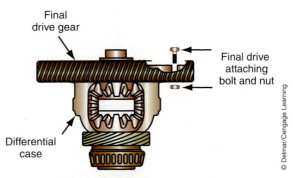

FIGURE 8-47 The ring gear should be fastened to the differential case with special nuts and bolts.

Ring Gear Replacement

The ring gear of many transaxles is riveted to the differential case. The rivets must be drilled then driven out with a hammer and drift to separate the ring gear from the case (Figure 8-46). To install a new ring gear to the case, new special nuts and bolts are used (Figure 8-47). These nuts and bolts must be of the specified hardness and should be tightened in steps and to the specified torque.

Speedometer Gear Replacement

The final drives of transaxle differential cases may be fitted with a speedometer gear pressed onto the differential case and under one side bearing. These gears are pulled off and pressed onto the case.

HOUSING AND CASE SERVICE

Critical to all functions of a transmission is the condition of the housing. The housing may actually comprise several units, sealed together by gaskets, sealant, or seals (Figure 8-48). Each case is part of the transmission housing should be carefully inspected.

While removing the parts from inside the case, make sure you note the location of the check balls that may be in the hydraulic network. Also, many housings have a small filtering screen placed into a hydraulic pathway. These are typically easily removed and should be removed before cleaning the case. Low air pressure (approximately 30 psi) applied to the back of the screen will typically remove them. Keep in mind that some cleaning solvents will destroy the plastic screens. Once removed, the screens should be inspected for debris and cleaned with hot water and soap.

Signs of leakage through parts of the housing can be caused by the porosity of the metal or by a crack. If the case is cracked, it should be replaced. If the problem is porosity, the problem may be corrected by applying an epoxy sealer over the spot of leakage after the case has been thoroughly cleaned.

After all parts contained in the housing are removed, the housing should be thoroughly cleaned and all passages (Figure 8-49) blown out with compressed air. These passages can be checked for restrictions by applying compressed air to each one. If air flows out the other end, there is no restriction. Compressed air can also be used to check for internal leaks. Plug off one end of the passage and apply air to the other. If pressure builds in that passage, there are probably no leaks in it.

Check the passages in the case for cross-tracking from one circuit to another. Fill the circuit with solvent and watch to see if the solvent disappears or leaks away. If the level of the solvent changes, you should check each part of the circuit to find where the leak is. If there is evidence of cross-tracking, the case should be replaced.

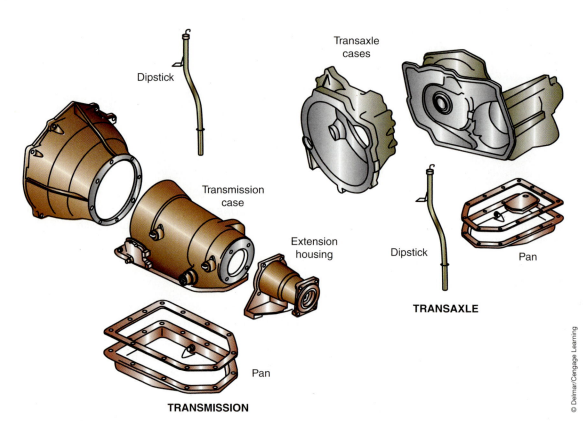

FIGURE 8-48 The transmission/transaxle housing may be comprised of several different cases, each sealed to one another.

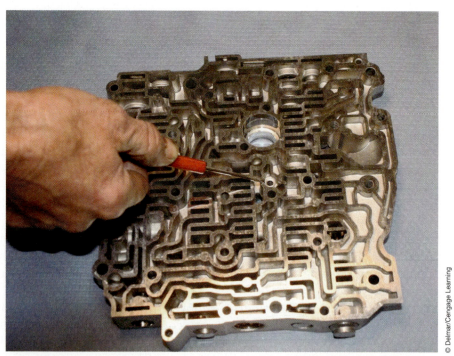

FIGURE 8-49 The oil passages in the housing should be cleaned and checked for blockage and cross-tracking. Make sure to remove all check balls before cleaning.

Vents are located in the oil pump body or transmission case to equalize pressure in the transmission. These vents can be checked by blowing low-pressure air through them, squirting solvent or brake-cleaning spray through them, or pushing a small diameter wire through the vent passage.

All sealing surfaces must be flat, smooth, and without damage that would lead to an oil leak. To keep all of the rotating parts in alignment, the surfaces must also be true. The flatness of the surfaces can be checked by laying a straightedge across them in several places.

Check the valve body mounting area for warpage with a straightedge and feeler gauge. This should be done in several locations. If there is a slight burr or high spot, it can be removed by flat-filing the surface. A straightedge should be laid across the lower flange of the case to check for distortion. Any warpage found here may result in circuit leakage, causing any number of hydraulically related problems.

All bores where a seal is installed should be inspected for surface roughness, nicks, or scratches. Imperfections in steel or cast iron parts can usually be polished out with crocus cloth.

All other internal and external bores must also be inspected. Check the fit of the servo pistons in their bore without their seals to be sure they have free travel. There should be no tight spots or binding over the whole range of travel. Any deep scratches or gouges that cause binding of the piston will require case replacement. Accumulator bores are checked the same as servo bores.

Case-mounted hydraulic clutch bores are prone to the same problems as servo bores. Look for any scratches or gouges in the sealing area that would affect the rubber seals. It is possible to damage these areas during disassembly, so be careful with tools used during overhaul.

Check all bell-housing bolt holes and dowels. Cracks around the bolt holes indicate that the case bolts were tightened with the case out of alignment with the engine block.

All internal threads should be carefully inspected. If the threads are distorted or stripped, they should be replaced with a threaded insert, or the case should be replaced. Sometimes a slightly damaged thread can be restored by running a thread chaser through the bore (Figure 8-50). Always make sure to check the pitch and size of the threads (Figure 8-51) and use a matching chaser or tap. After the threads have been repaired, make sure you thoroughly clean the case.

All nuts, bolts, and studs should also be inspected for damage and replaced if necessary. Check all of the bolts that were removed during disassembly for aluminum on the

FIGURE 8-50 A thread chaser should be used to clean out internal threads.

© Delmar/Cengage Learning

FIGURE 8-51 A thread pitch gauge ought to be used to determine the size and pitch of a bolt or screw before the threads are tapped or inserted.

threads. If aluminum is present, you can assume the thread bore is damaged and should be repaired.

Bushings in a transmission case are normally found in the rear of the case and require the same inspection and replacement techniques as other bushings in the transmission. Always be sure that the oil passage to a pressure-fed bushing or bearing is open and free of dirt and foreign material.

CASE STUDY

A customer had his 1998 Ford Taurus towed into the shop. He stated that nothing happened when the transmission was in any gear. The technician looked things over and attempted to verify the complaint. Sure enough, the car wouldn't move in forward or in reverse. He did notice a slight change in noise when he moved the lever into drive and reverse, so he knew something was happening.

He proceeded to do the normal checks with the scan tool and found nothing unusual except for the codes that said there was no signal from the output sensor. That didn't alarm him because he knew there wouldn't be a signal since the car wouldn't move. He then checked the fluid. The level was correct and smelled fine.

He then raised the car and checked the front axle shafts to see if they were stuck. He could turn the wheels easily; in fact, he noticed how easy it was to rotate them. But he wasn't sure what to make of that. He then ran a pressure test and found the pressure to be normal at idle speed and observed pressure changes as he selected drive and reverse. From this he knew it was not a hydraulic problem.

Since the problem didn't appear to be electronic or hydraulic, he knew it had to be mechanical. Thinking about what he had observed, he came to the conclusion that the problem must be in the input or the output of the transaxle. He strongly suspected the output because of the change of noise he noticed. Giving it more thought, he was convinced the problem was the output because of the little effort it took to rotate the wheels.

He proceeded to pull the transaxle and began his disassembly and inspection. He carefully checked the drive chain and sprocket assembly and found no evidence of a problem. He then pulled the output shaft from the transaxle. As soon as he had it out, he saw the problem; the splines were stripped and rounded off.

He proceeded to overhaul the transmission and clean every part thoroughly. He replaced the output shaft and transfer gear, then put the transaxle back together.

After the transaxle was in, he put it in drive and it moved. During the road test he found no problems. He told the customer it was now working fine and the customer was pleased.

TERMS TO KNOW

Etching

Selective thrust washers

Thrust load

ASE-STYLE REVIEW QUESTIONS

1. While inspecting a planetary gearset:

   Technician A says blackened pinion shafts indicate severe overloading.

   Technician B says bluish gears indicate overheating.

   Who is correct?
   A. A only   C. Both A and B
   B. B only   D. Neither A nor B

2. *Technician A* says loose planetary pinion gear bearings will cause the gear to whine when it is loaded.

   Technician B says damaged teeth on a planetary pinion gear will cause the gear to whine.

   Who is correct?
   A. A only   C. Both A and B
   B. B only   D. Neither A nor B

3. *Technician A* removes bushings by carefully cutting one side of the bushing and collapsing it. Once collapsed, the bushing can be easily removed with a pair of pliers.

   Technician B removes small-bore bushings by tapping the inside bore of the bushing with threads that match a selected bolt, which fits into the bushing. After tapping the bushing, the bolt is inserted and a slide hammer is used to pull the bolt and bushing out of its bore.

   Who is correct?
   A. A only   C. Both A and B
   B. B only   D. Neither A nor B

4. *Technician A* says bushings can be heated with a torch to remove them easily.

   Technician B says bushings can be removed with a slide hammer and the correct attachment.

   Who is correct?
   A. A only   C. Both A and B
   B. B only   D. Neither A nor B

5. While checking a planetary gearset:

   Technician A says the end clearance of the pinion gears should be checked with a feeler gauge.

   Technician B says the end clearance of the long pinions in a Ravigneaux gearset should be checked at both ends.

   Who is correct?
   A. A only   C. Both A and B
   B. B only   D. Neither A nor B

6. *Technician A* says most FWD final drive units require that ring and pinion backlash and pinion depth be set to specifications.

   Technician B says ring gear and side gear bearing adjustments are required on all FWD final drive units.

   Who is correct?
   A. A only   C. Both A and B
   B. B only   D. Neither A nor B

7. *Technician A* says a blocked oil delivery passage will cause a shaft to score.

   Technician B says that if a shaft is fitted with a check ball and the check ball does not seat properly, low oil pressure will result.

   Who is correct?
   A. A only   C. Both A and B
   B. B only   D. Neither A nor B

8. *Technician A* says thrust washers should be inspected for scoring, flaking, and wear through to the base material.

   Technician B says plastic thrust washers will not wear unless they are damaged.

   Who is correct?
   A. A only   C. Both A and B
   B. B only   D. Neither A nor B

9. While inspecting a parking pawl assembly:

   Technician A says the pivot pin must fit loosely in its bracket and tightly in the pawl.

   Technician B says the engagement lug on the pawl must be square in order to fully engage into the parking gear.

   Who is correct?
   A. A only   C. Both A and B
   B. B only   D. Neither A nor B

10. While checking the endplay of a transfer gear assembly:

    Technician A says that if the endplay of the output transfer shaft is incorrect, a different size thrust washer should be installed behind the output shaft transfer gear.

    Technician B says that after the endplay of the output shaft has been corrected, the turning torque of the shaft should be measured. If the turning torque is too high, a slightly thinner shim should be installed. If the turning torque is too low, a slightly thicker shim should be installed.

    Who is correct?
    A. A only   C. Both A and B
    B. B only   D. Neither A nor B

ASE CHALLENGE QUESTIONS

1. While servicing a final drive unit:

   Technician A checks gear and bearing endplay whenever new bearings are installed in the unit.

   Technician B reuses the bearings, seals, and thrust washers if the bearing preload and endplay are fine, as is the condition of the bearings.

   Who is correct?

   A. A only   C. Both A and B

   B. B only   D. Neither A nor B

2. *Technician A* says a drive chain that is too loose should be shortened by removing a pair of links in the chain.

   Technician B says the drive sprockets should be replaced if the gear teeth are polished or show any other signs of wear.

   Who is correct?

   A. A only   C. Both A and B

   B. B only   D. Neither A nor B

3. *Technician A* checks bushing wear by observing the lateral movement of the shaft that fits into the bushing. Any noticeable lateral movement indicates wear and therefore the bushing should be replaced.

   Technician B checks for bushing wear by measuring the inside diameter of the bushing and the outside diameter of the shaft with a dial caliper or micrometer.

   Who is correct?

   A. A only   C. Both A and B

   B. B only   D. Neither A nor B

4. *Technician A* drills out the ring gear retaining rivets to remove the ring gear from the carrier.

   Technician B installs the ring gear nuts and bolts of the specified hardness.

   Who is correct?

   A. A only   C. Both A and B

   B. B only   D. Neither A nor B

5. *Technician A* says all hubs, drums, and shells should be carefully examined for wear and damage.

   Technician B says minor scoring or burrs on band application surfaces of a drum can be removed by lightly polishing the surface with a 200-grit crocus cloth.

   Who is correct?

   A. A only   C. Both A and B

   B. B only   D. Neither A nor B

Name _____ Date _____

CHECKING THRUST WASHERS, BUSHINGS, AND BEARINGS

Upon completion of this job sheet, you should be able to inspect, measure, and replace thrust washers and bearings and inspect the bushings in a transmission/transaxle.

ASE Correlation

This job sheet is related to the Automatic Transmission and Transaxles Test's Content Area *Off-Vehicle Transmission and Transaxle Repair, Gear Train, Shafts, Bushings, and Case.*

Tasks: Inspect, measure, and replace thrust washers and bearings and inspect bushings; replace as needed.

Tools and Materials

Wire-type feeler gauge set
Bushing driver set
Bushing puller tool

Describe the vehicle being worked on:

Year _____ Make _____ Model _____

VIN _____ Engine type and size _____

Model and type of transmission _____

Procedure

_____

Task Completed

1. The best time to inspect thrust washers, bearings, and bushings is during disassembly. The bushings should be inspected for pitting and scoring. Describe their condition.

   _____

   _____

2. Check the depth that bushings are installed to and the direction of the oil groove, if so equipped, before you remove them. Many bushings that are used in the planetary gearing and output shaft areas have oiling holes in them. Be sure to line these up correctly during installation or you may block off oil delivery and destroy the geartrain. Describe their condition.

   _____

   _____

3. Observe the lateral movement of the shaft that fits into the bushing. Any noticeable lateral movement indicates wear and the bushing should be replaced. Describe your findings.

   _____

   _____

4. The amount of clearance between the shaft and the bushing can be checked with a wire-type feeler gauge. Insert the wire between the shaft and the bushing; if the gap is greater than the maximum allowable, the bushing should be replaced. What are the specifications for this gap and how do they compare to your measurement?

_____

_____

5. Measure the inside diameter of the bushing and the outside diameter of the shaft with a Vernier-type caliper or micrometer. Compare the two and state your conclusions.

_____

_____

6. Most bushings are press-fit into a bore. Remove them by driving them out of the bore with a properly sized bushing tool. Some bushings can be removed with a slide hammer fitted with an expanding or threaded fixture that grips the inside of the bushing. Another way to remove bushings is to carefully cut one side of the bushing and collapse it. Once collapsed, the bushing can be easily removed with a pair of pliers. What did you use to remove the bushing?

_____

_____

☐ 7. Small-bore bushings located in areas where it is difficult to use a bushing tool can be removed by tapping the inside bore of the bushing with threads that match a bolt that fits into the bushing. After the bushing has been tapped, insert the bolt and use a slide hammer to pull the bolt and bushing out of its bore.

☐ 8. All new bushings should be pre-lubed during transmission assembly and installed with the proper bushing driver. Make sure they are not damaged and are fully seated in their bores.

9. The purpose of a thrust washer is to support a thrust load and keep parts from rubbing together. Selective thrust washers come in various thicknesses to take up clearances and adjust shaft endplay. Flat thrust washers and bearings should be inspected for scoring, flaking, and wear through to the base material. Describe their condition.

_____

_____

10. Flat thrust washers should also be checked for broken or weak tabs. These tabs are critical for holding the washer in place. On metal flat thrust washers, the tabs may appear cracked at the bend of the tab; however, this is a normal appearance. Describe their condition.

_____

_____

11. Only damaged plastic thrust washers will show wear. The only way to check their wear is to measure the thickness and compare it to a new part. Describe their condition.

_____

_____

12. Proper thrust washer thicknesses are important to the operation of an automatic transmission. Always follow the recommended procedure for selecting the proper thrust plate. Use transjel or similar lubricant to hold thrust washers in place during assembly.

□

13. All bearings should be checked for roughness before and after cleaning.

□

14. Carefully examine the inner and outer races, and the rollers, needles, or balls for cracks, pitting, etching, or signs of overheating. Describe their condition.

_____

_____

15. Give a summary of your inspection.

_____

_____

_____

Instructor's Response _____

_____

_____

⚠️

CAUTION:
Never use white lube or chassis lube. These greases will not mix with the fluid and can plug up orifices and passages and hold check balls off their seats.

Name _____   **Date** _____

SERVICING OIL DELIVERY SEALS

Upon completion of this job sheet, you should be able to inspect oil delivery seal rings, ring grooves, and sealing surface areas.

ASE Correlation

This job sheet is related to the Automatic Transmission and Transaxles Test's Content Area *Off-Vehicle Transmission and Transaxle Repair, Gear Train, Shafts, Bushings, and Case.*
Tasks: Inspect oil delivery seal rings, ring grooves, and sealing surface areas.

Tools and Materials

Transjel	Crocus cloth
Seal driver tools	Feeler gauge set

Procedure Guidelines

- Three types of seals are used in automatic transmissions: O-ring and square-cut (lathe-cut), lip, and sealing rings. These seals are designed to stop fluid from leaking out of the transmission and to stop fluid from moving into another circuit of the hydraulic circuit.

- O-ring and square-cut seals are used to seal nonrotating parts. When installing a new O-ring or square-cut seal, coat the entire surface of the seal with assembly lube or petroleum jelly. Make sure you don't stretch or distort the seal while you are working it into its holding groove. After a square-cut seal is installed, double-check it to make sure it is not twisted. The flat surface of the seal should be parallel with the bore. If it is not, fluid will easily leak past the seal.

- Lip seals that are used to seal a shaft typically have a metal flange around their outside diameter. The shaft rides on the lip seal at the inside diameter of the seal assembly. The rigid outer diameter provides a mounting point for the lip seal and is pressed into a bore. Once pressed into the bore, the outer diameter of the seal prevents fluid from leaking into the bore, while the inner lip seal prevents leakage past the shaft.

- Piston lip seals are set into a machined groove on the piston. This type of lip seal is not housed in a rigid metal flange. They are designed to be flexible and provide a seal while the piston moves up and down. While the piston moves, the lip also flexes up and down. The most important thing to keep in mind while installing a lip seal is to make sure the lip is facing the correct direction. The lip should always be aimed toward the source of pressurized fluid. If installed backward, fluid under pressure will easily leak past the seal. Also remember to make sure the surfaces to be sealed are clean and not damaged.

- Teflon or metal sealing rings are commonly used to seal servo pistons, oil pump covers, and shafts. These rings may be designed to provide for a seal, but they may also be designed to allow a controlled amount of fluid leakage. Sealing rings are either solid rings or cut. Cut sealing rings are of one of three designs: open-end, butt-end, or locking-end.

- Solid sealing rings are made of a Teflon-based material and are never reused. To remove them, carefully cut the seal after it has been pried out of its groove.

- Installing a new solid sealing ring requires special tools. These tools allow you to stretch the seal while pushing it into position. Never attempt to install a solid seal without the proper tools. Because these seals are soft, they are easily distorted and damaged.
- Open-end sealing rings fit loosely into a machined groove. The ends of the rings do not touch when they are installed. This type of ring is typically removed and installed with a pair of snap-ring pliers. The ring should be expanded just enough to move it off or onto the shaft.
- Butt-end sealing rings are designed so that their ends butt up or touch each other once the seal is in place. This type of seal can be removed with a small screwdriver. The blade of the screwdriver is used to work the ring out of its groove. To install this type of ring, use a pair of snap-ring pliers and expand the ring to move it into position.
- Locking-end rings may have either hooked ends that connect or ends that are cut at an angle to hold the ends together. These seals are removed and installed in the same way as butt-end rings. After these rings are installed, make sure the ends are properly positioned and touching.
- All seals should be checked in their own bores prior to installation. They should be slightly smaller or larger (±3%) than their groove or bore. If a seal is not the proper size, find one that is. Do not assume that because a particular seal came with the overhaul kit it is the correct one.
- Never install a seal when it is dry. The seal should slide into position and allow the part it seals to slide into it. A dry seal is easily damaged during installation.
- Install only genuine seals recommended by the manufacturer of the transmission.

Describe the vehicle being worked on:

Year _____ Make _____ Model _____

VIN _____ Engine type and size _____

Transmission type and model _____

Task Completed	Procedure

Procedure

☐   1. Before installing seals, clean the shaft and/or bore area.

     2. Carefully inspect these areas for damage. File or stone away any burrs or bad nicks and polish the surfaces with a fine crocus cloth, then clean the area to remove the metal particles. Describe your findings.

_____

_____

☐   3. Lubricate the seal, especially any lip seals, to ease installation.

     4. All metal sealing rings should also be checked for proper fit. Since these rings seal on their outer diameter, the seal should be inserted in its bore and should feel tight there. If the seal has some form of locking ends, these should be interlocked prior to trying the seal in its bore. Describe your findings.

_____

_____

     5. Check the fit of the sealing rings in their shaft groove. Describe your findings.

_____

_____

6. Check the side clearance of the ring by placing the ring into its groove and measuring the clearance between the ring and the groove with a feeler gauge. Describe your findings.

_____

_____

7. While checking the clearance, look for nicks in the grooves and for evidence of groove taper or stepping. Describe your findings.

_____

_____

8. Use the correct driver when installing a seal and be careful not to damage the seal during installation.   □

Instructor's Response _____

_____

_____

Name _____ **Date** _____

Servicing Planetary Gear Assemblies

Upon completion of this job sheet, you should be able to inspect and measure planetary gear assemblies.

ASE Correlation

This job sheet is related to the Automatic Transmission and Transaxles Test's Content Area *Off-Vehicle Transmission and Transaxle Repair, Gear Train, Shafts, Bushings, and Case.*

Tasks: Inspect and measure planetary gear assembly (includes sun, ring gear, thrust washers, planetary gears, and carrier assembly); replace as needed.

Tools and Materials

Snap-ring pliers

Feeler gauge set

Describe the vehicle being worked on:

Year _____ Make _____ Model _____

VIN _____ Engine type and size _____

Transmission type and model _____

Procedure

_____

Task Completed

1. The planetary gears used in automatic transmissions are the helical type and all gear teeth should be inspected for chips or stripped teeth. Describe their general condition before disassembling the gearset.

    _____

    _____

2. Any gear that is mounted to a splined shaft needs the splines checked for mutilation or shifted splines. Record your findings.

    _____

    _____

3. Note any discoloration of the parts and explain the cause for it.

    _____

    _____

4. Check the planetary pinion gears for loose bearings. Record your findings.

    _____

    _____

5. Check each gear individually by rolling it on its shaft to feel for roughness or binding of the needle bearings. Wiggle the gear to be sure it is not loose on the shaft. Looseness will cause the gear to whine when it is loaded. Record your findings.

_____

_____

6. Inspect the gears' teeth for chips or imperfections, as these will also cause whine. Record your findings.

_____

_____

7. Check the gear teeth around the inside of the front planetary ring gear. Record your findings.

_____

_____

8. Check the fit between the front planetary carrier to the output shaft splines. Record your findings.

_____

_____

9. Remove the snap ring and thrust washer from the front planetary ring gear. Record your findings.

_____

_____

10. Examine the thrust washer and the outer splines of the front drum for burrs and distortion. Record your findings.

_____

_____

11. With the snap ring removed, the front planetary carrier can be removed from the ring gear. Check the planetary carrier gears for endplay by placing a feeler gauge between the planetary carrier and the planetary pinion gear. Compare the endplay to specifications. Record your findings.

_____

_____

12. Check the splines of the sun gear. Record your findings.

_____

_____

13. Sun gears should have their inner bushings inspected for looseness on their respective shafts. Record your findings.

_____

_____

14. Check the fit of the sun shell to the sun gear and inspect the shell for cracks, especially at the point where the gears mate with the shell. Record your findings.

_____

_____

15. Check the sun shell for a bell-mouthed condition where it is tabbed to the clutch drum. Any variation from a true round should be considered junk and should not be used. Record your findings.

_____

_____

16. Look at the tabs and check for the best fit into the clutch drum slots. This involves trial fitting the shell and drum at all the possible combinations and marking the point where they fit the tightest. Record your findings.

_____

_____

17. Check the gear carrier for cracks and other defects. Record your findings.

_____

_____

18. Check the thrust bearings for excessive wear and, if required, correct the input shaft thrust clearance by using a washer with the correct thickness. Record your findings.

_____

_____

19. To determine the correct thickness, measure the thickness of the existing thrust washer and compare it to the measured endplay. All the pinions should have about the same endplay. Record your findings.

_____

_____

20. Replace all defective parts and reassemble the gearset. ☐

Instructor's Response _____

_____

_____

Name _____ Date _____

SERVICING INTERNAL TRANSAXLE DRIVES

Upon completion of this job sheet, you should be able to inspect the transaxle drive, link chains, sprockets, gears, bearings, and bushings.

ASE Correlation

This job sheet is related to the Automatic Transmission and Transaxles Test's Content Area *Off-Vehicle Transmission and Transaxle Repair, Gear Train, Shafts, Bushings, and Case.*

Tasks: Inspect transaxle drive, link chains, sprockets, gears, bearings, and bushings; perform necessary action and inspect, measure, repair, adjust, or replace transaxle final drive components.

Tools and Materials

Marking tool
Machinist's rule

Describe the vehicle being worked on:

Year _____ Make _____ Model _____

VIN _____ Engine type and size _____

Model and type of transmission _____

Procedure

Task Completed

Chain Drives

1. This inspection is done with the transaxle on a bench and partially disassembled and should be repeated as a double check during reassembly. Begin by checking chain deflection between the centers of the two sprockets. Deflect the chain inward on one side until it is tight. ☐

2. Mark the housing at the point of maximum deflection. ☐

3. Deflect the chain outward on the same side until it is tight. ☐

4. Again mark the housing in line with the outside edge of the chain at the point of maximum deflection. ☐

5. Measure the distance between the two marks. If this distance exceeds specifications, replace the drive chain. Describe your findings.

   _____

   _____

6. Be sure to check for an identification mark on the chain during disassembly. The mark may be a painted or dark-colored link, and may indicate either the top or the bottom of the chain, so be sure you remember which side was up. How was your chain marked?

_____

_____

7. The sprockets should be inspected for tooth wear and for wear at the point where they ride. If the chain was found to be too slack, it may have worn sprockets in the same manner that engine timing gears wear when the timing chain stretches. A slightly polished appearance on the face of the gears is normal. Describe your findings.

_____

_____

8. Check the bearings and bushings used on the sprockets for damage. Describe your findings.

_____

_____

9. The radial needle thrust bearings must be checked for any deterioration of the needles and cage. Describe your findings.

_____

_____

10. The running surface in the sprocket must also be checked, as the needles may pound into the gear's surface during abusive operation. Describe your findings.

_____

_____

11. The bushings should be checked for any signs of scoring, flaking, or wear. Describe your findings.

_____

_____

12. Based on the above, what parts need to be replaced?

_____

_____

_____

_____

Final Drive Units

1. Final drive units may be helical or planetary gear units. A careful inspection of the assembly is done with the transaxle disassembled. The helical type should be checked for worn or chipped teeth, overloaded tapered roller bearings, and excessive differential side gear and spider gear wear. Describe your findings.

_____

_____

2. Measure the clearance between the side gears and the differential case. Compare your measurement to specifications. Describe your findings.

_____

_____

3. Check the fit of the spider gears on the spider gear shaft. Describe your findings.

_____

_____

4. Check the assembly's endplay. How do your measurements compare to specifications?

_____

_____

5. What is used to preload the side bearings on this transaxle?

_____

_____

6. With a torque wrench, measure the amount of rotating torque. Compare your readings against specifications. Describe your findings.

_____

_____

7. If the bearing preload, endplay, and the condition of the bearings are fine, the parts can be reused. However, always install new seals during assembly.

☐

8. Planetary-type final drives are checked for the same problems as helical-type ones. Check for worn or chipped teeth, overloaded tapered roller bearings, and excessive differential side gear and spider gear wear. Describe your findings.

_____

_____

9. The planetary pinion gears need to be checked for looseness or roughness on their shafts and for endplay. Describe your findings.

_____

_____

10. Check the endplay of the assembly. How did you do this and how do your measurements compare to specifications?

_____

_____

11. What is used to preload the side bearings on this transaxle?

_____

_____

12. With a torque wrench, measure the amount of rotating torque. Compare your readings against specifications. Describe your findings.

_____

_____

13. What are your conclusions about the final drive unit?

_____

_____

_____

_____

Instructor's Response _____

_____

_____

Name _____ **Date** _____

SERVICING FINAL DRIVE COMPONENTS

Upon completion of this job sheet, you will be able to inspect, measure, and adjust transaxle final drive units.

ASE Correlation

This job sheet is related to the Automatic Transmission and Transaxle Test's Content Area *Off-Vehicle Transmission and Transaxle Repair, Gear Train, Shafts, Bushings, and Case.* Tasks: Inspect, measure, repair, adjust, or replace transaxle final drive components.

Tools and Materials

Basic hand tools

Describe the vehicle being worked on:

Year _____ Make _____ Model _____

VIN _____ Engine type and size _____

Transmission type and model _____

Procedure

1. Final drive units may be helical gear or planetary gear units. A careful inspection of the assembly is done with the transaxle disassembled. The helical type should be checked for worn or chipped teeth, overloaded tapered roller bearings, and excessive differential side gear and spider gear wear. Describe your findings.

   _____

   _____

2. Measure the clearance between the side gears and the differential case. Compare your measurement to specifications. Describe your findings.

   _____

   _____

3. Check the fit of the spider gears on the spider gear shaft. Describe your findings.

   _____

   _____

4. Check the assembly's endplay. How do your measurements compare to specifications?

   _____

   _____

5. What is used to preload the side bearings on this transaxle?

   _____

   _____

6. With a torque wrench, measure the amount of rotating torque. Compare your readings against specifications. Describe your findings.

_____

_____

☐

7. If the bearing preload and endplay are fine, as is the condition of the bearings, the parts can be reused. However, always install new seals during assembly.

8. Planetary-type final drives are checked for the same problems as helical types. Check for worn or chipped teeth, overloaded tapered roller bearings, and excessive differential side gear and spider gear wear. Describe your findings.

_____

_____

9. The planetary pinion gears need to be checked for looseness or roughness on their shafts and for endplay. Describe your findings.

_____

_____

10. Check the endplay of the assembly. How did you do this and how do your measurements compare to specifications?

_____

_____

11. What is used to preload the side bearings on this transaxle?

_____

_____

12. With a torque wrench, measure the amount of rotating torque. Compare your readings against specifications. Describe your findings.

_____

_____

13. What are your conclusions about the final drive unit?

_____

_____

_____

_____

Instructor's Response _____

_____

_____

Name _____ **Date** _____

TRANSMISSION CASE SERVICE

Upon completion of this job sheet, you will be able to inspect case bores, passages, vents, bushings, and mating surfaces.

ASE Correlation

This job sheet is related to the Automatic Transmission and Transaxle Test's Content Area *Off-Vehicle Transmission and Transaxle Repair, Gear Train, Shafts, Bushings, and Case.*
Tasks: Inspect case bores, passages, bushings, vents, and mating surfaces; determine necessary action.

Tools and Materials

Air nozzle   Straightedge
Crocus cloth   Feeler gauge set

Describe the vehicle being worked on:

Year _____ Make _____ Model _____

VIN _____ Engine type and size _____

Model and type of transmission _____

Procedure   Task Completed

1. The transmission case should be thoroughly cleaned and all passages blown out.   ☐

2. The passages can be checked for restrictions by applying compressed air to each one. If air flows from the other end, there is no restriction. Describe your findings.

    _____

    _____

3. To check for leaks, plug off one end of the passage and apply air to the other. If pressure builds up in that passage, there are probably no leaks in it. Describe your findings.

    _____

    _____

4. Check the fit of the servo piston in the bore without the seal to be sure it has free travel. There should be no tight spots or binding over the whole range of travel. Any deep scratches or gouges that cause binding of the piston will require case replacement. Describe your findings.

    _____

    _____

5. Accumulator bores are checked the same as servo bores. Describe your findings.

_____

_____

6. Check the oil pump bore at the front of the case. Describe your findings.

_____

_____

7. Case mounted hydraulic clutch bores are prone to the same problems as servo bores. Look for any scratches or gouges in the sealing area that would affect the rubber seals. It is possible to damage these areas during disassembly, so be careful with tools used during overhaul. Describe your findings.

_____

_____

8. Sealing surfaces of the case should be inspected for surface roughness, nicks, or scratches where the seals ride. Imperfections in steel or cast iron parts can usually be polished out with crocus cloth. Describe your findings.

_____

_____

9. Check the passages in the case for cross-tracking of one circuit to another. Fill the circuit with solvent and watch to see if the solvent disappears or leaks away. If the solvent goes down, you should check each part of the circuit to find where the leak is. Describe your findings.

_____

_____

☐   10. Make sure all necessary check balls were in position during disassembly.

11. Check the valve body mounting area for warpage with a straightedge and feeler gauge. This should be done in several locations. If there is a slight burr or high spot, it can be removed by flat filing the surface. Describe your findings.

_____

_____

12. A long straightedge should be laid across the lower flange of the case to check for distortion. Any warpage found here may result in circuit leakage, causing any number of hydraulically related problems. Describe your findings.

_____

_____

13. Check all bell-housing bolt holes and dowel pins. Cracks around the bolt holes indicate that the case bolts were tightened with the case out of alignment with the engine block. Describe your findings.

_____

_____

14. Check all of the bolts that were removed during disassembly for aluminum on the threads. If so, the thread bore is damaged and should be repaired. Thread repair entails the installation of a thread insert or by retapping the bore. After the threads have been repaired, make sure you thoroughly clean the case. Describe your findings.

_____

_____

15. The small screens found during teardown should be inspected for foreign material. Describe your findings.

_____

_____

16. Most screens can be removed easily. Care should be taken when cleaning because some cleaning solvents will destroy the plastic screens. Low air pressure (approximately 30 psi) can be used to blow the screens out in a reverse direction.

☐

17. Bushings in a transmission case are normally found in the rear of the case and require the same inspection and replacement techniques as other bushings in the transmission. Always be sure that the oil passage to a pressure-fed bushing or bearing is open and free of dirt and foreign material. Describe your findings.

_____

_____

18. Vents are located in the pump body or transmission case and provide for equalization of pressures in the transmission. These vents can be checked by blowing low-pressure air through them, squirting solvent or brake cleaning spray through them, or by pushing a small diameter wire through the vent passage. Describe your findings.

_____

_____

Instructor's Response _____

_____

_____

Chapter 9

FRICTION AND REACTION UNIT SERVICE

UPON COMPLETION AND REVIEW OF THIS CHAPTER, YOU SHOULD BE ABLE TO:

- Inspect and replace bands and drums.

- Adjust bands, internally and externally.

- Inspect and repair or replace servos, including bore, piston, seals, pin, spring, and retainers as necessary.

- Inspect an accumulator bore, piston, seals, spring, and retainers while the transmission is in or out of the vehicle.

- Inspect and service clutch drums, pistons, check balls, springs, retainers, seals, and friction and pressure plates.

- Measure and adjust clutch pack clearance.

- Air test the operation of the clutch pack and servo assemblies.

- Inspect and service roller and sprag clutch, races, rollers, sprags, springs, cages, and retainers.

FRICTION AND REACTION UNITS

To provide for the various forward speed gears and a reverse, transmissions use compound gears controlled by friction and reaction devices. Although diagnosis can lead you to the device that is not holding or driving properly, recognition of those suspected parts becomes difficult when the transmission is apart. Clutch and brake application charts are a must during diagnosis and transmission overhaul.

During inspection and disassembly of friction units and their hub or drum, pay close attention to which planetary gearset member each is attached. Keep in mind that some planetary members are connected to each other and as one member rotates, it rotates the other at the same speed. Sometimes the drum or housing for a clutch is also the drum for a brake band (Figure 9-1). Often this means the same planetary member can be driven or held to achieve different gears.

Servos and accumulators control the application of bands and clutches to achieve a desirable shift feel. There are many different types and designs of each. Each has an important function to the operation of the various designs of transmissions and transaxles. Fortunately, service for all servos and accumulators is very similar.

BAND SERVICE

Servicing bands and their components includes inspection of the bands as well as the drums that the bands are wrapped around. Before the introduction of overdrive automatic transmissions, most bands operated in a free condition during most driving conditions. This means the band was not applied in the cruising gear range. However, many overdrive automatic transmissions use a band in the overdrive cruise range, which puts an additional load on the band and subsequently causes additional wear on the band. For this reason, a thorough inspection of the bands is very important (Figure 9-2).

BASIC TOOLS

Mechanic's basic tool set

Appropriate service manual

Torque wrench

Clean ATF

Clean drain pan

Lint-free rags

Classroom Manual

Chapter 9, page 281

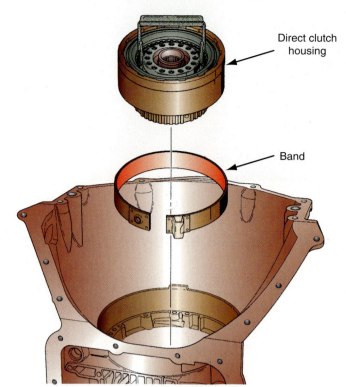

Direct clutch housing

Band

© Delmar/Cengage Learning

FIGURE 9-1 The drum for this clutch also serves as the clamping surface for the band. The tool shown in the clutch assembly is used to pull the assembly out of the transmission housing.

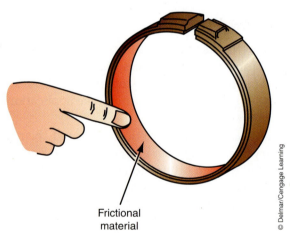

Frictional material

© Delmar/Cengage Learning

FIGURE 9-2 The friction material of the bands must be carefully inspected.

Classroom Manual

Chapter 9, page 283

The bands in a transmission are either single or double wrap, depending on the application. Both types can be the heavy-duty cast iron type or the normal strap type. The frictional material used on clutches and bands is quite absorbent. This characteristic can be used to tell if there is much life left in the lining. Simply squeeze the lining with your fingers to see if any fluid appears. If fluid appears, this tells you the lining can still hold fluid and has some life left in it. It is hard to tell exactly how long the band will last, but at least you have an indication that it is still useable. Strap- or flex-type bands should never be twisted or flattened out. This may crack the lining and lead to flaking of the lining.

Inspection

Band failures found during overhaul are easy to spot. Look for chipping, cracks, burn marks, glazing, and nonuniform wear patterns and flaking. If any of these defects are apparent, the band should be replaced.

Also inspect brake band frictional material for wear. If the linings show wear, carefully check the band struts (Figure 9-3), levers, and anchors for wear. Replace any worn or damaged parts. Look at the linings of heavy-duty bands to see if the lining is worn evenly. A twisted band will show tapered wear on the lining. If the frictional material is blackened, this is caused by an excessive buildup of heat. High heat may weaken the bonding of the lining and allow the lining to come loose from the metal portion of the band. On double-wrap bands, check both segments of the lining for cracks and other damage. Make sure you carefully check the band lugs for damage as well.

The drum surface should be checked for discoloration, scoring, glazing, and distortion. The drums will be either iron castings or steel stampings. Cast iron drums that are not scored can generally be restored to service by sanding the running surface with 180-grit emery paper in the drum's normal direction of rotation. A polished surface is not desirable on cast iron drums.

The surface of the drum must be flat (Figure 9-4). This is not usually a problem with a cast iron drum, but it can affect a stamped steel drum. It is possible for the outer surface of the drum to dish outward during its normal service life. This is a common problem on the GM 4L60 and should be inspected on any transmission that has a stamped steel band surface. Check the drum for flatness across the outer surface where the band runs. Any dishing here will cause the band to distort as it attempts to get a full grip on the drum. Distortion of the band weakens the bond of the frictional material to the band and will cause early failure due to flaking of the frictional lining. A dished stamped steel drum should be replaced. Check the service manual for maximum allowable tolerances.

Band Adjustments

After the band assembly has been installed in the transmission housing and around its drum, the band needs to be adjusted. Band adjustment is also part of a **transmission tune-up** on some models. Many transmissions have provisions for externally adjusting the band running

> A **transmission tune-up** usually includes a fluid and filter change, an inspection, and the adjustment of bands, if possible.

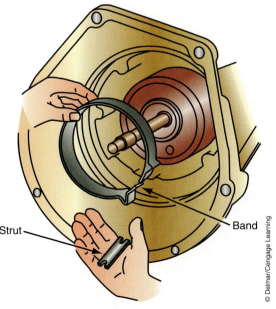

FIGURE 9-3 While removing the band, carefully inspect the band strut.

Strut — Band

© Delmar/Cengage Learning

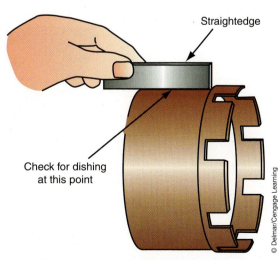

Straightedge

Check for dishing at this point

© Delmar/Cengage Learning

FIGURE 9-4 Check the flatness of the drum with a straightedge.

clearance (Figure 9-5). Some transmissions have no provisions for band adjustment other than selectively sized servo apply pins and struts.

To set the running clearance on transmissions that have an adjustment screw, loosen the locknut on the adjuster screw and back it off about five turns. Backing off the locknut allows for tightening the screw to a specified torque, which simulates a fully applied band (Figure 9-6). The amount of required torque varies with different manufacturers, but it gives the same result.

After torquing, the adjuster screw is backed off a number of turns as specified by the manufacturer. To hold the adjustment, the locknut is generally tightened to 30–35 ft.-lbs., while the adjuster screw is held stationary. The timing of band application has a lot to do with how the shift feels to the driver. This is one of the reasons there are so many different tightening torques on the various transmissions. The torque setting and number of turns the adjuster is backed off provides the proper clearance and grip for the many different types of

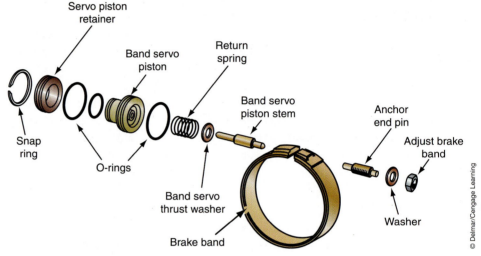

FIGURE 9-5 A complete band/servo assembly with an adjusting screw.

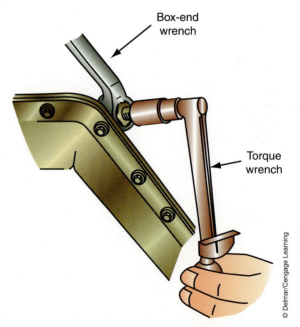

FIGURE 9-6 Adjusting a band with a torque wrench and box-end wrench.

bands used. Additionally, the pitch of the threads on the adjuster screws varies even on the same type of transmission. This is why you cannot use a single adjustment sequence for all transmissions.

Not all screw-type band adjusters are on the exterior of the transmission case. Torqueflite transmissions have the low/reverse band adjustment inside the oil pan. This requires the removal of the pan to make band adjustments. This may seem inconvenient, but since the low/reverse band is normally applied at idle conditions, not much lining wear occurs. Therefore, the low/reverse band does not require adjustment as often as the intermediate band, and it can be done during a fluid and filter change when the pan is removed.

As you can see, band adjustment is not a difficult task. However, gaining access to the adjuster screw may require the removal of some other linkages, wiring, or even cooler lines that could be in your way. Be sure any components you moved out of the way are restored to their original position after the band adjustment is finished.

SERVO AND ACCUMULATOR SERVICE

On some transmissions, the servo and accumulator assemblies are serviceable with the transmission in the vehicle (Figure 9-7). Others require the disassembly of the transmission and/or removal of the valve body. Internal leaks at the servo or clutch seal will cause excessive pressure drops during gear changes.

Before disassembling a servo or an accumulator, carefully inspect the area to determine the exact location of any leak. Do this before cleaning the area around the seal. Look at the path of the fluid and identify possible sources. These sources could be worn gaskets, loose bolts, cracked housings, or loose line connections.

Band servos and accumulators are basically pistons with seals in a bore held in position by springs and retaining bolts or snap rings (Figure 9-8). Some pistons have cast iron seal rings that may not need replacement but rubber and elastomer seals should always be replaced. Most technicians will replace all seals.

Removal

On many vehicles it is possible to remove a servo or an accumulator while the transmission is in the vehicle. To do this, the ATF needs to be drained and the oil pan removed. Some transmissions require that the valve body be removed to gain access to the accumulators or

CAUTION:
It is important that band adjustment procedures be followed exactly as outlined by the manufacturer. Serious damage to the transmission can result if the manufacturer's procedure is not followed.

Classroom Manual
Chapter 9, page 284

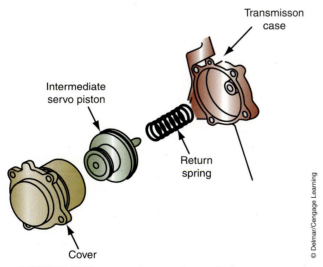

Intermediate servo piston

Transmisson case

Return spring

Cover

© Delmar/Cengage Learning

FIGURE 9-7 The servos in some transmissions are serviceable while the transmission is in the vehicle and are contained in its own bore.

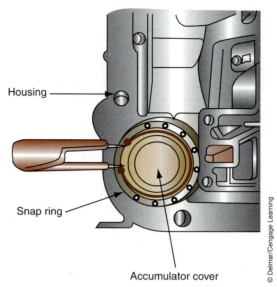

Housing

Snap ring

Accumulator cover

© Delmar/Cengage Learning

FIGURE 9-8 Servos and accumulators may be retained by a snap ring.

SERVICE TIP:
Depending on the transmission, a special tool may be required to keep the spring compressed so the cover can be removed.

CAUTION:
If the appropriate spring compressor is not available, be very careful while pushing down the piston with something else. The spring pressure can force the servo piston assembly out of case and at you. Also, the bore in the transmission case can be damaged by trying to pry on the retaining ring to remove it.

servos. Removal of the valve body, on some models, may require the removal of the transmission first.

To disassemble an accumulator, depress the piston to remove the accumulator cover snap ring. If the piston assembly is retained by a cover, unbolt the cover (Figure 9-9). After removing the accumulator cover, remove the springs and pistons.

Many accumulator pistons can be installed upside down. This results in free travel of the piston, or excessive compression of the accumulator spring. Note the direction of the piston during the teardown as you will not always find a good reference to follow during reassembly. It is quite common for manufacturers to mate servo piston assemblies with accumulators. This takes up less space in the transmission case and, because they have the same basic shape, can reduce some of the machining during manufacture.

A servo (Figure 9-10) is disassembled in a similar fashion. Be careful when disassembling a servo, some have two or more pistons and four or five fluid passages and fluid control orifices or check balls inside them. Make sure you do not mix the parts. Also note the exact location of the check balls.

In some servos, a strong return spring is under the piston; therefore, it is advisable to use the appropriate compressing tool (Figure 9-11). The tool will push down on the piston to allow removal of the retaining ring. Often, the servo assembly will contain more than one spring and two or more seals.

Inspection

Inspect the outside area of the seal. If it is wet, determine if the oil is leaking out or if it is merely a lubricating film of oil. When removing the servo, continue to look for causes of the leak. Check both the inner and outer parts of the seal for wet oil, which means leakage. When removing the seal, inspect the sealing surface, or lips before washing. Look for unusual wear, warping, cuts and gouges, or particles embedded in the seal.

The accumulator or servo's piston, spring, piston rod, and guide should be cleaned and dried. Check the servo piston for cracks, burrs, scores, and wear. Aluminum servo pistons should be carefully checked for cracks and their fit on the guide pins. Cracked pistons will allow for a pressure loss and being loose on the guide pin may allow the piston to bind in its

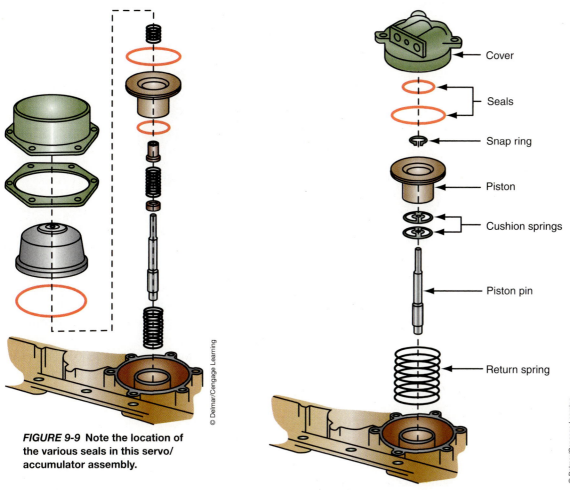

FIGURE 9-9 Note the location of the various seals in this servo/accumulator assembly.

Cover

Seals

Snap ring

Piston

Cushion springs

Piston pin

Return spring

© Delmar/Cengage Learning

FIGURE 9-10 A servo assembly.

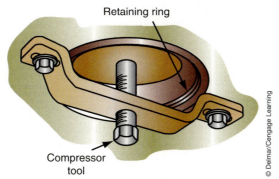

Retaining ring

Compressor tool

© Delmar/Cengage Learning

FIGURE 9-11 On some units, a spring compressor must be used to depress the piston enough to remove the retaining ring.

bore. The seal groove should be free of nicks or any imperfection that might pinch or bind the seal. Clean up any problems with a small file. The piston ring should rotate freely in its groove. If it does not, clean and inspect the grooves. Replace the piston ring during reassembly.

Also, check the condition of the servo/accumulator pins. Look for signs of wear and damage. Also, check the fit of the pins and pistons in the case. The bores in the case should not allow the pins and pistons to wobble. If they do, they are worn or the bores in the case are worn.

SERVICE TIP:
Some specialty shops have equipment that bores the worn out servo pin bore and installs a bushing with the correct inside diameter. This corrects any leakage past the pin and makes case replacement unnecessary.

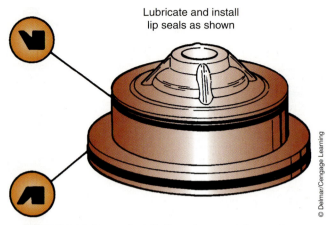

Lubricate and install
lip seals as shown

FIGURE 9-12 Proper installation of servo piston seals.

Inspect the servo or accumulator spring for cracks. Also, check the area where the spring rests against the case or piston. The spring may wear a groove, make sure the piston or case material has not worn too thin. Some servos have a steel chafing plate in order to eliminate this problem.

Inspect the servo cylinder for scores or other damage. Move the piston rod through the piston rod guide and check for freedom of movement. Check band servo components for wear and scoring. Replace all other components as necessary, and then reassemble the servo assembly.

Seals

Classroom Manual

Chapter 9, page 284

A faulty servo piston or cover seal can cause a loss in servo apply pressure which results in band slippage. A faulty accumulator piston or cover seal can cause abnormal shift quality. A cover seal is either a gasket or an O-ring, while the piston seal is either a metal or rubber rings (Figure 9-12). Molded rubber seals are usually replaced during overhaul as the rubber is subject to deterioration.

Most original equipment servo seals are of the Teflon type. These seals will exhibit no feeling of drag in the bore due to the slipperiness of the Teflon. A majority of replacement transmission gasket sets will supply cast iron hook-end seal rings, in place of the Teflon seals. When installed, the cast iron seals will have a noticeable drag as the servo piston is moved through the bore. This is not a problem, and in some cases may even improve operation.

Check the cast iron seal rings to make sure they are able to turn freely in the piston groove. These seal rings are not typically replaced unless they are damaged.

Installation

The springs used in accumulator and servo assemblies are specific to that application. Also, to see if one of the springs is weaker or stronger (both of which will change shift quality), compare the springs in the transmission you are working on to the specifications given in the service manual (Figure 9-13).

Servos and accumulators must be assembled in the correct sequence. Incorrect assembly can result in dragging bands and harsh shifting. If the servo pin has Teflon seals, these will need to be cut off with a knife and a new Teflon seal installed with the correct sleeve and sizing tools.

If rubber seal rings are installed on the piston, replace them whenever you are servicing an accumulator or a servo. When reassembling the servo or accumulator, lubricate the seal with ATF and install it on the piston rod. On spring-loaded lip seals, make sure the spring is seated around the lip, and that the lip is not damaged during installation. Lubricate and install

SPRING SPECIFICATIONS

Spring	Standard (New)–Unit: mm (in.)			
	Wire dia.	O.D.	Free length	No. of coils
1st accumulator spring	2.1 (0.083)	16.0 (0.630)	89.1 (3.508)	16.2
4th accumulator spring A	2.6 (0.102)	17.0 (0.669)	88.4 (3.480)	14.2
4th accumulator spring B	2.3 (0.091)	10.2 (0.402)	51.6 (2.031)	13.8
3rd accumulator spring	2.9 (0.114)	17.5 (0.689)	89.2 (3.512)	13.4
2nd accumulator spring A	2.4 (0.094)	14.5 (0.571)	68.0 (2.677)	12.5
3rd subaccumulator spring	2.7 (0.106)	17.0 (0.669)	39.0 (1.535)	6.3

© Delmar/Cengage Learning

FIGURE 9-13 Specifications for some of the springs used in servos and accumulators.

the piston rod guide with its snap ring into the piston. Lubricate the servo or accumulator cylinder walls. Then, install the piston assembly and return spring into the cylinder. Use a spring compressor tool to depress the piston enough to install the piston retaining ring. If the assembly had a cover plate, tighten it to specifications.

Selective Servo Apply Pins

Transmissions without a band adjusting screw use **selective servo apply pins**, which maintain the correct clearance between the band and the drum. These servo apply pins must be checked for correct length to assure proper stroking action of the servo piston, as well as shift timing. The travel needed to apply a band relates to the timing of band application. Although valving and orificing determines true shift timing, the adjustment of the servo pin completes the job by providing the correct clearance of the band to the drum. To select the correct apply pin, most transmissions require the use of special tools that simulate a fully applied band.

To select the correct apply pin for some Ford transmissions, follow this procedure: (This procedure does not compensate for any band wear and should only be used with new bands.)

1. Lubricate the piston seal.
2. Install the servo return spring and piston. Do not install the piston cover.
3. Install the special tool over the top of the piston (Figure 9-14).
4. Tighten the center bolt of the tool to the specified torque (normally 50 lbs.-in. or 5.6 Nm)
5. Attach the dial indicator holding tool to the transmission.
6. Position the stem of the dial indicator on the flat portion of the servo piston and zero the indicator.
7. Loosen the apply bolt of the piston tool until the piston stops against the tool.
8. Measure the amount of piston travel on the dial indicator and compare that to specifications (Figure 9-15).
9. If piston travel is not within specification, select and install the correct servo piston assembly to bring the servo piston travel within specification. The lengths are identified by the number of grooves near the piston (Figure 9-16).
10. Remove the dial indicator and servo selection tool.
11. Install the servo assembly. Using the special tool, install the servo retaining ring.

Classroom Manual
Chapter 9, page 286

Servo apply pins are called "selective" when they are available in a variety of lengths. The correct length must be used to prevent transmission damage and ensure proper band operation.

SPECIAL TOOLS
Dial indicator gauge with holding fixture
Servo piston remover/installer

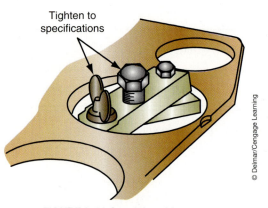

Tighten to specifications

FIGURE 9-14 Special tool for checking servo piston movement.

FIGURE 9-15 Measure the total movement of the piston after the piston stop on the special tool is loosened.

1 groove	2 groove	3 groove
2.936 in.	2.989 in.	3.043 in.
(74.56 mm)	(75.92 mm)	(77.29 mm)

FIGURE 9-16 The length of some servo pistons is identifiable by the grooves cut into the stem.

To select the servo apply pin in a 4L80-E transmission, two special tools and a torque wrench are needed. A gauge pin (Figure 9-17) is placed into the bore for the servo. Then the checking tool (Figure 9-18) is positioned over the bore with its hex nut facing the linkage for the parking pawl. The checking tool is fastened with two servo cover bolts that are torqued to a specified amount. The hex nut on the checking tool is then torqued to 25 lbs.-ft. and the exposed part of the pin gauge is studied to determine the proper apply pin length. A chart in the service manual (Figure 9-19) relates the appearance of the exposed part of the pin gauge to the correct pin length.

If a replacement band, drum, or even case has been used, an apply pin check must be made. Since these pins are selective, you will have to start by checking the length of the pin already in the transmission. If it is too long, or too short, a new length pin must be installed.

FIGURE 9-17 A special gauge pin that is placed into the bore for the servo to determine which selective servo pin should be used.

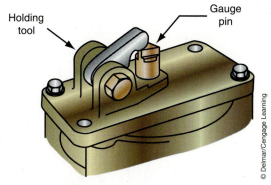

FIGURE 9-18 The checking tool is positioned over the bore and gauge pin with its hex nut facing the linkage for the parking pawl.

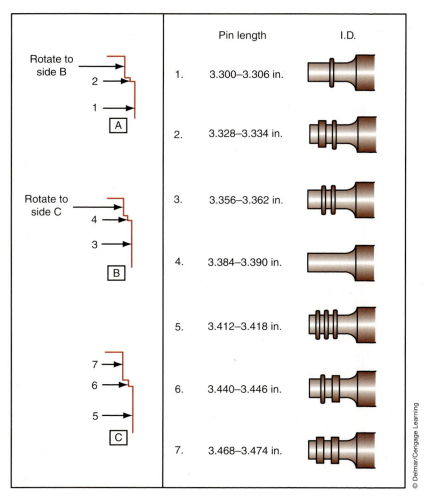

	Pin length	I.D.
1.	3.300–3.306 in.	
2.	3.328–3.334 in.	
3.	3.356–3.362 in.	
4.	3.384–3.390 in.	
5.	3.412–3.418 in.	
6.	3.440–3.446 in.	
7.	3.468–3.474 in.	

FIGURE 9-19 The exposed part of the gauge pin should be compared to the chart given in the service manual to determine the correct pin length.

MULTIPLE-FRICTION DISC ASSEMBLIES

Two types of multiple-friction disc assemblies are found in transmissions: rotating drum (Figure 9-20) and case grounded (Figure 9-21). Both are serviced in the same way but require slightly different inspection procedures.

All friction disc and steel plate packs are held in place by snap rings. These snap rings may be selective in thickness and must be kept with the clutch pack during disassembly. It is

Classroom Manual

Chapter 9, page 289

SERVICE TIP:
New friction discs should not be used with used steel plates unless the steel plates are deglazed.

417

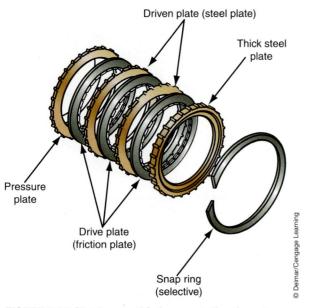

FIGURE 9-20 Clutch assembly from a rotating drum-type clutch.

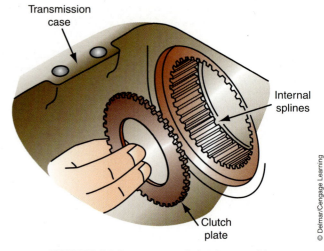

FIGURE 9-21 A case ground clutch assembly.

SPECIAL TOOLS

Snap-ring pliers
Feeler gauge
OSHA-approved
air nozzle

Many transmission rebuilders replace all friction discs and steel plates as a part of the overhaul process. This can actually save time because it eliminates the inspection time and there is no doubt about whether any discs or plates were of questionable condition.

common to use the same diameter snap ring on more than one clutch in a single transmission. However, the rings can be of differing thicknesses. This thickness variation can be used to set the clearance in the clutch pack. The snap rings may also have a distinct shape that will be effective only when used in the correct groove.

An example of complete disassembly, inspection, assembly, and clearance checks is shown in Photo Sequence 15. Always refer to the manufacturer's recommendations for a particular transmission.

Disassembly

Using a screwdriver, remove the large clutch-retaining-plate snap ring and remove the thick steel clutch pressure plate (Figure 9-22). Now remove the clutch pack. Some clutch packs have a wave snap ring that must be looped through the clutch pack until the steel plates can be removed.

Once the snap ring is removed, the backing plate will be the first steel plate you will take out. This plate, like the snap ring, may be a selective part. Since the backing plate is thicker than the other steel plates in the clutch pack, it is easy to spot. Now remove the remainder of the friction discs and steel plates, keeping them in the order in which they were in the drum (Figure 9-23).

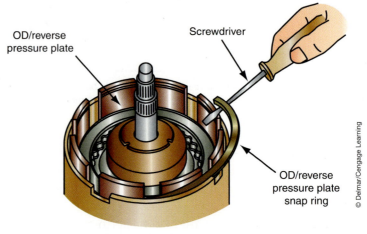

FIGURE 9-22 Removing the large snap ring to disassemble clutch pack.

PROPER PROCEDURE FOR DISASSEMBLING, INSPECTING, ASSEMBLING, AND CLEARANCE CHECKING A DIRECT CLUTCH

All photos in this sequence are © Delmar/Cengage Learning.

P15-1 Set the direct clutch on the bench.

P15-2 Pry out the snap ring.

P15-3 Remove the pressure plate and the clutch pack assembly.

P15-4 Install the piston compressor and compress the piston.

P15-5 Remove the snap ring.

P15-6 Remove the piston compressor, the spring retainer assembly, and the piston.

P15-7 Install new seals on the piston.

P15-8 Check the movement of the check ball.

P15-9 Lubricate the outer piston seal with transjel.

P15-10 Reinstall the piston.

P15-11 Place the piston spring retainer into the drum.

P15-12 Install the compressor and snap ring.

P15-13 Alternately install friction and steel discs.

P15-14 Install the pressure plate and snap ring.

P15-15 Check the clearance with a feeler gauge.

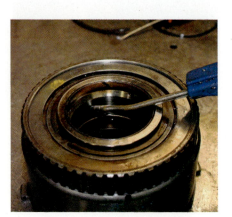

P15-16 Air test the assembly.

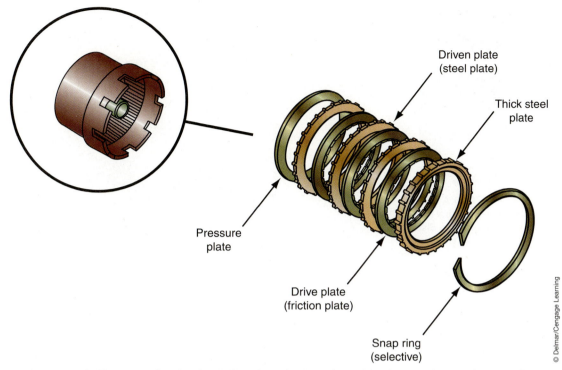

FIGURE 9-23 When removing the clutch discs from the drum, keep them in the order that they were in.

Using a clutch spring compressor tool, compress the clutch return springs. Then, using snap-ring pliers, remove the clutch hub retainer snap ring, retaining plate, and springs. Most retainers have small tabs that prevent you from removing the snap ring without using a compressor (Figure 9-24).

There are many types of clutch spring compressor tools available. The many different locations and depths of the clutch springs in their drums dictate having more than one simple compressor (Figures 9-25, 9-26, and 9-27).

With the compressor installed on the spring retainer (Figure 9-28), compress the spring and retainer just enough to allow the snap ring to be removed. Pushing the retainer down too much may bend or distort it. Make sure the snap ring is not partially caught in its groove before releasing the spring compressor. Also, be very careful when releasing the spring compressor tool; some springs have very high tensions and can injure you if they get a chance to fly out.

If the return spring assembly is installed upside down, the unit will pass an air check but will not apply the clutches.

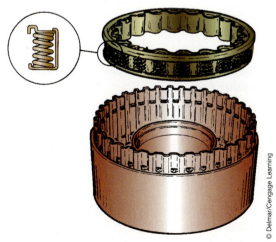

FIGURE 9-24 Retainer ring for captive return springs.

FIGURE 9-25 A spring compressor tool for a multiple-friction disc assembly in a GM 4T80E transaxle.

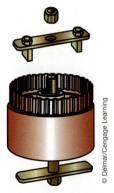

FIGURE 9-26 A spring compressor tool for a multiple-friction disc assembly in Honda transaxles.

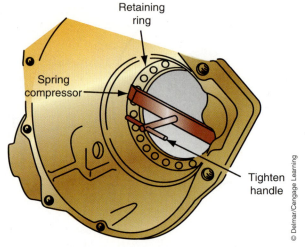

FIGURE 9-27 A spring compressor tool for a multiple-friction disc assembly that set into the transmission housing.

Retaining ring

Spring compressor

Tighten handle

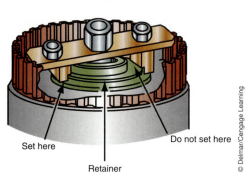

Set here

Do not set here

Retainer

FIGURE 9-28 Proper position of the spring compressor will prevent damage to the piston and seals.

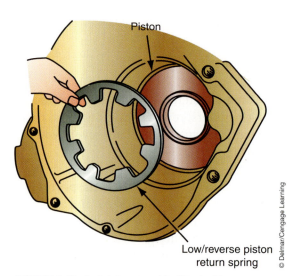

Piston

Low/reverse piston return spring

FIGURE 9-29 A clutch assembly fitted with a Belleville spring.

Classroom Manual

Chapter 9, page 291

Belleville-type return springs are usually found in the forward clutch.

⚠️ **WARNING:** Be careful when compressing the springs. Careless procedures can allow the springs and retainer to fly into your face.

With the retainer removed, a single large coil spring or multiple small coil springs will be exposed. Note the number and placement of the multiple springs for assembly.

If no coil springs were encountered when the clutch pack was removed, then a Belleville or disc-type return spring is used (Figure 9-29). This type of clutch will have a heavy steel plate with one rounded side in the bottom of the clutch pack. This heavy plate is a pressure plate. The released position of these springs has very little or no outward force on the retaining snap ring; therefore, the snap ring can be removed without the use of a compressor. The snap ring for this application may be selective, so remember which snap ring goes where.

Some clutches can have two or more snap rings, so pay attention to their placement in the clutch. Additionally, a wavy snap ring is sometimes used to retain the Belleville spring. This is used to give some cushion to the application of the clutch.

For easy removal of the piston from its drum, mount the clutch on the oil pump. Then lift the piston out of the bore. If it will not come out, the drum can be inverted and slammed squarely against a hardwood bench top. This will dislodge a tight piston, but it will not damage the drum.

Another way to remove the piston is to charge the apply circuit of the clutch with compressed air. Use an air nozzle with a rubber tip and apply air pressure to pop out the piston from the drum. Air is blown in either at the feed hole in the drum or by placing the drum on the clutch support and using its normal feed passage (Figure 9-30). Wrap a rag around the clutch drum to catch the piston and the fluid overspray. Make sure your fingers are out of the way of the moving piston.

 WARNING: **Air pressure should be reduced to 25–30 lbs. to avoid expelling the piston at a high rate of speed, which could cause injury.**

With the piston removed, take note of the types of seals used and their position. This is important with lip seals, as the lip will face the direction the fluid comes from. Lip seals installed in the wrong direction will not hold pressure. Seals are generally replaced during overhaul. The reuse of old seals is not recommended.

After the clutch pack has been disassembled, check the clutch drum, clutch housing, or transmission housing for defects. The splined area in the transmission housing should be carefully inspected for broken or chipped splines and other damage. If these splines are damaged, the case should be replaced. Clutch drums and housings should be checked for worn or damaged bushings. These should be replaced if they show signs of wear or damage. The splines inside the drum should be inspected for broken or chipped splines and other damage. If these splines are damaged, the drum or housing should be replaced. The outside surface of the housing should be inspected for signs of overheating and other defects. The flatness of the surface should also be checked by placing a straightedge across the surface (Figure 9-31). Look straight at the housing and the straightedge. If light shines between these two surfaces, the surface is distorted and the housing should be replaced.

Pistons located in the transmission case can be removed by blowing compressed air into the apply passage in the case.

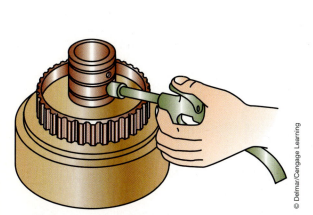

FIGURE 9-30 Use an air nozzle with a rubber tip and apply air pressure at the feed hole in the drum to pop out the piston from the drum.

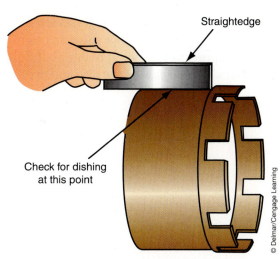

FIGURE 9-31 Checking the flatness of a housing with a straightedge.

423

Inspection and Cleaning

Once a clutch assembly has been taken apart, you may wish to inspect the clutch components or continue to disassemble the remainder of the clutch units in the transmission. If you choose the latter, make sure you keep the parts of each clutch separate from the others.

Clean the components of the clutch assembly. Make sure all clutch parts are free of any residue of varnish, burned disc facing material, or steel filings. Take special care to wash out any foreign material from the inside of drums and the hub disc splines. If left in, the material can be washed out by the fresh transmission fluid and sent through the transmission. This can ruin the rebuild.

The clutch splines must be in good shape with no excessively rounded corners or shifted splines. Test their fit by trial fitting each new clutch disc on the splines. Move the discs up and down the splines to check for binding. If they bind, this can cause dragging of the discs during a time when they should be free floating. Replace the hubs if the discs drag during this check. Check the spring retainer. It should be flat and not distorted at its inner circumference. Check all springs for height, cracks, and straightness (Figure 9-32). Any springs that are not the correct height or that are distorted should be replaced. Many retainers have the springs attached to them by crimping. This speeds up production at the assembly line. Turning this type of retainer upside down is a quick check of spring length. Closely examine the Belleville spring for signs of overheating or cracking, and replace it if it is damaged.

The steel plates should be checked to be sure they are flat and not worn too thin. Check all **steels** against the thickest one in the pack or a new one. Most steels will have an identifying notch or mark on the outer tabs. If the plates pass inspection, remove the polished surface finish and the steels are ready for reuse. The steel plates should also be checked for flatness by placing one plate on top of the other and checking the space on the inside and outside diameters. Clutch plates must not be warped or cone-shaped. Also, check the steel plates for burning and scoring and for damaged driving lugs. Check the grooves inside the clutch drum and check the fit of the steel plates, which should travel freely in the grooves.

Close inspection of the friction discs is simple. The discs will show the same types of wear as bands will. Disc facings should be free of chunking, flaking, and burned or blackened surfaces. Discs that are stripped of their facing have been overheated and subjected to abuse. In some cases, the friction discs and steel plates can be welded together. This occurs when the facing comes off the disc due to a loosening of the facing's bonding because of extreme heat. As the facing comes off, metal-to-metal contact is made between the discs and steel plates. The friction involved as the clutch tries to engage causes them to fuse together. This may lock the clutch in an engaged condition. Depending on which clutch is affected, driveability problems can occur, including driving in neutral, binding up in reverse (forward clutch seized), starting in direct drive, binding up in second (high/reverse clutch seized), and other problems that are less common.

If the discs do not show any signs of deterioration, squeeze each disc to see if fluid is still trapped in the facing material. If fluid comes to the surface, the disc is not glazed. Glazing seals the surface of the disc and prevents it from holding fluid. Holding fluid is basic to proper

> The steel plates are commonly called **steels** by the trade.

> **Classroom Manual**
> Chapter 9, page 290

FIGURE 9-32 Check the height of each one of the return springs.

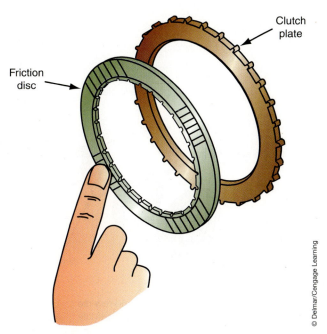

Clutch
plate

Friction
disc

© Delmar/Cengage Learning

FIGURE 9-33 Carefully examine the clutch discs.

disc operation. It allows the disc to survive engagement heat, which would burn the facing and cause glazing. Fluid stored in the frictional material cools and lubricates the facings as it transfers heat to the steel plates and also carries heat away as some oil is spun out of the clutch pack by centrifugal force. This helps avoid the scorching and burning of the disc.

Clutch discs must not be charred, glazed, or heavily pitted. If a disc shows signs of flaking frictional material or if the frictional material can be scraped off easily, replace the disc.

A black line around the center of the friction surface also indicates that the disc should be replaced. Examine the teeth on the inside diameter of each friction disc for wear and other damage (Figure 9-33).

Wave plates are used in some clutch assemblies to cushion the application of the clutch (Figure 9-34). These should be inspected for cracks and other damage. Never mix wave plates from one clutch assembly to another. As an aid in assembly, most wave plates have identifying marks.

The clutch pistons (Figure 9-35) are checked for cracks, warpage, and fit in their bores. Also check the check ball and the check ball bore for damage. Carefully examine the seal ring grooves and inside diameter of the piston for cracks, nicks, and burrs. Groove wear can be accelerated by excessive pump pressures. The excess pressure forces the seal rings against the sides of the grooves so hard that fluid cannot get between the ring and groove to lubricate the ring. If the bores of the clutch are severely grooved, a stuck pressure regulator valve could be the problem.

The reverse side of the pump cover is the clutch support that incorporates seal rings for the fluid circuits leading to the clutch drums. The seal rings fit loosely into the grooves in the clutch support and rely on pump oil pressure to push them against the side of the groove to make the seal. The seal rings should be checked for side play in the grooves and for proper fit into the drum. Check the grooves for burrs, step wear, or pinched groove conditions. It should be noted that these seals rotate with the drum. Any condition that hinders rotation will cause the ring seals to bind, resulting in drum wear. This can destroy the drum if it is not corrected.

Carefully inspect aluminum pistons, which may have hairline cracks that will cause pressure leakage during use. This could cause clutch slipping as a result of lost hydraulic force pushing the piston against the discs. If burned discs were found in the clutch pack, be sure to check for cracks in the piston.

Classroom Manual
Chapter 9, page 291

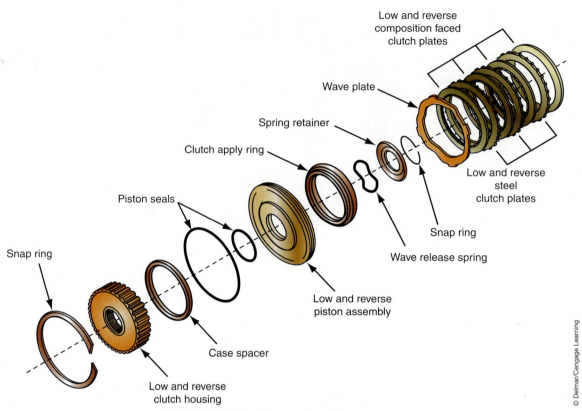

Low and reverse
composition faced
clutch plates

Wave plate

Spring retainer

Clutch apply ring

Piston seals

Snap ring

Low and reverse
steel
clutch plates

Snap ring

Wave release spring

Low and reverse
piston assembly

Case spacer

Low and reverse
clutch housing

FIGURE 9-34 **A clutch assembly with a wave plate.**

© Delmar/Cengage Learning

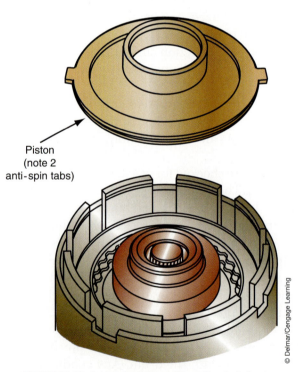

Piston
(note 2
anti-spin tabs)

FIGURE 9-35 **A clutch piston with anti-spin tabs.**

© Delmar/Cengage Learning

Stamped steel pistons have replaced most aluminum pistons because they are cheaper to produce. Aluminum pistons require a casting process followed by machining, whereas stamped pistons can be made by the thousands and may only require a simple spot-weld operation to prepare them for use. Stamped pistons show cracking more easily than aluminum. Also, look for any separation where spot welding is used on these pistons.

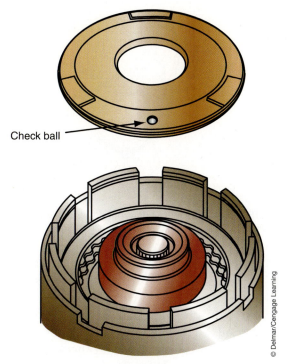

Check ball

FIGURE 9-36 A clutch piston with an annular check ball.

© Delmar/Cengage Learning

It is common to find annular check balls in clutch pistons (Figure 9-36). These balls allow for an air release from the bore while it is being filled with ATF and for a quick release of pressure when the clutch is released. Inspect each check ball to be sure it is free in its bore. Even after a thorough cleaning in solvent, the check balls may not be free. A fine wire can be pushed into the bore, followed by a spray of carburetor cleaner, to remove any stubborn deposits. The bore should then be blown out with compressed air.

A check ball may also be located in the drum and should be checked. However, it is often very difficult to use the same cleaning techniques as used on pistons. A quick way to determine if the check ball is free is to shake the clutch drum to hear the relief check ball rattle. If the check ball does not rattle, replace the drum.

The ability of the check ball to seal is also important. To check how well the ball seats in its bore, pour clean solvent into the bore. Observe the other end of the bore. If fluid leaks out, the ball is not seating and the piston or drum should be replaced.

Examine the outside surface of the drum for glazing. Glazing can be removed with emery cloth. Also check the drum's cylinder walls for deep scratches and nicks.

Inspect the front clutch bushing for wear and scores. If the bushing is worn, replace it. Also inspect any bushings found in the clutch drums for excessive wear, scoring, or looseness in the drum bore. Replace as needed.

Clutch Pack Reassembly

Begin assembly of a clutch unit by gathering the new seals and other new parts that may be necessary. Prior to installation, all clutch discs and bands are to be soaked in the type of transmission fluid that will be used in the transmission. The minimum soak time is 15 minutes. Be sure that all discs are submerged in the fluid and that both sides are coated.

Before fitting the new rubber seals to the piston, check them against the old seals. This will ensure correct sizing and shape of the new seal. Most overhaul kits include more seals than are required to complete the job. This is because changes in transmission design may dictate the use of a seal or gasket of different design or size. Therefore, both the old design seal and the new design seal are included in the kit. The rebuilder must check to be sure the correct seal is being used. This also holds true for cast iron and Teflon ring seals.

CAUTION:
Use caution when installing the piston to prevent damage to the seals. Be careful not to stretch the seals during installation.

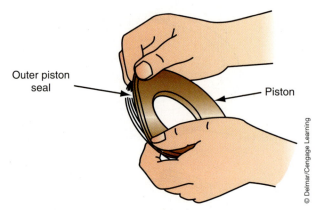

FIGURE 9-37 Install new seals around the outside of the piston before reassembling the clutch.

CAUTION:
Never assemble a clutch assembly when it is dry. Always lubricate its components thoroughly with clean ATF.

Some rebuilders use an old credit card in place of the feeler gauge. Its wide surface and plastic edge will not cut into the rubber seals.

SERVICE TIP:
Some rebuilders use a wax stick to coat lathe-cut seals for installation. This is available under the trade name Door-Ease. Its original use was to stop squeaks on rubber door bumpers and latches. If Door-Ease is used, do not coat the drum bore with ATF. The Door-Ease works fine by itself.

Once the correct seals are chosen from the kit, they can be installed on the piston (Figure 9-37). Remember to position lip seals so they face the same direction that the fluid pressure comes into the drum.

Seals should be lubricated with automatic transmission fluid or transjel. Never use chassis lube, "white lube," or motor oil. These will not melt into the transmission fluid as it heats up, but will clog filters and orifices or cause valves and check balls to stick.

Manufacturers switched to Teflon seals for these positions because they helped reduce the wear at the bore in the drum. During an overhaul, they may be replaced with hook-end-type steel rings.

Care must be taken to be sure the ends of scarf-cut Teflon seals are installed correctly. These rings must also be checked for fit into the drum. They should have a snug, but not too tight, fit into the bore.

Some manufacturers use an endless type of Teflon seal. These seals must be installed with special sleeves and pushing tools to avoid overstretching the seal. First the seals are pushed over the installing sleeve to their location in the groove, then a sizing tool is slipped over the sleeve and seal ring to fit the seal to the groove.

Assemble the piston, being careful not to allow the seal to kink or become damaged during installation. The piston can now be installed. Several methods may be used to aid piston installation. Lathe-cut seals can be helped into their bores by using a thin feeler gauge mounted on a handle. These are available from most automotive tool suppliers. Pistons with lathe-cut seals are installed by positioning the piston in the bore of the clutch drum. Then slowly work the piston down in the bore until resistance is felt. Using the feeler-type seal installer, work your way around the outer circumference of the seal, using a downward action followed by a clockwise pulling motion as you push the seal back into the groove in the piston.

Occasionally, a piston will not allow access to the outer seal area. A large chamfered edge is at the top of the seal bore to allow the seal to be worked into the bore without the help of any special tools. The piston can be installed by rotating the piston as you push down. Use even pressure to avoid binding the piston or cocking the piston in its bore. Uneven pressure can also cause the ring seal to be pushed out or to tear.

Pistons with lip seals require a more delicate installation. The lips can be bent back or torn unless proper caution is taken during installation. The basic shape of a lip seal makes it necessary to use an installation tool (Figure 9-38). The lip must be pushed back toward the piston body in order to allow the seal to enter the bore. Lip seals will often stick in snap-ring grooves as you try to slip the piston and seals into the drum. Piano wire installers can be used to roll lip seals back away from the snap-ring grooves or the bore of the drum (Figure 9-39). The round cross section of the wire prevents cutting or tearing the seal lip during installation.

While holding the piston as squarely in the drum as you can, work the tool around the lip to allow the seal to enter the bore or around the center of the drum. Do not apply too much

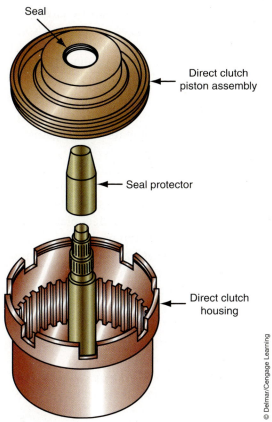

Seal

Direct clutch piston assembly

Seal protector

Direct clutch housing

© Delmar/Cengage Learning

FIGURE 9-38 Using a seal protector to install a piston into a drum without damaging the seals.

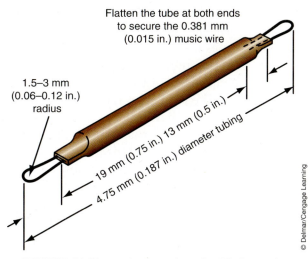

Flatten the tube at both ends to secure the 0.381 mm (0.015 in.) music wire

1.5–3 mm (0.06–0.12 in.) radius

19 mm (0.75 in.) 13 mm (0.5 in.)

4.75 mm (0.187 in.) diameter tubing

© Delmar/Cengage Learning

FIGURE 9-39 Piano wire-type piston installation tool.

downward force as the seal lip is worked into the bore. The piston will fall into place after the lip is fully inserted.

Pistons with multiple seals require special care to avoid damaging the other seals as you work on one seal. Multiple-seal piston installation is made simple by using plastic ring seal installers. These rings compress the seals back into the piston grooves so they will not hang up. Two installers are often used at the same time.

Regardless of the type of tool used to install the piston seals, always take your time to avoid tearing or rolling the new seals during installation.

When working with stamped steel pistons, install the ring spacer on the top side of the piston, making sure it has the correct thickness for the application.

Once the piston seals enter the bore, push the piston all the way down until it stops. Then lift the piston up slightly. Push it back down to the bottom of the bore to be sure it is all the way down. After the piston is installed, rotate the piston by hand to ensure that there is no binding.

Reassemble the springs and retainer after the piston has been installed. Place the single or multiple spring set onto the piston using the pockets provided to locate them. Be sure the loose spring sets are spaced as they were during teardown. Set the retainer plate over the springs. Make sure the retainer is facing in the correct direction. On some transmissions, if the retaining ring and springs are faced in the wrong direction, the clutch assembly will work well during an air check but it will not apply the clutches during use. Captive spring retainers are simply set on the piston and require no other setup.

Position the spring compressor on the retainer plate and compress the springs. Be careful not to allow the retainer to catch in the snap ring groove while compressing the spring. This will bend or distort the plate, making it unsuitable for use. Remember to compress the springs only enough to get the snap ring into its groove. Use snap-ring pliers to expand or contract the snap ring. Once the snap ring is installed and fully seated, release the compressor.

Loose springs are used mostly with aluminum pistons.

SERVICE TIP:
Used steels can be deglazed by sanding their faces with 320-grit emery paper, which can produce a dull surface on the plates.

Clutches with a Belleville return spring may not require a compressor for assembly. The Belleville spring is merely laid on the piston and centered in the bore. A wire ring is sometimes inserted on the piston where the Belleville spring touches the piston. This is used to prevent the steel spring from chafing the aluminum piston. If left out, there will be too much endplay and piston damage will occur.

The large snap ring that retains the Belleville spring is now inserted in its groove. This snap ring may be either flat or of wavy construction. There may also be a plastic spacer ring used under the snap ring to center the spring. Be certain the snap ring seats firmly against the drum. Then install the pressure plate on top of the Belleville spring.

Continue clutch assembly by stacking the clutch pack. Begin by installing the dish plate, with the dish facing outward. If a wavy plate or cushion spring is used, it will normally be installed next to the piston. Alternately stack the steels and friction discs until the correct number of plates have been installed. Place the backing plate (the thickest steel plate) on top and install the retainer ring in its groove. Assemble the remaining clutch packs.

As mentioned before, all models of a transmission do not always have the same number of discs and steels in their clutches. Most overhaul kits supply enough discs and steels to rebuild all models. They are likely to have more discs and steels than are required. This is why a technician should always note how many discs and steels are in each pack while disassembling the transmission.

Trying to fit all of the supplied discs and steels may result in clutches with no freeplay or no room for the snap ring. If these problems come up while stacking the pack, refer to a service manual. It may list the number of discs and steels for the model you are building. After the plates are installed, position the retainer plate and install the retaining snap ring. Proceed by measuring the clearance between the plate and snap ring.

Clearance Checks

The clearance check of a multiple-disc pack is critical for correct transmission operation. Excessive clearance causes delayed gear engagements, while too little clearance causes the clutch to drag. Adjusting the clearance of multiple-disc packs can be done with the large outer snap ring in place.

With the clutch pack and pressure plate installed, use a feeler gauge to check the distance between the pressure plate and the outer snap ring (Figure 9-40). Clearance can also be measured between the backing plate and the uppermost friction disc. If the clutch pack has a waved snap ring, place the feeler gauge between the flat pressure plate and the wave of the snap ring farthest away from the pressure plate. Compare the distance to specifications. Attempt to set pack clearance to the smallest dimension shown in the chart.

Clearances can also be checked with a dial indicator and hook tool (Figure 9-41). The hook tool is used to raise one disc from its downward position and the amount that it is able to move is recorded on the dial indicator. This represents the clearance.

Another way to measure clearance is to mount the clutch drum on the clutch support and use 25–35 psi of compressed air through the oil pump body channels to charge the clutch. Clearance can be measured by mounting a dial indicator so that it reads backing plate movement as the air forces the piston to apply the clutch (Figure 9-42).

If the clearance is greater than specified, install a thicker snap ring to take up the clearance. If the clutch clearance is insufficient, install a thinner snap ring.

Another way to adjust clutch clearance is to vary the thickness of the clutch pressure plate. By using a pressure plate of the desired thickness, you can obtain adequate clutch clearance (Figure 9-43).

Proper clearance can also be achieved through the use of special steel clutch plates. These are available from some manufacturers of transmission overhaul kits.

If specifications are not available, a general clutch pack clearance can be used. Allow 0.01 inch clearance for each friction disc in the clutch pack. For example, if the pack has four friction discs, set the minimum clearance to 0.04 inch. Since this is the minimum clearance,

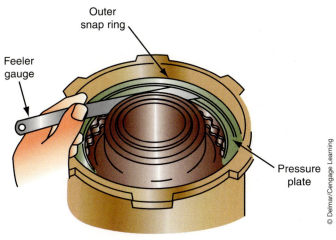

FIGURE 9-40 Measuring clutch clearance with a feeler gauge.

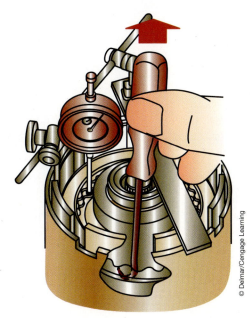

FIGURE 9-41 Using a hook tool to measure clutch clearance.

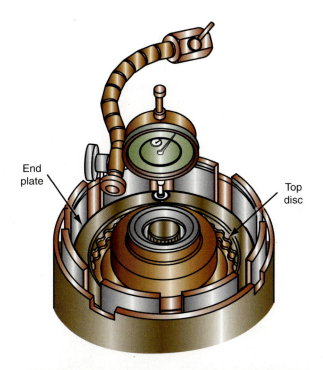

FIGURE 9-42 Dial indicator setup for measuring clutch clearances.

Service limit		
1st	0.65–0.85 mm	(0.026–0.033 in.)
2nd	0.65–0.85 mm	(0.026–0.033 in.)
3rd	0.40–0.60 mm	(0.016–0.024 in.)
4th	0.40–0.60 mm	(0.016–0.024 in.)
Low–hold	0.80–1.00 mm	(0.031–0.039 in.)

P/N	Plate no.	Thickness mm (in.)
22551–PX4–003	1	2.1 (0.082)
22552–PX4–003	2	2.2 (0.086)
22553–PX4–003	3	2.3 (0.090)
22554–PX4–003	4	2.4 (0.094)
22555–PX4–003	5	2.5 (0.098)
22556–PX4–003	6	2.6 (0.102)
22557–PX4–003	7	2.7 (0.106)
22558–PX4–003	8	2.8 (0.110)
22559–PX4–003	9	2.9 (0.114)

FIGURE 9-43 A typical chart of the various pressure plate thicknesses available to correct clutch clearances.

The 4T60-E and AX4S transmissions are common transaxles that have a chain link and an off-center input shaft.

Only use compressed air that is free of dirt and moisture.

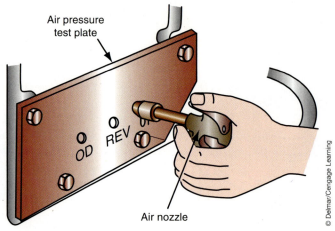

FIGURE 9-44 To test a clutch, apply air to the hole in the test plate designated for that clutch.

some transmission specialists then add 0.03 inch to the clearance. Therefore, in the example given, the maximum clearance would be 0.07 inch.

Air Testing

After the clearance of the clutch pack is set, perform an air test on each clutch. This test will verify that all the seals and check balls in the hydraulic component are able to hold and release pressure.

Air checks can also be made with the transmissions assembled. This is the absolute best way to check the condition of the circuit, because there are very few components missing from the circuit. The manufacturers of different transmissions have designed test plates that are available to test different hydraulic circuits. Testing with the transmission assembled also allows for testing of the servos.

To test a clutch assembly, install the oil pump assembly with its reaction shaft support over the input shaft and slide it into place on the front clutch drum. When the clutch drums are mounted on the oil pump, all components in the circuit can be checked. If the clutch cannot be checked in this manner, blocking off apply ports with your finger and applying air pressure through the other clutch apply port will work.

To conduct an air test with the transmission assembled, invert the entire assembly and place it in an open vise or transmission support tool. Pour clean ATF into the circuits that will be tested. Try to get as much fluid into the circuit as you can. Then, install the air test plate. On many transmissions it is necessary to remove the valve body in order to mount the test plate. Then, air test the circuit using the test hole designated for that clutch (Figure 9-44). Be sure to use low pressure compressed air (25–35 psi) to avoid damage to the seals. Higher pressures may blow the rubber seals out of the bore or roll them on the piston.

While applying air pressure, you may notice some escaping air at the metal or Teflon seal areas. This is normal, but not necessarily desirable, as these seals have a controlled amount of leakage designed into them. Basically, the less a seal leaks, the better it is. There should be no air escaping from the piston seals. The clutch should apply with a dull but positive thud. It should also release quickly without any delay or binding. Examine the check ball seat for evidence of air leakage.

CLUTCH VOLUME INDEX

Many electronically controlled transmissions regulate shift feel by regulating the volume of fluid used to apply or release a clutch or brake. The computer monitors gear ratio changes by monitoring the input and output speed of the transmission. By comparing the signal from the turbine

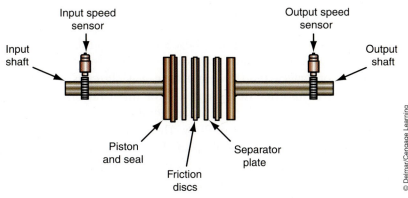

The TCM times how long it takes to compress the clutch pack to change the gear ratio.

Input speed sensor

Output speed sensor

Input shaft

Output shaft

Piston and seal

Friction discs

Separator plate

© Delmar/Cengage Learning

FIGURE 9-45 Things considered by the system's computer (TCM) when it calculates CVI.

or input speed sensor to the signal from the output speed sensor, the computer can determine the operating gear ratio. The computer also monitors the **Clutch Volume Index (CVI)**.

CVIs represent the volume of fluid needed to compress a clutch pack or apply a brake band. Based on its constant calculation of gear ratio, the computer can monitor how long it takes to change a gear (Figure 9-45). If the time is too long, more fluid is sent to the apply device. The volume of the fluid required to apply the friction elements is continuously updated as part of the transmission's adaptive learning. As friction material wears, the volume increases.

Clutch volume updates, as shown on a scan tool, are valuable in diagnosis of slippage problem and are typically updated, as follows.

- The volume of the low/reverse clutch volume is updated during a 2–1 or 3–1 manual downshift to low gear with a throttle angle below 5 degrees with the transmission's temperature above 110 °F. The clutch volume should be between 82 and 134.
- The volume of the 2nd clutch is updated during a 3–2 kickdown shift with the a throttle angle between 20 degrees and 54 degrees and the transmission's temperature above 110 °F. The clutch volume should be between 25 and 64.
- The volume of the overdrive clutch is updated when doing a 2–3 upshift with a throttle angle between 5 degrees and 54 degrees, and the temperature above 110 °F. The clutch volume should be between 30 and 64.
- The volume of the 4th clutch is updated when doing a 3–4 upshift with a throttle angle between 5 degrees and 54 degrees with a transmission temperature above 110 °F. The clutch volume should be between 30 and 64.
- The volume of the underdrive clutch is updated during a 4–3 kickdown shift with the throttle angle between 20 degrees and 54 degrees and the temperature above 110 °F. The clutch volume should be between 44 and 92.

For the correct CVI specifications for the vehicle you are working on, refer to the appropriate service manual.

GM's Transmission Adaptive Pressure

Similar to the CVI, GM's Transmission Adaptive Pressure (TAP) control system uses a line pressure control system during upshifts to compensate for the normal wear of the transmission. As the transmission's parts wear, shift times change. To compensate for these changes, the TCM adjusts its commands to the various pressure control solenoids, to maintain ideal

CVIs can be monitored with a scan tool and the readings compared to charts given in the service manual. Keep in mind that the **Clutch Volume Index (CVI)** lets you know how much fluid is needed to apply or release a friction component. When it is out of spec, a problem is indicated.

shift timing. The TCM monitors the signals from the input speed sensor (ISS) and the output speed sensor (OSS) during shifts. This data is used to determine if the shifts occurred too quickly or slowly. If the shifts did not take place according to the original calibration of the TCM, the TCM will change its commands to the solenoids.

ONE-WAY CLUTCH ASSEMBLIES

One-way clutches are holding devices. They allow a gearset member to rotate in one direction only and are used to ground or effectively hold a member by preventing it from rotating in the direction it would otherwise tend to rotate. One-way clutches are also used to transmit torque from one member to another. In these cases, the inner race is splined to one member of the gearset. The outer race is splined to another member.

An example of this is shown in Figure 9-46. The inner race of this overdrive clutch is splined to the coast clutch cylinder, which is splined to the overdrive sun gear. The one-way clutch's outer race is splined to the overdrive ring gear. When the overdrive ring gear has a counterclockwise torque from the movement of the vehicle, the gearset attempts to rotate the sun gear clockwise. The attempted counterclockwise rotation of the ring gear locks the one-way clutch, which in turn locks the ring gear with the sun gear. This action provides for direct drive through the gearset.

When overdrive gear is selected, the coast clutch cylinder is held by the inner race of the one-way clutch, which in turn holds the overdrive sun gear. This causes the planet gear assembly to walk around the sun and drive the ring gear clockwise with an overdrive. During coast, the clutch releases and allows the ring gear to rotate faster than the sun gear.

Because they are purely mechanical in nature, one-way clutches are relatively simple to inspect and test. They are not always easy to remove and the proper procedure must be followed. The following is given as an example and is the recommended procedure for removing and installing a reverse input and second roller clutch in a GM 4T45E transaxle:

1. Remove the reverse clutch housing.
2. Remove the roller clutch and the roller clutch retainer by prying up on the assembly with two screwdrivers (Figure 9-47). Note that the retainer is pressed onto the housing and should not be reused.
3. Rotate the roller clutch clockwise until it lines up with the roller clutch cam ramps (Figure 9-48).
4. Remove the roller clutch from the inside of the roller clutch cam. Inspect the roller clutch for excessive wear and damage.
5. Clean and dry each of the components.
6. To reassemble, reassemble the roller clutch assembly.
7. Press the roller clutch assembly onto the housing.

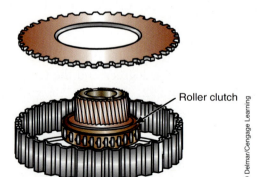

FIGURE 9-46 The inner race of this overdrive one-way clutch is splined to the coast clutch cylinder, which is splined to the overdrive sun gear and the outer race is splined to the overdrive ring gear.

Classroom Manual

Chapter 9, page 287

CAUTION:

Be careful not to damage the inner race of the roller clutch while removing it.

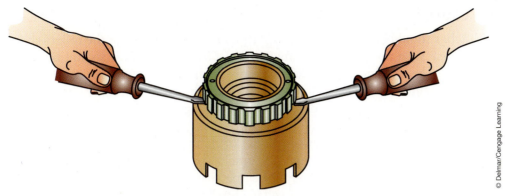

FIGURE 9-47 Remove the roller clutch and the roller clutch retainer by prying up on the assembly with two screwdrivers.

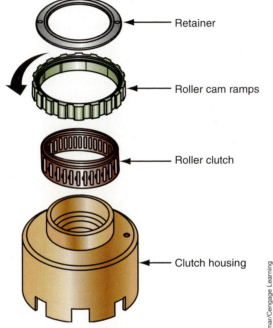

Retainer

Roller cam ramps

Roller clutch

Clutch housing

FIGURE 9-48 To separate the cam ramp assembly from the rollers, rotate the roller clutch clockwise until it lines up with the roller clutch cam ramps.

The durability of these clutches relies on constant fluid flow during operation. If a one-way clutch has failed, a thorough inspection of the hydraulic fluid feed circuit to the clutches must be made to determine if the failure was due to fluid starvation. The rollers and sprags ride on a wave of fluid, when they overrun. Since most of these clutches spend most of their time in the overrunning state, any loss of fluid can cause rapid failure of the components.

Sprags, by design, produce the fluid wave effect as they slide across the inner and outer races making them somewhat less prone to damage. Rollers, due to their spinning action, tend to throw off fluid, which allows more chance for damage during fluid starvation. During the check of the hydraulic circuit, take a look at the feedholes in the races of the clutch. Use a small diameter wire and spray brake cleaner to be certain the feedholes are clear. Push the wire through the feedholes and spray the cleaner into them. Blowing through them with compressed air after cleaning is recommended.

One-way clutches should be inspected for wear or damage, spline damage, surface finish damage, and damaged retainers and grooves. All smooth surfaces should be checked for any imperfections. If there is damage, the clutch should be replaced. Some clutch units have a plastic race that need be to carefully inspected for chafing or other damage. Many one-way clutches are retained by a snap ring that must be removed to remove the clutch.

Roller Clutches

Roller clutches should be disassembled for inspection. The surface of the rollers should have a smooth finish with no evidence of any flatness. Likewise, the race should be smooth and show no sign of brinnelling, as this indicates severe impact loading. This condition may also cause the roller clutch to "buzz" as it overruns. The roller springs should be carefully checked for signs of overheating, wear, or damage. A bad spring may allow a roller to fall out of its race.

All rollers and races that show any type of damage or surface irregularities should be replaced. Check the folded springs for cracks, broken ends, or flattening out. All of the springs should have approximately the same shape. Replace all distorted or otherwise damaged springs. The cam race of a roller clutch may show the same brinnelled wear due to impact overloading. The cam surface, like the smooth race, must be free of all irregularities.

When reassembling a roller clutch, make sure the rollers and springs are facing the correct direction (Figure 9-49). If they are reversed, the clutch will not lock in either direction.

Sprag Clutches

Sprag clutches (Figure 9-50) cannot easily be disassembled; therefore, a complete and thorough inspection of the assembly is necessary. Pay particular attention to faces of the sprags. If the faces are damaged, the clutch unit should be replaced. Sprags and races with scored or torn faces are an indication of dry running and require the replacement of the complete unit. Also check the sprags for looseness in their race. When assembling a sprag-type clutch, it is easy to place the sprags in the wrong direction. Make sure the unit is assembled correctly.

Mechanical Diodes

Some transmissions use a mechanical diode, rather than a sprag or roller clutch. These units use rectangular struts that engage a race when the unit is rotating in the direction of the torque. The struts are positioned between a plate with pockets for the retracted struts and a second plate with notches for the engagement of the struts. These units are often called

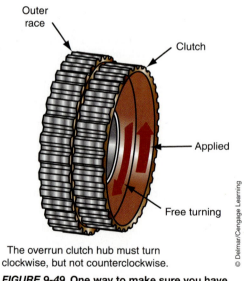

Outer race

Clutch

Applied

Free turning

© Delmar/Cengage Learning

The overrun clutch hub must turn clockwise, but not counterclockwise.

FIGURE 9-49 One way to make sure you have installed the clutch in the correct direction is to determine the direction of lockup before installing the clutch. The operation of a one-way clutch is checked in the same way.

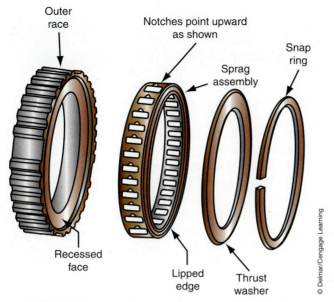

FIGURE 9-50 A typical sprag one-way clutch assembly.

"ratcheting" units because they make a ratcheting or clicking sound when they are freewheeling. This is how these are best checked. In one direction, the sound should be heard and the unit should not rotate in the opposite direction.

Installation

Once the one-way clutches are ready for installation, verify that they overrun in the proper direction. In some cases, it is possible to install these clutches backwards; this would allow them to lock in wrong direction. This would result in some definite driveability problems. One way to make sure you have installed the clutch in the correct direction is to determine the direction of lockup before installing the clutch, then study the transmission's powerflow and match the direction of the clutch with it. Most one-way clutches have some marking that indicates which direction the clutch should be set (Figure 9-51). During installation, use a new retaining snap ring.

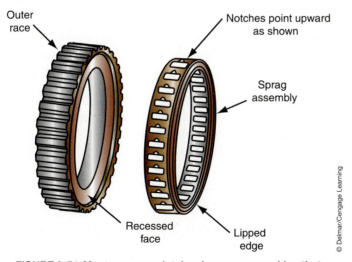

FIGURE 9-51 Most one-way clutches have some marking that indicates in which direction the clutch should be set.

CASE STUDY

A car dealer called to see if the automotive department at the college could help with a diagnostic problem. His customer had a new Ford pickup. The owner complained of a buzzing noise while driving. The technician at the dealership disassembled the transmission and installed an overhaul kit. The next day, the customer picked up his truck. He returned the next day, slightly annoyed, with the same complaint of buzzing while driving. The same technician once again disassembled the transmission and installed another overhaul kit. This time the technician road tested the truck after installing the kit. The noise was still there. By this time, both the owner of the truck and the technician were not very happy. After two rebuilds, the transmission still had the same noise.

The transmission was brought to the college and was checked out on the transmission "dyno." In the forward ranges, the buzzing was only apparent in intermediate and high gear ranges. There was no buzzing in low or reverse gear ranges. The students, using a power flow chart, determined that the only geartrain member not in motion in low and reverse was the rear carrier. Because the rear carrier is held by the one-way clutch, we decided to disassemble the transmission to check the clutch. The inner race of the roller clutch had a flat spot from an apparent machining mishap during manufacture. Every time a roller went past the flat spot, it clicked into the flat, producing the buzz at road speeds. We installed a new roller inner race, which was supplied by the dealer, and, like magic, the buzz was gone.

Why were students able to determine the problem while an experienced technician could not find it? Very simply, the technician did not understand the basic theories of transmission operation. The road test very clearly showed that the buzzing occurred only in second and third gears. A basic knowledge of the Simpson geartrain and knowing how to read and use a flow chart would have helped the technician diagnose the problem. Unfortunately, this technician, like many others, only knew how to "put a kit in it" and hope that the noise would go away. Just as a point of interest, according to the service manager, the technician loaded up her tools and quit when she learned that students, armed with a power flow chart and a shop manual, had diagnosed and repaired the problem that had her stumped.

TERMS TO KNOW

Clutch Volume Index (CVI)

Selective servo apply pins

Steels

Transmission tune-up

ASE-STYLE REVIEW QUESTIONS

1. While visually inspecting a roller-type one-way clutch:
   Technician A replaces the clutch because the rollers have a smooth finish.
   Technician B replaces the clutch because the race shows signs of brinnelling.
   Who is correct?
   A. A only
   B. B only
   C. Both A and B
   D. Neither A nor B

2. *Technician A* says most original equipment servo seals are of the Teflon type.
   Technician B says a majority of replacement transmission gasket sets will supply cast iron hook-end seal rings in place of Teflon seals, and these are acceptable replacements.
   Who is correct?
   A. A only
   B. B only
   C. Both A and B
   D. Neither A nor B

3. Which of the following is *not* a good reason to replace a band?
   A. The band is chipped.
   B. The friction material is glazed.
   C. The friction material is soaked with fluid.
   D. The friction material has random wear patterns.

4. While conducting an air test on a transaxle:
   Technician A notices escaping air at the metal or Teflon seals. He proceeds to replace them.
   Technician B says the clutch should apply with a slight delay, then a dull thud.
   Who is correct?
   A. A only
   B. B only
   C. Both A and B
   D. Neither A nor B

5. While inspecting a clutch assembly that has an aluminum piston:

   Technician A carefully inspects the piston for hairline cracks that will cause pressure leakage, resulting in clutch slippage.

   Technician B says that if burned discs were found in the clutch pack, the piston should be checked for defects and cracks.

   Who is correct?

   A. A only   C. Both A and B
   B. B only   D. Neither A nor B

6. *Technician A* says all adjustable bands have their locknut and adjusting screw on the outside of the transmission case.

   Technician B says some bands do not have adjusting screws.

   Who is correct?

   A. A only   C. Both A and B
   B. B only   D. Neither A nor B

7. *Technician A* says transmissions originally equipped with Teflon seals must be refitted with Teflon seals during an overhaul.

   Technician B says a press is needed for the installation of Teflon seals.

   Who is correct?

   A. A only   C. Both A and B
   B. B only   D. Neither A nor B

8. While assembling a multiple-friction disc pack:

   Technician A alternately stacks the steels and friction discs until the correct number of plates has been installed.

   Technician B installs all of the discs contained in the overhaul kit and corrects the clearance with a selective pressure plate or snap ring.

   Who is correct?

   A. A only   C. Both A and B
   B. B only   D. Neither A nor B

9. *Technician A* says an air test can be used to check servo action.

   Technician B says an air test can be used to check for internal fluid leaks.

   Who is correct?

   A. A only   C. Both A and B
   B. B only   D. Neither A nor B

10. While checking clutch discs:

   Technician A says the steel plates should be replaced if they are worn flat.

   Technician B says the friction discs should be squeezed to see if they can hold fluid. If they hold fluid and look okay, they are serviceable.

   Who is correct?

   A. A only   C. Both A and B
   B. B only   D. Neither A nor B

ASE CHALLENGE QUESTIONS

1. *Technician A* says friction discs that have blackened surfaces have been subject to overheating.

   Technician B says that if the friction facing has been burned or worn off, the discs can become welded together and can cause the vehicle to creep when the transmission is in neutral.

   Who is correct?

   A. A only   C. Both A and B
   B. B only   D. Neither A nor B

2. *Technician A* says excessive groove wear in a clutch piston can be caused by excessive pump pressures.

   Technician B says excessive wear, scoring, and grooving in the bore of a clutch can be caused by a stuck pressure regulator valve.

   Who is correct?

   A. A only   C. Both A and B
   B. B only   D. Neither A nor B

3. *Technician A* says incorrect assembly of servos and accumulators can result in dragging bands and harsh shifting.

   Technician B says internal leaks at the servo or clutch seal will cause excessive pressure drops during gear changes.

   Who is correct?

   A. A only   C. Both A and B
   B. B only   D. Neither A nor B

4. *Technician A* says that if a one-way clutch has failed, a thorough inspection of the hydraulic fluid feed circuit to the clutch must be made to determine if the failure was due to fluid starvation.

   Technician B says that if the race of a roller-type one-way clutch has random hard spots, it is likely that the roller clutch will "buzz" as it overruns.

   Who is correct?

   A. A only
   B. B only
   C. Both A and B
   D. Neither A nor B

5. While inspecting a servo:

   Technician A says the servo needs new seals because both the inside and outside of the cover seal are wet.

   Technician B says that whenever the servo is disassembled, all rubber, cast iron, and Teflon seals should be replaced.

   Who is correct?

   A. A only
   B. B only
   C. Both A and B
   D. Neither A nor B

Name _____ **Date** _____

INSPECTING APPLY DEVICES

Upon completion of this job sheet, you should be able to inspect various apply devices of a transmission.

ASE Correlation

This job sheet is related to the ASE Automatic Transmission and Transaxle Test's Content Area *Off-Vehicle Transmission and Transaxle Repair; Friction and Reaction Units.*

Tasks: Inspect clutch drum, piston, check balls, springs, retainers, seals, and friction and pressure plates; replace as needed. Inspect roller and sprag clutch, races, rollers, sprags, springs, cages, and retainers; replace as needed.

Tools and Materials

Compressed air and air nozzle   Lint-free shop towels
Supply of clean solvent   Service manual

Describe the vehicle being worked on:

Year _____ Make _____ Model _____
VIN _____ Engine type and size _____
Model and type of transmission _____

Procedure

Task Completed

1. Disassemble the transmission into major units. Set each unit aside until this job sheet refers to it. Describe any problems you encountered while disassembling the transmission. Be sure to follow the procedures given in the appropriate service manual while taking the transmission apart.

   _____

   _____

   _____

   _____

2. Clean each overrunning clutch assembly in fresh solvent. Allow the assemblies to air dry.   ☐

3. Check the rollers and sprags for signs of wear or damage. Describe your findings.

   _____

   _____

4. Check the springs for distortion, distress, and damage. Describe your findings.

   _____

   _____

5. Check the inner race and the cam surfaces for scoring and other damage. Describe your findings.

_____

_____

6. Check the condition of the snap rings. Describe your findings.

_____

_____

7. Summarize the condition of the overrunning clutch units.

_____

_____

_____

_____

☐   8. Wipe the transmission's bands clean with a dry, lint-free cloth.

9. Check the bands for damage, wear, distortion, and lining faults. Describe your findings.

_____

_____

10. Inspect the band apply struts for distortion and other damage. Describe your findings.

_____

_____

11. Summarize the condition of the bands.

_____

_____

_____

_____

☐   12. Clean servo and accumulator parts in fresh solvent and allow them to air dry.

13. Inspect their bores for scoring and other damage. Describe their condition.

_____

_____

14. Check the piston and piston rod for wear, nicks, burrs, and scoring. Describe your findings.

_____

_____

15. Inspect all springs for damage and distortion. Describe their condition.

_____

_____

16. Check the mating surfaces between the piston and the walls of their bores for scoring, wear, nicks, and other damage. Describe your findings.

_____

_____

17. Check the movement of each piston in its bore. Describe that movement.

_____

_____

18. Check the fluid passages for restrictions and clean out any dirt present in the passages. Describe your findings.

_____

_____

19. Summarize the condition of the servos and accumulators in this transmission.

_____

_____

_____

_____

Instructor's Response _____

_____

_____

Name _____ **Date** _____

CHECKING AND OVERHAULING A MULTIPLE-FRICTION DISC ASSEMBLY

Upon completion of this job sheet, you should be able to air test clutch packs and servos and disassemble, inspect, and reassemble a multiple-friction disc assembly.

ASE Correlation

This job sheet is related to the ASE Automatic Transmission and Transaxle Test's Content Area *Off-Vehicle Transmission and Transaxle Repair; Friction and Reaction Units.*
Tasks: Measure clutch pack clearance; adjust as needed. Air test operation of clutch and servo assemblies.

Tools and Materials

Feeler gauge set   Lint-free shop towels
Special tools for the assigned transmission   OSHA-approved air nozzle
Emery cloth   Air test plate
Crocus cloth   Service manual
Clean solvent

Describe the transmission being worked on:

Model and type of transmission _____
Vehicle the transmission is from:
Year _____ Make _____ Model _____
VIN _____ Engine type and size _____

Procedure

Task Completed

1. With the transmission disassembled, set aside each clutch pack for inspection. ☐

2. Disassemble each clutch pack. Keep each assembly separate from the others. Record the number of steel and friction plates in each assembly.

   _____

   _____

   _____

3. Check the steel plates for discoloration, scoring, distortion, and other damage. Describe their condition.

   _____

   _____

   _____

4. If the steel plates are in good condition, rough up the shiny surface with emery (80-grit) cloth. ☐

5. Inspect the friction plates for wear, damage, and distortion. Describe their condition.

_____

_____

_____

☐   6. Soak all new and reusable friction plates in ATF for at least 15 minutes before reassembling the clutch pack.

7. Check the pressure plate for discoloration, scoring, and distortion. Describe its condition.

_____

_____

_____

8. Inspect the coil springs for distortion and damage. Describe their condition.

_____

_____

_____

9. Inspect the Belleville spring for wear and distortion. Make sure to check the inner fingers. Describe your findings.

_____

_____

10. Check the clutch drums for damaged lugs, grooves, and splines. Also check them for score marks. Describe their condition.

_____

_____

_____

11. Check the movement of the check balls in the drums. Describe your findings.

_____

_____

_____

12. Summarize the condition of the clutch packs.

_____

_____

_____

13. Reassemble the clutch packs with the required new parts and with new seals. Follow the recommended procedure for doing this.

_____

_____

_____

14. Insert a feeler gauge between the pressure plate and the snap ring. What is the measured clearance? _____

15. What is the specified clearance? _____

16. If the measured clearance is not the same as the specified clearance, what should you do?

_____

_____

_____

17. Do what is necessary to correct the clearance. ☐

18. Tip the clutch assembly on its side and insert a feeler gauge between the pressure plate and the adjacent friction plate. What is the measured clearance? _____

19. What is the specified clearance? _____

20. If the measured clearance is not the same as what was specified, what should you do?

_____

_____

21. Do what is necessary to correct the clearance. ☐

22. After the clearance is set, check the service manual for the proper procedure for air testing the clutch pack. Summarize the procedure.

_____

_____

_____

23. Place the pack or the partially assembled transmission in a vise or on a stand. ☐

24. Apply low air pressure to the designated test port. Pay attention to the sound of the air leaking and the activation of the disc pack. Describe what you heard.

_____

_____

25. Release the air and listen for the release of the pack. Describe what happened.

_____

_____

_____

26. Based on the above, what are your conclusions?

_____

_____

_____

27. When the transmission is assembled, install the air test plate to the designated area on the transmission. What does the plate attach to?

_____

_____

28. What components can be checked with the test plate?

_____

_____

_____

29. Apply low pressure to each of the test ports and record the results.

_____

_____

_____

30. What is indicated by the results of this test?

_____

_____

_____

Instructor's Response _____

_____

_____

Name _____ **Date** _____

CHECKING END CLEARANCE OF A MULTIPLE-FRICTION DISC PACK

Upon completion of this job sheet, you should be able to use a feeler gauge and a service manual to check end clearance of a multiple-friction disc pack.

ASE Correlation

This job sheet is related to the ASE Automatic Transmission and Transaxle Test's content area *Off-Vehicle Transmission and Transaxle Repair: Friction and Reaction Units.*
Task: Measure clutch pack clearance; adjust as needed.

Tools and Materials

Feeler gauge set
A multiple-friction disc pack
Service manual
Goggles or safety glasses with side shields

Describe the transmission the pack is from:

Model and type of transmission _____
Vehicle the transmission is from:
Year _____ Make _____ Model _____
VIN _____ Engine type and size _____

Procedure

1. Insert a feeler gauge between the pressure plate of the pack and the snap ring. What is the measured clearance? _____

2. What is the specified clearance? _____

3. If the measured clearance is not the same as the specified clearance, what should be done?

    _____

    _____

4. Tip the clutch assembly on its side and insert a feeler gauge between the pressure plate and the adjacent friction plate. What is the measured clearance? _____

5. What is the specified clearance? _____

6. If the measured clearance is not the same as what was specified, what should be done?

    _____

    _____

7. What other measuring instrument can be used to measure the end clearance of this pack?

_____

_____

Instructor's Response _____

_____

_____

Chapter 10

REBUILDING COMMON TRANSMISSIONS

UPON COMPLETION AND REVIEW OF THIS CHAPTER, YOU SHOULD BE ABLE TO:

- Disassemble, service, and reassemble a Chrysler Torqueflite 36RH transmission.

- Disassemble, service, and reassemble a Chrysler 41TE transaxle.

- Disassemble, service, and reassemble a Chrysler 42LE transaxle.

- Disassemble, service, and reassemble a Ford 4R44E transmission.

- Disassemble, service, and reassemble a Ford 4EAT/ Mazda GF4A-EL transaxle.

- Disassemble, service, and reassemble a Ford AX4S transaxle.

- Disassemble, service, and reassemble a GM 4L60 transmission.

- Disassemble, service, and reassemble a GM 3T40 transaxle.

- Disassemble, service, and reassemble a GM 4T60 transaxle.

- Disassemble, service, and reassemble a Honda F4 transaxle.

- Disassemble, service, and reassemble a Nissan L4N71B/E4N71B transmission.

- Disassemble, service, and reassemble a Toyota A541E transaxle.

This chapter contains many photo sequences that show the step-by-step procedures for overhauling common transmissions and transaxles. Although a particular model of transmission was chosen for each set of photographs, variations of the same transmission can be used. The procedures in this chapter are for educational purposes and do not include the many changes that a different model would require.

In addition, please note that these sequences do not include detailed procedures for rebuilding subassemblies such as the valve body and clutch packs. Detailed information on these, as well as other subassemblies, is given in other chapters of this manual.

CHRYSLER TRANSMISSIONS

Chrysler Corporation introduced the Torqueflite transmission in 1956. This transmission was the first modern three-speed automatic transmission with a torque converter and the first to use the Simpson two-planetary compound geartrain. Nearly all Torqueflite-based transmissions and transaxles use a rotor-type oil pump and all use a Simpson geartrain. Late-model Torqueflites have a one-piece aluminum housing and a bolt-on extension housing.

There are two basic versions of the three-speed Torqueflite transmission: the 30RH (old A-904) and the 36RH (old A-727). The difference between the two is the intended use. The 36RH is designed for heavier-duty use.

The 42RE is a commonly found four-speed transmission and is used in pickups and SUVs. The 42RE is based on the 36H, which is a three-speed transmission. The primary difference

Mechanic's basic tool set

Appropriate service manual

Transmission holding fixture

Torque wrench

Rubber-tipped air nozzle

Pan of clean ATF

Feeler gauge

Classroom Manual

Chapter 10, page 310

between the two transmission models is that the 42RE has an additional planetary gearset at the rear of the compound planetary gearset. Hydraulic control of the first three forward speeds is the same as the 36RH. Overdrive is controlled by a direct clutch, an overdrive clutch, and an overdrive one-way clutch. The output from the compound gearset passes through the overdrive planetary gearset. In all gears, except overdrive, torque flows through the direct clutch to the output shaft. When overdrive is selected, the torque passes through the ring gear to the carrier of the overdrive planetary gearset.

The 42RE and 36RH are nearly identical, as are the service procedures for both. Like all transmissions, modifications have been made each year to include more electronic controls. This is true of the 36RH and especially true of the 42RE. Photo Sequence 16 goes through the procedure for overhauling a 36RH transmission. These overhaul procedures are similar to those for other Chrysler transmissions.

PHOTO SEQUENCE 16

TYPICAL PROCEDURE FOR OVERHAULING A 36RH TRANSMISSION

All photos in this sequence are © Delmar/Cengage Learning.

P16-1 Check and record the endplay of the input shaft, then unbolt and remove the oil pan and gasket.

P16-2 Unbolt and remove the oil filter. Then unbolt and remove the valve body by first disconnecting the parking lock rod from the manual lever and unbolting the TCC solenoid.

P16-3 Remove the accumulator piston spring and lift piston from the case.

P16-4 Remove parking lock rod. Then unbolt and remove the extension housing. There is a cover plate in the bottom of the extension housing that once removed will allow access to the output shaft's snap ring. This ring must be expanded while pulling the extension housing off the case.

P16-5 Tighten the front band adjustment screw to prevent damage to the discs during disassembly.

P16-6 Remove the oil pump bolts and pull the oil pump out of the case with special pullers.

P16-7 Loosen the front band adjustment screw and remove the band strut and anchor. Then remove the band.

P16-8 Remove the retaining or snap ring, then the front clutch assembly, with the input shaft, from the case.

P16-9 Remove front planetary gearset.

P16-10 Loosen rear band adjustment screw, then remove the band strut and band. Then remove the output shaft with the governor attached to it.

P16-11 Compress the return spring for the front servo and remove the retaining snap ring.

P16-12 Remove servo assembly from the case.

P16-13 Compress the return spring for the rear servo and remove the retaining snap ring.

P16-14 Remove servo assembly from the case.

P16-15 Inspect all servo piston seals.

P16-16 Reinstall the servo assemblies after careful inspection.

P16-17 Inspect the planetary gearset and drive shells. Replace components as needed.

P16-18 Inspect the bands and their struts and anchors. Replace as needed.

P16-19 Disassemble and inspect the clutch units. Replace parts as needed. Remember to allow the discs to soak in ATF before assembling the clutch pack.

P16-20 Replace the pistons' oil seal rings in the clutch assemblies.

P16-21 Reassemble the clutch packs, and then check the clearances.

P16-22 Unbolt the stator support from the oil pump housing. Separate the support from the oil pump.

P16-23 Check the wear of the gears and the pump itself. Replace parts as needed.

P16-24 Replace the pump seals and reassemble the pump. Torque the bolts to specifications.

P16-25 Disassemble, clean, and inspect the valve body.

P16-26 Bolt valve body together and tighten bolts to specifications.

P16-27 Install the output shaft with the governor into the case. Then install the rear band and anchor assembly and tighten the adjustment screw.

P16-28 Install the rear planetary gearset and drum. Make sure all thrust washers are being installed in their correct locations.

P16-29 Install the front clutch assembly and the front band assembly, and then tighten the adjustment screw.

P16-30 Assemble servos with new seals. Then install the oil pump and torque the bolts in the proper sequence and to the proper torque.

P16-31 Adjust both bands according to specifications.

P16-32 Air test the entire transmission before installing the oil filter and the oil pan.

SPECIAL TOOLS

Dial indicator and holding fixture

Clutch compressor tool

Oil pump puller

Seal remover/installer

Output shaft holding tool

Classroom Manual

Chapter 10, page 311

Classroom Manual

Chapter 10, page 315

SPECIAL TOOLS

Dial indicator and holding fixture

Clutch compressor tool

Oil pump puller

Seal remover/ installer

Output shaft holding tool

Chrysler Torqueflite Transaxles

In 1978, the basic Torqueflite transmission was modified for use as a FWD transaxle. Torqueflite transaxles contain the same basic parts as the 36RH and 42RE transmissions, with the addition of a transfer shaft, final drive gears, and differential unit.

Many different transaxle models have been used since then. The 41TE is commonly found in minivans and some late-model cars. This is a four-speed unit with electronic controls. The compound planetary gearset in the 41TE does not have a common member; rather, the two planetary units are connected in tandem, with the front carrier connected to the rear ring gear and the rear carrier connected to the front ring gear.

To control the forward gear ranges, the 41TE relies on an input clutch assembly, a 2nd/4th clutch, and a low/reverse clutch. The input clutch assembly houses the three different input clutches: the underdrive, overdrive, and reverse clutches.

The underdrive clutch is applied in first, second, and third gears. When applied, the underdrive hub drives the rear sun gear. The overdrive clutch is applied in third and overdrive gears. When the overdrive clutch is applied, it drives the front planet carrier. The underdrive clutch and the overdrive clutch are applied in third gear, giving two inputs to the gearset, which provides for direct drive. The reverse clutch is only applied in reverse and drives the front sun gear assembly.

The 2nd/4th gear and low/reverse units are holding devices and are placed at the rear of the transmission case. The 2nd/4th clutch is engaged in second and fourth gears and grounds the front sun gear to the case. The low/reverse clutch is applied in park, reverse, neutral, and first gear. When this clutch is engaged, the front planet carrier and rear ring gear assembly is grounded to the case. The overhaul procedure for this transaxle is shown in Photo Sequence 17.

With the introduction of new mid-size FWD cars in 1993, Chrysler introduced the 42LE, which is a longitudinally mounted transaxle. Although this transaxle is based on the 41TE and in most ways is very similar to the 41TE, its location in the vehicle requires different overhaul procedures. The 42LE uses a chain drive to transfer the output of the planetary gearset to a hypoid-type final drive unit. As a result, its appearance and construction is different. Therefore, Photo Sequence 18 is included to show the step-by-step procedures for overhauling this transaxle.

PHOTO SEQUENCE 17

TYPICAL PROCEDURE FOR OVERHAULING A 41TE TRANSAXLE

All photos in this sequence are © Delmar/Cengage Learning.

P17-1 Remove all electrical switches from the outside of the transaxle case.

P17-2 Remove the solenoid assembly and gasket.

P17-3 Remove the torque converter.

P17-4 Loosen and remove the oil pan's bolts. Then remove the oil pan.

P17-5 Remove the oil filter.

P17-6 Loosen and remove the valve body attaching bolts. Then push the park rod rollers away from the guide bracket and remove the valve body assembly. Pull straight up on the valve body as the manual shaft is attached to it.

P17-7 Remove the retaining snap ring for the accumulator, then remove the accumulator assembly.

P17-8 Remove the oil pump seal, using the correct seal puller.

P17-9 Loosen and remove the oil pump's attaching bolts. Then, using the correct puller, remove the oil pump. While pulling the pump, push in on the input shaft.

P17-10 Pull out the input shaft and clutch assembly.

P17-11 Remove and discard the oil pump gasket. Then remove the caged needle bearing assembly from the input shaft.

P17-12 Remove the front sun gear assembly.

P17-13 Twist and pull on the front carrier and rear ring gear assembly to remove it.

P17-14 Remove the rear sun gear with its thrust bearing.

P17-15 Using the proper clutch compressor, remove the 2–4 clutch snap ring and clutch retainer. Mark the alignment of the retainer with the return spring located below it. Then remove the return spring.

P17-16 Remove the clutch pack and tag it as the 2–4 pack.

P17-17 Remove the low/reverse tapered snap ring from the case. Follow the recommended sequence while prying the snap ring out of the case.

P17-18 Remove the low/reverse reaction plate and one friction disc.

P17-19 Remove the low/reverse reaction plate flat snap ring, and then remove the clutch pack. Tag the pack as the low/reverse clutch.

P17-20 Loosen and remove the rear cover bolts. Then remove the rear cover.

P17-21 Remove the transfer shaft gear nut and washer. Then, using a gear puller, remove the transfer gear.

P17-22 Remove the transfer gear selective shim and the bearing retainer. Then remove the transfer shaft bearing snap ring.

P17-23 Remove the transfer shaft and the output shaft gear bolt and washer. Then, using a gear puller, remove the output gear and selective shim.

P17-24 Remove the rear carrier assembly.

P17-25 After careful inspection of all parts, begin reassembly by installing the rear carrier assembly.

P17-26 Install the output gear, selective shim, washer, and bolt. While holding the shaft with a special holding tool, tighten the bolt to specifications.

P17-27 Install the transfer shaft.

P17-28 Install the transfer shaft bearing snap ring, selective shim, and bearing retainer. Then install the transfer gear, washer, and nut. While holding the shaft with a special tool, tighten the nut to specifications.

P17-29 Apply a 1/8-inch bead of RTV on the rear cover. Then install the rear cover and tighten the rear cover bolts to specifications.

P17-30 Install the low/reverse clutch pack, reaction plate flat snap ring, friction disc, and low/reverse reaction plate. Then install the low/reverse tapered snap ring into the case. Follow the recommended sequence while installing the snap ring into the case.

P17-31 Install the 2–4 clutch pack snap ring and return spring.

P17-32 Using the proper clutch compressor, install the 2–4 clutch snap ring, and clutch retainer. Make sure the alignment marks on the spring and retainer are matched.

P17-33 Install the rear sun gear with its thrust bearing. Then, install the front carrier and rear ring gear assembly by twisting and pushing on it.

P17-34 Install the front sun gear assembly.

P17-35 Install the new oil pump gasket. Then, install the oil pump. Tighten the oil pump's attaching bolts to specifications.

P17-36 Install the oil pump seal, using the correct seal installing tool.

P17-37 Install the accumulator assembly and retaining snap ring. Then install the valve body assembly. Push the park rod rollers away from the guide bracket while positioning the valve body. Install and tighten the valve body retaining bolts to specifications.

P17-38 Install a new oil filter. Then, install the oil pan with a new gasket. Tighten the oil pan bolts to specifications.

P17-39 Install the torque converter, the solenoid assembly with a new gasket, and all electrical switches located on the outside of the transaxle case.

TYPICAL PROCEDURE FOR OVERHAULING A 42LE TRANSAXLE

All photos in this sequence are © Delmar/Cengage Learning.

P18-1 Remove all electrical switches from the outside of the transaxle case.

P18-2 Remove the torque converter and measure the endplay of the transaxle.

P18-3 Loosen and remove the oil pan's bolts. Then remove the oil pan.

P18-4 Remove the oil filter.

P18-5 Loosen and remove the valve body attaching bolts. Then remove the valve body assembly. Care should be taken not to lose the overdrive and underdrive accumulator springs, which may fall out when removing the valve body.

P18-6 Remove the retaining snap ring for the accumulators, and then remove the accumulator assemblies. Be careful to keep the parts of each accumulator separated from each other.

P18-7 Unbolt and remove the solenoid assembly from the valve body.

P18-8 Remove the long stub shaft from the transaxle.

P18-9 Index the inner and outer differential adjusters at the case.

P18-10 Remove the lock bracket and back out the adjuster one complete revolution. Then remove the differential cover.

P18-11 Remove the inner adjuster lock bracket, then remove the inner adjuster.

P18-12 Remove the transfer shaft nut, rear cone, rear cup, oil baffle, rear shim, transfer shaft, and transfer shaft seals.

P18-13 Remove the oil pump seal, using the correct seal puller.

P18-14 Loosen and remove the oil pump's attaching bolts. Then, using the correct puller, remove the oil pump. While pulling the pump, push in on the input shaft.

P18-15 Remove and discard the oil pump gasket. Then remove the by-pass valve from the case.

P18-16 Remove the caged needle bearing assembly from the input shaft.

P18-17 Pull out the input shaft and clutch assembly.

P18-18 Remove the front sun gear assembly and thrust washer.

P18-19 Twist and pull on the front carrier and rear ring gear assembly to remove it. Then remove the rear sun gear with its thrust bearing.

P18-20 Using the proper clutch compressor, remove the 2–4 clutch snap ring and clutch retainer. Mark the alignment of the retainer with the return spring located below it. Then remove the return spring.

P18-21 Remove the clutch pack and tag it as the 2–4 pack.

P18-22 Remove the low/reverse tapered snap ring from the case. Follow the recommended sequence while prying the snap ring out of the case.

P18-23 Remove the low/reverse reaction plate and one friction disc. Then remove the low/reverse reaction plate flat snap ring and the clutch pack. Tag the pack as the low/reverse clutch.

P18-24 Remove the output shaft gear bolt and washer. Then, using a gear puller, remove the output gear and selective shim.

P18-25 Press the output shaft from the case.

P18-26 After careful inspection of all parts, begin reassembly by installing the output shaft.

P18-27 Install the output gear, selective shim, washer, and bolt. While holding the shaft with a special holding tool, tighten the bolt to specifications.

P18-28 Install the low/reverse clutch pack and the reaction plate flat snap ring.

P18-29 Install the friction disc and low/reverse reaction plate. Then install the low/reverse tapered snap ring into the case. Follow the recommended sequence while installing the snap ring into the case.

P18-30 Install the 2–4 clutch pack and return spring.

P18-31 Using the proper clutch compressor, install the 2–4 clutch snap ring and clutch retainer. Make sure the alignment marks on the spring and retainer are matched.

P18-32 Install the rear sun gear with its thrust bearing. Then, install the front carrier and rear ring gear assembly by twisting and pushing on it.

P18-33 Install the front sun gear assembly.

P18-34 Install the input shaft and clutch assembly. Then install the input shaft's caged needle bearing.

P18-35 Install the new oil pump gasket. Then, install the oil pump. Tighten the oil pump's attaching bolts to specifications.

P18-36 Install the oil pump seal, using the correct seal installing tool.

P18-37 Install the transfer shaft, transfer shaft seals, oil baffle, rear cup, rear cone, and a new shaft nut.

P18-38 Install a new O-ring onto the inner adjuster. Lube the inner adjuster threads and O-ring with gear oil and reinstall according to the index marks made during disassembly.

P18-39 Reinstall the inner adjuster locking bracket.

P18-40 Install the differential carrier. Then install a stub shaft seal protector.

P18-41 Install the differential cover/outer adjuster assembly with silicone sealant. Install and tighten differential cover bolts.

P18-42 Tighten the outer adjuster 3/4 of a turn. Seat the bearings by turning the differential carrier three or four turns in each direction. Finish tightening the adjuster 1/4 turn to its index mark. Then, reinstall the adjuster bracket and long stub shaft.

P18-43 Install the solenoid assembly onto the valve body.

P18-44 Install the accumulator assemblies and retaining snap rings. Then install the valve body assembly. Install and tighten the valve body retaining bolts to specifications.

P18-45 Install a new oil filter. Then, install the oil pan with a new gasket. Tighten the oil pan bolts to specifications.

P18-46 Install the torque converter and all electrical switches on the outside of the transaxle case.

P18-47 Install a new oil pump gasket on the oil pump.

P18-48 Install the bypass valve and oil pump, then tighten the oil pump's attaching bolts to specifications.

P18-49 Install the transfer shaft nut, transfer shaft seals, oil baffle, rear cup, rear cone, rear shim, transfer shaft, and a new shaft nut.

P18-50 Install the chain and sprockets as a unit using the chain spreader tool.

P18-51 Install the wave washers and snap rings on the output and transfer shafts.

P18-52 Install the chain snubber guide.

P18-53 Install the rear chain cover and torque the bolts to specifications.

P18-54 Install a new O-ring onto the inner adjuster. Lube the inner adjuster threads and O-ring with gear oil and reinstall according to the index marks made during disassembly. Then install the inner adjuster and locking bracket. Torque the locking bracket to specifications.

P18-55 Install the differential carrier and ring gear assembly.

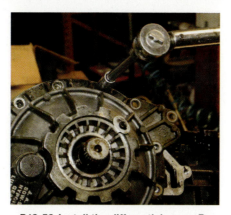

P18-56 Install the differential cover. Be careful to protect the seals. Torque the attaching bolts to specifications.

P18-57 Install the outer adjuster lock and torque its bolt to specifications.

P18-58 Install the long stub shaft and snap ring.

P18-59 Install the solenoid assembly on the valve body.

P18-60 Install the accumulator spring, accumulator, O-ring, and snap ring.

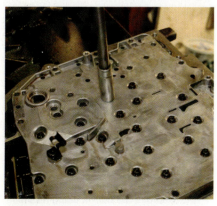

P18-61 Install the valve body assembly. Install and tighten the valve body retaining bolts to specifications.

P18-62 Install a new oil filter.

P18-63 Install the oil pan with a new gasket. Tighten the oil pan bolts to specifications.

P18-64 Attach the wiring harness to the case.

P18-65 Install the electrical switches on the outside of the transaxle case. Tighten them to specifications.

SPECIAL TOOLS

Dial indicator and holding fixture

Clutch compressor tool

Oil pump puller

Oil pump alignment tool

Seal remover/ installer

Tapered punch

Classroom Manual
Chapter 10, page 322

FORD MOTOR COMPANY TRANSMISSIONS

Ford Motor Company began to use the Simpson geartrain with the introduction of the C-4 transmission in 1964. Previous transmissions were based on the Ravigneaux design, as are some of their current models. The most commonly used transmissions in cars and trucks are the 4R70W, 4R100, and 4R44E. Another commonly used transmission is a heavy-duty version of the 4R44E, referred to as the 4R55E. The 4R44E (old A4LD) is a Simpson gearset-based four-speed transmission. It has full electronic controls and relies on three multiple-friction disc packs, three brake bands, and two one-way clutches to provide the various gear ratios. The overhaul procedures for this transmission are shown in Photo Sequence 19.

PHOTO SEQUENCE 19

TYPICAL PROCEDURE FOR OVERHAULING A 4R44E TRANSMISSION

All photos in this sequence are © Delmar/Cengage Learning.

P19-1 Remove the torque converter.

P19-2 Remove the input shaft. Note that the two splined ends of the shaft are different.

P19-3 Remove the bell-housing retaining bolts, and then the bell housing and oil pump as an assembly. Also remove the thrust washer and gasket.

P19-4 Separate the oil pump from the bell housing.

P19-5 Remove the steel plate from the housing.

P19-6 Remove the oil pan bolts, then the oil pan.

P19-7 Remove the retaining bolt for the filter screen. Remove filter screen, then remove the detent spring under the screen.

P19-8 Disconnect the wires at the solenoids.

P19-9 Remove the valve body retaining bolts. While moving the valve body away, unlock and remove the selector lever connecting link. Then remove the valve body and gasket. Note and mark the valve body bolts, they are different lengths.

P19-10 Remove the Allen head retaining bolt holding the center support from within the case's worm tracks.

P19-11 Remove solenoid harness and connector from case.

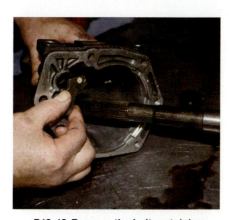

P19-12 Remove the bolts retaining the extension housing to the case, then remove the extension housing, parking pawl, and pawl spring.

P19-13 Loosen overdrive band locknut and back off adjustment screw. Discard the locknut.

P19-14 Remove the anchor and apply struts.

P19-15 Lift out clutch assembly and band. Mark the band as "Overdrive" and label the anchor end of the band.

P19-16 Remove the overdrive one-way clutch.

P19-17 Pull the overdrive planetary assembly out of the case.

P19-18 Remove the center support retaining snap ring.

P19-19 Remove the overdrive apply lever and shaft. Then remove the overdrive control bracket from the valve body side of the case. The overdrive apply lever shaft is longer than the intermediate apply lever shaft.

P19-20 Remove and mark the thrust washer from the top of the center support.

P19-21 Carefully remove the center support bearing by prying on it evenly and upwardly.

P19-22 Remove and mark the thrust washer from below the center support.

P19-23 Loosen the intermediate band locknut and back off the adjusting screw. Discard the locknut. Remove the anchor and apply struts. Then remove the reverse/high and forward clutch assembly.

P19-24 Remove the intermediate band. Mark the band as "Intermediate" and label the anchor end of the band.

P19-25 Remove the forward planetary gear assembly with its thrust washer. Mark the thrust washer.

P19-26 Remove the sun gear shell.

P19-27 Remove the planet and ring gear assembly.

P19-28 Remove the large snap ring from the case to release the reverse planetary gear assembly.

P19-29 Remove the reverse planetary gear assembly with its thrust washer. Mark the thrust washer.

P19-30 Remove the small snap ring on the output shaft and remove the output shaft ring gear.

P19-31 Remove the low/reverse drum and one-way clutch assembly.

P19-32 Remove the low/reverse servo from the valve body side of the case.

P19-33 Remove the low/reverse band. Then remove and mark the thrust washer.

P19-34 Remove the intermediate band apply lever and shaft.

P19-35 Turn transmission so that the output shaft points upward. Then lift out the output shaft.

P19-36 Remove the park gear/collector body assembly from the rear of the case. Then remove and mark the thrust washer.

P19-37 Remove the intermediate and overdrive servos' cover snap rings, covers, pistons, and springs. Mark the covers as to which is the overdrive servo cover.

P19-38 Remove the neutral/safety switch.

P19-39 Remove the linkage centering pin, manual lever nut, manual lever, internal park pawl rod, and detent plate assembly. Then remove lever shaft oil seal.

P19-40 Remove TCC and 3-4 shift solenoid connector. To remove the connector, a tab on the outside of the case must be depressed while the connector is pulled.

P19-41 Before installing the center support into the case, install new high clutch seals onto the support hub. You must size these seals. Failure to do so will cause the seals to

be cut or rolled over during installation. Use the overdrive brake drum for sizing. Carefully rotate the center support while inserting it into the drum housing. Apply a liberal amount of transjel to the center support hub and seals.

P19-42 Make sure the center support is seated fully into the overdrive drum. Set the assembly aside to allow the assembly to stand for several minutes, this will allow the seals to seat in their grooves.

P19-43 Position the thrust washer in the rear of the case. Install the collector body and output shaft. Tighten the bolts to specifications.

P19-44 Position the thrust washer into the case from the front. Then install the low-reverse brake drum, output shaft ring gear and snap ring. Position the reverse planet assembly and thrust washers.

P19-45 Install the snap ring to retain the center support. Ends of the snap ring should be positioned in the wide shallow cavity located in the five o'clock position. Then install the Allen head bolt that retains the center support to the case. Note: Two types of snap rings are used. One has no notches and the other has a notch on its outer and inner diameters. The outer diameter notch should be positioned on the left.

P19-46 Replace the piston's O-ring and install the intermediate servo piston assembly, piston cover, and snap ring.

P19-47 Install the intermediate servo apply lever and shaft into the case. Then install the complete forward clutch and reverse-high clutch assemblies. Rotate the transmission so that the output shaft points downward.

P19-48 Install the intermediate band, apply strut, and anchor strut. Temporarily install the input shaft.

P19-49 Insert the input shaft with its short splines facing down, through the center support, and into the splines of the forward clutch cylinder. Carefully place the center support into the case. Do not seat it into the intermediate brake drum but make sure it is in line with the retaining bolthole in the worm tracks.

P19-50 Install the sun gear and support into the overdrive planet assembly and one_way clutch. Make sure the needle bearing race is centered inside the planetary assembly.

P19-51 Install the overdrive planet assembly and one_way clutch into the case. Install the overdrive drum assembly, overdrive bracket, apply lever, and shaft. Then install the overdrive band, apply strut, and anchor strut.

P19-52 Make sure the needle bearing race in the overdrive planetary is centered and the overdrive clutch is fully seated. Then place the selective washer on top of the overdrive clutch drum and temporarily install the pump assembly (without its gasket) into the case.

P19-53 Using a dial indicator check the endplay. If the endplay exceeds limits, replace the selective washer with one that will allow for proper endplay. After the endplay is correct, remove the oil pump and the selective thrust washer.

P19-54 Install a new oil pump seal and position the separator plate onto the converter housing. Position the pump assembly onto the separator plate and converter housing. Install the retaining bolts finger tight.

P19-55 With the recommended tool, align the pump in the converter housing to prevent seal leakage, pump gear breakage, or bushing failure.

P19-56 Install and tighten the retaining bolts. Then remove the alignment tool.

P19-57 Install the input shaft into the pump and install the converter into the pump gears. Rotate the converter to check for free movement.

P19-58 Install a new lock nut on the overdrive band adjusting screw. Tighten the adjusting screw to 10 ft. lbs. and back it off exactly 2 turns. Then tighten the lock nut to 35_45 ft. lbs. Repeat this procedure for the intermediate band. Perform air pressure tests to ensure proper transmission operation.

P19-59 Install the shift lever oil seal. Then install the internal shift linkage, external manual control lever, and centering pin. Then tighten the nut to specifications. Install the O-ring and retaining nut. Install the Neutral/safety switch.

P19-60 Using tapered punches, align the valve body to the separator plate and gasket. Use transjel to hold gasket in place, then tighten the retaining bolts to specifications. Attach and lock the link to the manual valve. Carefully ease the valve body into case and install and tighten the retaining bolts. Make sure the correct length bolts are installed in their proper location.

P19-61 Connect the wires to the various solenoids.

P19-62 Then install the reverse servo piston assembly and spring. Make sure the piston rod is correctly seated into the reverse band apply end. Then install the correct servo rod and cover, using a new servo cover gasket. Then tighten the attaching bolts. Install new O-rings on the filter screen. Then install the filter screen and the oil pan with a new gasket. Tighten the pan bolts to specifications in two steps.

P19-63 Install the parking pawl and return spring into the extension housing. Then install the extension housing with a new gasket. Make sure the parking pawl rod is fully seated. Then install and tighten the extension housing bolts.

P19-64 Finish assembly by installing the input (pump drive) shaft and torque converter. Make sure the torque converter is properly seated.

SPECIAL TOOLS

Dial indicator and holding fixture

Clutch compressor tool

Oil pump puller

Seal remover/ installer

Servo cover compressor tool

Depth micrometer and fixture

Seal protector

Ford Transaxles

The AX4S (old AXOD) four-speed transaxle is used in many FWD vehicles and is commonly found in Taurus, Windstar, and Sable models. It relies on two simple planetary units that operate in tandem. The planet carriers of each planetary unit are locked to the other planetary unit's ring gear. Each planetary set has its own sun gear and set of planet pinion gears. The AXOD uses four multiple-friction disc assemblies, two band assemblies, and two one-way clutches to control the operation of the planetary gearset. All the multiple-disc packs are applied hydraulically and are released by several small coil springs when hydraulic pressure is diverted from the clutch's piston. The overhaul procedure for this transaxle is shown in Photo Sequence 20.

Classroom Manual
Chapter 10, page 329

PHOTO SEQUENCE 20

TYPICAL PROCEDURE FOR OVERHAULING AN AX4S TRANSAXLE

All photos in this sequence are © Delmar/Cengage Learning.

P20-1 Remove the torque converter and mount the transaxle in a vertical position on a holding fixture. Then drain the fluid.

P20-2 Move the transaxle to a horizontal position. Remove the two governor cover bolts, and then remove the cover and seal. Discard the seal. Lift the governor, speedometer drive gear assembly, and bearing from case.

P20-3 Remove the bolts from the overdrive servo cover. Then mark the alignment of the cover and remove it, the piston assembly, and spring. Discard the O ring from the cover. Then remove the low-intermediate servo cover bolts. Mark the alignment of the cover and remove it, the piston assembly, and the spring. Remove and discard the gasket.

P20-4 Remove the neutral safety switch bolts and remove the switch.

P20-5 Remove the dipstick tube attaching bolt and pull the tube from the case. Then, remove the chain cover bolts from inside torque converter housing.

P20-6 Remove the valve body cover bolts, valve body cover, and gasket. Discard the gasket.

P20-7 Unplug the electrical connectors from the pressure switches and solenoid. Use both hands to do this and do not pull on the wires, pull on the connector. Compress the tabs on both sides of the five-pin bulkhead connector from inside of the chain cover and remove the connector and wiring.

P20-8 Using a 9 mm wrench on the flats of the manual shaft, rotate the shaft clockwise so that the manual valve is all the way in. Note the location, according to length, of the oil pump and valve body bolts. Then remove the bolts. Do not remove the two bolts that hold the oil pump and valve body together. The oil pump cover bolts should also not be removed at this time.

P20-9 Push in the TV plunger. Pull the pump and valve body assembly outward. Rotate the valve body clockwise and remove the manual valve link from the manual valve and disconnect it from the detent lever. Now remove the oil pump and valve body assembly.

P20-10 Remove the throttle valve bracket bolts and remove the bracket. Pull the oil pump drive shaft out of the case and remove and discard the Teflon seals from the shaft.

P20-11 Place transaxle in a vertical position. Then remove and discard the circlip for the left-hand output shaft. Then remove the shaft's seal with a seal remover.

P20-12 Mark the location and length of the chain cover bolts. Then remove them and the chain cover. Discard the gasket.

P20-13 Mark and remove the accumulator springs from the chain cover.

P20-14 Simultaneously lift out both sprockets with the chain assembly. Mark and remove the thrust washers from the drive and driven sprocket supports. Inspect the drive sprocket support bearing to determine if it needs replacing.

P20-15 Remove the lock pin and roll pins from the manual shaft. Be careful not to damage the machined surfaces. Slide the shaft out of the case, and then pry the seal from the case.

P20-16 Use a straightedge and note whether the machined bolthole surface of the driven sprocket support is above or below the case's machined surface. Remember this for reassembly. Then remove the driven sprocket support assembly. It may be necessary to back out the reverse clutch anchor bolt.

P20-17 Remove the Teflon seals from the support assembly. Mark and remove the thrust washers and needle bearing. Then remove the plastic overdrive band retainer and the overdrive band.

P20-18 Lift the front sun gear and shell assembly out of the case.

P20-19 Remove the oil pan cover bolts and the cover. Discard the gasket. Remove the reverse apply tube/oil filter bolt, bracket, and oil filter screen.

P20-20 Remove the tube retaining bracket bolts and brackets. NOTE: For complete transaxle disassembly, the reverse apply tube MUST be removed prior to removing the reverse clutch. The rear lube tube must also be removed and the seal replaced whenever the differential is removed.

P20-21 Remove the park rod abutment bolts. Remove the roll pin for the park pawl shaft. Then using a magnet, remove the park pawl shaft, park pawl, and return spring.

P20-22 Place the transaxle in a horizontal position. Grasp the outer diameter of the reverse clutch cylinder with your fingertips and slide the assembly out of case.

P20-23 Place the transaxle in a vertical position. Grasp the front planetary shaft and lift out both the front and rear planetary assemblies. Lift out the low-intermediate drum and sun gear assembly. Then, remove the low-intermediate band.

P20-24 Remove the snap ring for the final drive assembly from the case using a screwdriver inserted through side of case. Lift out the final drive assembly using the output shaft.

P20-25 Remove the final drive ring gear, thrust washer, and needle bearing. Remove and discard the rear lube tube seal by tapping it toward the inside of the case. Then, install the converter oil seal and right-hand output shaft seal using a seal installer.

P20-26 Place the transaxle in a vertical position. Install the needle bearing over the case boss with its flat side facing up and outer lip facing down.

P20-27 Install the final drive ring gear with its external splines up. Lightly tap the ring gear to fully seat it in the case.

P20-28 Reassemble the governor drive gear, differential assembly, final drive sun gear, parking gear, needle bearings, rear planetary support, and thrust washer.

P20-29 Lower the final drive assembly into the case. Install the snap ring and align it with the low-intermediate band anchor pin.

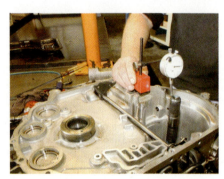

P20-30 Mount a dial indicator with its plunger on the end of the output shaft. Check end clearance. If the clearance is not within specifications, replace the thrust washer with one that will bring the clearance to specifications. Available thicknesses are: 0.045–0.049" (1.15–1.25 mm) Orange, 0.055–0.059" (1.40–1.50 mm) Purple, and 0.064–0.069" (1.65–1.75 mm) Yellow.

P20-31 Install the park pawl, return spring, park pawl shaft, and locator pin. Make sure the park pawl engages park gear and returns freely. Install the park rod actuating lever and park rod into the case. Install the park rod abutment and start the abutment bolts. Push in the park pawl and locate the rod between the pawl and the abutment.

P20-32 Using a 3/8" drift, gently install lube tube seal flush against the rear case support. Install the low-intermediate band and align the anchor pin pocket with the anchor pin. Install the low-intermediate drum and sun gear assembly.

P20-33 Reassemble the ring gear and shell assembly, rear planetary, needle bearing, front planetary and snap ring. Carefully slide the planetary assembly over the output shaft.

P20-34 Lower the reverse clutch into the case and start clutch plate engagement. Align the clutch cylinder anchor pin pocket with the anchor pin case hole. Use the intermediate clutch hub to complete clutch plate engagement and fully seat the reverse clutch. Rotate the planet with the hub to engage splines.

P20-35 Start the reverse anchor pin bolt, but do not tighten. Reassemble the forward, direct, and intermediate clutch assembly. Lower the assembly into the case. Align the shell and sun gear splines with those in the forward planetary.

P20-36 Install the overdrive band into the case. Install the plastic retainer with the crosshairs facing up.

P20-37 Check the drive sprocket end clearance to determine required thrust washer thicknesses. To do this, you must first determine if the machined bolthole surfaces on the driven sprocket support are above or below the case machined surface. If they are above the surface, place a depth micrometer on the machined bolthole surface and measure the distance to the case's machined surface. If they are below the case's surface, place the depth micrometer on the case's surface and measure the distance to the machined bolthole surface. Install the correct thrust washer. Repeat this procedure for the drive sprocket.

P20-38 Tap the seal for the manual shaft into the case. Start the manual shaft through the seal and slide the manual detent lever onto the shaft. Then slide the shaft through the park rod actuating lever and tap it into the bore in the case. Install a new lock pin through the case hole. Make sure it is aligned with the groove in the shaft. Install new roll pins.

P20-39 Align the tabs of the thrust washers and install them onto the drive and driven sprocket supports. Lubricate and install the cast iron sealing ring onto the input shaft. Install the chain over the sprockets.

P20-40 Lower the chain assembly onto sprocket supports. Rotate the supports while lowering the chain assembly to ensure that the sprockets are fully seated on the supports.

P20-41 Install and align the thrust washers into the chain cover. Install a new chain cover gasket onto the cover. Then install the correct accumulator springs in their proper location. Carefully apply downward pressure, to overcome accumulator spring pressure, and start two of the chain cover bolts. Position these bolts 180 degrees apart.

P20-42 Start the remaining chain cover bolts and tighten them in sequence and to specifications. The input shaft should now have little endplay and should rotate freely. If it does not rotate freely, remove the chain cover and check for a damaged cast iron seal.

P20-43 Tighten the park rod abutment bolts, reverse anchor pin bolt, and locknut to specifications. Lightly tap the lube tubes until they are fully seated in their bores. Install the tube retaining brackets.

P20-44 Install the O-rings onto the oil filter and press the filter into the case. Install the reverse apply tube/oil filter bracket and bolt. Install the oil pan with a new gasket. Tighten the bolts to specification.

P20-45 Install new Teflon seals onto the pump drive shaft and install the shaft.

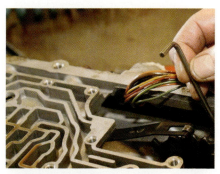

P20-46 Install the TV bracket and tighten bracket bolts to specification. Connect the manual valve link to the detent lever.

P20-47 Start the oil pump and valve body assembly over the pump shaft and connect the manual valve link to the manual valve.

P20-48 Check the alignment of the valve body to the pump and make sure the gasket is the correct one for the application. Then, install the oil pump and valve body assembly.

P20-49 Install the bulkhead connector and other electrical connectors. Install the neutral start switch. With the manual shaft in neutral, align the switch using a .089" drill bit. Then tighten the switch to specification.

P20-50 Install the valve body cover with a new gasket. Then tighten the bolts to specification. Install the dipstick tube grommet and dipstick tube in the case.

SPECIAL TOOLS

Dial indicator and holding fixture

Clutch compressor tool

Oil pump puller

Seal remover/installer

Servo cover compressor tool

Output shaft support tool

GENERAL MOTORS TRANSMISSIONS

Most General Motors transmissions and transaxles are based on the Simpson geartrain. Many different versions of the same basic design can be found. For many designs, the differences lie in the electronic controls and their intended use. The 4L60-E is used in most full-size GM cars and trucks and in some heavy-duty and high-performance vehicles. It is the four-speed unit that features full electronic control. The available gear ratios are provided by two planetary gearsets. The ring gear of one gearset is splined to the carrier of the other. One gearset is referred to as the input gearset, the other is called the reaction gearset. To control gearset activity, five multiple-friction disc assemblies, one sprag one-way clutch, one roller one-way clutch, and a band are used. The 4L60-E has a one-piece case casting that incorporates the bell housing. The extension housing is a separate casting bolted to the rear of the case. The overhaul procedures for this transmission are outlined in Photo Sequence 21.

PHOTO SEQUENCE 21

TYPICAL PROCEDURE FOR OVERHAULING A 4L60 TRANSMISSION

All photos in this sequence are © Delmar/Cengage Learning.

P21-1 Remove torque converter.

P21-2 Mount the transmission in a holding fixture and position it so the oil pan is facing up. Position a servo cover compressor on two oil pan bolts. Then, compress the servo cover and remove the retaining ring, servo cover, and O-ring.

P21-3 Remove the 2–4 servo assembly. Servo pin length should be checked prior to disassembly. This helps to identify any wear on the 2–4 band and/or reverse input drum.

P21-4 Remove the governor cover and O-ring, then remove the governor assembly.

P21-5 Remove the speed sensor assembly or the speedometer driven gear. This type transmission will be equipped with one or the other.

P21-6 Remove the extension housing bolts, then pull the housing away from the case.

P21-7 Remove the oil pan bolts, pan, and gasket.

P21-8 Remove the oil filter.

P21-9 Remove the solenoid retaining bolts. Then remove the solenoid and O-ring.

P21-10 Remove the wiring harness. Note the location and position for reference during reassembly.

P21-11 Loosen and remove the valve body and auxiliary valve body attaching bolts. Mark the location of each bolt, as they are different lengths. Also mark the location of any check balls you encounter.

P21-12 Carefully remove the 1–2 accumulator cover retaining bolts, cover and pin assembly, piston, seal, and spring.

P21-13 Remove spacer plate, and note the location of the check balls and filter.

P21-14 Remove the 3–4 accumulator spring, piston, and pin. Note the location of the spacer plate, gasket, check balls, and filters.

P21-15 Remove oil pump retaining bolts. Then using the correct puller, remove the oil pump.

P21-16 Remove the oil pump seal and gasket. Then remove the reverse input clutch-to-pump thrust washer from the pump.

P21-17 Remove the reverse and input clutch assembly by lifting it out with the input shaft.

P21-18 Remove the 2–4 band anchor pin. Then remove the band from the case.

P21-19 Remove the input sun gear. Install an Output Shaft Support tool onto the output shaft, then remove the input carrier to output shaft retaining ring.

P21-20 Remove the input carrier and output shaft. Then remove the input carrier thrust washer from the reaction carrier shaft.

P21-21 Remove the input ring gear and the reaction carrier shaft.

P21-22 Remove the reaction sun shell and thrust washer. Then, remove the sun shell to clutch race thrust washer and support-to-case retaining ring.

P21-23 Remove the spring retainer from the low/reverse support, then remove the reaction sun gear, low/reverse clutch race, clutch rollers, support assembly, and reaction carrier assembly.

P21-24 Remove the low/reverse clutch assembly.

P21-25 Remove the reaction ring gear and bearing assembly. Then remove the reaction ring gear support-to-case bearing assembly.

P21-26 Remove the parking lock bracket retaining bolts and the lock bracket.

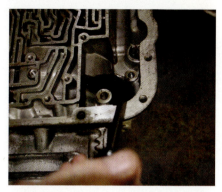

P21-27 Remove the parking pawl shaft, parking pawl, and return spring by using a screw extractor to remove the shaft plug from the case.

P21-28 Using the correct spring compressor, compress the low/reverse clutch spring retainer and remove the retaining ring and spring assembly.

P21-29 Remove the low/reverse piston assembly by applying air to the apply passage.

P21-30 Remove the manual shaft nut, shaft, and retainer. Then remove the parking lock actuator assembly and inner detent lever.

P21-31 Pry the shaft seal from the case.

P21-32 Install a new shaft seal into the case. Then install the parking lock actuator assembly and detent lever. Install the manual shaft, nut, and retainer.

P21-33 Install new seals onto the low/reverse piston and coat them with transjel.

P21-34 With the transmission in a vertical position, install the piston into the transmission case. Make sure the piston is properly aligned and fully seated.

P21-35 Install the clutch springs. Using the spring compressor, compress the springs past the ring groove in the case. Then install the retaining ring.

P21-36 Coat the bearing assembly with transjel and install it into the case. Then install the reaction ring gear and support.

P21-37 Install the bearing onto the support, then install the oil deflector and reaction carrier assembly into the case.

P21-38 Install the clutch pack with the correct number of plates.

P21-39 Install the low/reverse support into the case, then install the inner race by pushing down while rotating it until it is fully engaged.

P21-40 Install the spring retainer, then install the low/reverse retainer ring.

P21-41 Install the snap ring onto the reaction sun gear, then install the sun gear into the reaction carrier.

P21-42 Install the thrust washer onto the low/reverse clutch race and install the reaction sun gear shell onto the sun gear.

P21-43 Install the thrust washer onto the reaction sun gear shell, make sure the thrust washer tangs are positioned in the slots in the shell.

P21-44 Install the input ring gear and reaction carrier shaft into the sun gear shell. The carrier shaft splines must engage with the reaction carrier.

P21-45 Install the thrust washer onto the reaction carrier shaft. Then install the output shaft into the transmission case.

P21-46 Install the Output Shaft Support tool so that the shaft is positioned as upward as possible. Then install the input carrier assembly onto the output shaft.

P21-47 Install a new retaining ring onto the output shaft, then remove the Output Shaft Support tool.

P21-48 Install the input sun gear.

P21-49 Install the selective thrust washer and bearing assembly onto the input housing.

P21-50 Install new input shaft seals, and then size them according to the procedures given in the service manual.

P21-51 Position the reverse input assembly onto the input clutch assembly. Then install the reverse and input clutch assemblies into the case as a single unit. Align the 3–4 clutch plates of the input assembly with the input ring gear.

P21-52 Install the 2–4 band into the case. Align the anchor pin end with the case pin hole and install the anchor pin into the case. Make sure the anchor pin lines up with the end of the band. Then install the 2–4 servo assembly into the case and index the apply pin onto the end of the band. Install the servo cover with a new O-ring.

P21-53 Check endplay. Install the correct thickness of thrust washer at the rear of the oil pump, and then install the oil pump after installing a new pump O-ring and gasket. Make sure all holes; especially the filter and pressure regulator holes are lined up.

P21-54 Install the retaining bolts and torque them to specifications.

P21-55 Install the 3–4 accumulator piston pin into the case. Then install the piston seal and piston onto the pin. Install the accumulator spring; check balls, filters, retainer, and the spacer plate and gasket.

P21-56 Install the 1–2 accumulator spring, oil seal ring, piston, cover, and bolts. Tighten the bolts to specifications.

P21-57 Install the valve body and auxiliary valve body. Then install the retaining bolts and tighten them to specifications.

P21-58 Install the speedometer gear or speed sensor rotor onto the output shaft. Then install a new O-ring into the output shaft sleeve. Install the extension housing and torque the bolts to specifications. Using a seal installer, install a new oil seal in the extension housing.

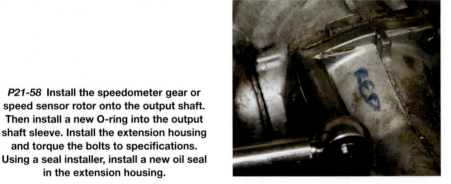

P21-59 Install the speedometer driven gear or the speed sensor into the extension housing. Then install all external electrical connectors. Install the manual shift lever and the torque converter.

Allison Series 1000

The Allison series 1000 transmission is found in many GM trucks, such as the Silverado, GMC Sierra, and Hummer H1. They are built to handle heavy loads and high torque engines. They also are very heavy weighing more than 350 lbs (160 kg).

The 1000 series is very similar to the 2000 series. These transmissions are either five- or six-speed units and are electronically controlled. Shifting is controlled by a TCM and up to six different solenoids. They are manufactured by the Allison Division of GM.

Some 1000 series units had a sixth gear, which was an additional overdrive. The unit in Photo Sequence 22 is a five speed.

CAUTION:
This transmission is very heavy. Always use caution when installing or removing it. Use lifting equipment and an assistant while moving the transmission. Also, make sure the lifting equipment can support the weight of the transmission.

PHOTO SEQUENCE 22

TYPICAL PROCEDURE FOR OVERHAULING AN ALLISON SERIES 1000 TRANSMISSION

All photos in this sequence are © Delmar/Cengage Learning.

P22-1 Mount the unit on a suitable stand. If the unit has PTO covers, remove them.

P22-2 Remove the various speed sensors and the Park/Neutral Position Switch.

P22-3 Remove the vent and torque converter assembly. Gently pry the vent out of the housing. Then rotate the transmission so the converter is facing up. Using the correct tools, remove the torque converter and place it on a work bench. The converter is heavy so make sure you are using the proper lifting equipment.

P22-4 Lift and remove the 1–2–3–4, 4–5 clutch piston assembly. Then remove the thrust bearing assembly from the 1–2–3–4, 4–5 clutch piston assembly or the input carrier assembly.

P22-5 Install the special main shaft holding tool or an equivalent tool to keep the main shaft in the housing. Then rotate the transmission's pan is facing up. Remove the oil pan bolts. Then remove the pan with its gasket. Now remove the internal oil filter and seal.

P22-6 Depress the tabs on internal wiring harness connector and gently push the connector into and through the transmission case. Disconnect the wiring harness from the various solenoids. Then remove the wiring harness from the valve body.

P22-7 Remove the manual shift detent spring bolts and its spring. Unbolt the valve body but do not attempt to remove it. Lift the valve body until its locating pins are disengaged from the transmission. The slide the valve body to the side and disconnect the manual valve pin from the manual shift shaft detent lever. Remove the control valve body. Then remove the shift valve body and manual valve as a unit.

P22-8 Rotate the transmission so that the rear is facing up. Remove the flange from the output shaft. Then remove the oil seal assembly.

P22-9 Using the correct tool, remove the retaining ring from the low and reverse clutch housing. Then remove the bearings from the end of the output shaft, along with the selective spacer.

P22-10 Unbolt the low and reverse clutch housing and remove the gasket.

P22-11 Remove the park pawl and its return spring from its support pin. Then pull the output shaft and output carrier assembly out of the transmission. Separate the shaft from the carrier. Then remove the low and reverse clutch piston return spring assembly.

P22-12 Keeping them as a single unit, remove the main shaft, intermediate sun gear, and sun gear spacer. Then separate the spacer and sun gear from the main shaft.

P22-13 Remove the intermediate carrier assembly. Then remove the thrust bearing from the intermediate carrier assembly.

P22-14 Remove the low and reverse clutch selective steel plate. Then remove the friction and steel clutch plates. Now remove the clutch backing plate. Set these aside and carefully inspect them for wear and damage.

P22-15 Remove the input carrier assembly. Then separate the thrust bearing from the carrier.

P22-16 Using the correct tool, compress the second clutch return spring and remove the low and reverse clutch retaining ring. Then remove the second clutch retaining ring. Remove the tool and remove the second clutch backing plate.

P22-17 Remove the friction and steel clutch plates. Then, as a unit, remove the second clutch spring plate and the piston return spring assemblies. Remove the second clutch piston assembly. Now remove the input ring gear.

P22-18 Rotate the transmission case so the front is facing up. Install the correct compressing tool and attachments and compress the third, fifth, and reverse clutch piston return spring assemblies. Then remove the retaining ring. Remove the tools and then remove the third, fifth, and reverse clutch backing plate assembly.

P22-19 Remove the friction and steel clutch plates. Then, as a unit, remove the third, fifth, and reverse clutch spring plate and the piston return spring assemblies. Remove the third, fifth, and reverse clutch piston assembly.

P22-20 Remove the oil seal assembly from the input shaft. Unbolt and remove the converter housing from the oil pump cover. Remove the channel plate gasket and the seals at the rear hub of the oil pump cover and the stator shaft. Remove the thrust washer and unbolt and remove the oil pump assembly.

P22-21 On a work bench, position the 1–2–3–4, 4–5 clutch assembly so the rear of the assembly is up. Remove the retaining ring from the clutch assembly. Then remove the input sun gear with the input drive flange attached. Remove the bronze thrust washer from the 4–5 drive hub and remove the hub. Remove the thrust bearing assembly from the rear of the 1–2–3–4 clutch drive hub and remove the hub.

P22-22 Remove the friction and steel clutch from the housing. Then, remove the retaining ring from the 1–2–3–4 clutch backing plate. Now remove the retaining ring from the turbine shaft and the 1–2–3–4 clutch backing plate from the turbine shaft.

P22-23 Remove the friction and steel clutch plates from the turbine shaft. Remove the retaining ring from the turbine shaft. Then remove the clutch housing and seals from the turbine shaft. Set the turbine shaft aside.

P22-24 Install the correct spring compressor onto the 1–2–3–4 clutch piston return spring assembly. Using the correct attachments, compress the spring assembly so the retaining ring can be removed.

P22-25 Remove the piston return spring assembly from the 1–2–3–4 clutch piston. Then remove the 1–2–3–4 clutch piston assembly. Using the correct tools, compress the return ring assembly enough to remove the retaining ring from the 1–2–3–4, 4–5 clutch housing. Then separate the 1–2–3–4 clutch piston housing from the 1–2–3–4, 4–5 clutch housing.

P22-26 Remove the 4–5 clutch piston return spring assembly from the 4–5 clutch piston. Then remove the retaining ring from the 1–2–3–4, 4–5 clutch housing, which will allow the removal of the 4–5 clutch piston assembly.

P22-27 To begin reassembling the 1–2–3–4, 4–5 clutch housing, install the piston seal rings onto the 4–5 clutch piston. Lubricate all seals and the piston. Install the O-ring into the bottom groove on the inner hub of the housing. Install the 4–5 clutch assembly into the housing. Make sure the piston is properly seated on the housing. Push the balance piston below the retaining ring groove on the housing and install the retaining ring.

P22-28 Align the 4–5 clutch piston return spring assembly so that its notches are facing and are aligned with the tangs of the piston. Then install return spring assembly into the piston. Install the 1–2–3–4 clutch piston housing into the 4–5 clutch piston.

P22-29 Position the 1–2–3–4, 4–5 clutch housing on the base of the special compressing tool. Compress the 4–5 clutch piston return spring assembly until the retaining ring can be installed. Remove the tools. Lubricate and install the clutch piston seals onto the clutch piston.

P22-30 Align the 1–2–3–4 clutch piston so that the spring indents are facing up. Install the clutch piston into the housing. Align the piston return spring assembly so that the springs are facing and are aligned with the spring pockets of piston and install the spring assembly.

P22-31 Turn the housing over so its rear is facing up. Then set the retaining ring into the inner hub. Now set up the spring compressor tool and compress the 1–2–3–4 clutch spring assembly until the retaining ring is able to be seated. Remove the tools. Then lubricate the seals and piston, and install them in the housing. Insert the retaining ring. It may be necessary to press on the assembly to do this.

P22-32 Install the overlap seal rings onto the turbine shaft. Position the turbine shaft so the rear of the shaft is facing up. Align the 1–2–3–4, 4–5 clutch housing so the rear of the housing is facing up. Install the housing onto the rear of turbine shaft and then install the retaining ring.

P22-33 Alternately install the friction and steel clutch discs onto the turbine shaft. Align the pressure plate with the turbine shaft so that the countersunk holes of the pressure plate are facing up. Then install it onto the turbine shaft. Align the ridge plate with the turbine shaft so its inner ridge is facing down and then slide the inner ridge into the turbine shaft's retaining ring groove. Align the holes in the ridge plate with the holes of the pressure plate and install the retaining bolts. Tighten these specifications.

P22-34 Select the correct clutch backing plate according to the manufacturer's guidelines. Align the backing plate so that its beveled edge is facing up. Then install it onto the turbine shaft. Install the retaining ring onto the turbine shaft. Lift the backing plate and install the spiral retaining ring into the backing plate.

P22-35 Position the thrust bearing assembly according to specifications and install it onto the turbine shaft. Center the clutch plates and install the drive hub over the clutch plates. Install the 4–5 clutch drive hub so that the smaller end is facing up. Now, alternately install the friction and steel clutch discs into the housing.

P22-36 Align the tangs of the bronze thrust washer with the holes in the 4-5 clutch drive hub and install it. Install the input sun gear into the input sun gear flange with its teeth facing the concave face of the flange. Hold the sun gear and align the double wide splines of the flange with the double wide spaces in the housing. Then install the sun gear and flange into the housing. Install the sun gear external retaining ring into the housing.

P22-37 Position the oil pump assembly so the rear of the pump is facing up. Align the locating pin holes in the channel plate with the locating pins in the oil pump assembly. Install the channel plate onto the oil pump assembly.

P22-38 Position the oil pump cover assembly so that the stator shaft is facing up. Align the oil pump assembly with the channel plate and install the oil pump assembly and channel plate onto the oil pump cover. Tighten the retaining bolts to specifications.

P22-39 Align the tangs of the thrust washer with the slots at the base of the oil pump cover's rear hub. Then install it onto the oil pump cover. Install new overlap seal rings onto the rear hub and the stator shaft.

P22-40 Install two guide bolts into the oil pump cover module and align the converter housing bolt holes with the bolt holes in the oil pump cover. Install the converter housing onto the guide bolts and onto the cover. Remove the guide bolts and install new seals the pump retaining bolts. Install and tighten the bolts in a criss-cross pattern and to specifications. Using the correct tool, install the seal over the oil pump's hub.

P22-41 Rotate the transmission so that the front of the case is facing up. Apply lubricant to the inside and outside diameter of the third, fifth, and reverse clutch piston assembly. Align the piston with the transmission case so the seal is facing down and the lubrication/cooling orifice is at the top of the case. Gently tap the piston assembly into position.

P22-42 Install the piston return spring assemblies onto the third, fifth, and reverse spring plate. Then install the spring plate into the transmission case. Alternately install the friction and steel clutch discs into the case. Install the third, fifth, and reverse clutch backing plate assembly with gear thrust plates down. With the compressor tool, compress the clutch pack until the retaining ring is able to be installed.

P22-43 Rotate the transmission so the rear of the case is facing up. Apply lubricant to the inside and outside diameters of the second clutch piston assembly. Align the piston with the transmission case so the seal is facing down and the lubrication/cooling orifice is at the top of the case. Gently tap the piston assembly into position. Install the input ring gear so that the outside splines are facing down. Then rotate the ring gear until the friction clutch plates are engaged.

P22-44 Install the input carrier module, rotate it while doing this until the carrier drops into place. Install the thrust bearing onto the input carrier. Install the piston return spring assemblies onto the second spring plate. Then install the spring plate into the transmission case. Alternately install the friction and steel clutch discs into the case. Install the third, fifth, and reverse clutch backing plate assembly with gear thrust plates down. With the compressor tool, compress the clutch pack until the retaining ring is able to be installed. Then install the low and reverse clutch retaining ring with its gap 180 degrees from the second clutch retaining ring gap. Remove the compressor tools.

P22-45 Locate low and reverse clutch housing, alternately install the friction and steel clutch discs into the housing. Then install the correct low and reverse clutch steel plate.

P22-46 Install the intermediate carrier module. Rotate it until it drops into place. Install the thrust washer onto the carrier. Then install the main shaft holding tool.

P22-47 Install the intermediate sun gear and its spacer onto the main shaft. Then install this assembly into the intermediate carrier. Install the output sun gear onto the main shaft with its flat side facing the spacer. Install the thrust bearing assembly into the sun gear.

P22-48 Install the low and reverse clutch piston return spring assembly, make sure the tang aligns with the notch in the transmission case. Install the park pawl cam guide and park pawl support pin into the transmission case. Install the park pawl return spring onto the park pawl and install the assembly onto the park pawl support pin. Then connect the park pawl return spring to the case.

P22-49 Install the output shaft into the output carrier assembly. While holding the output shaft, place the carrier assembly into the top of the output ring gear. Rotate the output shaft until the carrier drops into the ring gear. Gently place the output shaft onto the main shaft and rotate the carrier until it seats.

P22-50 Prepare to install the low and reverse clutch housing. Install a new gasket onto the transmission case. Install the low and reverse clutch housing making sure it is aligned with locating pin in the case and the park pawl pin. Tighten four of the retaining bolts by hand to seat the housing in the case. Then install the rest of the bolts and tighten all of them to specifications.

P22-51 Rotate the transmission so the valve body area of the case is facing up. Align the manual valve pin with the manual selector valve. Install the manual valve pin into the manual selector valve and then install the valve into shift valve body. Align the control valve assembly with the case while installing the manual valve pin into the manual shift shaft detent lever. Then install the control valve body assembly onto the transmission case, using the alignment pins as a guide. Install the retaining bolts and tighten them to specifications. Install the manual shift shaft detent spring onto the valve body. Then install and tighten the manual shift shaft detent spring bolts.

P22-52 With your hand, install the internal wiring harness connector into the transmission case. Then using the special tool make sure the connector is securely in place and the four tabs on the connector are snapped into place. Then fasten the harness to the valve body and connect the individual connectors to the appropriate solenoid. Then install the oil filter and pan. Make sure to tighten the bolts to specifications.

P22-53 To install the 1–2–3–4, 4–5 clutch housing assembly, place the transmission so that the front of the case is facing up. Then remove the main shaft holding tool. Apply lubricant to the thrust bearing assembly and install it into the input carrier assembly. Install the 1–2–3–4, 4–5 clutch assembly. Rotate the assembly until it seats.

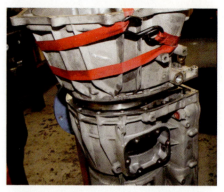

P22-54 The torque converter housing is very heavy. Make sure you have assistance while installing it. Install two guide bolts into the transmission case. Place a new converter housing to case gasket onto the transmission case. Apply lubrication to the seal rings on the rear of the oil pump cover assembly. Lower the converter housing/ oil pump cover assembly over the turbine shaft and onto the case. Tap the housing until it seats on the case. Remove the guide bolts from the transmission case and install the retaining bolts. Make sure the correct length is placed in its appropriate bore. Then tighten them to specifications.

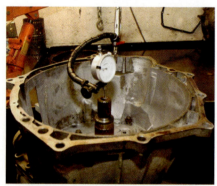

P22-55 Rotate the transmission so its rear is facing up. Using the proper tool, install the rear bearing. Install the retaining ring with its beveled edge facing up. Now rotate the transmission so the front is facing up. Set a dial indicator onto the end of the turbine shaft. Zero the dial indicator. Then lift the rear of the 1–2–3–4, 4–5 clutch housing with a screwdriver through the PTO hole in the case. This is the amount of endplay and your measurement should be compared to specifications. If the end play is not within the specification, a different selective spacer must be installed.

P22-56 Raise the torque converter over the torque converter housing and lower it over the turbine and stator shafts and into the oil pump assembly. Make sure the torque converter is seated in the oil pump assembly.

P22-57 Set the special switch alignment tool onto the Park/Neutral position switch. Align the PNP switch with the case so that the tool is facing outward. While maintaining the correct switch-to-manual-shift-shaft alignment, slide the switch over the manual shift shaft and install the retaining bolts. Remove the tool and install the switch's splash shield over the manual shift shaft. Install the shift lever, and then install the shift lever nut by hand, on the end of manual shift shaft. Tighten the shift lever nut to specification. Reconnect the shift selector linkage/cable to the shift lever and the external wiring harness to the PNP switch.

P22-58 Using the correct driver, install a new drive shaft slip yoke oil seal. Make sure the seal is installed to the correct depth below the rear face of the low and reverse clutch housing. Then install the vent by gently tapping on it.

P22-59 Install the ISS, TSS, and OSS in their designated bores. Make sure to install them with new seals that are lubricated. Tighten the retaining bolts to specifications.

P22-60 Attach a sling to the transmission. Fasten the sling to a hoist. Use the hoist to remove the weight of the transmission off the transmission stand. Then unbolt the transmission from the stand. Place the transmission on a work table with the oil pan facing down. Install the PTO covers with new gaskets. Tighten the cover bolts to specification.

GM Transaxles

Through the years, GM has used many different transaxles. Early units were mostly based on the Simpson gear train. Later, the transaxles relied on two simple planetary gears connected in series. The commonly used series of this type of transaxle is the 4Txx series. The 4T60 was introduced in 1984 and was the replacement for the 440-T4. In 1991, new electronic controls were added and the unit was referred to as the 4T60E. The latter was used in nearly all GM vehicles. It was replaced with an updated version, the 4T65E, which is still found in some vehicles.

The geartrain of the 4T60 is based on two simple planetary gearsets operating in tandem. The combination of the two planetary units functions much like a compound unit. The two tandem units do not share a common member; rather, certain members are locked together or are integral with each other. The front planet carrier is locked to the rear ring gear and the front ring gear is locked to the rear planet carrier. The transaxle houses a third planetary unit, which is used only as the final drive unit and not for overdrive.

The 4T60 uses a variable displacement vane-type pump, and four multiple-friction disc assemblies, two bands, and two one-way clutches to provide the various gear ranges. One of the one-way clutches is a roller clutch, the other is a sprag. The multiple-disc packs are released by several small coil springs when hydraulic pressure is diverted from the clutch's piston. Photo Sequence 23 covers the overhaul procedures for this transaxle.

The 6T70/75 is a fully automatic, six-speed, electronic-controlled transaxle. The planetary gearsets provide the six forward gear ratios and reverse. Changing gear ratios is fully automatic and is accomplished through the use of a TCM inside the transmission. The TCM receives and monitors various electronic sensor inputs.

The TCM commands shift solenoids and variable bleed pressure control solenoids to control shift timing and quality. The TCM also controls the torque converter clutch. All of the solenoids and the TCM are packaged into a single unit, called the control solenoid valve assembly.

Classroom Manual
Chapter 10, page 337

SPECIAL TOOLS
Dial indicator and holding fixture
Clutch compressor tool
Oil pump puller
Seal remover/installer
Servo cover compressor tool
Output shaft support tool

TYPICAL PROCEDURE FOR OVERHAULING A 4T60 TRANSAXLE

All photos in this sequence are © Delmar/Cengage Learning.

P23-1 Remove torque converter, then place the transaxle in a holding fixture. Remove the speedometer sensor and governor assembly.

P23-2 Remove the bottom oil pan, oil filter, modulator, and modulator valve.

P23-3 Remove the accumulator cover with governor feed and return pipes and accumulator pistons, gaskets, retainers, and pipes from their bores.

P23-4 Mark the cover for the reverse servo so you know from where it came. Then, remove the cover by applying pressure to it and removing the retaining ring. Then, remove the servo assembly from the case.

P23-5 Mark the cover for the 1–2 servo so you know from where it came. Then, apply pressure to the cover and remove the servo cover retaining ring. Then remove the cover and servo assembly.

P23-6 Remove the side cover bolts, nuts, and washers. Then remove the side covers and gaskets. Disconnect and remove the wiring harness to the pressure switches, solenoid, and case connector.

P23-7 Remove the TV lever, linkage, and bracket from the valve body assembly. Remove the pump assembly cover bolts, then the pump cover. Remove the servo pipe retainer bolt, retainer plate, mounting bolts, and valve body.

P23-8 Remove the oil reservoir weir. Then mark the location of and remove the check balls between the spacer plate and the valve body and between the channel plate and the spacer plate.

P23-9 Disconnect the manual valve link from the manual valve. Place the detent lever in the park position and remove the retaining clip. Then remove the channel plate with its gaskets.

P23-10 Remove the oil pump drive shaft. Remove the input clutch accumulator and converter clutch piston assemblies. Remove the fourth clutch plates and the apply plate, then remove the fourth clutch's thrust bearing, hub, and shaft.

P23-11 Rotate the final drive unit until both ends of the output shaft retaining ring are showing.

P23-12 With a C-ring removal tool, loosen the ring from the output shaft. Rotate the shaft 180 degrees and remove the ring from the shaft. Then remove the output shaft.

P23-13 Remove and discard the O-ring from the input shaft, then remove the drive sprocket, driven sprocket, and chain as an assembly with the selective fit thrust washers. Place the chain assembly on a bench so that it is in the same direction and position it was in the transmission.

P23-14 Remove the driven sprocket support and thrust washer from between the sprockets and the channel plate. Remove the scavenging scoop and driven sprocket support with the second clutch thrust washer.

P23-15 Using the correct tool, remove the second clutch and input shaft clutch housings as an assembly. Remove the reverse band.

P23-16 Remove the thrust washers. Then measure the endplay of the input shaft clutch housing. Select the correct thickness of thrust washer and set aside for reassembly.

P23-17 Remove input clutch sprag assembly, third clutch assembly, and the input sun gear. Then remove the reverse reaction drum, input carrier assembly, and thrust washer.

P23-18 Remove the reaction carrier, reaction sun gear/drum assembly, and forward band.

P23-19 Remove the reaction sun gear thrust bearing and final drive sun gear shaft.

P23-20 Check final drive endplay and select the correct size thrust washer for the unit.

P23-21 Using the proper tool, remove the final drive assembly and selective thrust washers and bearings.

P23-22 Clean and inspect the transaxle case and all of the transaxle components. Then position the correct size thrust washers and bearings onto the final drive assembly and install the unit. Transjel can be used to hold the washers and bearings in place while positioning the unit.

P23-23 Install the final drive sun gear shaft through the final drive ring gear, the splines must engage with the parking gear and the final drive sun gear.

P23-24 Install the forward band into the case, make sure the band is aligned with the anchor pin.

P23-25 Install reaction sun gear-to-final drive ring gear thrust bearing. Then assemble the reaction sun gear and drum assembly onto the final drive ring gear.

P23-26 Check the endplay of the carriers in the reaction planetary gearset. Then install the thrust washer and carrier assembly into the case. Rotate the carrier until the pinions engage with the reaction sun gear.

P23-27 Install the input carrier with its thrust bearing into the case. Then install the reverse reaction drum, make sure its splines engage with the input carrier.

P23-28 Inspect the roller and sprag clutches, then put the spacer onto the sun gear followed by the input sprag retainer, sprag assembly, and roller clutch. Make sure the input sprag and third roller clutch hold and freewheel in opposite directions while holding the input sun gear.

P23-29 Lubricate the inner seal on the input clutch's piston, then install the seal and assemble the input piston into the input housing. Install a new O-ring onto the input shaft.

P23-30 Install spring retainer and guide into the piston, then install the third clutch piston housing into the input housing. Using the proper compressor, install the retaining snap ring.

P23-31 Install the inner seal for the third clutch, then install the third clutch piston into the housing. Compress the spring retainer and install the retaining snap ring.

P23-32 Install the wave plate, and then install the correct number of clutch plates in the correct sequence. Install the input clutch backing plate and the retaining snap ring. Air check the operation of the clutch.

P23-33 Assemble the second clutch piston in the housing. Install the apply ring, return spring, snap ring, and wave plate. Then assemble the correct number of clutch plates in the correct sequence. Install the backing plate and snap ring, then air check the clutch's operation.

P23-34 Install the thrust washers. Using the correct tool, install the second clutch and input shaft clutch housings as an assembly. Install the reverse band.

P23-35 Install the scavenging scoop and driven sprocket support with the second clutch thrust washer. Install the driven sprocket support and thrust washer between the sprockets and the channel plate.

P23-36 Install the drive sprocket, driven sprocket, and chain as an assembly.

P23-37 Install the output shaft. Start the C-ring onto the output shaft, then rotate the shaft 180 degrees and fully seat the ring onto the shaft.

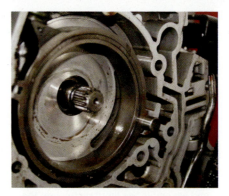

P23-38 Install the fourth clutch's thrust bearing, hub, and shaft. Then install the fourth clutch plates and the apply plate. Install the input clutch accumulator and converter clutch piston assemblies. Install the oil pump drive shaft.

P23-39 Install the channel plate with new gaskets; connect the manual valve link to the manual valve. Place the detent lever in the Park position and install the retaining clip.

P23-40 Install the oil reservoir weir. Install the check balls in their proper location between the spacer plate and the valve body and between the channel plate and the spacer plate.

P23-41 Install the servo pipe retainer bolt, retainer plate, mounting bolts, and valve body. Install thel pump assembly cover bolts, then the pump cover. Tighten the bolts to specifications. Install the TV lever, linkage, and bracket onto the valve body assembly.

P23-42 Connect and install the wiring harness to the pressure switches, solenoid, and case connector. Then install the side covers and gaskets. Install the side cover bolts, nuts, and washers, and then tighten them to specifications.

P23-43 Install the 1–2 servo cover, the servo assembly, and the retaining ring.

P23-44 Install the reverse servo cover, the servo assembly, and the retaining ring.

P23-45 Install the accumulator cover with governor feed and return pipes and accumulator pistons, gaskets, retainers, and pipes into their bores.

P23-46 Install the bottom oil pan, oil filter, modulator, and modulator valve.

P23-47 Install the torque converter, speedometer sensor, and governor assembly.

The hydraulic system primarily consists of a vane-type pump and two control valve body assemblies. These units rely on five multiple-disc clutches and a one-way clutch. The gearsets then transfer torque through the transfer drive gear, transfer driven gear and differential assembly.

This series of transaxle was jointly designed with Ford who fits a similar unit in many of their FWD vehicles. The basic construction of the GM and Ford units are similar. The most noticeable differences are in the application of electronic controls. Photo Sequence 24 covers the basic steps for overhauling this series of transaxle.

> **CUSTOMER CARE:** Because this model uses a sliding reverse gear, damage to the transaxle can occur if the car is towed or moved while the transaxle's gear selector is in neutral but the transaxle is still mechanically in reverse gear. A servo normally is used to delay forward gear engagement after the transaxle has been in reverse. Simply shifting into neutral may not disengage reverse gear. Any time the vehicle is to be moved, start the engine and place the transaxle into drive, then shift it into neutral. Turn off the engine. Now the vehicle is ready to be moved.

NISSAN MOTOR COMPANY TRANSMISSIONS

Widely used Nissan RWD transmissions are the 3N71B, which is a three-speed unit, and the L4N71B/E4N71B series, which provides four forward gears through the use of a Simpson gearset and an additional planetary unit mounted in front of the Simpson gearset. The primary difference between the "L" and the "E" is that the E model provides electronic control for shifting and torque converter clutch operation. Nissan transaxles are similar to other transaxles covered in this chapter.

These transmissions use four multiple-friction disc assemblies, two servos and bands, and a one-way clutch to provide for the different ranges of gears. All the clutches except the low/reverse unit are released by several small coil springs. The low/reverse unit utilizes a Belleville-type spring for greater clamping pressures. Photo Sequence 25 covers the typical overhaul procedures for this model transmission. Always follow the recommended procedure for the transmission or transaxle you are working on.

SPECIAL TOOLS

Dial indicator and holding fixture

Clutch compressor tool

Oil pump puller

Seal remover/installer

Classroom Manual

Chapter 10, page 353

TYPICAL PROCEDURE FOR OVERHAULING A 6T70 TRANSAXLE

All photos in this sequence are © Delmar/Cengage Learning.

P24-1 Remove the torque converter before mounting the transaxle on its holding fixture. Rotate it so that the valve cover body is facing up. Then unbolt the cover and remove it with its gasket. While lifting off the cover, make sure not to pill on the wires leading to the solenoids and TCM.

P24-2 Unbolt and remove the control solenoid valve assembly. Then remove the filter plate.

P24-3 Unbolt and remove the control valve body assembly. Keep track of the location of the shorter bolts. Then unbolt the retaining bolt for the OSS and remove the sensor.

P24-4 Rotate the unit so that the torque converter side is facing down. Unbolt and remove the case cover with its gasket. Set this assembly aside. Then remove the input shaft thrust bearing. Now, the 3–5–Reverse and 4–5–6 clutch housing can be lifted out of the case.

P24-5 Now, remove the reaction carrier hub assembly and its thrust washer. Then remove the 2–6 clutch spring, plate, friction and steel discs, and its backing plate.

P24-6 Remove the reaction carrier assembly. Then remove the thrust bearings for the input carrier and the input sun gear. Remove the input carrier assembly.

P24-7 Remove the input sun gear. Then remove the output carrier with its thrust washers.

P24-8 Remove the output sun gear assembly. Then remove the output carrier transfer drive gear hub assembly with its bearings.

P24-9 Lift the spring, apply plate, friction and steel discs, plate, and backing plate for the low and reverse clutch out of the case.

P24-10 Remove the low and reverse clutch retaining ring. Then remove the clutch assembly. Now the friction and steel discs for the 1–2–3–4 clutch can be removed. Then remove the clutch plate and the waved plate. Remove the clutch spring retainer ring and the spring. Remove the clutch piston by pulling on it with pliers.

P24-11 Rotate the transaxle so that the torque converter side is facing up. Unbolt and remove the torque converter and oil pump housing with its outer seal.

P24-12 Remove the lubrication tube for the differential's drive pinion. Then remove the differential's drive pinion gear assembly and set it aside.

P24-13 Remove the differential's carrier assembly and set it aside.

P24-14 Using the correct pullers, remove the differential carrier and drive pinion bearing cups from the case. Then remove the thrust washers and the drive pinion's lubricant dam.

P24-15 With the correct puller and holding tool, remove the bearings on the carrier assembly.

P24-16 Then install new bearings with the correct driver.

P24-17 Remove the transfer drive gear support assembly. Inspect it for damage or wear to the splines, bushings, machined surfaces, and threaded holes. Also, check the gear for damage or wear. Make sure the drive gear bearing rotates smoothly.

P24-18 Unbolt the fluid passage tube from the support. Check the tubes for cracks or other damage. Remove the old gasket and place a new one onto the support. Reinstall the passage tube and tighten the retaining bolts to specifications.

P24-19 Remove the old drive gear support seal. With the correct driving tools, install a new seal. Turn the transfer gear support assembly over and remove and install the seal on that side. Then, replace the torque converter fluid seal and install the support assembly into the housing.

P24-20 Place the new bearing cups over the lower bearings on the differential carrier and drive pinion assemblies. Carefully lower the gear assemblies into the case. Then place the bearing cups onto the upper bearings of the assemblies.

P24-21 Install the special selection sleeves over the top of the bearing cups on the gears. Place a housing gasket over the case. Then install the long guide pins and spacers that are part of the gauging set. Lower the T/C housing onto the case and over the guide pins. Make sure the spacers and gauges are lined up.

P24-22 Remove the guide pins and install spacers in their place. Make sure all bolt holes in the housing and case have a spacer. Then install the bolts that are part of the gauging set and tighten them in the prescribed sequence and to the specified torque. Rotate the differential assembly at least 10 times to allow the bearings to seat into the cups.

P24-23 Hold the top sleeve of the gauging tool up. Measure the gap between the gauging tool sleeves. To do this, use a feeler gauge to identify the largest leaf that will fit into the gap. Take two measurements, each 180 degrees apart, at each gear. The two measurements should be averaged and the correct thrust washer to install behind the bearing cups is the one that has a thickness that is closest to that average. Do this at both gears.

P24-24 Unbolt the cover from the housing and remove the special tools and spacers. Remove the bearing cups from the gearsets. Remove the housing gasket. Place the correct selective washers into the case's bores for the bearing cups and with the correct tool install the cups. Then install the cups in their bores in the housing.

P24-25 Place the gear assemblies into the case. Place a new housing gasket on the case. Install the retaining bolts and tighten them to specifications.

P24-26 Rotate the transaxle so that the T/C side is facing down. Gather the 1–2–3–4 clutch assembly and the proper tools for assembling it. Install a new seal on the clutch's piston and lubricate it. Place the seal protector into the piston's bore and push the piston in place.

P24-27 Install the clutch's spring and place its retainer ring into the bore. With the correct spring compressor, compress the spring enough to allow the retaining ring to be installed.

P24-28 Install the 1–2–3–4 clutch's wave plate, pressure plate, and friction and steel discs into the bore. Install the low and reverse clutch assembly and retaining ring. Make sure the gap of the ring is facing the valve body side of the case.

P24-29 Install the low and reverse clutch backing plate with the clutch plate facing the bottom of the case. Then install the friction and steel clutch discs, the apply plate, and the cushion spring.

P24-30 Install the input shaft thrust bearing, transfer gear hub assembly, and hub bearing.

P24-31 Install the output sun gear and carrier assemblies. Then install the thrust bearings.

P24-32 Install the input carrier assembly, followed by the input sun gear and thrust bearing. Then install the input carrier thrust bearing and the reaction carrier assembly.

P24-33 Now install the 2–6 clutch's backing plate, friction and steel discs, apply plate, and cushion spring. Then install the reaction sun gear and 2–6 clutch hub thrust bearing.

P24-34 Gather the 4–5–6 clutch housing and the tools required to service it. With the compressor tool, press down on the reluctor just enough to remove the retaining ring. Then remove the reluctor wheel. Remove the 2–5 Reverse clutch piston and check the seals on the piston. Replace them if necessary.

P24-35 Turn the input shaft over and remove the backing plate retaining ring. Then remove the reaction hub carrier assembly and the 4–5–6 clutch's backing plate, apply plate, friction and steel discs, and the thrust bearing.

P24-36 Remove the 4–5–6 clutch dam retaining ring. Then remove the dam and the 4–5–6 clutch spring assembly, piston, and seals. Replace the piston seals. Then reinstall the piston, spring assembly, and dam. Now install the retaining ring.

P24-37 Install the 4–5–6 clutch hub thrust bearing, reaction carrier hub assembly, 4–5–6 apply plate, 4–5–6 friction and steel plates, 4–5–6 backing plate, and retaining ring.

P24-38 Now install the 3–5 Reverse clutch's wave plate, apply plate, friction and steel discs, backing plate, and retaining ring.

P24-39 Turn the input shaft over and install the dam seal, 3–5 Reverse clutch spring, and the 3–5 reverse piston and seals. Then install the reluctor wheel. Using the correct tool, compress the assembly enough to install the reluctor wheel retaining ring.

P24-40 Put the case in an area suitable for work and remove the retainer for the ISS, then remove the sensor. Now remove the seals for the 3–5 Reverse and 4–5–6 clutches from the center of the case.

P24-41 Remove the low/reverse clutch spring retaining ring, spring, and piston. Then with the correct tool, compress the 2–6 spring assembly to remove the retaining ring, spring assembly, and piston.

P24-42 Now using the seal protector, install low and reverse clutch piston and spring assemblies. Make sure the air bleed on the piston is aligned with the mark on the cover.

P24-43 Then compress the springs to install the retaining ring.

P24-44 Insert the wiring from the ISS through its bore in the cover and fasten the ISS to the cover.

P24-45 Install the input shaft thrust washer. Then place a new case cover gasket on the housing. Route the ISS wire harness through its bore in the case. Using guide pins, lower the case cover onto the housing. Then remove the guide pins and install the cover bolts. Tighten them to specifications.

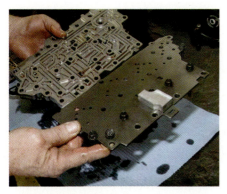

P24-46 Move the control valve body to a suitable working space. Separate the upper and lower assemblies by unbolting them. Make sure to keep track of the check ball locations. After doing this, inspect the upper channel plate for damage. Discard the old lower spacer plate assembly.

P24-47 Carefully clean and inspect the upper and lower valve body assemblies. If either of these is damaged, the complete assembly must be replaced.

P24-48 Assemble the upper and lower valve bodies with a new spacer plate. Make sure the check balls are in their proper locations. Tighten the bolts to specifications. Then install the channel plate.

P24-49 Install the OSS and tighten the retaining nut to specifications. Then align the manual valve to the detent lever assembly. Make sure the detent assembly is properly aligned before tightening the bolt.

P24-50 Place the valve body onto the case and insert the retaining bolts. Make sure the shorter bolts are in their proper bores. Then tighten the bolts in the correct sequence and to the specified torque.

P24-51 Place a new filter plate over the top of the valve body and make sure it is secured by the retaining clips.

P24-52 Place the control solenoid valve assembly in position over the filter plate. Then place the control solenoid valve spring in position and install the retaining bolts for the assembly. Make the different length bolts are in their proper location.

P24-53 Connect the wiring harness connectors to their appropriate place. Make sure the harness is correctly routed and secure.

P24-54 Install a new wiring connector seal into the valve body cover. Then place a new cover gasket onto the case. Make sure the connector is seated in its seal. Install the retaining nuts and bolts and tighten them to specifications.

Classroom Manual

Chapter 10, page 354

SPECIAL TOOLS

Dial indicator and holding fixture

Clutch compressor tool

Oil pump puller

Air nozzle

Seal remover/ installer

Gear pullers and drivers

Small chisel

Punch and drift set

TOYOTA TRANSAXLES

Toyota led the way with electronically controlled transmissions, and since the first model was released they have continued to refine their basic model. That base model, the A-140E transaxle, was a three-speed transmission with an overdrive assembly. The basis for operation is a Simpson gearset in line with a single overdrive planetary gearset. The transaxles uses four multiple-disc clutches, two band and servo assemblies, and three one-way clutches to provide the various gear ranges.

The A541E transaxle is a revised copy of the A-140E. The biggest change was in the electronic controls, which now has an adaptive learning capability. This transaxle has been used in many different Toyota and Lexus models, including the Toyota Camry, Toyota Avalon, and the Lexus ES-300. This transaxle uses six multiple-disc clutches, three one-way clutches, and one brake band. The operation of the transaxle and the lockup converter is totally controlled by the PCM. However, a throttle pressure cable is used to mechanically modulate line pressure. Photo Sequence 26 covers the typical overhaul procedures for this model transmission.

TYPICAL PROCEDURE FOR OVERHAULING A NISSAN L4N71B TRANSMISSION

All photos in this sequence are © Delmar/Cengage Learning.

P25-1 Place the transmission on a suitable workbench. Check endplay and record your readings.

P25-2 Remove the torque converter.

P25-3 Unscrew and remove the electrical kickdown solenoid and O-ring.

P25-4 Unscrew and remove the throttle modulator valves, its diaphragm rod and O-ring.

P25-5 Remove the speedometer drive assembly, with gear and O-ring.

P25-6 Remove the oil pan and inspect its contents. An analysis of foreign material will provide clues regarding the types of problems to look for during the overhaul procedure. Check the service manual for specific details.

P25-7 Remove valve body from the case.

P25-8 Remove manual valve from the valve body to prevent the valve from dropping out.

P25-9 Back off the servo piston stem locknut and tighten piston stem snug to prevent the front clutch drum from dropping out when removing the front pump.

P25-10 Use the correct puller and remove the front pump from the case.

P25-11 Remove front clutch thrust washer and bearing race.

P25-12 Remove the overdrive servo cover.

P25-13 Loosen the overdrive servo locknut, and back off the overdrive band's servo piston stem to release the band.

P25-14 Remove the front and rear clutch assemblies. Note the positions of the front pump thrust washers and rear clutch thrust washer.

P25-15 Remove the brake band strut and overdrive brake band.

P25-16 Remove the overdrive housing.

P25-17 Remove the extension housing. Take care not to lose the parking pawl, spring, pin, and parking actuator.

P25-18 Remove the governor valve assembly.

P25-19 Remove the front drum support and input shaft.

P25-20 Remove the second brake band strut and brake band.

P25-21 Remove the front clutch and planetary gear pack.

P25-22 Remove the output shaft snap ring.

P25-23 Remove the output shaft.

P25-24 Remove the snap ring, connecting drum, and one-way clutch assembly.

P25-25 Use a screwdriver to remove the large retaining snap ring. Then remove the low and reverse brake assembly.

P25-26 Install the low and reverse brake assembly starting with the steel dished plate and then alternating steel and friction plates.

P25-27 Install the retaining plate and snap ring. Check for proper clearance between the snap ring and the retaining plate. Select the proper thickness of retaining plate that will give the correct ring to plate clearance if the measurement does not meet the specified limits. Check the operation of the low and reverse brake by using compressed air.

P25-28 Install the one-way clutch assembly, connecting drum, and snap ring.

P25-29 Install the governor needle bearing, thrust washer, output shaft, and oil distributor in the case.

P25-30 Install the planetary gear pack and front clutch assembly.

P25-31 Install the snap ring on the output shaft.

P25-32 Install the second brake band, band strut, and band servo. Lubricate the O-ring seals with ATF.

P25-33 Lubricate the drum support gasket with ATF and install the drum support into the housing.

P25-34 Install the governor valve assembly and torque the retaining bolts to specifications.

P25-35 Install the parking actuator and parking pawl assemblies in the extension housing. Then install the extension housing. Torque its mounting bolts to specifications.

P25-36 Install the front drum support gasket and new O-ring onto the overdrive housing. Gently tap it into place using a rubber mallet. Make sure the boltholes are aligned.

P25-37 Install the needle bearing race and direct clutch thrust washer. Then install the overdrive brake band, strut, and servo assembly. Lubricate the servo O-ring before installing it.

P25-38 Install the overdrive pack on the drum support.

P25-39 Adjust the overdrive band. Tighten the piston stem to specifications. Then back off two full turns. Tighten the servo locknut to specifications. Test the overdrive servo with compressed air.

P25-40 Install the overdrive servo cover and tighten to specifications.

P25-41 Adjust the second brake band. Tighten the piston stem to specifications and then back off three full turns.

P25-42 Secure the piston stem while you tighten the servo locknut to specifications.

P25-43 Install the oil pump bearing and thrust washer.

P25-44 Install the oil pump assembly, using the special tool. Be sure to lubricate the gasket and O-ring with ATF. Align the mounting boltholes in the pump housing with the boltholes in the overdrive housing.

P25-45 Lubricate and install the manual valve into the control valve body.

P25-46 Install the control valve body and tighten the valve body attaching bolts to the specified torque.

P25-47 Install the oil pump gasket, oil pan, and retaining bolts. Tighten the bolts to specifications, while using a star pattern to tighten them.

P25-48 Install the speedometer drive assembly with its drive gear and lubricated O-ring.

P25-49 Install the vacuum diaphragm, diaphragm rod, and O-ring.

P25-50 Install the kickdown (or downshift) solenoid and O-ring.

P25-51 Install the torque converter housing. Torque the mounting bolts to specifications in a star pattern.

TYPICAL PROCEDURE FOR OVERHAULING A TOYOTA A541E TRANSAXLE

All photos in this sequence are © Delmar/Cengage Learning.

P26-1 With the transaxle mounted in a stand and its external electrical components, oil pan, and linkages removed, remove the upper cover of the transaxle case.

P26-2 Remove the oil pipe bracket.

P26-3 Remove the oil strainer.

P26-4 Remove the retaining bolts for the detent spring, and then remove the spring.

P26-5 Unbolt and remove the retaining bolts for the manual valve body and remove it. Then remove the retaining bolt for the oil pipes, and then remove the pipes.

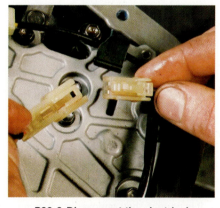

P26-6 Disconnect the electrical connectors to the solenoids, then remove the connector clamp.

P26-7 Remove the oil pipes' retaining bolts, then pry out the apply pipes.

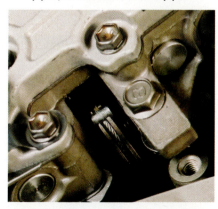

P26-8 Unbolt the valve body retaining bolts, and then disconnect the throttle cable from the valve body while removing the valve body.

P26-9 Remove the throttle cable, solenoid wiring.

P26-10 Remove the accumulator pistons and springs.

P26-11 Measure the piston stroke for the second coast brake. Use a divider to transfer the distance to a vernier caliper or micrometer for measurement.

P26-12 Remove the second coast brake piston's retaining ring and use air pressure to remove the second coast brake piston.

P26-13 Position the transaxle so that it is upright, then remove the oil pump with a puller.

P26-14 Remove the direct and forward clutch assembly.

P26-15 Separate the direct clutch from the forward clutch.

P26-16 Remove the guide for the second/coast brake band.

P26-17 Remove the second/coast brake band.

P26-18 Remove the ring gear for the front planetary gearset.

P26-19 Remove the front planetary gearset with its bearings.

P26-20 Remove the sun gear and the sun gear input drum.

P26-21 Remove the second brake drum, and then remove the second brake piston return spring, and the #1 one-way clutch.

P26-22 Remove the friction discs and plates from the housing.

P26-23 Remove the snap ring and the #2 one-way clutch.

P26-24 Then remove the thrust washer and the rear planetary ring gear.

P26-25 Remove the flange, discs, and plates of the first/reverse brake.

P26-26 Rotate the transaxle and remove the retaining bolts for the overdrive unit.

P26-27 Remove the overdrive cover.

P26-28 Then pull the overdrive planetary gearset from the transaxle.

P26-29 Remove the output gear and nut.

P26-30 Remove the retaining ring, output shaft, and bearing race.

P26-31 Install the output shaft with a new race and retaining ring.

P26-32 Install the output gear with a new nut and tighten the nut to specifications.

P26-33 Stake the output gear's retaining nut to the output shaft.

P26-34 Remove the overdrive clutch pack by compressing the first/reverse brake piston return spring and remove the snap ring. Then remove the piston.

P26-35 Remove the parking lock pawl bracket and guide.

P26-36 Remove the manual shaft and seal.

P26-37 Loosen the final drive and axle seal covers.

P26-38 Remove the covers, then remove and service the differential unit.

P26-39 Begin reassembly by installing the differential and drive axle unit. Make sure the unit is properly setup.

P26-40 Install new seals and gaskets, then position the final drive housing and tighten the bolts to specifications.

P26-41 Install the parking pawl lock, pawl bracket and guide and the manual shaft. Use a new seal and collar when doing this. Then check the operation of the parking lock pawl.

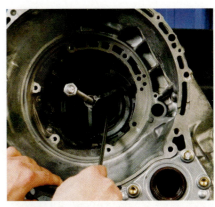

P26-42 Install the first/reverse brake piston into the case and install the piston return spring and retaining snap ring.

P26-43 Install the overdrive brake band and overdrive planetary gearset into the case.

P26-44 Install the overdrive cover with a new gasket. Tighten the retaining bolts to specifications.

P26-45 Check the endplay of the intermediate shaft. If is within specs, coat the thrust washer with transjel and install the rear planetary gearset.

P26-46 Install the discs, plates, and flange for the first/reverse brake. Then install the retaining snap ring.

P26-47 Check the operation of the brake with compressed air.

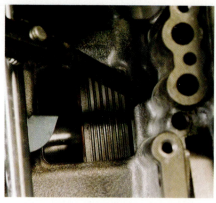

P26-48 Check the clearance of the disc pack with a feeler gauge.

P26-49 Install the #2 one-way clutch into the case with the shiny side of the flange facing out.

P26-50 Install the second/coast brake band guide.

P26-51 Then coat the thrust washer for the #1 one-way clutch and install it with the clutch.

P26-52 Install the discs and flange for the second brake, then install the snap ring.

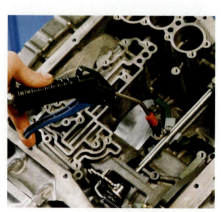

P26-53 Check the operation of the second brake with compressed air.

P26-54 Coat the thrust washer with transjel, then install the sun gear with its drum.

P26-55 Turn the drum clockwise until it seats into the #1 one-way clutch.

P26-56 Install a new O-ring onto the intermediate shaft, then install the front planetary gearset.

P26-57 Coat the thrust washers with transjel and assemble the forward and direct clutch assemblies.

P26-58 Install the second/coast brake band and the forward and direct clutch assembly.

P26-59 Position the transaxle with the oil pump opening up. Coat the new pump O-ring with ATF and install it on the pump.

P26-60 Install the pump into the transmission housing.

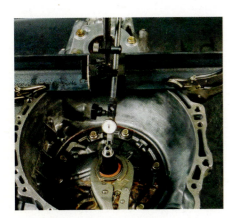

P26-61 Check the endplay of the input shaft.

P26-62 Install the piston for the second/coast brake. Use a new snap ring to retain it.

P26-63 Identify the correct accumulator and piston spring for each of the accumulator bores.

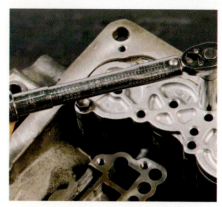

P26-64 Install the accumulator pistons and springs. Use a new gasket and torque the bolts to specifications.

P26-65 Connect the wires to the solenoids and connect the throttle cable.

P26-66 Put the oil pipes into their proper position and then mount the valve body to the case. Install and tighten the retaining bolts to specifications.

P26-67 Connect the solenoids to the electrical harness. Then install the manual valve body and detent spring. Be sure to tighten the retaining bolts to specifications.

P26-68 Install and properly tighten the oil pipe bracket and retaining bolts.

P26-69 Install the oil strainer.

P26-70 Complete assembly by installing the upper cover of the transaxle case and its external electrical components, oil pan, and linkages.

CASE STUDY

A very irate customer had his late-model Honda towed back to the shop. Just three weeks had passed since the transmission had been rebuilt and now it had failed again.

The technician who had done the overhaul was assigned the car. He conducted a visual inspection and found nothing obviously wrong with the electrical system or the transaxle itself. However, the fluid had a burned smell. He conducted an oil pressure test and found extremely low pressures. He suspected that the pump or pressure regulator had failed.

He pulled the transaxle and checked the pressure regulator and oil pump. He found the pump's gears had seized together. This was the first Honda transaxle he had ever rebuilt and he was disappointed that it had failed. Wondering how this could happen, he reviewed the service manual and Technical Service Bulletins. In bold print in the service manual he found the answer. He had failed to correctly align the oil pump shaft during assembly. Based on his experiences with other transmissions he had rebuilt, he hadn't known this was critical and had assumed that the shaft would only fit one way.

He aligned the shaft, replaced the oil pump gears, and reassembled the transmission. He road tested the car and conducted a pressure test. Everything was fine. He then gave the car back to the customer, with an apology, knowing that the transaxle was now right. The customer appreciated his honesty and left with confidence that the transmission was okay. The technician learned a lesson: always refer to and follow the directions given in the service manual.

ASE-STYLE REVIEW QUESTIONS

1. *Technician A* says some seals must be cut to size before they are installed.
   Technician B says square-cut seals are designed to roll over when a part is fit over them.
   Who is correct?
   A. A only
   B. B only
   C. Both A and B
   D. Neither A nor B

2. While assembling a transmission:
   Technician A coats the steel clutch discs with transjel.
   Technician B soaks the friction discs in clean ATF before installing them.
   Who is correct?
   A. A only
   B. B only
   C. Both A and B
   D. Neither A nor B

3. *Technician A* says endplay is often corrected by selective snap rings.
   Technician B says endplay is often corrected by selective thrust washers.
   Who is correct?
   A. A only
   B. B only
   C. Both A and B
   D. Neither A nor B

4. *Technician A* says all Chrysler transmissions and transaxles are Simpson gear based.
   Technician B says the 36RH and 42LE are longitudinally mounted and are very similar in con struction and operation. The major difference between the two is that one is a transaxle.
   Who is correct?
   A. A only
   B. B only
   C. Both A and B
   D. Neither A nor B

5. *Technician A* says all transaxles that have a drive chain and use the chain to transfer transmission output to the final drive unit.
   Technician B says tandem planetary gearsets do not have a common member; rather, a member of one gearset is connected to a member of the other gearset. Sometimes there are two of these connections.
   Who is correct?
   A. A only
   B. B only
   C. Both A and B
   D. Neither A nor B

6. A transaxle slips in all gear ranges. Which of the following is the *least* likely cause?
   A. A defective one-way clutch
   B. Low fluid level
   C. Faulty electronic controls
   D. A clogged fluid filter

7. While assembling a transaxle:
   Technician A reuses all seals unless they are damaged.
   Technician B lubricates all seals and bearings with clean bearing grease before installing them.
   Who is correct?
   A. A only   C. Both A and B
   B. B only   D. Neither A nor B

8. A transmission abruptly makes unwanted downshifts at high speeds. Which of the following is the *most* likely cause?
   A. Throttle cable out of adjustment
   B. Defective oil pump
   C. Linkage out of adjustment
   D. Sticking valves in the valve body

9. While diagnosing incorrect shift points:
   Technician A says a disconnected shift solenoid could be the cause.
   Technician B says a dirty valve body could be the cause.
   Who is correct?
   A. A only   C. Both A and B
   B. B only   D. Neither A nor B

10. While checking transmission endplay:
    Technician A measures the movement of the shaft with a dial indicator.
    Technician B uses a clutch compressor tool to get the maximum movement reading.
    Who is correct?
    A. A only   C. Both A and B
    B. B only   D. Neither A nor B

ASE CHALLENGE QUESTIONS

1. The customer complains of harsh automatic downshifts.
   Technician A says the anticlunk spring may be broken or positioned incorrectly.
   Technician B says line pressure may be entering the governor assembly.
   Who is correct?
   A. A only   C. Both A and B
   B. B only   D. Neither A nor B

2. All of the following may cause a rough initial engagement in forward and reverse *except*:
   A. Excessive backlash in final drive/differential assembly
   B. Retarded ignition timing
   C. Missing check ball
   D. Leaking transmission oil filter

3. During disassembly, the low/reverse band is found to be very worn with some frictional material missing.
   Technician A says the damage may be the result of improper band adjustment.
   Technician B says high oil pump pressure may be the cause.
   Who is correct?
   A. A only   C. Both A and B
   B. B only   D. Neither A nor B

4. The 3–4 gear switch winding tested open.

    Technician A says this would prevent the torque converter from lock up in fourth gear.

    Technician B says the PCM may place the system in default under this condition.

    Who is correct?

    A. A only
    B. B only
    C. Both A and B
    D. Neither A nor B

5. The vehicle experiences an intermittent second-gear start.

    Technician A says a bad one-way clutch may be the cause.

    Technician B says low governor pressure may be the cause.

    Who is correct?

    A. A only
    B. B only
    C. Both A and B
    D. Neither A nor B

Name _____ **Date** _____

IDENTIFICATION OF SPECIAL TOOLS AND PROCEDURES

Upon completion of this job sheet, you should be able to use the service manual to determine what special tools and procedures are required to correctly service a particular transmission or transaxle.

Tools and Materials

Appropriate service manual

Procedure

Task Completed
_____

Your instructor will assign you to a particular transmission or transaxle. You will use the service manual to identify and describe the special tools recommended for the complete overhaul of the transmission. You will also identify any tools your shop has that are acceptable substitutes for the factory-recommended special tools. After you have assembled the list of special tools, go through the steps for complete overhaul and identify those procedures that are more than remove and install types (such as checking input shaft endplay).

1. Describe the transmission that was assigned to you.
   Model and type of transmission _____
   Describe the vehicle it is from: Year _____ Make _____
   Model _____ VIN _____

2. What service manual or electronic database are you using for this assignment? Be specific.

   _____

   _____

3. List the special tools referenced in the overhaul section of the manual for this transmission. Include in your list the part or subsystem the tool is used on.

   _____

   _____

   _____

   _____

   _____

   _____

   _____

   _____

   _____

   _____

   _____

4. List the required special tools that are available in your shop.

_____

_____

_____

_____

_____

_____

_____

5. List the available tools that are acceptable substitutes for the special tools you don't have in your shop.

_____

_____

_____

_____

_____

_____

_____

6. Review the overhaul procedures and describe the special procedures that must be known in order to properly service the transmission assigned to you. Include in your description the parts or subsystems that require this procedure. Do not simply say, "check endplay." Describe what needs to be done to check the endplay.

_____

_____

_____

_____

_____

_____

_____

_____

Instructor's Response _____

_____

_____

Name _____ Date _____

CONTINUOUSLY VARIABLE TRANSMISSIONS

Upon completion of this job sheet, you will be able to explain how a normal CVT works.

Tools and Materials

Service Information

Describe the vehicle being worked on:

Model and type of transmission _____

Year _____ Make _____ Model _____

VIN _____ Engine type and size _____

Procedure

Task Completed

_____

1. Why is this type of transmission called a continuously variable transmission?

   _____

   _____

2. What is the basic goal of a CVT?

   _____

   _____

3. Describe the major components of this transmission.

   _____

   _____

   _____

   _____

4. Explain how this CVT works.

   _____

   _____

   _____

   _____

5. Does the driver feel a change in ratio as the vehicle is driven? Why or why not?

   _____

   _____

   _____

   _____

6. Using the service information as a guide, what inputs does the transmission rely on to change ratios?

_____

_____

_____

_____

Instructor's Response _____

_____

_____

Name _____ Date _____

HYBRID TRANSMISSIONS

Upon completion of this job sheet, you will be able to explain how a transmission in a hybrid vehicle works.

Tools and Materials

Service Information

Describe the vehicle being worked on:

Model and type of transmission _____

Year _____ Make _____ Model _____

VIN _____ Engine type and size _____

Procedure

Task Completed

1. Using the service information, briefly describe the hybrid system used on this vehicle.

   _____

   _____

   _____

   _____

2. Describe the key characteristics that make this a hybrid vehicle.

   _____

   _____

   _____

   _____

3. Where is the electric motor(s) located?

   _____

   _____

   _____

   _____

4. What are principle purposes of the electric motor?

   _____

   _____

   _____

   _____

5. What is the voltage rating of the hybrid battery?

_____

_____

6. Does this vehicle use a conventional transmission?

_____

_____

7. If the vehicle has a unique transmission, describe why it is unique.

_____

_____

_____

_____

8. Can this vehicle run on electric power only? If so, when?

_____

_____

_____

_____

9. Briefly explain the action of the electric motor(s), when are they turned on and when do they function as a generator.

_____

_____

10. Briefly describe the precautions that must be followed when servicing these transmissions and vehicles.

_____

_____

_____

_____

Instructor's Response _____

_____

_____

Final Exam Automatic Transmission/Transaxle A2

1. Which of the following is the *least* likely cause for a buzzing noise from a transmission?
   A. Improper fluid level or condition
   B. Defective oil pump
   C. Defective flexplate
   D. Damaged planetary gearset

2. A vehicle experiences engine flare in low gear only.
   Technician A says the torque converter lockup clutch is slipping.
   Technician B says the transmission oil pump is not providing the required pressure.
   Who is correct?
   A. A only
   B. B only
   C. Both A and B
   D. Neither A nor B

3. The results of a hydraulic pressure test are being discussed:
   Technician A says low idle pressure may be caused by a defective exhaust gas recirculation system.
   Technician B says low neutral and park pressures may indicate a fluid leakage past the clutch and servo seals.
   Who is correct?
   A. A only
   B. B only
   C. Both A and B
   D. Neither A nor B

4. The vehicle creeps in neutral.
   Technician A says a too high engine idle speed could be the cause.
   Technician B says a too tight clutch pack may be the problem.
   Who is correct?
   A. A only
   B. B only
   C. Both A and B
   D. Neither A nor B

5. *Technician A* says overtorqued valve body fasteners may cause a lack of engine braking in manual low.
   Technician B says a lack of engine braking in manual third may be caused by a faulty overrunning clutch.
   Who is correct?
   A. A only
   B. B only
   C. Both A and B
   D. Neither A nor B

6. The transmission's output shaft and its sealing components are being discussed:
   Technician A says the shaft and all of its sealing components must be replaced if nicks and scratches are found in the shaft's sealing area.
   Technician B says all of the shaft's seals and rings must be replaced during a rebuild.
   Who is correct?
   A. A only
   B. B only
   C. Both A and B
   D. Neither A nor B

7. The vehicle will only upshift to second at full throttle. This could be caused by any of the following EXCEPT:
   A. Clogged oil passages
   B. Low fluid level
   C. Bad clutch pack
   D. Open upshift switch

8. The vehicle will not move in any gear.
   Technician A says a misadjusted TV cable could be the cause.
   Technician B says leakage at the oil pump and/or valve body could cause this condition.
   Who is correct?
   A. A only
   B. B only
   C. Both A and B
   D. Neither A nor B

9. Sensors are being discussed:
   Technician A says most speed sensors are AC generators.
   Technician B says most speed sensors use a stationary magnet, rotor, and a voltage sensor.
   Who is correct?
   A. A only
   B. B only
   C. Both A and B
   D. Neither A nor B

10. *Technician A* says the PCM monitors the amount of voltage generated by the speed sensor to calculate the vehicle's speed.
    Technician B says the output of a speed sensor is pulsed as an on/off voltage signal when displayed on a DSO.
    Who is correct?
    A. A only
    B. B only
    C. Both A and B
    D. Neither A nor B

11. Shift solenoids are being discussed:

    Technician A says engine flare during upshifts may be caused by high resistance in the solenoid's windings.

    Technician B says delay shifts may be caused by open solenoid windings.

    Who is correct?
    A. A only
    B. B only
    C. Both A and B
    D. Neither A nor B

12. When diagnosing the cause of no engine braking during manual low operation:

    Technician A checks for a defective oil pump.

    Technician B suspects a damaged drive link.

    Who is correct?
    A. A only
    B. B only
    C. Both A and B
    D. Neither A nor B

13. The customer complains of transmission noises in all gears except park and neutral.

    Technician A says the oil pump may be the cause.

    Technician B says the drive chain or sprocket may be the source of the noise.

    Who is correct?
    A. A only
    B. B only
    C. Both A and B
    D. Neither A nor B

14. Converter installation is being discussed:

    Technician A says converter depth must be measured before it is removed from the transmission.

    Technician B says the converter must be engaged to the turbine (input) shaft, stator shaft, and pump drive to be properly installed.

    Who is correct?
    A. A only
    B. B only
    C. Both A and B
    D. Neither A nor B

15. A vehicle with an EAT exhibits erratic shifting. The speed sensor is suspect.

    Technician A says the PM sensor can be checked on the vehicle with a voltmeter.

    Technician B says there should be the specified resistance between the leads when checked with an ohmmeter.

    Who is correct?
    A. A only
    B. B only
    C. Both A and B
    D. Neither A nor B

16. The vehicle will crank in reverse or drive, but not in any other gear.

    Technician A says the park/neutral switch is open.

    Technician B says the TV linkage is misadjusted.

    Who is correct?
    A. A only
    B. B only
    C. Both A and B
    D. Neither A nor B

17. While diagnosing the cause of sluggish acceleration:

    Technician A suspects a faulty governor.

    Technician B checks the condition of the fluid.

    Who is correct?
    A. A only
    B. B only
    C. Both A and B
    D. Neither A nor B

18. Which of the following is the *most* likely cause of slipping in all forward gear ranges?
    A. Faulty governor
    B. Clogged oil filter
    C. Faulty band or clutch
    D. Sticking valve in the valve body

19. A vehicle with a newly rebuilt EAT shifts directly to third from low. The most probable cause for this concern is:
    A. 1–2 servo is defective
    B. 2–3 shift valve is sticking
    C. Manual linkages are out of adjustment
    D. Throttle linkage is out of adjustment

20. While diagnosing the cause of no forced downshifts during full throttle operation:

    Technician A suspects a misadjusted manual linkage.

    Technician B suspects a dirty valve body and valves.

    Who is correct?
    A. A only
    B. B only
    C. Both A and B
    D. Neither A nor B

21. Which of the following is the *least* likely cause for a transmission not upshifting and only operating in first gear?
    A. Improper fluid level or condition
    B. Faulty valve body
    C. Defective oil pump
    D. Faulty governor

22. The vehicle exhibits a scraping or grating noise in every gear position anytime engine torque is changed up or down.

    Technician A says this could be caused by a damaged planetary gearset.

    Technician B says cracks around the flexplate mounting holes would cause this noise.

    Who is correct?

    A. A only   C. Both A and B
    B. B only   D. Neither A nor B

23. A MAP sensor test reveals below specified voltage signals at all engine speeds.

    Technician A says an EAT may have late upshifts based on the signals.

    Technician B says a vacuum modulator controlled TV pressure will cause late upshifts based on the signals.

    Who is correct?

    A. A only   C. Both A and B
    B. B only   D. Neither A nor B

24. The transmission will not shift and has high line pressure during testing. This could be caused by:

    A. A stuck regulator valve
    B. High governor pressure
    C. Low TV pressure
    D. A faulty manual lever position sensor

25. The valve body bores are being discussed:

    Technician A says the bores should be measured with small bore (hole) gauges.

    Technician B says the bores only need to be cleaned if the damage to the valves can be corrected by polishing.

    Who is correct?

    A. A only   C. Both A and B
    B. B only   D. Neither A nor B

26. During disassembly, some extension-to-case bolts were found to have aluminum strips in the threads.

    Technician A says the extension housing should be replaced.

    Technician B says to clean and repair the internal threads with a thread insert.

    Who is correct?

    A. A only   C. Both A and B
    B. B only   D. Neither A nor B

27. *Technician A* says use two flare nut wrenches to disconnect transmission cooler lines.

    Technician B says use vice grip pliers to hold a cooler line that is twisting as the flare nut is turned and then replace the line and fitting.

    Who is correct?

    A. A only   C. Both A and B
    B. B only   D. Neither A nor B

28. Air checking the transmission is being discussed:

    Technician A says using a special adapter plate is necessary on some transmissions.

    Technician B says a special style blow gun should be used.

    Who is correct?

    A. A only   C. Both A and B
    B. B only   D. Neither A nor B

29. During inspection, the valve body mounting surface of the case is found to be slightly warped.

    Technician A says the problem may be corrected by using a fine file on the valve body.

    Technician B says the case must be replaced if the mounting surface for the valve body is warped.

    Who is correct?

    A. A only   C. Both A and B
    B. B only   D. Neither A nor B

30. A bushing was nicked and scratched as the shaft was being removed. There is no other apparent wear or damage.

    Technician A says use fine emery cloth to repair the damage.

    Technician B says sanding or polishing the bushing may cause the shaft to wobble in the bushing bore.

    Who is correct?

    A. A only   C. Both A and B
    B. B only   D. Neither A nor B

31. While diagnosing the cause of gear slipping in second gear only:

    Technician A suspects a worn or damaged clutch assembly.

    Technician B suspects a damaged drive link.

    Who is correct?

    A. A only   C. Both A and B
    B. B only   D. Neither A nor B

32. While conducting a pressure test:

    Technician A says the cause of low pressure in all operating ranges could be a clogged filter.

    Technician B says the cause of high pressure in all operating ranges could be a defective throttle valve.

    Who is correct?

    A. A only   C. Both A and B
    B. B only   D. Neither A nor B

33. *Technician A* says pump side clearance is determined using a straightedge and feeler gauges.

    Technician B says the pump's gear side clearance may be measured with a dial indicator.

    Who is correct?

    A. A only   C. Both A and B
    B. B only   D. Neither A nor B

34. *Technician A* says the transmission fluid level may be checked when the transmission is hot or cold and in park with the engine running on some vehicles.

    Technician B says if the ATF has a dark color and a burned smell it may indicate a transmission rebuild may be needed.

    Who is correct?

    A. A only   C. Both A and B
    B. B only   D. Neither A nor B

35. There are traces of transmission fluid in the engine coolant.

    Technician A says the leak may be coming from a poor line connection to the radiator.

    Technician B says a cracked cooler line may be the source of the leak.

    Who is correct?

    A. A only   C. Both A and B
    B. B only   D. Neither A nor B

36. *Technician A* says when removing the drive chain, input gear, and output gear from certain transaxles, a special tool must be used to spread the gears.

    Technician B says a special tool must be used to draw the gears together in order to remove the drive chain from this type of transaxle.

    Who is correct?

    A. A only   C. Both A and B
    B. B only   D. Neither A nor B

37. The customer complains of delayed upshifts during hard acceleration.

    Technician A says the throttle valve linkage adjustment is the most probable cause.

    Technician B says a clogged catalytic converter may be the cause.

    Who is correct?

    A. A only   C. Both A and B
    B. B only   D. Neither A nor B

38. A shudder is noticeable at speeds around 40–55 mph in third or fourth gear.

    Technician A says the TCC solenoid may be partially blocked.

    Technician B says TCC oil pressure may be too high.

    Who is correct?

    A. A only   C. Both A and B
    B. B only   D. Neither A nor B

39. While diagnosing the cause of no torque converter clutch engagement on a transmission that seems to shift fine:

    Technician A suspects a damaged clutch pressure plate into all gears.

    Technician B suspects a severely worn input clutch.

    Who is correct?

    A. A only   C. Both A and B
    B. B only   D. Neither A nor B

40. Inspection of the torque converter is being discussed:

    Technician A says to replace the converter if the transmission's front pump is damaged.

    Technician B says to reweld or repair loose converter drive lugs.

    Who is correct?

    A. A only   C. Both A and B
    B. B only   D. Neither A nor B

41. *Technician A* says transaxle turning torque may be checked by turning the axle with a torque wrench.

    Technician B says the turning torque is measured by turning the final drive's axle (side) gear with a torque wrench.

    Who is correct?

    A. A only   C. Both A and B
    B. B only   D. Neither A nor B

42. An ohmmeter is being used to check several shift solenoids:

    Technician A says high resistance in the shift solenoids will cause engine flare during upshifts.

    Technician B says an infinite reading means the solenoid's windings have more resistance than the meter can measure.

    Who is correct?

    A. A only   C. Both A and B
    B. B only   D. Neither A nor B

43. *Technician A* says some OBD II systems have a TCC break-in mode or routine.

    Technician B says this OBD II break-in mode must be manually activated using a scan tool.

    Who is correct?

    A. A only   C. Both A and B
    B. B only   D. Neither A nor B

44. Transmission/transaxle removal is being discussed:

    Technician A says some electrical components of the ABS system must be disconnected at the wheels of RWD vehicles.

    Technician B says the converter access plate must be removed before removing a transmission or transaxle.

    Who is correct?

    A. A only   C. Both A and B
    B. B only   D. Neither A nor B

45. Which of the following is the *least* likely cause for ATF coming out of the transmission vent or the dipstick tube?

    A. Defective pressure regulator valve
    B. Plugged drain holes in valve body
    C. Fluid at too high of a level
    D. Defective oil the cooler

46. Transmission/transaxle installation is being discussed:

    Technician A says to always replace the pilot bearing/bushing for smooth torque converter movement on the flexplate.

    Technician B says the converter drive hub should be lubricated with ATF to help prevent damage to the pump's front seal.

    Who is correct?

    A. A only   C. Both A and B
    B. B only   D. Neither A nor B

47. While diagnosing the cause of transmission overheating:

    Technician A checks for contaminated fluid.

    Technician B suspects a damaged flexplate.

    Who is correct?

    A. A only   C. Both A and B
    B. B only   D. Neither A nor B

48. *Technician A* says slippage in only first gear can only be caused by a faulty vehicle speed sensor.

    Technician B says a defective one-way clutch can cause slippage in all gears.

    Who is correct?

    A. A only   C. Both A and B
    B. B only   D. Neither A nor B

49. *Technician A* cleans the valve body in carburetor cleaner, rinses it in water, and blow dries it with compressed air.

    Technician B says valves that are dragging in their bores should be polished with fine emery cloth until the drag is removed.

    Who is correct?

    A. A only   C. Both A and B
    B. B only   D. Neither A nor B

50. Transmission oil pumps are being discussed:

    Technician A says high line pressure would result if the relief valve spring is broken or missing.

    Technician B says the relief valve is designed to restrict the fluid flow until operating pressures are achieved.

    Who is correct?

    A. A only   C. Both A and B
    B. B only   D. Neither A nor B

APD Transmission Parts
Atlanta, GA

Atec Trans. Tool
San Antonio, TX

Automatic Transmission Rebuilders Association
Ventura, CA

Automotive Specialty Tools
Green Valley Lake, CA

Baum Tools Unlimited, Inc.
Longboat Key, FL

Big A Auto Parts, APC Inc.
Houston, TX

Carquest Corp.
Tarrytown, NY

Drivetrain Specialists
Las Vegas, NV

Hastings Manufacturing
Hastings, MI

KD Tools, Danaher Tool Group
Lancaster, PA

Kent-Moore, Div. SPX Corp.
Warren, MI

Lisle Corp.
Clarinda, IA

MacTools
Washington Courthouse, OH

MATCO Tool Co.
Stow, OH

NAPA Hand/Service Tools
Lancaster, PA

OTC, Div. SPX Corp.
Owatonna, MN

Parts Plus
Memphis, TN

Snap-on Tools Corp.
Kenosha, WI

Transtar Industries, Inc.
Cleveland, OH

To convert these	to these,	multiply by:
TEMPERATURE		
Centigrade Degrees	Fahrenheit Degrees	1.8 then +32
Fahrenheit Degrees	Centigrade Degrees	0.556 after −32
LENGTH		
Millimeters	Inches	0.03937
Inches	Millimeters	25.4
Meters	Feet	3.28084
Feet	Meters	0.3048
Kilometers	Miles	0.62137
Miles	Kilometers	1.60935
AREA		
Square Centimeters	Square Inches	0.155
Square Inches	Square Centimeters	6.45159
VOLUME		
Cubic Centimeters	Cubic Inches	0.06103
Cubic Inches	Cubic Centimeters	16.38703
Cubic Centimeters	Liters	0.001
Liters	Cubic Centimeters	1000
Liters	Cubic Inches	61.025
Cubic Inches	Liters	0.01639
Liters	Quarts	1.05672

To convert these	to these,	multiply by:
Quarts	Liters	0.94633
Liters	Pints	2.11344
Pints	Liters	0.47317
Liters	Ounces	33.81497
Ounces	Liters	0.02957
WEIGHT		
Grams	Ounces	0.03527
Ounces	Grams	28.34953
Kilograms	Pounds	2.20462
Pounds	Kilograms	0.45359
WORK		
Centimeter-Kilograms	Inch-Pounds	0.8676
Inch-Pounds	Centimeter-Kilograms	1.15262
Meter-Kilograms	Foot-Pounds	7.23301
Foot-Pounds	Newton-Meters	1.3558
PRESSURE		
Kilograms/Square Centimeter	Pounds/Square Inch	14.22334
Pounds/Square Inch	Kilograms/Square Centimeter	0.07031
Bar	Pounds/Square Inch	14.504
Pounds/Square Inch	Bar	0.06895

GLOSSARY
GLOSARIO

Note: **Terms are highlighted in color,** followed by **Spanish translation in bold.**

Abrasion Wearing or rubbing away of a part.

Abrasión El desgaste o consumo por rozamiento de una parte.

Acceleration An increase in velocity or speed.

Aceleración Un incremento en la velocidad.

Accumulator A device used in automatic transmissions to cushion the shock of shifting between gears, providing a smoother feel inside the vehicle.

Acumulador Un dispositivo que se usa en las transmisiones automáticas para suavizar el choque de cambios entre las velocidades, así proporcionando una sensación más uniforme en el interior del vehículo.

Adhesives Chemicals used to hold gaskets in place during the assembly of an engine. They also aid the gasket in maintaining a tight seal by filling in the small irregularities on the surfaces and by preventing the gasket from shifting due to engine vibration.

Adhesivo Los productos químicos que se usan para sujetar a los empaques en una posición correcta mientras que se efectua la asamblea de un motor. También ayuden para que los empaques mantengan un sello impermeable, rellenando a las irregularidades pequeñas en las superficies y previniendo que se mueva el empaque debido a las vibraciones del motor.

Aeration The process of mixing air into a liquid.

Aireación El proceso de mezclar el aire en un líquido.

ALDL Assembly line data link.

ALDL Siglas para una trasmisión de datos de la planta de fabricación.

Alignment An adjustment to a line or to bring into a line.

Alineación Un ajuste que se efectúa en una linea o alinear.

Ammeter A test instrument used to measure electrical current.

Amperímetro Un aparato que mide corriente eléctrica.

Amplitude The height of a waveform is called its amplitude.

Amplitud La altura de onda eléctrica se llama su amplitud.

Antifriction bearing A bearing designed to reduce friction. This type of bearing normally uses ball or roller inserts to reduce the friction.

Cojinetes de antifricción Un cojinete diseñado con el fin de disminuir la fricción. Este tipo de cojinete suele incorporar una pieza inserta esférica o de rodillos para disminuir la fricción.

Antiseize Thread compound designed to keep threaded connections from damage due to rust or corrosion.

Antiagarrotamiento Un compuesto para filetes diseñado para proteger a las conecciones fileteados de los daños de la oxidación o la corrosión.

Apply devices Devices that hold or drive members of a planetary gearset. They may be hydraulically or mechanically applied.

Dispositivos de aplicación Los dispositivos que sujeten o manejan los miembros de un engranaje planetario. Se pueden aplicar mecánicamente o hidráulicamente.

Arbor press A small, hand-operated shop press used when only a light force is required against a bearing, shaft, or other part.

Prensa para calar Una prensa de mano pequeña del taller que se puede usar en casos que requieren una fuerza ligera contra un cojinete, una flecha u otra parte.

ASE The National Institute for Automotive Service Excellence (ASE) has a voluntary certification program for automotive, heavy-duty truck, auto body repair, and engine machine shop technicians.

ASE El Instituto Nacional para la Excelencia en el Servicio Automotriz (ASE) tiene un programa voluntario de certificación para técnicos automotrices, de camiones de servicio pesado, de reparación de carrocerías y de talleres mecánicos.

ATF Automatic transmission fluid.

ATF Fluido de transmisión automática.

Automatic transmission A transmission in which gear or ratio changes are self-activated, eliminating the necessity of hand-shifting gears.

Transmisión automática Una transmisión en la cual un cambio deengranajes o los cambios en relación son por mando automático, así eliminando la necesidad de cambios de velocidades manual.

Automotive Service Excellence (ASE) The National Institute for Automotive Service Excellence (ASE) has established a voluntary certification program for technicians employed in the many related fields of the automotive industry.

Excelencia Automotora Del Servicio El instituto nacional para la excelencia automotora del servicio (ASE) ha establecido un programa voluntario de la certificación para los técnicos empleados en los muchos ramos relacionados de la industria del automóvil.

Axial Parallel to a shaft or bearing bore.

Axial Paralelo a una flecha o al taladro del cojinete.

Axis The centerline of a rotating part, a symmetrical part, or a circular bore.

Eje La linea de quilla de una parte giratoria, una parte simétrica, o un taladro circular.

Axle The shaft or shafts of a machine on which the wheels are mounted.

Semieje El eje o los ejes de una máquina sobre los cuales se montan las ruedas.

Axle ratio The ratio between the rotational speed (rpm) of the driveshaft and that of the driven wheel; gear reduction through the differential, determined by dividing the number of teeth on the ring gear by the number of teeth on the drive pinion.

Relación del eje La relación entre la velocidad giratorio (rpm) del árbol propulsor y la de la rueda arrastrada; reducción de los engranajes por medio del diferencial, que se determina por dividir el número de dientes de la corona por el número de los dientes en el piñón de ataque.

Axle shaft A shaft on which the road wheels are mounted.

Flecha del semieje Una flecha en la cual se monta las ruedas.

Backlash The amount of clearance or play between two meshed gears.

Juego La cantidad de holgura o juego entre dos engranajes endentados.

Balance Having equal weight distribution. The term is usually used to describe the weight distribution around the circumference and between the front and back sides of a wheel.

Equilibrio Lo que tiene una distribución igual de peso. El término suele usarse para describir la distribución del peso alrededor de la circunferencia y entre los lados delanteros y traseros de una rueda.

Balance valve A regulating valve that controls a pressure of just the right value to balance other forces acting on the valve.

Válvula niveladora Una válvula de reglaje que controla a la presión del valor correcto para mantener el equilibrio contra las otras fuerzas que afectan a la válvula.

Ball bearing An antifriction bearing consisting of a hardened inner and outer race with hardened steel balls that roll between the two races, and supports the load of the shaft.

Rodamiento de bolas Un cojinete de antifricción que consiste de una pista endurecida interior e exterior y contiene bolas de acero endurecidos que ruedan entre las dos pistas, y sostiene la carga de la flecha.

Ball joint A suspension component that attaches the control arm to the steering knuckle and serves as the lower pivot point for the steering knuckle. The ball joint gets its name from its ball-and-socket design. It allows both up-and-down motion as well as rotation. In a MacPherson strut FWD suspension system, the two lower ball joints are nonload carrying.

Articulación esférica Un componente de la suspensión que une el brazo de mando a la articulación de la dirección y sirve como un punto pivote inferior de la articulación de la dirección. La articulación esférica derive su nombre de su diseño de bola y casquillo. Permite no sólo el movimiento de arriba y abajo sino también el de rotación. En un sistema de suspensión tipo FWD con poste de MacPherson, las articulaciones esféricas inferiores no soportan el peso.

Ballooning A condition in which the torque converter has been blown up like a balloon caused by excessive pressure in the converter.

Inflación Una condición en el qual el convertidor del esfuerzo de torsión se ha inflado como un baloon, que es una condición causada por la presión excesiva dentro del convertidor.

Band A steel band with an inner lining of frictional material. A device used to hold a clutch drum at certain times during transmission operation.

Banda Una banda de acero que tiene un forro interior de una materia de fricción. Un dispositivo que retiene al tambor del embrague en algunos momentos durante la operación de la transmisión.

Bearing The supporting part that reduces friction between a stationary and rotating part or between two moving parts.

Cojinete La parte portadora que reduce la fricción entre una parte fija y una parte giratoria o entre dos partes que muevan.

Bearing cage A spacer that keeps the balls or rollers in a bearing in proper position between the inner and outer races.

Jaula del cojinete Un espaciador que mantiene a las bolas o a los rodillos del cojinete en la posición correcta entre las pistas interiores e exteriores.

Bearing caps In the differential, caps held in place by bolts or nuts which, in turn, hold bearings in place.

Tapones del cojinete En un diferencial, las tapas que se sujeten en su lugar por pernos o tuercas, los cuales en su turno, retienen y posicionan a los cojinetes.

Bearing cone The inner race, rollers, and cage assembly of a tapered roller bearing. Cones and cups must always be replaced in matched sets.

Cono del cojinete La asamblea de la pista interior, los rodillos, y el jaula de un cojinete de rodillos cónico. Se debe siempre reemplazar a ambos partes de un par de conos del cojinete y los anillos exteriores a la vez.

Bearing cup The outer race of a tapered roller bearing or ball bearing.

Anillo exterior La pista exterior de un cojinete cónico de rodillas o de bolas.

Bearing race The surface on which the rollers or balls of a bearing rotate. The outer race is the same thing as the cup, and the inner race is the one closest to the axle shaft.

Pista del cojinete La superficie sobre la cual rueden los rodillos o las bolas de un cojinete. La pista exterior es lo mismo que un anillo exterior, y la pista interior es la más cercana a la flecha del eje.

Belleville spring A tempered spring steel cone-shaped plate used to aid the mechanical force in a pressure plate assembly.

Resorte de tensión Belleville Un plato de resorte del acero revenido en forma cónica que aumenta a la fuerza mecánica de una asamblea del plato opresor.

Bell housing A housing that fits over the clutch components and connects the engine and the transmission.

Concha del embrague Un cárter que encaja a los componentes del embrague y conecta al motor con la transmisión.

Bloodborne pathogens Pathogenic microorganisms that are present in human blood and can cause disease. These pathogens include, but are not limited to, hepatitis B virus (HBV) and human immunodeficiency virus (HIV).

Patógenos transportados en la sangre Microorganismos patógenos que se encuentran en la sangre humana y que pueden causar enfermedades. Estos son, entre otros, el virus de la hepatitis B y el virus de inmunodeficiencia humana (VIH).

Bolt head The part of a bolt that the socket or wrench fits over to torque or tighten the bolt.

Cabeza del perno La pieza de un perno que el socket o la llave ajusta sobre la cabeza del perno para apretarlo.

Bolt shank The smooth area on a bolt from the bottom surface of the head to the start of the threads.

Asta del perno El área lisa entre el fondo de la cabeza y el principio de las roscas se llama la asta del perno.

Bolt torque The turning effort required to offset resistance as the bolt is being tightened.

Torsión del perno El esfuerzo de torsión que se requiere para compensar la resistencia del perno mientras que esté siendo apretado.

Brinnelling Rough lines worn across a bearing race or shaft due to impact loading, vibration, or inadequate lubrication.

Efecto brinel Líneas ásperas que aparecen en las pistas de un cojinete o en las flechas debido al choque de carga, la vibración, o falta de lubricación.

Burr A feather edge of metal left on a part being cut with a file or other cutting tool.

Rebaba Una lima espada de metal que permanece en una parte que ha sido cortado con una lima u otro herramienta de cortar.

Bus A common connector used as an information source for the vehicle's various control units.

Bus Un conector común que se usa como un fuente de información para los varios aparatos de control del vehículo.

Bushing A cylindrical lining used as a bearing assembly; can be made of steel, brass, bronze, nylon, or plastic.

Buje Un forro cilíndrico que se usa como una asamblea de cojinete que puede ser hecho del acero, del latón, del bronze, del nylon, o del plástico.

Butt-end locking ring A locking ring whose ends are cut to butt up against or contact each other once in place. There is no gap between the ends of a butt-end ring.

Anillo de enclavamiento a tope Un anillo de enclavamiento cuyos extremidades son cortadas para toparse o ajustarse una contra la otra en lugar. No hay holgura entre las extremidades de un anillo de retén a tope.

C-clip A C-shaped clip used to retain the drive axles in some rear axle assemblies.

Grapa de C Una grapa en forma de C que retiene a las flechas motrices en algunas asambleas de ejes traseras.

Cage A spacer used to keep the balls or rollers in proper relation to one another. In a constant-velocity joint, the cage is an open metal framework that surrounds the balls to hold them in position.

Jaula Una espaciador que mantiene una relación correcta entre los rodillos o las bolas. En una junta de velocidad constante, la jaula es un armazón abierto de metal que rodea a las bolas para mantenerlas en posición.

Cap An object that fits over an opening to stop flow.

Tapón Un objecto que tapa a una apertura para detener el flujo.

Case porosity Leaks caused by tiny holes that are formed by trapped air bubbles during the casting process.

Porosidad del cárter Las fugas que se causan por los hoyitos pequeños formados por burbújas de aire entrapados durante el proceso del moldeo.

Caustic Something that causes corrosion.

Cáustico Algo que provoca corrosión.

Chamfer A bevel or taper at the edge of a hole or a gear tooth.

Chaflán Un bisél o cono en el borde de un hoyo o un diente del engranaje.

Chamfer face A beveled surface on a shaft or part that allows for easier assembly. The ends of FWD driveshafts are often chamfered to make installation of the CV joints easier.

Cara achaflanada Una superficie biselada en una flecha o una parte que facilita la asamblea. Los extremos de los árboles de mando de FWD suelen ser achaflandos para facilitar la instalación de las juntas CV.

Chase To straighten up or repair damaged threads.

Embutir Enderezar o reparar a los filetes dañados.

Chasing To clean threads with a tap.

Embutido Limpiar a los filetes con un macho.

Chassis The vehicle frame, suspension, and running gear. On FWD cars, it includes the control arms, struts, springs, trailing arms, sway bars, shocks, steering knuckles, and frame. The driveshafts, constant-velocity joints, and transaxle are not part of the chassis or suspension.

Chasis El armazón de un vehículo, la suspensión, y el engranaje de marcha. En los coches de FWD, incluye los brazos de mando, los postes, los resortes (chapas), los brazos traseros, las estabilizadoras, las articulaciones de la dirección y el armazón. Los árboles de mando, las juntas de velocidad constante, y la flecha impulsora no son partes del chasis ni de la suspensión.

Circlip A split steel snap ring that fits into a groove to hold various parts in place. Circlips are often used on the ends of FWD driveshafts to retain the constant-velocity joints.

Grapa circular Un seguro partido circular de acero que se coloca en una ranura para posicionar a varias partes. Las grapas circulares se suelen usar en las extremidades de los árboles de mando en FWD para retener las juntas de velocidad constante.

Class A fire A fire in which wood, paper, and other ordinary materials are burning.

Fuego de la Clase A Un fuego en el cual la madera, el papel, y otros materiales ordinarios se están quemando se llama un fuego de la Clase A.

Class B fire A fire involving flammable liquids, such as gasoline, diesel fuel, paint, grease, oil, and other similar liquids.

Fuego de la Clase B Un fuego que quema líquidos inflamables, tales como gasolina, el combustible diesel, la pintura, la grasa, el aceite, y otros líquidos similares se llama un fuego de la clase B.

Class C fire An electrical fire.

Fuego de la Clase C Un fuego eléctrico se llama un fuego de la clase C.

Class D fire A unique type of fire because the material burning is a metal.

Fuego de la Clase D Un fuego de la clase D es un fuego inusual en el cual un metal se está quemando.

Clearance The space allowed between two parts, such as between a journal and a bearing.

Holgura El espacio permitido entre dos partes, tal como entre un muñon y un cojinete.

Clutch A device for connecting and disconnecting the engine from the transmission or for a similar purpose in other units.

Embrague Un dispositivo para conectar y desconectar el motor de la transmisión o para tal propósito en otros conjuntos.

Clutch packs A series of clutch discs and plates installed alternately in a housing to act as a driving or driven unit.

Conjuntos de embrague Una seria de discos y platos de embrague que se han instalado alternativamente en un cárter para funcionar como una unedad de propulsión o arrastre.

Clutch slippage A situation in which engine speed increases but increased torque is not transferred through to the driving wheels.

Resbalado del embrague Una situatión en el qual la velocidad del motor aumenta pero la torsión aumentada del motor no se transfere a las ruedas de marcha.

Clutch volume index (CVI) This represents the volume of fluid needed to compress a clutch pack or apply a brake band.

Índice del Volumen del Embrague Esto representa el volumen de líquido necesitado para comprimir un paquete del embrague o para aplicar una banda del freno.

Coefficient of friction The ratio of the force resisting motion between two surfaces in contact to the force holding the two surfaces in contact.

Coeficiente de la fricción La relación entre la fuerza que resiste al movimiento entre dos superficies que tocan y la fuerza que mantiene en contacto a éstas dos superficies.

Coil preload springs Coil springs are made of tempered steel rods formed into a spiral that resist compression; located in the pressure plate assembly.

Muelles de embrague Los muelles espirales son fabricadas de varillas de acero revenido y resisten la compresión; se ubican en el conjunto del plato opresor.

Coil spring A heavy wire-like steel coil used to support the vehicle weight while allowing for suspension motions. On FWD cars, the front coil springs are mounted around the MacPherson struts. On the rear suspension, they may be mounted to the rear axle, to trailing arms, or around rear struts.

Muelles de embrague Un resorte espiral hecho de acero en forma de alambre grueso que soporte el peso del vehículo mientras que permite a los movimientos de la suspensión. En los coches de FWD, los muelles de embrague delanteros se montan alrededor de los postes Macpherson. En la suspensión trasera, pueden montarse en el eje trasero, en los brasos traseros, o alrededor de los postes traseros.

Compound A mixture of two or more ingredients.

Compuesto Una combinación de dos ingredientes o más.

Concentric Two or more circles having a common center.

Concéntrico Dos círculos o más que comparten un centro común.

Constant-velocity joint A flexible coupling between two shafts that permits each shaft to maintain the same driving or driven speed regardless of operating angle, allowing for a smooth transfer of power. The constant-velocity joint (also called CV joint) consists of an inner and outer housing with balls in between, or a tripod and yoke assembly.

Junta de velocidad constante Un acoplador flexible entre dos flechas que permite que cada flecha mantenga la velocidad de propulsión o arrastre sin importar el ángulo de operación, efectuando una transferencia lisa del poder. La junta de velociadad constante (también llamado junta CV) consiste en un cárter interior e exterior entre los cuales se encuentran bolas, o de un conjunto de trípode y yugo.

Control arm A suspension component that links the vehicle frame to the steering knuckle or axle housing and acts as a hinge to allow up-and-down wheel motions. The front control arms are attached to the frame with bushings and bolts and are connected to the steering knuckles with ball joints. The rear control arms attach to the frame with bushings and bolts and are welded or bolted to the rear axle or wheel hubs.

Brazo de mando Un componente de la suspensión que une el armazón del vehículo al articulación de dirección o al cárter del eje y que se porta como una bisagra para permitir a los movimientos verticales de las ruedas. Los brazos de mando delanteros se conectan al armazón por medio de pernos y bujes y se conectan al articulación de dirección por medio de los articulaciones esféricos. Los brazos de mando traseros se conectan al armazón por medio de pernos y bujes y son soldados o empernados al eje trasero o a los cubos de la rueda.

Corrosion Chemical action, usually by an acid, that eats away (decomposes) a metal.

Corrosión Un acción químico, por lo regular un ácido, que corroe (descompone) un metal.

Corrosivity The characteristic of a material that enables it to dissolve metals and other materials or burn the skin.

Corrosividad Característica de un material que le permite disolver metales y otros materiales, o que quemar la piel.

Cotter pin A type of fastener, made from soft steel in the form of a split pin, that can be inserted in a drilled hole. The split ends are spread to lock the pin in position.

Pasador de chaveta Un tipo de fijación, hecho de acero blando en forma de una chaveta que se puede insertar en un hueco tallado. Las extremidades partidas se despliegen para asegurar la posición de la chaveta.

Counterclockwise rotation Rotating in the opposite direction of the hands on a clock.

Rotación en sentido inverso Girando en el sentido opuesto de las agujas de un reloj.

Coupling A connecting means for transferring movement from one part to another; may be mechanical, hydraulic, or electrical.

Acoplador Un método de conección que transfere el movimiento de una parte a otra; puede ser mecánico, hidráulico, o eléctrico.

Coupling phase Point in torque converter operation in which the turbine speed is 90 percent of impeller speed and there is no longer any torque multiplication.

Fase del acoplador El punto de la operación del convertidor de la torsión en el cual la velocidad de la turbina es el 90% de la velocidad del impulsor y no queda ningún multiplicación de la torsión.

Cover plate A stamped steel cover bolted over the service access to the manual transmission.

Cubrejuntas Un cubierto de acero estampado que se emperna en la apertura de servicio de la transmisión manual.

Crocus cloth A very fine polishing paper designed to remove very little metal; therefore it is safe to use on critical surfaces.

Tela de óxido férrico Un papel muy fino para pulir. Fue diseñado para raspar muy poco del metal; por lo tanto, se suele emplear en las superficies críticas.

CRT The acronym for a cathode ray tube. This term normally refers to the display of a computer.

CRT La sigla en inglés por un tubo de rayos catódicos. Este termino suele referirse a la presentación en una computadora.

Cycle One set of changes in a signal that repeats itself several times.

Ciclo Un conjunto de cambios en una señal que se repite varias veces.

Default mode A mode of operation that allows for limited use of the transmission in the case of electronic failure.

Modo del uso limitado Un modo de operación que permite el uso limitado de la transmisión en el caso del incidente electrónico.

Deflection Bending or movement away from normal due to loading.

Desviación Curvación o movimiento fuera de lo normal debido a la carga.

Degree A unit of measurement equal to 1/360th of a circle.

Grado Una uneda de medida que iguala al 1/360 parte de un círculo.

Density Compactness; relative mass of matter in a given volume.

Densidad La firmeza; una cantidad relativa de la materia que ocupa a un volumen dado.

Detent A small depression in a shaft, rail, or rod into which a pawl or ball drops when the shaft, rail, or rod is moved. This provides a locking effect.

Detención Un pequeño hueco en una flecha, una barra o una varilla en el cual cae una bola o un linguete al moverse la flecha, la barra o la varilla. Esto provee un efecto de enclavamiento.

Detent mechanism A shifting control designed to hold the manual transmission in the gear range selected.

Aparato de detención Un control de desplazamiento diseñado a sujetar a la transmisión manual en la velocidad selecionada.

Diagnosis A systematic study of a machine or machine parts to determine the cause of improper performance or failure.

Diagnóstico Un estudio sistemático de una máquina o las partes de una máquina con el fín de determinar la causa de una falla o de un operación irregular.

Dial indicator A measuring instrument with the readings indicated on a dial rather than on a thimble as on a micrometer.

Indicador de carátula Un instrumento de medida cuyo indicador es en forma de muestra en contraste al casquillo de un micrómetro.

Differential A mechanism between drive axles that permits one wheel to run at a different speed than the other while turning.

Diferencial Un mecanismo entre dos semiejes que permite que una rueda gira a una velocidad distincta que la otra en una curva.

Differential action An operational situation in which one driving wheel rotates at a slower speed than the opposite driving wheel.

Acción del diferencial Una situación durante la operación en la cual una rueda propulsora gira con una velocidad más lenta que la rueda propulsora opuesta.

Differential case The metal unit that encases the differential side gears and pinion gears, and to which the ring gear is attached.

Caja de satélites La unedad metálica que encaja a los engranajes planetarios (laterales) y a los satélites del diferencial, y a la cual se conecta la corona.

Differential drive gear A large circular helical gear that is driven by the transaxle pinion gear and shaft and drives the differential assembly.

Corona Un engranaje helicoidal grande circular que es arrastrado por el piñon de la flecha de transmisión y la flecha y propela al conjunto del diferencial.

Differential housing Cast iron assembly that houses the differential unit and the drive axles. Also called the rear axle housing.

Cárter del diferencial Una asamblea de acero vaciado que encaja a la unedad del diferencial y los semiejes. También se llama el cárter del eje trasero.

Differential pinion gears Small beveled gears located on the differential pinion shaft.

Satélites Engranajes pequeños biselados que se ubican en la flecha del piñon del diferencial.

Differential pinion shaft A short shaft locked to the differential case. This shaft supports the differential pinion gears.

Flecha del piñon del diferencial Una flecha corta clavada en la caja de satélites. Esta flecha sostiene a los satélites.

Differential ring gear A large circular hypoid-type gear enmeshed with the hypoid drive pinion gear.

Corona Un engranaje helicoidal grande circular endentado con el piñon de ataque hipoide.

Differential side gears The gears inside the differential case that are internally splined to the axle shafts and are driven by the differential pinion gears.

Planetarios (laterales) Los engranajes adentro de la caja de satélites que son acanalados a los semiejes desde el interior, y que se arrastran por los satélites.

Dipstick A metal rod used to measure the fluid in an engine or transmission.

Varilla de medida Una varilla de metal que se usa para medir el nivel de flúido en un motor o en una transmisión.

Direct drive System in which one turn of the input driving member corresponds to one complete turn of the driven member, such as when there is direct engagement between the engine and driveshaft in which the engine crankshaft and the driveshaft turn at the same rpm.

Mando directo Una vuelta del miembro de ataque o propulsión que se compara a una vuelta completa del miembro de arrastre, tal como cuando hay un enganchamiento directo entre el motor y el árbol de transmisión en el qual el cigueñal y el árbol de transmisión giran al mismo rpm.

Disengage When the operator moves the clutch pedal toward the floor to disconnect the driven clutch disc from the driving flywheel and pressure plate assembly.

Desembragar Cuando el operador mueva el pedal de embrague hacia el piso para desconectar el disco de embrague del volante impulsor y del conjunto del plato opresor.

Distortion A warpage or change in form from the original shape.

Distorción Abarquillamiento o un cambio en la forma original.

DLC The data link connector. This is the connector used to connect into a vehicle's computer system for the purpose of diagnostics. Prior to J1930, this was commonly referred to as the ALDL.

DLC El conectador de enlaces de datos. Este es el conectador que se usa en conectarse al sistema computerizado del vehículo con el propósito de efectuar los diagnósticos. Antes del J1930, esto solía referirse como el ALDL.

DMM The acronym for a digital multimeter.

DMM La sigla en inglés por un multímeter digital.

Dowel A metal pin attached to one object that, when inserted into a hole in another object, ensures proper alignment.

Espiga Una clavija de metal que se fija a un objeto, que al insertarla en el hoyo de otro objeto, asegura una alineación correcta.

Dowel pin A pin inserted in matching holes in two parts to maintain those parts in fixed relation one to another.

Clavija de espiga Una clavija que se inserte en los hoyos alineados en dos partes para mantener ésos dos partes en una relación fijada el uno al otro.

Downshift To shift a transmission into a lower gear.

Cambio descendente Cambiar la velocidad de una transmision a una velocidad más baja.

Driveline torque Relates to rear-wheel driveline; the transfer of torque between the transmission and the driving axle assembly.

Potencia de la flecha motríz Se relaciona a la flecha motríz de las ruedas traseras y transfere la potencia de la torsión entre la transmisión y el conjunto del eje trasero.

Driven gear The gear meshed directly with the driving gear to provide torque multiplication, reduction, or a change of direction.

Engranaje de arrastre El engranaje endentado directamente al engranaje de ataque para proporcionar la multiplicación, la reducción, o los cambios de dirección de la potencia.

Drive pinion gear One of the two main driving gears located within the transaxle or rear driving axle housing. Together the two gears multiply engine torque.

Engranaje de piñon de ataque Uno de dos engranajes de ataque principales que se ubican adentro de la flecha de transmisión o en el cárter del eje de propulsión. Los dos engranajes trabajan juntos para multiplicar la potencia.

Driveshaft An assembly of one or two universal joints connected to a shaft or tube; used to transmit power from the transmission to the differential. Also called the propeller shaft.

Árbol de mando Una asamblea de uno o dos uniones universales que se conectan a un árbol o un tubo; se usa para transferir la potencia desde la transmisión al diferencial. También se le refiere como el árbol de propulsión.

Dry friction The friction between two dry solids.

Fricción seca Fricción entre dos s-lidos secos.

DSO A common acronym for a digital storage oscilloscope.

DSO Siglas comunes para un osciloscopio con memoria digital.

DTC The acronym for diagnostic trouble code.

DTC La sigla en inglés por un código diagnóstico de averías.

Dynamic In motion.

Dinámico En movimiento.

EEPROM An electrically erasable programmable read only memory chip.

EEPROM Una placa de memoria, leer solamente, erasible y programable.

Elastomer Any rubber-like plastic or synthetic material used to make bellows, bushings, and seals.

Elastómero Cualquiera materia plást parecida al hule o una materia sintética que se utiliza para fabricar a los fuelles, los bujes y las juntas.

End clearance Distance between a set of gears and their cover, commonly measured on oil pumps.

Holgura del extremo La distancia entre un conjunto de engranajes y su placa de recubrimiento, suele medirse en las bombas de aceite.

Endplay The amount of axial or end-to-end movement in a shaft due to clearance in the bearings.

Juego de las extremidades La cantidad del movimiento axial o del movimiento de extremidad a extremidad en una flecha debido a la holgura que se deja en los cojinetes.

Engage When the vehicle operator moves the clutch pedal up from the floor, this engages the driving flywheel and pressure plate to rotate and drive the driven disc.

Accionar Cuando el operador del vehículo deja subir el pedal del embrague del piso, ésto acciona la volante de ataque y el plato opresor para impulsar al disco de arrastre.

Engagement chatter A shaking, shuddering action that takes place as the driven disc makes contact with the driving members. Chatter is caused by a rapid grip and slip action.

Chasquido de enganchamiento Un movimiento de sacudo o temblor que resulta cuando el disco de ataque viene en contacto con los miembros de propulsión. El chasquido se causa por una acción rápida de agarrar y deslizar.

Engine torque A turning or twisting action developed by the engine, measured in foot-pounds or kilogram-meters.

Torsión del motor Una acción de girar o torcer que crea el motor, ésta se mide en librasópie o kilosómetros.

EP toxicity The characteristic of a material that enables it to leach one or more of eight heavy metals in concentrations greater than 100 times standard drinking water concentrations.

Toxicidad por el procedimiento de extracción Característica de un material que permite que se filtren uno o más de ocho metales pesados en concentraciones mayores a 100 veces las concentraciones estándares en el agua potable.

Essential tool kit A set of special tools designed for a particular model of car or truck.

Estuche de herramientas principales Un conjunto de herramientas especiales diseñadas para un modelo específico de coche o camión.

Etching A discoloration or removal of some material caused by corrosion or some other chemical reaction.

Grabado por ácido Una descoloración o remueva de una materia que se efectua por medio de la corrosión u otra reacción química.

Extension housing An aluminum or iron casting of various lengths that encloses the transmission output shaft and supporting bearings.

Cubierta de extensión Una pieza moldeada de aluminio o acero que puede ser de varias longitudes que encierre a la flecha de salida de la transmisión y a los cojinetes de soporte.

External gear A gear with teeth across the outside surface.

Engranaje exterior Un engranaje cuyos dientes estan en la superficie exterior.

Externally tabbed clutch plates Clutch plates that are designed with tabs around the outside periphery to fit into grooves in a housing or drum.

Placas de embrague de orejas externas Las placas de embrague que se diseñan de un modo para que las orejas periféricas de la superficie se acomoden en una ranura alrededor de un cárter o un tambor.

Extreme-pressure lubricant A special lubricant for use in hypoid-gear differentials; needed because of the heavy wiping loads imposed on the gear teeth.

Lubricante de presión extrema Un lubricante especial que se usa en las diferenciales de tipo engranaje hipóide; se requiere por la carga de transmisión de materia pesada que se imponen en los dientes del engranaje.

Face The front surface of an object.

Cara La superficie delantera de un objeto.

Fatigue The buildup of natural stress forces in a metal part that eventually causes it to break. Stress results from bending and loading the material.

Fatiga El incremento de tensiones y esfuerzos normales en una parte de metal que eventualmente causen una quebradura. Los esfuerzos resultan de la carga impuesta y el doblamiento de la materia.

Feeler gauge A metal strip or blade finished accurately with regard to thickness used for measuring the clearance between two parts; such gauges ordinarily come in a set of different blades graduated in thickness by increments of 0.001 inch.

Calibrador de laminillas Una lámina o hoja de metal que ha sido acabado precisamente con respecto a su espesor que se usa para medir la holgura entre dos partes; estas galgas típicamente vienen en un conjunto de varias espesores graduados desde el 0.001 de una pulgada.

Fillet The small, smooth curve on a bolt, where the shank flows into the bolt head.

Filete La curva pequeña, lisa en un perno, adonde la asta fluye en la cabeza del perno se llama un filete.

Final drive ratio The ratio between the drive pinion and ring gear.

Relación del mando final La relación entre el piñón de ataque y la corona.

Fit The contact between two machined surfaces.

Ajuste El contacto entre dos superficies maquinadas.

Fixed-type constant-velocity joint A joint that cannot telescope or plunge to compensate for suspension travel. Fixed joints are always found on the outer ends of the driveshafts of FWD cars. A fixed joint may be of either Rzeppa or tripod type.

Junta tipo fijo de velocidad constante Una junta que no tiene la capacidad de los movimientos telescópicos o repentinos que sirven para compensar en los viajes de suspensión. Las juntas fijas siempre se ubican en las extremidades exteriores de los árboles de mando en los coches de FWD. Una junta tipo fijo puede ser de un tipo Rzeppa o de trípode.

Flammability A statement of how well a substance supports combustion.

Inflamabilidad Declaración acerca de cuánto resiste a la combustión una sustancia.

Flange A projecting rim or collar on an object that keeps it in place.

Reborde Una orilla o un collar sobresaliente de un objeto cuyo función es de mantenerlo en lugar.

Flexplate A lightweight flywheel used only on engines equipped with an automatic transmission. A flexplate is equipped with a starter ring gear around its outside diameter and also serves as the attachment point for the torque converter.

Placa articulada Un volante ligera que se usa solamente en los motores que se equipan con una transmisión automática. El diámetro exterior de la placa articulada viene equipado con un anillo de engranajes para arrancar y también sirve como punto de conección del convertidor de la torsión.

Fluid coupling A device in the powertrain consisting of two rotating members; transmits power from the engine, through a fluid, to the transmission.

Acoplamiento de fluido Un dispositivo en el tren de potencia que consiste de dos miembros rotativos; transmite la potencia del motor, por medio de un fluido, a la transmisión.

Fluid drive A drive in which there is no mechanical connection between the input and output shafts, and power is transmitted by moving oil.

Dirección fluido Una dirección en la cual no hay conecciones mecánicas entre las flechas de entrada o salida, y la potencia se transmite por medio del aceite en movimiento.

Flywheel A heavy metal wheel that is attached to the crankshaft and rotates with it; helps smooth out the power surges from the engine power strokes; also serves as part of the clutch and engine-cranking system.

Volante Una rueda pesada de metal que se fija al cigueñal y gira con ésta; nivela a los sacudos que provienen de la carrera de fuerza del motor; también sirve como parte del embrague y del sistema de arranque.

Flywheel ring gear A gear fitted around the flywheel that is engaged by teeth on the starting motor drive to crank the engine.

Engranaje anular del volante Un engranaje, colocado alrededor del volante que se acciona por los dientes en el propulsor del motor de arranque y arranca al motor.

Foot-pound (ft.-lb.) A measure of the amount of energy or work required to lift 1 pound a distance of 1 foot.

Pie libra Una medida de la cantidad de energía o fuerza que requiere mover una libra a una distancia de un pie.

Force Any push or pull exerted on an object; measured in pounds and ounces, or in newtons (N) in the metric system.

Fuerza Cualquier acción empujado o jalado que se efectua en un objeto; se mide en pies y onzas, o en newtones (N) en el sistema métrico.

Four-wheel-drive On a vehicle, driving axles at both front and rear, so that all four wheels can be driven.

Tracción a cuatro ruedas En un vehículo, se trata de los ejes de dirección fronteras y traseras, para que cada uno de las ruedas puede impulsar.

Frame The main understructure of the vehicle to which everything else is attached. Most FWD cars have only a subframe for the front suspension and drivetrain. The body serves as the frame for the rear suspension.

Armazón La estructura principal del vehículo al cual todo se conecta. La mayoría de los coches FWD sólo tiene un bastidor auxiliar para la suspensión delantera y el tren de propulsión. El carrocería del coche sirve de chassis par la suspensión trasera.

Freewheel To turn freely and not transmit power.

Volante libre Da vueltas libremente sin transferir la potencia.

Freewheeling clutch A mechanical device that will engage the driving member to impart motion to a driven member in one direction but not the other. Also known as an "overrunning clutch."

Embrague de volante libre Un dispositivo mecánico que acciona el miembro de tracción y da movimiento al miembro de tracción en una dirección pero no en la otra. También se conoce bajo el nombre de un "embrague de sobremarcha."

Friction The resistance to motion between two bodies in contact with each other.

Fricción La resistencia al movimiento entre dos cuerpos que están en contacto.

Friction bearing A bearing in which there is sliding contact between the moving surfaces. Sleeve bearings, such as those used in connecting rods, are friction bearings.

Rodamientos de fricción Un cojinete en el cual hay un contacto deslizante entre las superficies en movimiento. Los rodamientos de manguitos, como los que se usan en las bielas, son rodamientos de fricción.

Friction disc In the clutch, a flat disc, faced on both sides with frictional material and splined to the clutch shaft. It is positioned between the clutch pressure plate and the engine flywheel. Also called the clutch disc or driven disc.

Disco de fricción En el embrague, un disco plano al cual se ha cubierto ambos lados con una materia de fricción y que ha sido estriado a la flecha del embrague. Se posiciona entre el plato opresor del embrague y el volante del motor. También se llama el disco del embrague o el disco de arrastre.

Friction facings A hard-molded or woven asbestos or paper material that is riveted or bonded to the clutch driven disc.

Superficie de fricción Un recubrimiento remachado o aglomerado al disco de arrastre del embrague que puede ser hecho del amianto moldeado o tejido o de una materia de papel.

Front pump Pump located at the front of the transmission. It is driven by the engine through two dogs on the torque converter housing. It supplies fluid whenever the engine is running.

Bomba delantera Una bomba ubicado en la parte delantera de la transmisión. Se arrastre por el motor al través de dos álabes en el cárter del convertidor de la torsión. Provee el fluido mientras que funciona el motor.

Front-wheel-drive (FWD) Describes a vehicle with all drivetrain components located at the front.

Tracción de las ruedas delanteras (FWD) El vehículo tiene todos los componentes del tren de propulsión en la parte delantera.

Galling Wear caused by metal-to-metal contact in the absence of adequate lubrication. Metal is transferred from one surface to the other, leaving behind a pitted or scaled appearance.

Desgaste por fricción El desgaste causado por el contacto de metal a metal en la ausencia de lubricación adecuada. El metal se transfere de una superficie a la otra, causando una aparencia agujerado o con depósitos.

Gasket A layer of material, usually made of cork, paper, plastic, composition, or metal, or a combination of these, placed between two parts to make a tight seal.

Empaque Una capa de una materia, normalmente hecho del corcho, del papel, del plástico, de la materia compuesta o del metal, o de cualquier combinación de éstos, que se coloca entre dos partes para formar un sello impermeable.

Gasket cement A liquid adhesive material or sealer used to install gaskets.

Mastique para empaques Una substancia líquida adhesiva, o una substancia impermeable, que se usa para instalar a los empaques.

Gear A wheel with external or internal teeth that serves to transmit or change motion.

Engranaje Una rueda que tiene dientes interiores o exteriores que sirve para transferir o cambiar el movimiento.

Gear lubricant A type of grease or oil blended especially to lubricate gears.

Lubricante para engranaje Un tipo de grasa o aceite que ha sido mezclado específicamente para la lubricación de los engranajes.

Gear ratio The number of revolutions of a driving gear required to turn a driven gear through one complete revolution. For a pair of gears, the ratio is found by dividing the number of teeth on the driven gear by the number of teeth on the driving gear.

Relación de los engranajes El número de las revoluciones requeridas del engranaje de propulsión para dar una vuelta completa al engranaje arrastrado. En una pareja de engranajes, la relación se calcula al dividir el número de los dientes en el engranaje de arrastre por el número de los dientes en el engranaje de propulsión.

Gear reduction When a small gear drives a large gear, there is an output speed reduction and a torque increase that results in a gear reduction.

Velocidad descendente Cuando un engranaje pequeño impulsa a un engranaje grande, hay una reducción en la velocidad de salida y un incremento en la torsión que resulta en una cambio descendente de los velocidades.

Gearshift A linkage-type mechanism by which the gears in an automobile transmission are engaged and disengaged.

Varillaje de cambios Un mecanismo tipo eslabón que acciona y desembraga a los engranajes de la transmisión.

Gear whine A high-pitched sound developed by some types of meshing gears.

Ruido del engranaje Un sonido agudo que proviene de algunos tipos de engranajes endentados.

Glitches Abnormal, slight movements of a waveform on a lab scope. These can be caused by circuit problems or noise in the circuit.

Irregularidades espontáneas ("glitch") Los movimientos pequeños e abnormales en una onda en una pantalla de laboratorio. Estos pueden ser causados por los problemas de los circuitos o el ruido dentro del circuito.

Governor pressure The transmission's hydraulic pressure; directly related to output shaft speed. It is used to control shift points.

Regulador de presión La presión hidráulica de una transmisión se relaciona directamente a la velocidad de la flecha de salida. Se usa para controlar los puntos de cambios de velocidad.

Governor valve A device used to sense vehicle speed. The governor valve is attached to the output shaft.

Válvula reguladora Un dispositivo que se usa para determinar la velocidad de un vehículo. La válvula reguladora se monta en la flecha de salida.

Grade marks Marks on fasteners that indicate strength.

Marcas del grado Las marcas en los sujetadores que indican fuerza se llaman las marcas del grado.

Hazardous waste Any used material or any byproduct that can be classified as potentially hazardous to one's health and/or the environment.

Desechos Peligrosos Cualquier material usado, o cualquier subproducto, que se pueda clasificar como potencialmente dañoso al medio ambiente o parjudicial para salud.

Hub The center part of a wheel, to which the wheel is attached.

Cubo La parte central de una rueda, a la cual se monta la rueda.

Hydraulic press A piece of shop equipment that develops a heavy force by use of a hydraulic piston-and-jack assembly.

Prensa hidráulica Una herramienta del taller que provee una fuerza grande por medio de una asamblea de gato con un pistón hidráulico.

Hydraulic pressure Pressure exerted through the medium of a liquid.

Presión hidráulica La presión esforzada por medio de un líquido.

Hydrocarbons Particles of gasoline, present in the exhaust and in crankcase vapors, that have not been fully burned.

Hidrocarburos Partículas de la gasolina, presentes en los vapores del cárter del cigüeñal y el tubo de escape, que no han sido completamente quemadas.

ID Inside diameter.

DI Diámetro interior.

Idle Engine speed when the accelerator pedal is fully released and there is no load on the engine.

Marcha lenta La velocidad del motor cuando el pedal accelerador esta completamente desembragada y no hay carga en el motor.

Ignitability The characteristic of a solid that enables it to spontaneously ignite. Any liquid with a flash point below 140°F is also said to possess ignitability.

Incendiabilidad Característica de un sólido que le permite encenderse espontáneamente. Cualquier líquido con un punto de inflamabilidad por debajo de los 140°F también se dice que posee incendiabilidad.

Impedance The operating resistance of an electrical device.

Impedancia La resistencia operativa de un dispositivo eléctrico.

Impeller The pump or driving member in a torque converter.

Impulsor La bomba o el miembro impulsor en un convertidor de torsión.

Increments Series of regular additions from small to large.

Incrementos Una serie de incrementos regulares que va de pequeño a grande.

Index To orient two parts by marking them. During reassembly the parts are arranged so the index marks are next to each other. Used to preserve the orientation between balanced parts.

Índice Orientar a dos partes marcándolas. Al montarlas, las partes se colocan para que las marcas de índice estén alinieadas. Se usan los índices para preservar la orientación de las partes balanceadas.

Input shaft The shaft carrying the driving gear by which the power is applied, as to the transmission.

Flecha de entrada La flecha que porta el engranaje propulsor por el cual se aplica la potencia, como a la transmisión.

Inspection cover A removable cover that permits entrance for inspection and service work.

Cubierta de inspección Una cubierta desmontable que permite a la entrada para inspeccionar y mantenimiento.

Integral Built into, as part of the whole.

Integral Incorporado, una parte de la totalidad.

Internal gear A gear with teeth pointing inward, toward the hollow center of the gear.

Engranaje internal Un engranaje cuyos dientes apuntan hacia el interior, al hueco central del engranaje.

International System (SI) The International System of weights and measures. The system is normally called the metric system.

Sistema Internacional (SI) El sistema internacional de pesos y de medidas. Este sistema normalmente se llama la sistema métrico.

Jack (safety) stands Used to hold a vehicle up while using a floor jack.

Gato de seguridad Se usa para sostener un vehículo mientras se está usando un gato patín.

Jam nut A second nut tightened against a primary nut to prevent it from working loose. Used on inner and outer tie-rod adjustment nuts and on many pinion-bearing adjustment nuts.

Contra tuerca Una tuerca secundaria que se aprieta contra una tuerca primaria para prevenir que ésta se afloja. Se emplean en las tuercas de ajustes interiores e exteriores para las barras de acoplamiento y también en muchas de las tuercas de ajuste de portapiñones.

Journal A bearing with a hole in it for a shaft.

Manga de flecha Un cojinete que tiene un hoyo para una flecha.

Key A small block inserted between the shaft and hub to prevent circumferential movement.

Chaveta Un tope pequeño que se meta entre la flecha y el cubo para prevenir un movimiento circunferencial.

Keyway A groove or slot cut to permit the insertion of a key.

Ranura de chaveta Un corte de ranura o mortaja que permite insertar una chaveta.

Knock A heavy metallic sound usually caused by a loose or worn bearing.

Golpe Un sonido metálico fuerte que suele ser causado por un cojinete suelto o gastado.

Lapping The process of fitting one surface to another by rubbing them together with an abrasive material between the two surfaces.

Pulido El proceso de ajustar a una superficie con otra por frotarlas juntas con una materia abrasiva entre las dos superficies.

Lash The amount of free motion in a geartrain, between gears, or in a mechanical assembly, such as the lash in a valve train.

Juego La cantidad del movimiento libre en un tren de engranajes, entre los engranajes o en una asamblea mecánica, tal como el juego en un tren de vávulas.

LED The common acronym for a light-emitting diode.

LED La sigla común en inglés por un diódo emisor de luz.

Linkage Any series of rods, yokes, levers, and so on used to transmit motion from one unit to another.

Biela Cualquiera serie de barras, yugos, palancas, y todo lo demás, que se usa para transferir los movimientos de una unedad a otra.

Locking ring A type of sealing ring that has ends that meet or lock together during installation. There is no gap between the ends when the ring is installed.

Anillo de enclavamiento Un tipo de anillo obturador que tiene las extremidades que se tocan o se enclavan durante la instalación. No hay holgura entre las extremidades cuando se ha instalado el anillo.

Locknut A second nut turned down on a holding nut to prevent loosening.

Contra tuerca Una tuerca segundaria apretada contra una tuerca de sostén para prevenir que ésta se afloja.

Lock pin Used in some ball sockets (inner tie-rod end) to keep the connecting nuts from working loose. Also used on some lower ball joints to hold the tapered stud in the steering knuckle.

Clavija de cerrojo Se usan en algunas rótulas (las extremidades interiores de la barra de acoplamiento) para prevenir que se aflojan las tuercas de conexión. También se emplean en algunas juntas esféricas inferiores para retener al perno cónico en la articulación de dirección.

Lockplates Metal tabs bent around nuts or bolt heads.

Placa de cerrojo Chavetas de metal que se doblan alrededor de las tuercas o las cabezas de los pernos.

Lockwasher A type of washer which, when placed under the head of a bolt or nut, prevents the bolt or nut from working loose.

Arandela de freno Un tipo de arandela que, al colocarse bajo la cabeza de un perno, previene que el perno o la tuerca se aflojan.

Low speed The gearing that produces the highest torque and lowest speed of the wheels.

Velocidad baja La velocidad que produce la torsión más alta y la velocidad más baja a las ruedas.

Lubricant Any material, usually a petroleum product such as grease or oil, that is placed between two moving parts to reduce friction.

Lubricante Cualquier substancia, normalmente un producto de petróleo como la grasa o el aciete, que se coloca entre dos partes en movimiento para reducir la fricción.

Machinist's rule A ruler used to measure short distances.

Regla de mecánico Regla que se usa para medir distancias cortas.

Mainline pressure The hydraulic pressure that operates apply devices and is the source of all other pressures in an automatic transmission. It is developed by pump pressure and regulated by the pressure regulator.

Línea de presión La presión hidráulica que opera a los dispositivos de applicación y es el orígen de todas las presiones en la transmisión automática. Proviene de la bomba de presión y es regulada por el regulador de presión.

Main oil pressure regulator valve Regulates the line pressure in a transmission.

Válvula reguladora de la linea de presión Regula la presión en la linea de una transmisión.

Manifold absolute pressure (MAP) sensor The sensor that measures changes in the intake manifold pressure that result from changes in engine load and speed.

Sensor de presión absoluta del colector de escape Un sensor que mide los cambios en la presión del colector de escape que resultan de cambios en carga y velocidad del motor.

Manual control valve A valve used to manually select the operating mode of the transmission. It is moved by the gearshift linkage.

Válvula de control manual Una válvula que se usa para escojer a una velocidad de la transmisión por mano. Se mueva por la biela de velocidades.

Material safety data sheets (MSDS) Information sheets containing chemical composition and precautionary information for all products that can present a health or safety hazard.

Hoja de datos de seguridad de materiales Hoja informativa que contiene la composición química e información de precaución para todos los productos que puedan presentar un riesgo para la salud o la seguridad.

Meshing The mating, or engaging, of the teeth of two gears.

Engrane Embragar o endentar a los dientes de dos engranajes.

Meter 1/10,000,000 of the distance from the North Pole to the Equator, or 39.37 inches.

Metro Un 1/10,000,000 de la distancia del polo del norte al ecuador.

Micrometer A precision measuring device used to measure small bores, diameters, and thicknesses. Also called a mike.

Micrómetro Un dispositivo de medida precisa que se emplea a medir a los taladros pequeños y a los espesores. También se llama un mike (mayk).

MIL The malfunction indicator lamp for a computer control system. Prior to J1930, the MIL was commonly called a Check Engine or Service Engine Soon lamp.

MIL La lámpara de indicación de averías para un sistema de control de computadora. Antes del J1930, la MIL se llamaba la lámpara de Revise Motor o Servicia el Motor Pronto.

Millimeter (mm) The third and most commonly used division of a meter.

Milímetro (mm) Un milímetro es el tercero y la división lo más comúnmente posible usada de un metro, una medida lineal en la sistema métrico, conteniendo la milésima pieza de un metro; igual al 0.03934 de una pulgada.

Mineral spirits An alcohol-based liquid that leaves no surface residue after it dries.

Alcoholes mineral Un líquido a base de alcohol que no sale de ningún residuo superficial después de que se seque.

Misalignment When bearings are not on the same centerline.

Desalineamineto Cuando los cojinetes no comparten la misma linea central.

Modulator A vacuum diaphragm device connected to a source of engine vacuum. It provides an engine load signal to the transmission.

Modulador Un dispositivo de diafragma de vacío que se conecta a un orígen de vacío en el motor. Provee un señal de carga del motor a la transmisión.

Mounts Made of rubber to insulate vibrations and noise while they support a powertrain part, such as engine or transmission mounts.

Monturas Hecho de hule para insular a las vibraciones y a los ruidos mientras que sujetan una parte del tren de propulsión, tal como las monturas del motor o las monturas de la transmisión.

Multimeter A tool that combines the voltmeter, ohmmeter, and ammeter together in a diagnostic instrument.

Multímetro Una herramienta que combina el voltímetro, el ohmímetro, y el amperímetro junta en uno instrumento de diagnóstico.

Multiple disc A clutch with a number of driving and driven discs as compared to a single plate clutch.

Discos múltiples Un embrague que tiene varios discos de propulsión o de arraste al contraste con un embrague de un sólo plato.

Needle bearing An antifriction bearing using a great number of long, small-diameter rollers. Also known as a quill bearing.

Rodamiento de agujas Un rodamiento (cojinete) antifricativo que emplea un gran cantidad de rodillos largos y de diámetro muy pequeños.

Needle deflection Distance of travel from zero of the needle on a dial gauge.

Desviación de la aguja La distancia del cero que viaja una aguja de un indicador.

Neoprene A synthetic rubber that is not affected by the various chemicals that are harmful to natural rubber.

Neoprene Un hule sintético que no se afecta por los varios productos químicos que pueden dañar al hule natural.

Neutral In a transmission, the setting in which all gears are disengaged and the output shaft is disconnected from the drive wheels.

Neutral En una transmisión, la velocidad en la cual todos los engranajes estan desembragados y el árbol de salida esta desconectada de las ruedas de propulsión.

Neutral-start switch A switch wired into the ignition switch to prevent engine cranking unless the transmission shift lever is in neutral or the clutch pedal is depressed.

Interruptor de arranque en neutral Un interruptor eléctrico instalado en el interruptor de encendido que previene el arranque del motor al menos de que la palanca de cambio de velocidad esté en una posición neutral o que se pisa en el embrague.

Newton-meter (Nm) Metric measurement of torque or twisting force.

Metro newton (Nm) Una medida métrica de la fuerza de torsión.

Nominal shim A shim with a designated thickness.

Laminilla fina Una cuña de un espesor especificado.

Nonhardening A gasket sealer that never hardens.

Sinfragua Un cemento de empaque que no endurece.

Nut A removable fastener used with a bolt to lock pieces together; made by threading a hole through the center of a piece of metal that has been shaped to a standard size.

Tuerca Un retén removable que se usa con un perno o tuerca para unir a dos piezas; se fabrica al filetear un hoyo taladrado en un pedazo de metal que se ha formado a un tamaño específicado.

Occupational Safety and Health Administration (OSHA) The government agency charged with ensuring safe work environments for all workers.

Administraion de la seguridad y de la salud ocupacionales (OSHA) La agencia de estatal cargada con asegurar los ambientes seguros del trabajo para todos los trabajadores.

OD Outside diameter.

DE Diámetro exterior.

Ohmmeter A test meter used to measure electrical resistance and/or continuity.

Contador de ohmios Un aparato que mide resistencia y/o continuidad eléctrica.

Oil seal A seal placed around a rotating shaft or other moving part to prevent leakage of oil.

Empaque de aciete Un empaque que se coloca alrededor de una flecha giratoria para prevenir el goteo de aceite.

One-way clutch See Sprag clutch.

Embrague de una via Vea Sprag clutch.

Open An open is a break in the circuit that stops current flow.

Abierto Una rotura en el circuito que para el flujo de la corriente eléctrica.

O-ring A type of sealing ring, usually made of rubber or a rubber-like material. In use, the O-ring is compressed into a groove to provide the sealing action.

Anillo en O Un tipo de sello anular, suele ser hecho de hule o de una materia parecida al hule. Al usarse, el anillo en O se comprime en una ranura para proveer un sello.

Oscillate To swing back and forth like a pendulum.

Oscilar Moverse alternativamente en dos sentidos contrarios como un péndulo.

Outer bearing race The outer part of a bearing assembly on which the balls or rollers rotate.

Pista exterior de un cojinete La parte exterior de una asamblea de cojinetes en la cual ruedan las bolas o los rodillos.

Out-of-round Wear of a round hole or shaft which, when viewed from an end, will appear egg-shaped.

Defecto de circularidad Desgaste de un taladro o de una flecha circular, que al verse de una extremidad, tendrá una forma asimétrica, como la de un huevo.

Output shaft The shaft or gear that delivers the power from a device, such as a transmission.

Flecha de salida La flecha o la velocidad que transmite la potencia de un dispositivo, tal como una transmisión.

Overall ratio The product of the transmission gear ratio multiplied by the final drive or rear axle ratio.

Relación global El producto de multiplicar la relación de los engranajes de la transmisión por la relación del impulso final o por la relación del eje trasero.

Overdrive Any arrangement of gearing that produces more revolutions of the driven shaft than of the driving shaft.

Sobremultiplicación Un arreglo de los engranajes que produce más revoluciones de la flecha de arrastre que los de la flecha de propulsión.

Overdrive ratio Identified by the decimal point indicating less than one driving input revolution compared to one output revolution of a shaft.

Relación del sobremultiplicación Se identifica por el punto decimal que indica menos de una revolución del motor comparado a una revolución de una flecha de salida.

Overrun coupling A freewheeling device to permit rotation in one direction but not in the other.

Acoplamiento de sobremarcha Un dispositivo de marcha de rueda libre que permite las giraciones en una dirección, pero no en la otra dirección.

Overrunning clutch A device consisting of a shaft or housing linked together by rollers or sprags operating between movable and fixed races. As the shaft rotates, the rollers or sprags jam between the movable and fixed races. This jamming action locks together the shaft and housing. If the fixed race should be driven at a speed greater than the movable race, the rollers or sprags will disconnect the shaft.

Embrague de sobremarcha Un dispositivo que consiste de una flecha o un cárter eslabonados por medio de rodillos o palancas de detención que operan entre pistas fijas y movibles. Al girar la flecha, los rodillos o palancas de detención se aprietan entre las pistas fijas y movibles. Este acción de apretarse enclava el cárter con la flecha. Si la pista fija se arrastra en una velocidad más alta que la pista movible, los rodillos o palancas de detención desconectarán a la flecha.

Oxidation Burning or combustion; the combining of a material with oxygen. Rusting is slow oxidation, and combustion is rapid oxidation.

Oxidación Quemando o la combustión; la combinación de una materia con el oxígeno. El orín es una oxidación lenta, la combustión es la oxidación rápida.

Pascal's law The law of fluid motion.

Ley de pascal La ley del movimiento del fluido.

Parallel The quality of two items being the same distance from each other at all points; usually applied to lines and, in automotive work, to machined surfaces.

Paralelo La calidad de dos artículos que mantienen la misma distancia el uno del otro en cada punto; suele aplicarse a las líneas y, en el trabajo automotívo, a las superficies acabadas a máquina.

Pawl A lever that pivots on a shaft. When lifted, it swings freely and when lowered, it locates in a detent or notch to hold a mechanism stationary.

Trinquete Una palanca que gira en una flecha. Levantado, mueve sín restricción, bajado, se coloca en una endentación o una muesca para mantener sín movimiento a un mecanismo.

PCM The powertrain control module of a computer control system. Prior to J1930, the PCM was commonly called a ECA, ECM, or one of many acronyms used by the various manufacturers.

PCM El módulo de control del sistema de transmisión de fuerza de un sistema de control de una computadora. Antes d3l J1930, el PCM se llamaba un ECA, un ECM, o una de varias siglas usadas por los varios fabricantes.

Pitch The number of threads per inch on any threaded part.

Paso El número de filetes por pulgada de cualquier parte fileteada.

Pivot A pin or shaft on which another part rests or turns.

Pivote Una chaveta o una flecha que soporta a otra parte o sirve como un punto para girar.

Planetary gearset A system of gearing that is modeled after the solar system. A pinion is surrounded by an internal ring gear and planet gears are in mesh between the ring gear and pinion around which all revolve.

Conjunto de engranajes planetarios Un sistema de engranaje cuyo patrón es el sistema solar. Un engranaje propulsor (la corona interior) rodea al piñon de ataque y los engranajes satélites y planetas se endentan entre la corona y el piñon alrededor del cual todo gira.

Planet carrier The carrier or bracket in a planetary gear system that contains the shafts on which the pinions or planet gears turn.

Perno de arrastre planetario El soporte o la abrazadera que contiene las flechas en las cuales giran los engranajes planetarios o los piñones.

Planet gears The gears in a planetary gearset that connect the sun gear to the ring gear.

Engranages planetarios Los engranajes en un conjunto de engranajes planetario que connectan al engranaje propulsor interior (el engranaje sol) con la corona.

Planet pinions In a planetary gear system, the gears that mesh with, and revolve about, the sun gear; they also mesh with the ring gear.

Piñones planetarios En un sistema de engranajes planetarios, los engranajes que se endentan con, y giran alrededor, el engranaje propulsor (sol); también se endentan con la corona.

Plug Anything that will fit into an opening to stop fluid or airflow.

Tapón Cualquier cosa que se ajuste en una apertura para prevenir el goteo o el escape de un corriente del aire.

Pneumatic tools Power tools that rely on compressed air for power.

Herramientas neumáticas Las herramientas de motor cuyo energía proviene del aire comprimido.

Porosity A statement of how porous or permeable to liquids a material is.

Porosidad Una expresión de lo poroso o permeable a los líquidos es una materia.

Powertrain The mechanisms that carry the power from the engine crankshaft to the drive wheels; these include the clutch, transmission, driveline, differential, and axles.

Tren impulsor Los mecanismos que transferen la potencia desde el cigueñal del motor a las ruedas de propulsión; éstos incluyen el embrague, la transmisión, la flecha motríz, el diferencial y los semiejes.

Preload A load applied to a part during assembly so as to maintain critical tolerances when the operating load is applied later.

Carga previa Una carga aplicada a una parte durante la asamblea para asegurar sus tolerancias críticas antes de que se le aplica la carga de la operación.

Press-fit Forcing a part into an opening that is slightly smaller than the part itself to make a solid fit.

Ajustamiento a presión Forzar a una parte en una apertura que es de un tamaño más pequeño de la parte para asegurar un ajustamiento sólido.

Pressure Force per unit area, or force divided by area. Usually measured in pounds per square inch (psi) or in kilopascals (kPa) in the metric system.

Presión La fuerza por unedad de una area, o la fuerza divida por la area. Suele medirse en libras por pulgada cuadrada (lb/pulg2) o en kilopascales (kPa) en el sistema métrico.

Pressure plate That part of the clutch that exerts force against the friction disc; it is mounted on and rotates with the flywheel.

Plato opresor Una parte del embraque que aplica la fuerza en el disco de fricción; se monta sobre el volante, y gira con éste.

Propeller shaft See Driveshaft.

Flecha de propulsion Vea Flecha motríz.

Prussian blue A blue pigment; in solution, useful in determining the area of contact between two surfaces.

Azul de Prusia Un pigmento azul; en forma líquida, ayuda en determinar la area de contacto entre dos superficies.

psi Abbreviation for *pounds per square inch,* a measurement of pressure.

Lb/pulg2 Una abreviación de libras por pulgada cuadrada, una medida de la presión.

Puller Generally, a shop tool used to separate two closely fitted parts without damage. Often contains one screw, or several screws that can be turned to apply a gradual force.

Extractor Generalmente, una herramienta del taller que sirve para separar a dos partes apretadas sin incurrir daños. Suele tener una tuerca o varias tuercas, que se pueden girar para aplicar la fuerza gradualmente.

Pulsation To move or beat with rhythmic impulses.

Pulsación Moverse o batir con impulsos rítmicos.

Race A channel in the inner or outer ring of an antifriction bearing in which the balls or rollers roll.

Pista Un canal en el anillo interior o exterior de un cojinete antifricción en el cual ruedan las bolas o los rodillos.

Radial The direction moving straight out from the center of a circle. Perpendicular to the shaft or bearing bore.

Radial La dirección al moverse directamente del centro de un círculo. Perpendicular a la flecha o al taladro del cojinete.

Radial clearance Clearance within the bearing and between balls and races perpendicular to the shaft. Also called radial displacement.

Holgura radial La holgura en un cojinete entre las bolas y las pistas que son perpendiculares a la flecha. También se llama un desplazamiento radial.

Radial load A force perpendicular to the axis of rotation.

Carga radial Una fuerza perpendicular al centro de rotación.

Radio frequency interference (RFI) An unwanted voltage signal that rides on another voltage signal.

Interferencia de la radiofrecuencia Una señal indeseada del voltaje que interfiere con otra señal del voltaje.

Ratio The relation or proportion that one number bears to another.

Relación La correlación o proporción de un número con respeto a otro.

Reactivity The characteristic of a material that enables it to react violently with water or other materials. Materials that release cyanide gas, hydrogen sulfide gas, or similar gases when exposed to low pH acid solutions are also said to possess reactivity, as are materials that generate toxic mists, fumes, vapors, and flammable gases.

Reactividad Característica de un material que le permite reaccionar violentamente con agua u otros materiales. Los materiales que emiten gas de cianuro, sulfuro de hidrógeno o gases similares cuando se exponen a soluciones ácidas de bajo pH también se dice que poseen reactividad, como los materiales que generan vapores tóxicos o gases inflamables.

Rear-wheel-drive A term associated with a vehicle in which the engine is mounted at the front and the driving axle and driving wheels are at the rear of the vehicle.

Tracción trasera Un término que se asocia con un vehículo en el cual el motor se ubica en la parte delantera y el eje propulsor y las ruedas propulsores se encuentran en la parte trasera del vehículo.

Relief valve A valve used to protect against excessive pressure in the case of a malfunctioning pressure regulator.

Válvula de seguridad Una válvula que se usa para guardar contra una presión excesiva en caso de que malfulciona el regulador de presión.

Repair order A legal document written for every vehicle brought into the shop for service. It gives detailed information about the customer, the vehicle, the customer's concern or request, and an estimate of the cost of the services and when the services should be completed.

Orden de reparación Documento legal que se emite para cada vehículo que se lleva a un taller para reparación. Da información detallada sobre el cliente, el vehículo, la preocupación o el pedido del cliente, y una estimación del costo de los servicios y el plazo en el que se completarán los mismos.

Retaining ring A removable fastener used as a shoulder to retain and position a round bearing in a hole.

Anillo de retén Un seguro removible que sirve de collarín para sujetar y posicionar a un cojinete en un agujero.

RFI Radio frequency interference. This acronym is used to describe a type of electrical interference that may affect voltage signals in a computerized system.

RFI Interferencia de frecuencias de radio. Esta sigla se usa en describir un tipo de interferencia eléctrica que puede afectar los señales de voltaje en un sistema computerizado.

Right-to-know laws Laws requiring employers to provide employees with a safe work place as it relates to hazardous materials, and information about any and all hazards the employees face while performing their job.

Prerrogativa a saber leyes Los leyes que requieren a patrones proveer de empleados un lugar de trabajo seguro, relacionada con los materiales peligrosos y la información sobre cualesquiera y todos todos los peligros que los empleados pudieron encontrar mientras que realizaban su trabajo.

Rivet A headed pin used for uniting two or more pieces by passing the shank through a hole in each piece and securing it by forming a head on the opposite end.

Remache Una clavija con cabeza que sirve para unir a dos piezas o más al pasar el vástago por un hoyo en cada pieza y asegurarlo por formar una cabeza en el extremo opuesto.

Roller bearing An inner and outer race on which hardened steel rollers operate.

Cojinete de rodillos Una pista interior y exterior en la cual operan los rodillos hecho de acero endurecido.

Rollers Round steel bearings that can be used as the locking element in an overrunning clutch or as the rolling element in an antifriction bearing.

Rodillos Articulaciones redondos de acero que pueden servir como un elemento de enclavamiento en un embrague de sobremarcha o como el elemento que rueda en un cojinete antifricción.

Rotary flow A fluid force generated in the torque converter that is related to vortex flow. The vortex flow leaving the impeller is not only flowing out of the impeller at high speed but is also rotating faster than the turbine. The rotating fluid striking the slower turning turbine exerts a force against the turbine that is defined as rotary flow.

Flujo rotativo Una fuerza fluida producida en el convertidor de torsión que se relaciona al flujo torbellino. El flujo torbellino saliendo del rotor no sólo viaja en una alta velocidad sino también gira más rápidamente que el turbino. El fluido rotativo chocando contra el turbino que gira más lentamente, impone una fuerza contra el turbino que se define como flujo rotativo.

rpm Abbreviation for revolutions per minute, a measure of rotational speed.

RPM Abreviación de revoluciones por minuto, una medida de la velocidad rotativa.

RTV sealer Room-temperature vulcanizing gasket material that cures at room temperature; a plastic paste squeezed from a tube to form a gasket of any shape.

Sellador RTV Una materia vulcanizante de empaque que cura en temperaturas del ambiente; una pasta plástica exprimida de un tubo para formar un empaque de cualquiera forma.

Runout Deviation of the specified normal travel of an object. The amount of deviation or wobble a shaft or wheel has as it rotates. Runout is measured with a dial indicator.

Corrimiento Una desviación de la carrera normal e especificada de un objeto. La cantidad de desviación o vacilación de una flecha o una rueda mientras que gira. El corrimiento se mide con un indicador de carátula.

RWD Abbreviation for rear-wheel-drive.

RWD Abreviación de tracción trasera.

SAE Society of Automotive Engineers.

SAE La Sociedad de Ingenieros Automotrices.

Scan tool A microprocessor designed to communicate with a vehicle's onboard computer in order to perform diagnosis and troubleshooting.

Herramienta de la exploración Una herramienta con un microprocesador diseñado para comunicarse con el ordenador a bordo de un vehículo para realizar diagnosis y la localización de averías.

Score A scratch, ridge, or groove marring a finished surface.

Entalladura Una raya, una arruga o una ranura que desfigure a una superficie acabada.

Scuffing A type of wear in which there is a transfer of material between parts moving against each other; shows up as pits or grooves in the mating surfaces.

Erosión Un tipo de desgaste en el cual hay una tranferencia de una materia entre las partes que estan en contacto mientras que muevan; se manifesta como hoyitos o muescas en las superficies apareadas.

Seal A material shaped around a shaft, used to close off the operating compartment of the shaft, preventing oil leakage.

Sello Una materia, formado alrededor de una flecha, que sella el compartimiento operativo de la flecha, preveniendo el goteo de aceite.

Sealer A thick, tacky compound, usually spread with a brush, that may be used as a gasket or sealant to seal small openings or surface irregularities.

Sellador Un compuesto pegajoso y espeso, comœnmente aplicado con una brocha, que puede usarse como un empaque o un obturador para sellar a las aperturas pequeñas o a las irregularidades de la superficie.

Seat A surface, usually machined, on which another part rests or seats; for example, the surface on which a valve face rests.

Asiento Una superficie, comúnmente maquinada, sobre la cual yace o se asienta otra parte; por ejemplo, la superficie sobre la cual yace la cara de la válvula.

Selective washer This type of washer is available in different thicknesses to provide the correct endplay or clearance.

Arandela selectiva Este tipo de arandela está disponible en diversos espesores, para proporcionar a la separación correcta o amplitud correcta.

Serial data The communications to and from the computer are commonly referred to as the system's serial data.

Datos seriales del sistema Las comunicaciones al ordenador y del ordenador se llama comúnmente los datos seriales del sistema.

Servo A device that converts hydraulic pressure into mechanical movement, often multiplying it. Used to apply the bands of a transmission.

Servo Un dispositivo que convierte la presión hidráulica al movimiento mecánico, frequentemente multiplicándola. Se usa en la aplicación de las bandas de una transmisión.

Shank The portion of a shoe that protects the ball of your foot. It is the narrow part of a shoe between the heel and sole.

Cambrillón Porción del zapato que protege la bola del pie. Es la parte angosta del zapato entre la suela y el tacón.

Shift lever The lever used to change gears in a transmission. Also the lever on the starting motor that moves the drive pinion into or out of mesh with the flywheel teeth.

Palanca del cambiador La palanca que sirve para cambiar a las velocidades de una transmisión. También es la palanca del motor de arranque que mueva al piñon de ataque para engranarse o desegranarse con los dientes del volante.

Shift valve A valve that controls the shifting of the gears in an automatic transmission.

Válvula de cambios Una válvula que controla a los cambios de las velocidades en una transmisión automática.

Shim Thin sheets used as spacers between two parts, such as the two halves of a journal bearing.

Laminilla de relleno Hojas delgadas que sirven de espaciadores entre dos partes, tal como las dos partes de un muñón.

Short An unwanted path for current.

Cortocircuito Un cortocircuito se describe lo más mejor posible como un camino indeseado para la corriente eléctrica.

Side clearance The clearance between the sides of moving parts when the sides do not serve as load-carrying surfaces.

Holgura lateral La holgura entre los lados de las partes en movimiento mientras que los lados no funcionan como las superficies de carga.

Sinusoidal The wave shaped like a sine wave.

Sinusoidal El término sinusoidal quiere decir que la onda tiena una formaparecida a una onda senoidal.

Sliding-fit Where sufficient clearance has been allowed between the shaft and journal to allow freerunning without overheating.

Ajuste corredera Donde se ha dejado una holgura suficiente entre la flecha y el muñon para permitir una marcha libre sin sobrecalentamiento.

Snap ring Split spring-type ring located in an internal or external groove to retain a part.

Anillo de seguridad Un anillo partido tipo resorte que se coloca en una muesca interior o exterior para retener a una parte.

Spalling A condition in which the material of a bearing surface breaks away from the base metal.

Escamación Una condición en la cual una materia de la superficie de un rodamiento se separa del metal base.

Spindle The shaft on which the wheels and wheel bearings mount.

Husillo La flecha en la cual se montan las ruedas y el conjunto del cojinete de las ruedas.

Spline Slot or groove cut in a shaft or bore; a splined shaft onto which a hub, wheel, gear, and so on, with matching splines in its bore is assembled so that the two must turn together.

Acanaladura (espárrago) Una muesca o ranura cortada en una flecha o en un taladro; una flecha acanalada en la cual se asamblea un cubo, una rueda, un engranaje, y todo lo demás que tiene un acanaladura pareja en el taladro de manera de que las dos deben girar juntos.

Split lip seal Typically, a rope seal sometimes used to denote any two-part oil seal.

Sello hendido Típicamente, un sello de cuerda que se usa a veces para demarcar cualquier sello de aceite de dos partes.

Split pin A round split spring steel tubular pin used for locking purposes; for example, locking a gear to a shaft.

Chaveta hendida Una chaveta partida redonda y tubular hecho de acero para resorte que sirve para el enclavamiento; por ejemplo, para enclavar un engranaje a una flecha.

Spool valve A cylindrically shaped valve with two or more valleys between the lands. Spool valves are used to direct fluid flow.

Válvula de carrete Una válvula de forma cilíndrica que tiene dos acanaladuras de cañón o más entre las partes planas. Las válvulas de carrete sirven para dirigir el flujo del fluido.

Sprag clutch A member of the overrunning clutch family using a sprag to jam between the inner and outer races used for holding or driving action.

Embrague de puntal Un miembro de la familia de embragues de sobremarcha que usa a una palanca de detención trabada entre las pistas interiores e exteriores para realizar una acción de asir o marchar.

Spring A device that changes shape when it is stretched or compressed, but returns to its original shape when the force is removed; the component of the automotive suspension system that absorbs road shocks by flexing and twisting.

Resorte Un dispositivo que cambia de forma al ser estirado o comprimido, pero que recupera su forma original al levantarse la fuerza; es un componente del sistema de suspensión automotívo que absorba los choques del camino al doblarse y torcerse.

Spring retainer A steel plate designed to hold a coil or several coil springs in place.

Retén de resorte Una chapa de acero diseñado a sostener en su posición a un resorte helicoidal o más.

Squeak A high-pitched noise of short duration.

Chillido Un ruido agudo de poca duración.

Squeal A continuous high-pitched noise.

Alarido Un ruido agudo continuo.

Stall A condition in which the engine is operating and the transmission is in gear, but the drive wheels are not turning because the turbine of the torque converter is not moving.

Paro Una condición en la cual opera el motor y la transmisión esta embragada pero las ruedas de impulso no giran porque no muevea el turbino del convertidor de la torsión.

Stall test A test of the one-way clutch in a torque converter.

Prueba de paro Una prueba del embrague de una vía en un convertidor de la torsión.

Static A form of electricity caused by friction.

Estático Una forma de la electridad causada por la fricción.

Stress The force to which a material, mechanism, or component is subjected.

Esfuerzo La fuerza a la cual se somete a una materia, un mecanísmo o un componente.

Sun gear The central gear in a planetary gear system around which the rest of the gears rotate. The innermost gear of the planetary gearset.

Engranaje principal (sol) El engranaje central en un sistema de engranajes planetarios alrededor del cual giran los otros engranajes. El engranaje más interno del conjunto de los engranajes planetarios.

Sweep rate The rate at which the trace moves across the display of an oscilloscope.

Velocidad de barrido Velocidad a la cual la señal se mueve por la pantalla de un osciloscopio.

Tap To cut threads in a hole with a tapered, fluted, threaded tool.

Roscar con macho Cortar las roscas en un agujero con una herramienta cónica, acanalada y fileteada.

Teardown A term often used to describe the process of disassembling a transmission.

Desmontaje Un término común que describe el proceso de desarmar a una transmisión.

Technical Service Bulletin (TSB) A special report detailing the repair of a particular customer concern or updated changes in specifications and repair procedures for a particular model vehicle.

Boletín del servicio técnico (TSB) Informe especial que detalla la reparación hecha a un cliente en particular o la actualización de cambios en las especificaciones y los procedimientos de reparación para un modelo de vehículo en particular.

Temper To change the physical characteristics of a metal by applying heat.

Templar Cambiar las características físicas de un metal mediante una aplicación del calor.

Tension Effort that elongates or "stretches" a material.

Tensión Un esfuerzo que alarga o "estira" a una materia.

Thickness gauge Strips of metal made to an exact thickness, used to measure clearances between parts.

Calibre de espesores Las tiras del metal que se han fabricado a un espesor exacto, sirven para medir las holguras entre las partes.

Thread chaser A device similar to a die that is used to clean threads.

Peine de roscar Un dispositivo parecido a una terraja que sirve para limpiar a las roscas.

Thread pitch The number of threads in 1 inch of threaded bolt length. In the metric system, thread pitch is the distance in millimeters between two adjacent threads.

Paso del filete La cantidad de filetes en una pulgada de la rosca de un tornillo. En el sistema métrico decimal, el paso del filete es la distancia en milímetros entre dos filetes adyacentes.

Threaded insert A threaded coil that is used to restore the original thread size to a hole with damaged threads.

Pieza inserta roscada Una bobina roscada que sirve para restaurar a su tamaño original una rosca dañada.

Throttle position (TP) sensor A potentiometer used to monitor changes in throttle plate opening. The position of the throttle plate determines the voltage drop at the sensor's resistor and the resultant voltage signal is sent to a computer system.

Sensor de posición de la válvula reguladora Un potenciómetro usado para vigilar cambios en la apertura de la placa de la válvula reguladora. La posición de la placa de la válvula reguladora determina la caída de voltaje en el resistor del sensor y la señal resultante del voltaje se envía a un sistema informático.

Thrust bearing A bearing designed to resist or contain side or end motion as well as reduce friction.

Cojinete de empuje Un cojinete diseñado a detener o reprimir a los movimientos laterales o de las extremidades y también reducir la fricción.

Thrust load A load that pushes or reacts through the bearing in a direction parallel to the shaft.

Carga de empuje Una carga que empuja o reacciona por el cojinete en una dirección paralelo a la flecha.

Thrust washer A washer designed to take up end thrust and prevent excessive endplay.

Arandela de empuje Una arandela diseñada para rellenar a la holgura de la extremidad y prevenir demasiado juego en la extremidad.

Tolerance A permissible variation between the two extremes of a specification or dimension.

Tolerancia Una variación permisible entre dos extremos de una especificación o de un dimensión.

Torque A twisting motion, usually measured in ft.-lb. (Nm).

Torsión Un movimiento giratorio, suele medirse en pies/libra (Nm).

Torque capacity The ability of a converter clutch to hold torque.

Capacidad de la torsión La abilidad de un convertidor de embraque a sostener a la torsión.

Torque converter A turbine device utilizing a rotary pump, one or more reactors (stators), and a driven circular turbine or vane, whereby power is transmitted from a driving to a driven member by hydraulic action. It provides varying drive ratios; with a speed reduction, it increases torque.

Convertidor de la torsión Un dispositivo de turbino que utilisa a una bomba rotativa, a un reactor o más, y un molinete o turbino circular impulsado, por cual se transmite la energía de un miembro de impulso a otro arrastrado mediante la acción hidráulica. Provee varias relaciones de impulso; al descender la velocidad, aumenta la torsión.

Torque curve A line plotted on a chart to illustrate the torque personality of an engine. When the engine operates on its torque curve, it is producing the most torque for the quantity of fuel being burned.

Curva de la torsión Una linea delineada en una carta para ilustrar las características de la torsión del motor. Al operar un motor en su curva de la torsión, produce la torsión óptima para la cantidad del combustible que se consuma.

Torque multiplication The result of meshing a small driving gear and a large driven gear to reduce speed and increase output torque.

Multiplicación de la torsión El resultado de engranar a un engranaje pequeño de ataque con un engranaje más grande arrastrado para reducir la velocidad y incrementar la torsión de salida.

Torque steer An action felt in the steering wheel as the result of increased torque.

Dirección la torsión Una acción que se nota en el volante de dirección como resultado de un aumento de la torsión.

Torque wrench A tool used to measure how tight a nut or bolt is.

Llave dinamométrica Herramienta que se usa para medir cuán ajustado está un tornillo o una tuerca.

Traction The gripping action between the tire tread and the road's surface.

Tracción La acción de agarrar entre la cara de la rueda y la superficie del camino.

Transaxle Type of construction in which the transmission and differential are combined in one unit.

Flecha de transmisión Un tipo de construcción en el cual la transmisión y el diferencial se combinan en una unidad.

Transaxle assembly A compact housing most often used in front-wheel-drive vehicles that houses the manual transmission, final drive gears, and differential assembly.

Asamblea de la flecha de transmisión Un cárter compacto que se usa normalmente en los vehículos de tracción delantera que contiene la transmisión manual, los engranajes de propulsión, y la asamblea del diferencial.

Transfer case An auxiliary transmission mounted behind the main transmission. Used to divide engine power and transfer it to both front and rear differentials, either full time or part time.

Cárter de la transferencia Una transmisión auxiliar montada detrás de la transmisión principal. Sirve para dividir la potencia del motor y transferirla a ambos diferenciales delanteras y traseras todo el tiempo o la mitad del tiempo.

Transmission The device in the powertrain that provides different gear ratios between the engine and drive wheels as well as reverse.

Transmisión El dispositivo en el trén de potencia que provee las relaciones diferentes de engranaje entre el motor y las ruedas de impulso y también la marcha de reversa.

Transverse Powertrain layout in a front-wheel-drive automobile extending from side to side.

Transversal Una esquema del tren de potencia en un automóvil de tracción delantera que se extiende de un lado a otro.

U-joint A four-point cross connected to two U-shaped yokes that serves as a flexible coupling between shafts.

Junta de U Una cruceta de cuatro puntos que se conecta a dos yugos en forma de U que sirven de acoplamientos flexibles entre las flechas.

Underdrive The condition that exists when there is speed reduction through the gears.

Bajamarcha Bajamarcha es la condición que existe cuando hay reducción de velocidad a través de los engranajes.

United States Customary System (USCS) A system of weights and measures used in places that do not use the metric system, such as the United States. The USCS system is also known as the English or inch system.

Sistema acostumbrado de Estados Unidos (USCS) Un sistema de los pesos y de las medidas usados en los países que no utilizan la sistema métrico, tal como los Estados Unidos. El sistema de USCS también se conoce como el sistema del inglés o de la pulgada.

Universal joint A mechanical device that transmits rotary motion from one shaft to another shaft at varying angles.

Junta universal Un dispositivo mecánico que transmite el movimiento giratorio desde una flecha a otra flecha en varios ángulos.

Upshift To shift a transmission into a higher gear.

Cambio ascendente Cambiar a la velocidad de una transmisión a una más alta.

Valve body Main hydraulic control assembly of a transmission containing the components necessary to control the distribution of pressurized transmission fluid throughout the transmission.

Cuerpo de la válvula Asamblea principal del control hidráulico de una transmisión que contiene los componentes necessarios para controlar a la distribución del fluido de la transmisión bajo presión por toda la transmisión.

Vehicle identification number (VIN) The number assigned to each vehicle by its manufacturer, primarily for registration and identification purposes.

Número de identificacíon del vehículo El número asignado a cada vehículo por su fabricante, primariamente con el propósito de la registración y la identificación.

Vehicle speed sensor (VSS) A sensor used to track the current speed and the total miles traveled by a vehicle. This input is used by many computer systems in the vehicle.

Sensor de velocidad del vehículo Un sensor utilizado para medir la velocidad actual y las millas totales viajaron por un vehículo. Muchos de los sistemas informáticos del vehículo utilizan estos datos como entrada de información.

Vibration A quivering, trembling motion; felt in a vehicle at different speed ranges.

Vibración Un movimiento de estremecer o temblar que se siente en el vehículo en varios intervalos de velocidad.

Viscosity The resistance to flow exhibited by a liquid. A thick oil has greater viscosity than a thin oil.

Viscosidad La resistencia al flujo que manifiesta un líquido. Un aceite espeso tiene una viscosidad mayor que un aceite ligero.

Volatility The tendency for a fluid to evaporate rapidly or pass off in the form of vapor. For example, gasoline is more volatile than kerosene because it evaporates at a lower temperature.

Volatilidad Tendencia de un fluido a evaporarse rápidamente o a transformarse en vapor. Por ejemplo, la gasolina es más volátil que el queroseno porque se evapora a menor temperatura.

Voltmeter A test instrument used to measure voltage.

Voltímetro Un aparato utilizado para medir voltaje.

Vortex Path of fluid flow in a torque converter. The vortex may be high, low, or zero depending on the relative speed between the pump and turbine.

Vórtice La vía del flujo de los fluidos en un convertido de torsión. El vórtice puede ser alto, bajo, o cero depende de la velocidad relativa entre la bomba y la turbina.

Vortex flow Recirculating flow between the converter impeller and turbine that causes torque multiplication.

Flujo del vórtice El fluyo recirculante entre el impulsor del convertidor y la turbina que causa la multiplicación de la torsión.

Wet-disc clutch A clutch in which the friction disc (or discs) is operated in a bath of oil.

Embrague de disco flotante Un embrague en el cual el disco (o los discos) de fricción opera en un baño de aceite.

Wheel A disc or spokes with a hub at the center that revolves around an axle, and a rim around the outside on which the tire is mounted.

Rueda Un disco o rayo que tiene en su centro un cubo que gira alrededor de un eje, y tiene un rim alrededor de su exterior en la cual se monta el neumático.

Yoke In a universal joint, the drivable torque-and-motion input and output member attached to a shaft or tube.

Yugo En una junta universal, el miembro de la entrada y la salida que transfere a la torsión y al movimiento, que se conecta a una flecha o a un tubo.

Note: Figures and tables are denoted by f and t, respectively.